Functional Genomics in Aquaculture

Functional Genomics in Aquaculture

Editor:

Linda Lait

Delve Publishing

www.delvepublishing.com

Functional Genomics in Aquaculture

Editor: Linda Lait

708 3rd Avenue, 6th Floor
New York, NY 10017
United States of America
www.delvepublishing.com

ISBN: 978-1-68095-851-5

Library of Congress Control Number: 2016951604

About the Editor

Linda Lait

Linda obtained her MSc from the University of Lethbridge in 2011 and will be completing her PhD at Memorial University of Newfoundland in 2016. Her interests include the evolutionary history of species, particularly those of conservation concern, which she studies with molecular markers. Increasingly complex questions can be answered with the improved technologies that are emerging. She is currently finishing up her PhD in molecular genetics, and will be starting as a Postdoctoral Fellow studying phylogeography and molecular evolution in insects.

Contents

List of Contributors

Cristiano De Pittà
CRIBI Biotechnology Centre, Università degli Studi di Padova, Via U. Bassi, 58/B, 35121, Padova, Italy,
Dipartimento di Biologia, Università degli Studi di Padova, Via U. Bassi, 58/B, 35121, Padova, Italy

Cristiano Bertolucci
Dipartimento di Biologia ed Evoluzione, Università degli Studi di Ferrara, Via L. Borsari, 46, 44100 Ferrara, Italy

Gabriella M Mazzotta
Dipartimento di Biologia, Università degli Studi di Padova, Via U. Bassi, 58/B, 35121, Padova, Italy

Filippo Bernante
CRIBI Biotechnology Centre, Università degli Studi di Padova, Via U. Bassi, 58/B, 35121, Padova, Italy

Giorgia Rizzo
Dipartimento di Biologia, Università degli Studi di Padova, Via U. Bassi, 58/B, 35121, Padova, Italy

Barbara De **Nardi**
Dipartimento di Biologia, Università degli Studi di Trieste, P.le Valmaura, 9, 34148 Trieste, Italy

Alberto Pallavicini
Dipartimento di Biologia, Università degli Studi di Trieste, P.le Valmaura, 9, 34148 Trieste, Italy

Gerolamo Lanfranchi
CRIBI Biotechnology Centre, Università degli Studi di Padova, Via U. Bassi, 58/B, 35121, Padova, Italy
Dipartimento di Biologia, Università degli Studi di Padova, Via U. Bassi, 58/B, 35121, Padova, Italy

Rodolfo Costa
Dipartimento di Biologia, Università degli Studi di Padova, Via U. Bassi, 58/B, 35121, Padova, Italy

Keyi Ma
Key laboratory of Freshwater Aquatic Genetic Resources Certificated by Ministry of Agriculture, College of Fisheries and Life Science, Shanghai Ocean University, Shanghai, P. R. China

Gaofeng Qiu
Key laboratory of Freshwater Aquatic Genetic Resources Certificated by Ministry of Agriculture, College of Fisheries and Life Science, Shanghai Ocean University, Shanghai, P. R. China
E-Institute of Shanghai Universities, Shanghai Ocean University, Shanghai, P. R. China Auburn University, United States of America

Jianbin Feng
Key laboratory of Freshwater Aquatic Genetic Resources Certificated by Ministry of Agriculture, College of Fisheries and Life Science, Shanghai Ocean University, Shanghai, P. R. China

Jiale Li
Key laboratory of Freshwater Aquatic Genetic Resources Certificated by Ministry of Agriculture, College of Fisheries and Life Science, Shanghai Ocean University, Shanghai, P. R. China
E-Institute of Shanghai Universities, Shanghai Ocean University, Shanghai, P. R. China Auburn University, United States of America

Jiang H
College of Animal Science and Technology, Northwest A&F University, Shaanxi Key Laboratory of Molecular Biology for Agriculture, Yangling, Shaanxi, People's Republic of China.
Agriculture Ministry Key Laboratory of Healthy Freshwater Aquaculture, Key Laboratory of Freshwater Aquatic Animal Genetic and Breeding of Zhejiang province, Zhejiang Institute of Freshwater Fisheries, Huzhou, Zhejiang, People's Republic of China.
College of Fisheries, Henan Normal University, Xinxiang, Henan, People's Republic of China.

Li X
Agriculture Ministry Key Laboratory of Healthy Freshwater Aquaculture, Key Laboratory of Freshwater Aquatic Animal Genetic and Breeding of Zhejiang province, Zhejiang Institute of Freshwater Fisheries, Huzhou, Zhejiang, People's Republic of China.

Sun Y
College of Animal Science and Technology, Northwest A&F University, Shaanxi Key Laboratory of Molecular Biology for Agriculture, Yangling, Shaanxi, People's Republic of China.

Hou F
College of Animal Science and Technology, Northwest A&F University, Shaanxi Key Laboratory of Molecular Biology for Agriculture, Yangling, Shaanxi, People's Republic of China.

Zhang Y
Agriculture Ministry Key Laboratory of Healthy Freshwater Aquaculture, Key Laboratory of Freshwater Aquatic Animal Genetic and Breeding of Zhejiang province, Zhejiang Institute of Freshwater Fisheries, Huzhou, Zhejiang, People's Republic of China.

Li F
Agriculture Ministry Key Laboratory of Healthy Freshwater Aquaculture, Key Laboratory of Freshwater Aquatic Animal Genetic and Breeding of Zhejiang province, Zhejiang Institute of Freshwater Fisheries, Huzhou, Zhejiang, People's Republic of China.

Gu Z
Agriculture Ministry Key Laboratory of Healthy Freshwater Aquaculture, Key Laboratory of Freshwater Aquatic Animal Genetic and Breeding of Zhejiang province, Zhejiang Institute of Freshwater Fisheries, Huzhou, Zhejiang, People's Republic of China.

Liu X
College of Animal Science and Technology, Northwest A&F University, Shaanxi Key Laboratory of Molecular Biology for Agriculture, Yangling, Shaanxi, People's Republic of China.

Jianjian Lv
Key Laboratory for Sustainable Utilization of Marine Fisheries Resources, Ministry of Agriculture, Yellow Sea Fisheries Research Institute, Chinese Academy of Fishery Sciences, Qingdao, China

Ping Liu
Key Laboratory for Sustainable Utilization of Marine Fisheries Resources, Ministry of Agriculture, Yellow Sea Fisheries Research Institute, Chinese Academy of Fishery Sciences, Qingdao, China

Baoquan Gao
Key Laboratory for Sustainable Utilization of Marine Fisheries Resources, Ministry of Agriculture, Yellow Sea Fisheries Research Institute, Chinese Academy of Fishery Sciences, Qingdao, China

Yu Wang
Key Laboratory for Sustainable Utilization of Marine Fisheries Resources, Ministry of Agriculture, Yellow Sea Fisheries Research Institute, Chinese Academy of Fishery Sciences, Qingdao, China
College of Fisheries and Life Science, Shanghai Ocean University, Shanghai, China State Key Laboratory of Pathogen and Biosecurity, Beijing Institute of Microbiology and Epidemiology, China

Zheng Wang
Key Laboratory for Sustainable Utilization of Marine Fisheries Resources, Ministry of Agriculture, Yellow Sea Fisheries Research Institute, Chinese Academy of Fishery Sciences, Qingdao, China
College of Fisheries and Life Science, Shanghai Ocean University, Shanghai, China State Key Laboratory of Pathogen and Biosecurity, Beijing Institute of Microbiology and Epidemiology, China

Ping Chen
Key Laboratory for Sustainable Utilization of Marine Fisheries Resources, Ministry of Agriculture, Yellow Sea Fisheries Research Institute, Chinese Academy of Fishery Sciences, Qingdao, China

Jian Li
Key Laboratory for Sustainable Utilization of Marine Fisheries Resources, Ministry of Agriculture, Yellow Sea Fisheries Research Institute, Chinese Academy of Fishery Sciences, Qingdao, China

Hong Liu
The Fish Molecular Genetics and Biotechnology Laboratory, Department of Fisheries and Allied Aquacultures and Program of Cell and Molecular Biosciences, Aquatic Genomics Unit, Auburn University, Auburn, AL 36849, USA,
College of Fisheries, Huazhong Agricultural University, Wuhan, 430070, China

Yanliang Jiang
The Fish Molecular Genetics and Biotechnology Laboratory, Department of Fisheries and Allied Aquacultures and Program of Cell and Molecular Biosciences, Aquatic Genomics Unit, Auburn University, Auburn, AL 36849, USA,
Key Laboratory of Protein Chemistry and Developmental Biology of State Education Ministry of China, College of Life Sciences, Hunan Normal University, Changsha, 410081, China

Shaolin Wang
The Fish Molecular Genetics and Biotechnology Laboratory, Department of Fisheries and Allied Aquacultures and Program of Cell and Molecular Biosciences, Aquatic Genomics Unit, Auburn University, Auburn, AL 36849, USA

Parichart Ninwichian
The Fish Molecular Genetics and Biotechnology Laboratory, Department of Fisheries and Allied Aquacultures and Program of Cell and Molecular Biosciences, Aquatic Genomics Unit, Auburn University, Auburn, AL 36849, USA

Benjaporn Somridhivej
The Fish Molecular Genetics and Biotechnology Laboratory, Department of Fisheries and Allied Aquacultures and Program of Cell and Molecular Biosciences, Aquatic Genomics Unit, Auburn University, Auburn, AL 36849, USA

Peng Xu
The Fish Molecular Genetics and Biotechnology Laboratory, Department of Fisheries and Allied Aquacultures and Program of Cell and Molecular Biosciences, Aquatic Genomics Unit, Auburn University, Auburn, AL 36849, USA

Jason Abernathy
The Fish Molecular Genetics and Biotechnology Laboratory, Department of Fisheries and Allied Aquacultures and Program of Cell and Molecular Biosciences, Aquatic Genomics Unit, Auburn University, Auburn, AL 36849, USA

Huseyin Kucuktas
The Fish Molecular Genetics and Biotechnology Laboratory, Department of Fisheries and Allied Aquacultures and Program of Cell and Molecular Biosciences, Aquatic Genomics Unit, Auburn University, Auburn, AL 36849, USA

Zhanjiang Liu
The Fish Molecular Genetics and Biotechnology Laboratory, Department of Fisheries and Allied Aquacultures and Program of Cell and Molecular Biosciences, Aquatic Genomics Unit, Auburn University, Auburn, AL 36849, USA

Anti Vasemägi
Division of Genetics and Physiology, Department of Biology, University of Turku, 20014 Turku, Finland
Department of Aquaculture, Institute of Veterinary Medicine and Animal Science, Estonian University of Life Sciences, 51014 Tartu, Estonia

Jan Nilsson
Department of Wildlife, Fish, and Environmental Studies, Swedish University of Agricultural Sciences, 901 83 Umeå, Sweden

Philip McGinnity
Aquaculture and Fisheries Development Centre, School of Biological, Earth, and Environmental Sciences, University College Cork, Cork, Ireland
Marine Institute, Furnace, Newport, Co. Mayo, Ireland

Tom Cross
Aquaculture and Fisheries Development Centre, School of Biological, Earth, and Environmental Sciences, University College Cork, Cork, Ireland

Patrick O'Reilly
Population Ecology Division, Department of Fisheries and Oceans, Bedford Institute of Oceanography, Challenger Drive, Dartmouth, NS, Canada B2Y 4A2

Brian Glebe
Fisheries and Oceans Canada, Department of Fisheries and Oceans, St. Andrews Biological Station, St. Andrews, NB, Canada E0G 2X0

Bo Peng
Department of Genetics, The University of Texas M. D. Anderson Cancer Center, Houston, TX 77030, USA

Paul Ragnar Berg
Department of Animal and Aquacultural Sciences, Centre for Integrative Genetics, Norwegian University of Life Sciences, 1432 Aas, Norway

Craig Robert Primmer[1]
Division of Genetics and Physiology, Department of Biology, University of Turku, 20014 Turku, Finland

Diego Robledo
Departamento de Xenética, Facultade de Bioloxía (CIBUS), Universidade de Santiago de Compostela, Santiago de Compostela 15782, Spain

Carlos Fernández
Departamento de Xenética, Facultade de Veterinaria, Universidade de Santiago de Compostela, Lugo 27002, Spain

Miguel Hermida
Departamento de Xenética, Facultade de Veterinaria, Universidade de Santiago de Compostela, Lugo 27002, Spain

Andrés Sciara
Instituto de Biología Molecular y Celular de Rosario (IBR-CONICET), Rosario S2002LRK, Argentina

José Antonio Álvarez-Dios
Departamento de Matemática Aplicada, Facultade de Matemáticas, Universidade de Santiago de Compostela, Santiago de Compostela 15782, Spain

Santiago Cabaleiro
Cluster de Acuicultura de Galicia (Punta do Couso), Aguiño-Ribeira 15695, Spain

Rubén Caamaño
Cluster de Acuicultura de Galicia (Punta do Couso), Aguiño-Ribeira 15695, Spain

Paulino Martínez
Departamento de Xenética, Facultade de Veterinaria, Universidade de Santiago de Compostela, Lugo 27002, Spain

Carmen Bouza
Departamento de Xenética, Facultade de Veterinaria, Universidade de Santiago de Compostela, Lugo 27002, Spain

Serena Ferraresso
Department of Public Health, Comparative Pathology, and Veterinary Hygiene, Faculty of Veterinary Medicine, University of Padova, Viale dell'Università 16, 35020 Legnaro, Italy

Massimo Milan
Department of Public Health, Comparative Pathology, and Veterinary Hygiene, Faculty of Veterinary Medicine, University of Padova, Viale dell'Università 16, 35020 Legnaro, Italy
Department of Evolutionary Biology, University of Florence, 50125 Florence, Italy

Caterina Pellizzari
Department of Public Health, Comparative Pathology, and Veterinary Hygiene, Faculty of Veterinary Medicine, University of Padova, Viale dell'Università 16, 35020 Legnaro, Italy

Nicola Vitulo
CRIBI, University of Padova, Complesso Biologico Vallisneri, Via Ugo Bassi 58/B, Padova, Italy

Richard Reinhardt
Max Planck Institute for Molecular Genetics, Ihnestraße 63-73, 14195 Berlin, Germany

Adelino VM Canario
Centro de Ciências do Mar, Universidade do Algarve, Gambelas, 8005-139 Faro, Portugal

Tomaso Patarnello
Department of Public Health, Comparative Pathology, and Veterinary Hygiene, Faculty of Veterinary Medicine, University of Padova, Viale dell'Università 16, 35020 Legnaro, Italy

Luca Bargelloni
Department of Public Health, Comparative Pathology, and Veterinary Hygiene, Faculty of Veterinary Medicine, University of Padova, Viale dell'Università 16, 35020 Legnaro, Italy

Fabian Grammes
Institute of Animal and Aquaculture Sciences, Norwegian University of Life Sciences, N-1432 As-UMB,Norway
NOFIMA, N-1432 Aas, Norway.

Kjell-Arne Rørvik
Institute of Animal and Aquaculture Sciences, Norwegian University of Life Sciences, N-1432 As-UMB,Norway.
NOFIMA, , N-1432 Aas, Norway.

Magny S Thomassen
Institute of Animal and Aquaculture Sciences, Norwegian University of Life Sciences, N-1432 As-UMB,Norway.
NOFIMA, N-1432 Aas, Norway.

Rolf K Berge
Institute of Medicine, Haukeland University Hospital, University of Bergen, N-5021 Bergen, Norway.

Harald Takle
NOFIMA, N-1432 Aas, Norway.
AVS Chile SA, Casilla 300, Puerto Varas, Chile.

Jun Hong Xia

Molecular Population Genetics Group, Temasek Life Sciences Laboratory, National University of Singapore

Gen Hua Yue
Molecular Population Genetics Group, Temasek Life Sciences Laboratory, National University of Singapore

Elena Sarropoulou
Institute of Marine Biology and Genetics, Hellenic Center of Marine Research, 710 03 Iraklio, Crete, Greece,

Pilar Sepulcre
Department of Cell Biology and Histology, Faculty of Biology, University of Murcia, 30100 Murcia, Spain,

Laura Poisa-Beiro
Instituto de Investigaciones Marinas, Consejo Superior de Investigaciones Científicas (CSIC), Eduardo Cabello, 6 36208, Vigo, Spain and

Victoriano Mulero
Department of Cell Biology and Histology, Faculty of Biology, University of Murcia, 30100 Murcia, Spain,

José Meseguer
Department of Cell Biology and Histology, Faculty of Biology, University of Murcia, 30100 Murcia, Spain,

Antonio Figueras
Instituto de Investigaciones Marinas, Consejo Superior de Investigaciones Científicas (CSIC), Eduardo Cabello, 6 36208, Vigo, Spain

Beatriz Novoa
Instituto de Investigaciones Marinas, Consejo Superior de Investigaciones Científicas (CSIC), Eduardo Cabello, 6 36208, Vigo, Spain and

Vasso Terzoglou
Institute of Marine Biology and Genetics, Hellenic Center of Marine Research, 710 03 Iraklio, Crete, Greece,

Richard Reinhardt
Max-Planck Institute for Molecular Genetics, Ihnestrasse 63–73, 14195 Berlin-Dahlem, Germany

Antonios Magoulas
Institute of Marine Biology and Genetics, Hellenic Center of Marine Research, 710 03 Iraklio, Crete, Greece,

Georgios Kotoulas
Institute of Marine Biology and Genetics, Hellenic Center of Marine Research, 710 03 Iraklio, Crete, Greece

Michael J Leaver
Institute of Aquaculture, University of Stirling, Stirling FK9 4LA, UK

Laure AN Villeneuve
Institute of Aquaculture, University of Stirling, Stirling FK9 4LA, UK

Alex Obach
Skretting Aquaculture Research Centre (ARC), N- 4001 Stavanger, Norway

Linda Jensen
Institute of Aquaculture, University of Stirling, Stirling FK9 4LA, UK

James E Bron
Institute of Aquaculture, University of Stirling, Stirling FK9 4LA, UK

Douglas R Tocher
Institute of Aquaculture, University of Stirling, Stirling FK9 4LA, UK

John B Taggart
Institute of Aquaculture, University of Stirling, Stirling FK9 4LA, UK

Tiago S Hori
Ocean Sciences Centre, Memorial University of Newfoundland, St. John's A1C 5S7, NL, Canada

A Kurt Gamperl
Ocean Sciences Centre, Memorial University of Newfoundland, St. John's A1C 5S7, NL, Canada

Marije Booman
Ocean Sciences Centre, Memorial University of Newfoundland, St. John's A1C 5S7, NL, Canada

Gordon W Nash
Ocean Sciences Centre, Memorial University of Newfoundland, St. John's A1C 5S7, NL, Canada

Matthew L Rise
Ocean Sciences Centre, Memorial University of Newfoundland, St. John's A1C 5S7, NL, Canada

Robert W Li
USDA, Agricultural Research Service, Catfish Genetics Research Unit, Stoneville, MS 38776, USA
USDA, Agricultural Research Service, Bovine Functional Genomics Laboratory, Beltsville, MD 20705, USA

Geoffrey C Waldbieser
USDA, Agricultural Research Service, Catfish Genetics Research Unit, Stoneville, MS 38776, USA

Belén G Pardo
Departamento de Genética. Facultad de Veterinaria, Universidad de Santiago de Compostela, Campus de Lugo, 27002 Lugo, Spain,

Carlos Fernández
[1]Departamento de Genética. Facultad de Veterinaria, Universidad de Santiago de Compostela, Campus de Lugo, 27002 Lugo, Spain,

Adrián Millán
Departamento de Genética. Facultad de Veterinaria, Universidad de Santiago de Compostela, Campus de Lugo, 27002 Lugo, Spain,

Carmen Bouza
Departamento de Genética. Facultad de Veterinaria, Universidad de Santiago de Compostela, Campus de Lugo, 27002 Lugo, Spain,

Araceli Vázquez-López
Departamento de Genética. Facultad de Veterinaria, Universidad de Santiago de Compostela, Campus de Lugo, 27002 Lugo, Spain,

Manuel Vera
Departamento de Genética. Facultad de Veterinaria, Universidad de Santiago de Compostela, Campus de Lugo, 27002 Lugo, Spain,

José A Alvarez-Dios
Departamento de Matemática Aplicada, Facultad de Matemáticas, Universidad de Santiago de Compostela, 15782 Santiago de Compostela, Spain,

Manuel Calaza
Departamento de Matemática Aplicada, Facultad de Matemáticas, Universidad de Santiago de Compostela, 15782 Santiago de Compostela, Spain,

Antonio Gómez-Tato
Departamento de Geometría y Topología, Facultad de Matemáticas, Universidad de Santiago de Compostela, 15782 Santiago de Compostela, Spain

María Vázquez
Cluster de la Acuicultura de Galicia (CETGA), Punta de Couso s/n, 15965, Aguiño (Ribeira), A Coruña, Spain and

Santiago Cabaleiro
Cluster de la Acuicultura de Galicia (CETGA), Punta de Couso s/n, 15965, Aguiño (Ribeira), A Coruña, Spain and

Beatriz Magariños
Departamento de Microbiología y Parasitología. Universidad de Santiago de Compostela, 15782 Santiago de Compostela, Spain

Manuel L Lemos
Departamento de Microbiología y Parasitología. Universidad de Santiago de Compostela, 15782 Santiago de Compostela, Spain

José M Leiro
Departamento de Microbiología y Parasitología. Universidad de Santiago de Compostela, 15782 Santiago de Compostela, Spain

Paulino Martínez
Departamento de Genética. Facultad de Veterinaria, Universidad de Santiago de Compostela, Campus de Lugo, 27002 Lugo, Spain,

Li-xin Xiang
College of Life Science, Zhejiang University, Hangzhou 310058, China.
Key lab for Cell and Gene Engineering of Zhejiang province, Hangzhou 310058, China.

Ding He
College of Life Science, Zhejiang University, Hangzhou 310058, China.
Key lab for Cell and Gene Engineering of Zhejiang province, Hangzhou 310058, China.

Wei-ren Dong
College of Life Science, Zhejiang University, Hangzhou 310058, China.
Key lab for Cell and Gene Engineering of Zhejiang province, Hangzhou 310058, China.

Yi-wen Zhang
College of Life Science, Zhejiang University, Hangzhou 310058, China.
Key lab for Cell and Gene Engineering of Zhejiang province, Hangzhou 310058, China.

Jian-zhong Shao
College of Life Science, Zhejiang University, Hangzhou 310058, China.
Key lab for Cell and Gene Engineering of Zhejiang province, Hangzhou 310058, China.

Zexia Gao
Key Lab of Freshwater Animal Breeding, College of Fisheries, Ministry of Agriculture, Key Lab of Agricultural Animal Genetics, Breeding and Reproduction of Ministry of Education, Huazhong Agricultural University, Wuhan, People's Republic of China

Wei Luo
Key Lab of Freshwater Animal Breeding, College of Fisheries, Ministry of Agriculture, Key Lab of Agricultural Animal Genetics, Breeding and Reproduction of Ministry of Education, Huazhong Agricultural University, Wuhan, People's Republic of China

Hong Liu
Key Lab of Freshwater Animal Breeding, College of Fisheries, Ministry of Agriculture, Key Lab of Agricultural Animal Genetics, Breeding and Reproduction of Ministry of Education, Huazhong Agricultural University, Wuhan, People's Republic of China

Cong Zeng
Key Lab of Freshwater Animal Breeding, College of Fisheries, Ministry of Agriculture, Key Lab of Agricultural Animal Genetics, Breeding and Reproduction of Ministry of Education, Huazhong Agricultural University, Wuhan, People's Republic of China

Xiaolian Liu
Key Lab of Freshwater Animal Breeding, College of Fisheries, Ministry of Agriculture, Key Lab of Agricultural Animal Genetics, Breeding and Reproduction of Ministry of Education, Huazhong Agricultural University, Wuhan, People's Republic of China

Shaokui Yi
Key Lab of Freshwater Animal Breeding, College of Fisheries, Ministry of Agriculture, Key Lab of Agricultural Animal Genetics, Breeding and Reproduction of Ministry of Education, Huazhong Agricultural University, Wuhan, People's Republic of China

Weimin Wang
Key Lab of Freshwater Animal Breeding, College of Fisheries, Ministry of Agriculture, Key Lab of Agricultural Animal Genetics, Breeding and Reproduction of Ministry of Education, Huazhong Agricultural University, Wuhan, People's Republic of China

Preface

Aquaculture is a billion dollar industry. In 2010 more than 150 billion kg of fish, crustaceans, mollusks, and plants was harvested from aquaculture, almost twice that caught in the wild. As with any new venture, aquaculture maintains both advantages and weaknesses. With the global decline in fish and invertebrates, and an increasing demand for human consumption, farming of these species provides a potentially sustainable method to satiate demand and help slow or even reverse population declines. However, domestication comes with its own set of problems: low fecundity and growth rates, high levels of disease and parasitism, and the risk of introducing non-native species to an environment through escape. Recent studies have turned to functional genomics to address many of these problems. While some key commercial species have been studied extensively, we see here how comparative studies can be used to look at a non-model organism with few genomic resources. Through the use of next generation sequencing and microarray analysis, as well as many other techniques, the authors here address issues such as how captivity may affect the genome, whether there is evidence for growth-related genes, what genes are associated with infection, disease, or deformity, and how the use of dietary supplementation affects gene expression.

Editor
Linda Lait

INTRODUCTION

Functional genomics is a field of molecular biology that attempts to make use of the vast wealth of data produced by genomic and transcriptomic projects (such as genome sequencing projects and RNA-seq) to describe gene (and protein) functions and interactions. Unlike genomics, functional genomics focuses on the dynamic aspects such as gene transcription, translation, regulation of gene expression and protein–protein interactions, as opposed to the static aspects of the genomic information such as DNA sequence or structures. Functional genomics attempts to answer questions about the function of DNA at the levels of genes, RNA transcripts, and protein products.

The aim of functional genomics studies is to understand the complex relationship between genotype and phenotype on a global (genome-wide) scale. Studies investigate a range of processes such as transcription, translation and epigenetic regulation, in an attempt to answer relevant biological questions such as:

- When and where are genes expressed?
- How do gene expression levels differ in various cell types and states?
- What are the functional roles of different genes and in what cellular processes do they participate?
- How are genes regulated? Where are the active gene promoters in a particular cell type?
- How do genes and gene products interact?
- How does gene expression change in various diseases or following a treatment?

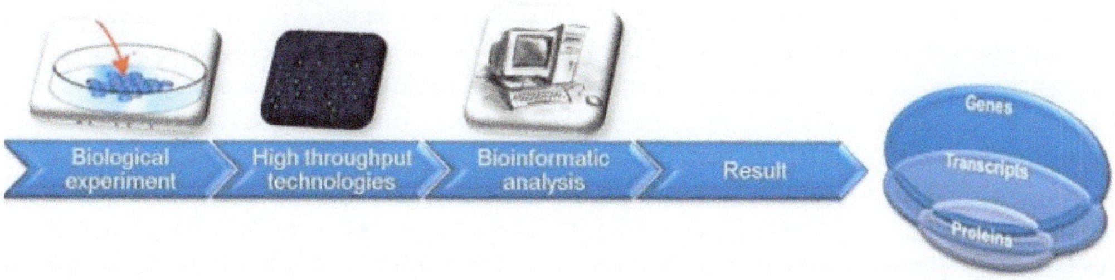

Figure 1. Functional genomics integrates information from various molecular methodologies to gain an understanding of how DNA sequence is translated into complex information in a cell (DNA → RNA → Proteins → biological process).

Experiments

Functional genomic experiments (Figure 1) typically utilize large-scale, high-throughput assays to measure and track many genes or proteins in parallel under different experimental or environmental conditions (e.g. with samples from patients and healthy individuals). This "genome-wide" approach allows the function of different parts of the genome to be discovered by combining information from genes, transcripts and proteins.

Goals of functional genomics

The goal of functional genomics is to understand the relationship between an organism's genome and its phenotype. The term functional genomics is often used broadly to refer to the many possible approaches to understanding the properties and function of the entirety of an organism's genes and gene products. This definition is somewhat variable; Gibson and Muse define it as "approaches under development to ascertain the biochemical, cellular, and/or physiological properties of each and every gene product",[1] while Pevsner includes the study of nongenic elements in his definition: "the genome-wide study of the function of DNA (including genes and nongenic elements), as well as the nucleic acid and protein products encoded by DNA". [2] Functional genomics involves studies of natural variation in genes, RNA, and proteins over time (such as an organism's development) or space (such as its body regions), as well as studies of natural or experimental functional disruptions affecting genes, chromosomes, RNA, or proteins.

The Concept of Functional Genomics

Genes and Gene Functions Most eukaryotic organisms harbor tens of thousands of genes, and in most cases, perhaps more than 20,000 but fewer than 40,000 genes. Each gene has its own functions. Historically, gene functions were determined by observations of a phenotypic mutation followed by genetic mapping of the mutated phenotype and eventually trace to the gene controlling the trait. Such an approach is usually considered to be a forward genetics approach. This approach is highly straightforward; however, mutations, whether as a result of spontaneous mutation, or induced mutation, are rare and mutated phenotypes are oftentimes difficult to be observed in the first place. With the rapid progress of DNA sequencing technologies, scientists started to know much quicker about a specific gene sequence and its protein structures than the functions of the gene. As a result, a new set of reverse genetics approaches was developed. In reverse genetics, the simple concept is to inactivate the gene, and then determine the changes of the phenotypes; alternatively, to add more of the gene products, and then determine the changes of the phenotypes. In the former, genes can be specifically targeted to "knockout" the gene. This worked really well with some model species where embryonic stem cell technology is available. For instance, with rat or mouse, a specific gene can be knocked out in an embryonic cell line, and then individuals can be developed from the cells, and phenotypes can be observed in the "knocked out" animals. However, this approach is to date not applicable to aquaculture species because embryonic stem cell technologies have not been developed for aquaculture species. In recent years, the discovery of RNA interference (RNAi) has lent research tools for the study of gene functions. RNAi is an RNAdependent gene silencing process that has been applied to knockdown gene expression (Hannon, 2002). This is particularly useful for some aquaculture species, as RNAi technologies have been applied to investigate gene function as well as to develop antiviral agents to combat various infections in some fish and crustaceans species (Acosta et al., 2005; Copf et al., 2006; Liu et al., 2006; Wargelius et al., 1999; Kelly and Hurlstone, 2011). For instance, the highly efficient gene knockdown was observed to result in similar embryonic defects to the known mutant phenotypes in zebrafish (Wargelius et al., 1999). In a study of genes expressed differentially in freshwater crayfish infected with the white spot syndrome virus (WSSV), the anti-lipopolysaccharide factor gene (AF) was shown to protect against WSSV infection, and knockdown of AF by RNAi specifically resulted in higher rates of viral propagation (Liu et al., 2006). In addition to gene knockout and knockdown technologies, transgenic technology has been widely used to demonstrate the functions of genes. The basic principle is that if a gene has certain functions, its over-expression should cause changes in phenotypes. Such an approach was best demonstrated by transgenic fish harboring the growth hormone gene that grow much faster and bigger than their nontransgenic controls (Du et al., 1992; Gross et al., 1992; Devlin et al., 1994; Rahman et al., 1998; Rahman et al., 2001). All the traditional approaches for the study of gene functions are effective, but they have fatal problems that include the following:

- Not all genes can cause a visible phenotype.
- Knockout of one gene may cause numerous other changes in genome expression; many of these are compensatory or consequential, making analysis of gene functions very difficult.
- Functional study of gene function by "one gene at a time" is too laborious, too expensive, and too slow. As a result, scientists have explored to study gene functions on the genomic scale that leads to the emergence of functional genomics.

Functional genomics can be defined as a discipline for the understanding of gene functions and regulation on a genome-wide scale. It is a field of molecular biology that attempts to make use of the vast wealth of data produced by genomic projects (such as genome sequencing projects) to describe gene (and protein) functions and their interactions. Unlike structural genomics and proteomics, functional genomics focuses on the dynamic aspects such as gene transcription, translation, and protein–protein interactions and interactions between proteins, DNA, and RNA. As opposed to the static aspects of the genomic information such as nucleotide and amino acid sequences or structures, functional genomics attempts to answer questions about the functions of DNA at the levels of genes, RNA transcripts, and protein products (Hackett and Clark, 2007). The ultimate goal for functional genomics is to bridge the gap between the blueprint (genome sequence or genotype) and the living organism (trait or phenotype) under various environmental conditions (Cogburn et al., 2007). Spawned by the technological revolution in genome sequencing (Venter et al., 1996; Rowen et al., 1997), functional genomics studies currently are greatly stimulated by the high-throughput sequencing and screening technologies. Therefore, a key characteristic of functional genomics studies is the genome-wide approach generally involving high-throughput technologies rather than the traditional "gene-by-gene" approach.

Approaches to Functional Genomics

Numerous approaches with sophisticated experimental techniques have been developed for functional genomics studies. The huge numbers of genes, and even larger numbers of transcripts contribute to the complexity of functional genomics analysis. Such complexities are amplified again by an even greater diversity of polypeptides and their post-translational modifications, the variation in expression from different tissues under different developmental stages, and under various environmental conditions (Hackett and Clark, 2007). High throughput methodologies are required to simultaneously examine the expression and functions of all genes of an organism. Clearly, high-speed computation is an essential feature of most functional genomics techniques.

We describe some of the general approaches currently used in functional genomics studies, with the understanding that more powerful methods are constantly emerging, especially as computational power continues to improve. The improvements include tools and instruments for digital recording of information and newer algorithms for sorting and analyzing information developed as our understanding of gene function increases. Because of the large quantity of data generated by these techniques and the desire to discover biologically meaningful patterns, bioinformatics is crucial to functional genomics data analysis. While whole genome sequencing is now a routine, functional analysis continues to be a great challenge. There are several general strategies for elucidating the functions and regulation of gene expression. In terms of strategies, functional genomics approaches can be divided into four general categories:

- Functional correlation of gene expression and thereby inference of gene functions;
- Functional correlation of gene positions and thereby inference of candidate gene functions;
- Functional assignment by association with gene pathways; and
- Direct testing of gene functions.

In breeding programmes, information on family relations of individuals is used when estimating genetic parameters (heritabilities and genetic correlations) and breeding values for traits, and when optimizing selection and mating in order to avoid inbreeding. Similarly, pedigrees are useful in the management of conservation programmes and wild populations. In addition to physical individual tagging, pedigrees can be determined using DNA markers. This procedure is quite straightforward using microsatellites. Typically 10–20 variable genetic markers are needed to assign >95% of individuals to single pairs of parents. To do this, one needs tissues samples from both potential parents and their offspring, and several freely available software exists for parental assignment. Physical individual markers are useful when large facilities exist where family groups can be held separately until fish are large enough to be individually tagged. For instance, in salmonid breeding programs fish are typically held in hundreds of family tanks until they are individually tagged at a weight of approximately 50g. Using DNA pedigrees is useful in many aquaculture breeding and conservation programmes when individual tagging is difficult or when facilities for family tanks do not exist. Three examples of using DNA pedigrees are presented below: walk-back selection, estimation of genetic parameters, and conservation programmes. Walk-back selection refers to a selection programme where a group of superior individuals are first selected, and then only the selected animals are genotyped for family relations. Using the established pedigree, only those superior animals that are not too closely related are used in matings. This is an improved mass selection scheme to obtain genetic improvement while simultaneously controlling for inbreeding. This is cost-effective because only some hundreds of individuals among the potentially (tens of) thousands of individuals reared need to be genotyped. This is especially useful for species for which no extensive resources are available, or for new species whose reproduction cannot be fully controlled. Furthermore, there are studies showing that microsatellite markers are useful for determining the effective number of parents and their individual reproductive success, and level of inbreeding in mass selection scheme in the Nile tilapia Oreochromis niloticus (Komen H., Animal Breeding and Genetics Group, Wageningen University, The Netherlands, pers. comm.) and the flat oyster, Ostrea edulis. A full population can be genotyped for parental analysis, which allows one to estimate heritabilities and genetic correlations to traits of interest. Such an approach has been used on common carp. Likewise, DNA pedigrees are useful in conservation programmes of wild fish and shellfish, especially when aiming at controlling inbreeding. Microsatellite genotyping in the induced mass spawns of lion-paw scallop Nodipecten subnodosus demonstrated that some parents contribute a much higher percent to the progeny than expected (Petersen J.L., Genomic Variation Laboratory, University of California Davis, Davis, CA USA, pers. comm.). Similarly, the impact of hatchery practices on the genetic variability of progenies can be monitored. That is, when the effective population size is reduced and inbreeding can decrease hatchery stocks. Microsatellite genotyping can be also useful in assessment of introgression to/from a natural population. Multiplexing (i.e. simultaneous PCR amplification) of microsatellites and SNP-based parentage assignment are now greatly facilitating these types of studies.

Aquaculture Genome Technologies

Genomics is a rapidly growing scientific field with applications ranging from improved disease resistance to increased rate of growth. Aquaculture Genome Technologies comprehensively covers the field of genomics and its applications to the aquaculture industry. This volume looks to bridge the gap between a basic understandings of genomic technology to its practical use in the aquaculture industry. Aquaculture Genome Technologies provides the reader with a tour through the aquaculture genomics process. Aquaculture Genome Technologies is edited by an international leader in the field of aquaculture genome research, with contributions from an international team of leading researchers. Aquaculture Genome Technologies is an essential reference for anyone with an interest in aquaculture genetics.

CHAPTER 1

SYSTEMATIC SEQUENCING OF MRNA FROM THE ANTARCTIC KRILL (EUPHAUSIA SUPERBA) AND FIRST TISSUE SPECIFIC TRANSCRIPTIONAL SIGNATURE

Cristiano De Pittà, Cristiano Bertolucci, Gabriella M Mazzotta, Filippo Bernante, Giorgia Rizzo, Barbara De Nardi, Alberto Pallavicini, Gerolamo Lanfranchi and Rodolfo Costa

BACKGROUND

Euphausiacea (krill) are small shrimplike crustaceans that are abundant in the pelagic ecosystems of all oceans. There are about 85 species of Euphausiacea, making this one of the smallest orders in the class of Malacostraca [1]. Phylogenetic analysis of the Eumalacostraca orders based on 28S rDNA sequences suggests that Euphausiacea are more closely related to Mysida than to the Decapoda [2].

In the Southern Ocean, krill is a critical link between primary productivity and most of the predators at higher trophic levels such as birds, fish, seals, squid and whales [3]. The krill biomass in the Southern Ocean has been estimated at 400–1550 million tons with sustainable annual harvest at around 70–200 million tons. Therefore, krill biomass that could be available for human food is comparable to the biomass of all the other aquatic species currently fished by humans, but only six species of krill are at present harvested commercially [4, 5]. Commercial fishing of krill is done in the Southern Ocean and around Japan. The global annual production amounts to 150 – 200.000 tons, most of this from the Scotia Sea [6, 7]. Most of the fished krill is used for aquaculture and aquarium feedings, as baits in sport fishing, or in the pharmaceutical industry.

The Antarctic krill (Euphausia superba, Dana 1852) has a circumpolar distribution with the highest concentrations in the Atlantic sector of the Southern Ocean. It is a key species of the Antarctic ecosystem and plays an important role both as feeder of algae, bacteria and micro-zooplancton and as a prey of vertebrates [8]. E. superba displays a large daily vertical migration that occurs generally within the upper 200 m water column making a significant amount of biomass available as food for predators near the surface at night and in deeper waters during the day [9]. Basic knowledge of crustacean biology is limited by the lack of information about their genomes. Considering all orders in the class of Malacostraca, no genome has yet been fully sequenced. At present Genbank carries 116,640 nucleotide and 11,932 protein sequences (Table 1), with a high rate of redundancy. Currently only 434 nucleotide and 310 protein sequences have been identified in Euphausiacea (GenBank source, release of November 2007). Specifically for E. superba only 69 nucleotide and 17 amino acid sequences have been obtained; they identify key proteins and enzymes of oxidative phosphorylation (NADH dehydrogenase subunit 1, 2, 3, 4, 4L, 5; Cytochrome oxidase subunit I, II, III; ATP synthase subunit 6; cytochrome b; cytochrome b apoenzyme and cytochrome c oxidase subunit I) and of phototransduction (opsin). In the subphylum Crustacea there are 33 complete (or nearly complete) mitochondrial DNA sequences: 4 Branchiopoda, 8 Maxillopoda, 18 Malacostraca, and one of Ostracoda, Cephalocarida and Remipedia (Table 2). In a previous investigation Machida et al. [10] determined the nearly complete DNA sequence of the mitochondrial genome of E. superba (14,606 bp).

Citation: Cristiano De Pittà, Cristiano Bertolucci, Gabriella M Mazzotta, Filippo Bernante, Giorgia Rizzo, Barbara De Nardi, Alberto Pallavicini, Gerolamo Lanfranchi and Rodolfo Costa, Systematic sequencing of mRNA from the Antarctic krill (Euphausia superba) and first tissue specific transcriptional signature, doi:10.1186/1471-2164-9-45

Table 1: Nucleotide and protein sequences belonging to all orders of the Malacostraca class available from public databases at November 2007.

Class	Subclass	Superorder	Order or Sub-order	Nucleotide sequences	Protein sequences
Malacostraca	Eumalacostraca	Eucarida	Decapoda	97,083	7,862
Malacostraca	Eumalacostraca	Eucarida	Euphausiacea	434	310
Malacostraca	Eumalacostraca	Hoplocarida	Stomatopoda	119	214
Malacostraca	Eumalacostraca	Peracarida	Amphipoda	14,732	1,745
Malacostraca	Eumalacostraca	Peracarida	Cumacea	93	82
Malacostraca	Eumalacostraca	Peracarida	Isopoda	3,355	1,155
Malacostraca	Eumalacostraca	Peracarida	Mictacea	1	0
Malacostraca	Eumalacostraca	Peracarida	Mysidacea	771	561
Malacostraca	Eumalacostraca	Peracarida	Spelaeogriphacea	1	0
Malacostraca	Eumalacostraca	Peracarida	Tanaidacea	18	1
Malacostraca	Eumalacostraca	Peracarida	Thermosbaenacea	4	0
Malacostraca	Eumalacostraca	Syncarida	Anaspidacea	27	2
Malacostraca	Eumalacostraca	Syncarida	Bathynellacea	2	0
Total				**116,640**	**11,932**

Table 2: List of complete or nearly complete mitochondrial DNA sequences of the subphylum Crustacea available from public databases at November 2007.

#	Class	Subclass	Organism	Refseq	Reference
1	Branchiopoda	Phyllopoda	*Triops longicaudatus*	NC_006079	[50]
2	Branchiopoda	Sarsostraca	*Artemia franciscana*	NC_001620	Unpublished
3	Branchiopoda	Phyllopoda	*Daphnia pulex*	NC_000844	[51]
4	Branchiopoda	Phyllopoda	*Triops cancriformis*	NC_004465	[52]
5	Cephalocarida	Brachypoda	*Hutchinsoniella macracantha*	NC_005937	[53]
6	Malacostraca	Eumalacostraca	*Halocaridina rubra*	NC_008413	[54]
7	Malacostraca	Eumalacostraca	*Geothelphusa dehaani*	NC_007379	[55]
8	Malacostraca	Eumalacostraca	*Ligia oceanica*	NC_008412	[56]
9	Malacostraca	Eumalacostraca	*Squilla mantis*	NC_006081	[50]
10	Malacostraca	Eumalacostraca	*Harpiosquilla harpax*	NC_006916	[57]
11	Malacostraca	Eumalacostraca	*Pseudocarcinus gigas*	NC_006891	[58]
12	Malacostraca	Eumalacostraca	*Macrobrachium rosenbergii*	NC_006880	[58]
13	Malacostraca	Eumalacostraca	*Callinectes sapidus*	NC_006281	Unpublished
14	Malacostraca	Eumalacostraca	*Squilla empusa*	NC_007444	Unpublished
15	Malacostraca	Eumalacostraca	*Lysiosquillina maculata*	NC_007443	Unpublished
16	Malacostraca	Eumalacostraca	*Gonodactylus chiragra*	NC_007442	Unpublished
17	Malacostraca	Eumalacostraca	*Marsupenaeus japonicus*	NC_007010	[59]
18	Malacostraca	Eumalacostraca	*Eriocheir sinensis*	NC_006992	[60]
19	Malacostraca	Eumalacostraca	*Cherax destructor*	NC_011243	[61]
20	Malacostraca	Eumalacostraca	*Panulirus japonicus*	NC_004251	[62]
21	Malacostraca	Eumalacostraca	*Portunus trituberculatus*	NC_005037	[63]
22	Malacostraca	Eumalacostraca	*Pagurus longicarpus*	NC_003058	[64]
23	Malacostraca	Eumalacostraca	*Penaeus monodon*	NC_002184	Unpublished
24	Maxillopoda	Copepoda	*Tigriopus californicus*	NC_008831	Unpublished
25	Maxillopoda	Cirripedia	*Pollicipes mitella*	NC_008742	Unpublished
26	Maxillopoda	Branchiura	*Argulus americanus*	NC_005935	[50]
27	Maxillopoda	Copepoda	*Tigriopus japonicus*	NC_003979	[65]
28	Maxillopoda	Copepoda	*Lepeophtheirus salmonis*	NC_007215	[66]
29	Maxillopoda	Cirripedia	*Megabalanus volcano*	NC_006293	Unpublished
30	Maxillopoda	Cirripedia	*Pollicipes polymerus*	NC_005936	[53]
31	Maxillopoda	Cirripedia	*Tetraclita japonica*	NC_008974	Unpublished
32	Ostracoda	Myodocopa	*Vargula hilgendorfii*	NC_005306	[67]
33	Remipedia	Nectiopoda	*Speleonectes tulumensis*	NC_005938	[53]

The identification of novel shrimp genes by systematic sequencing of genomic DNA is hindered by the dispersion of the genes among large non-coding regions and by the presence of introns within genes. Current genomics technologies, like SAGE [11], differential display [12] and systematic sequencing of expressed sequence tags [13, 14], are very useful approaches to identify protein coding genes rapidly on a large scale. Moreover, the frequency of a given sequence in the SAGE or cDNA libraries can be related to the relative abundance of the corresponding mRNA, giving an indication of the level of gene expression [15].

The aim of our study was to significantly increase the number of krill genes in the public database and to discover tissue specific genes. For this purpose we have produced and sequenced five cDNA libraries from different Antarctic krill tissues: head, abdomen, thoracopods and photophores. We have developed special cDNA libraries optimized to directionally cloning full-length cDNA in plasmid vectors without enzymatic digestion. We have identified 1,770 high-quality EST clones that have been grouped in 1,017 different clusters. Of these, 309 clusters were successfully annotated while 708 did not show a significant similarity with known genes from other organisms. Clusters, consensus and related similarity and gene ontology searches were organized in a dedicated E. superba database [16].

RESULTS AND DISCUSSION

Construction of cDNA libraries and EST analysis

Total RNA was independently isolated from tissues (head, abdomen, thoracopods and photophores) dissected from specimens of E. superba collected at five different time points during 24 hours. The analysis of total RNA samples, performed by capillary electrophoresis, showed absence of genomic DNA contamination and a peculiar electropherogram as shown in Fig. 1. In particular, total RNAs show low molecular weight from 200 bp to 1 kb, perhaps as result of a partial RNA degradation.

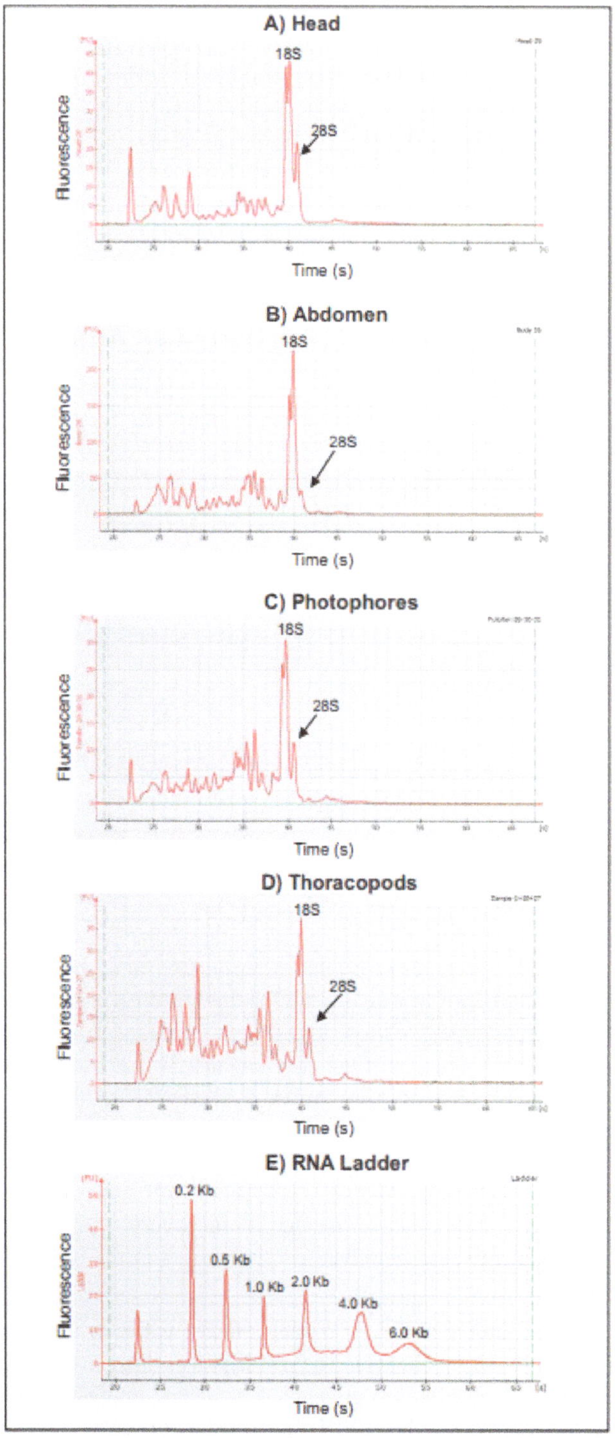

Figure 1: Electropherograms of E. superba tissue-specific total RNAs. (A-D) Electropherograms resulting from Agilent 2100 bioanalyzer analysis on total RNA extracted from head, abdomen, photophores and thoracopods. X-axis: time of ribosomal RNA peak appearance, corresponding to the size of the fragment; Y-axis: fluorescence of the peak, corresponding to its concentration. The size and the concentration of the sample peaks are calculated by the software via comparison with a RNA ladder at known concentration (**E**). E. superba RNA samples showed some products with a migration time between 22 and 35 seconds (from 200 bp to 1.000 bp) indicating a partial RNA degradation.

Five independent tissue-specific cDNA libraries, named K01 and K05 (head), K06 (abdomen), K07 (photophores) and K09 (thoracopods), were produced from total RNA pools. For head we have sequenced only K05 cDNA library because it presented more recombinant clones compared to K01. Recombinant bacterial clones from each library were randomly picked and the EST were sequenced from the 5'-end. The average insert size for all libraries was estimated to be 412 bp.

EST assembly and construction of an Antarctic krill transcript catalogue

A total of 2,046 ESTs were initially analyzed for sequence quality and vector sequences were recognized and deleted. Two-hundred-seventy-six low quality ESTs were removed and 1,770 (86.5%) high-quality ESTs were further processed. These ESTs assembled by similarity into 216 clusters and 801 singletons, resulting in a total of 1,017 non-redundant (consensus) sequences. A list of the sequencing trend for each cDNA library is presented in Table 3. Interestingly, we obtained a low percentage of clusters composed by ≥ 2 ESTs from cDNA libraries prepared from the krill head (K01, K05) showing that no specific transcripts are particularly enriched in this tissue. As expected from the above data, the head reveals a higher percentage (about 65%) of new putative transcripts, with respect to abdomen (52.3%), photophores (50.2%) and thoracopods (44.4%) in which the presence of enriched transcripts was instead revealed by EST sequencing.

Table 3: Results of EST assembly for each krill cDNA library.

Library	Tissue	Total ESTs	Discarded ESTs	Analyzed ESTs	# GenBank	# EST in cluster	# cluster	# singletons	# consensus	% redundancy	% discovery
K01	Head	35	9	26	ES542703–ES542728	12	10	14	24	38.46	61.54
K05	Head	579	124	455	ES542729–ES543183	225	108	230	338	35.16	64.84
K06	Abdomen	508	47	461	ES543184–ES543644	250	110	211	321	42.73	57.27
K07	Photophores	506	52	454	ES543645–ES544098	255	114	199	313	49.78	50.22
K09	Thoracopods	418	44	374	ES544099–ES544472	227	108	147	255	55.61	44.39
Overall		2046	276	1770		969	450	801	1251	44.35	55.65

Total ESTs = number of produced chromatograms; Discarded ESTs = number of low quality ESTs; Analyzed ESTs = number of sequences processed for clustering; # GenBank = accession numbers of ESTs deposited in GenBank; # EST in cluster = number of sequences in cluster; # clusters = total number of clusters; # singletons = number of putative transcripts identified by one EST; # consensus = number of non-redundant sequences; % redundancy = percentage of sequences not identified in an exclusive manner; % discovery = percentage of ESTs identifying putative new transcripts in the total EST analyzed.

The number of ESTs in each cluster varies from 2 (93 clusters) to 88 (1 cluster). The average length of a cluster is 436 bp with the longest assembled sequence being 1,417 bp and the shortest 153 bp. The 91% of cluster consensus contains the 3'-end region of mRNAs as demonstrated by the presence of a poly-adenylation signal.

Each non-redundant sequence was searched in the nucleotides database and UniProtKB database using Blast-N and Blast-X with an e-value cut off of < e^{-40} and < e^{-10}, respectively. These values were empirically chosen considering the low amount of sequences data available for Euphausiacea and similar shrimp species and the need of stringency in providing a reliable catalogue of Antarctic krill genes. All annotations were further manually examined, in order to assign the best describing text to the correspondent cluster.

Overall, 70% of non-redundant sequences (708 out of 1,017), identified by about 50% of total produced ESTs, showed no or poor similarity matches, and they probably represent completely unknown Antarctic krill transcripts that could be characterized in future studies. Additional file 1 lists the 309 (30%) non-redundant sequences identifying known Euphausia genes or sequences showing significant similarity to genes from arthropods (63.4%) and other species (36.6%), such as Homarus americanus (3.2%), Aedes aegypti (12.0%), Drosophila melanogaster(3.9%), Bombyx mori (3.6%), Brachydanio rerio (4.9%), Mus musculus (3.9%), Rattus norvegicus (3.9%), Homo sapiens (4.5%). Antarctic krill sequences generally show a greater similarity to genes of insects (about 36%) than to genes of Malacostraca (about 23%) and only 7% were similar to the known sequences of Euphausiacea. This result could be due to the limited number of Malacostraca gene and protein sequences available in the public databases (gene sequences: 116,666 and protein sequences: 11,941 at November 2007) respect to insects (gene sequences: 1,869,511 and protein sequences: 336,246 at November 2007). The total collection of 1,770 E. superba 3'-EST has been deposited in the EBI-GenBank-DBJ database (GenBank accession numbers from ES542703 to ES544472).

One of the most noticeable features of our EST catalogue is that mitochondrial transcripts are quite abundant (about 10% of the total ESTs): 84 ESTs (about 4.7%) matched with nine different E. superba mitochondrial transcripts (ATPase6, COI, COII, COIII, cyt b, ND1, ND2, ND3, ND6) and about 5% ESTs (88 out of 1,770) identified the large mitochondrial 16S rRNA gene. In future experiments, to avoid the repetitive sequencing of this abundant mRNA, we plan to introduce, during cDNA insert amplification, interference primers specifically designed for the 16S rRNA [17]. Moreover, we found a small percentage of ESTs (1.6%), showing similarity with 18S and 28S rRNA. About 2% of ribosomal RNAs contamination is common to other systematic sequencing projects.

The E. superba mitochondrial gene sequences are very similar to those commonly found in the mitochondrial genomes of other arthropods, including 13 protein-coding genes, 19 tRNA and 2 rRNA genes [10]. Machida et al. [10] have demonstrated that mitochondrial protein-coding genes are transcribed from the same DNA strand, left to right, except for ND1, ND4L, ND4 and ND5 genes, while two ribosomal RNA genes are encoded by L strands.

We did not find highly represented tissue-specific mRNAs, while some ribosomal proteins like L22 (31 ESTs), S25 (31), P1 (16), L37A (15), L24 (13), P2 (13), S14 (9), S14 (10), L34 (9) were expressed at the same level in all analyzed krill tissues. The only genuine tissue-specific mRNA appears to be myosin light chain (8 ESTs, cluster KRC00032) that is highly represented in a strictly committed tissue such as skeletal muscle localized in abdomen (library K06). Moreover, we found 3 unknown transcripts among the 20 most expressed genes: KRC00118 (10 ESTs), KRC00407 (9 ESTs) and KRC00101 (7 ESTs) that could be interesting for future functional studies.

Since very few abundant transcripts were found that could hamper the identification of rare transcripts, it seems plausible that random sequencing of our Antarctic krill libraries would continue to represent an effective strategy for identifying novel E. superba mRNAs.

Functional Categorization of *E. superba* ESTs

In order to facilitate functional genomic studies in Antarctic krill, 309 consensus sequences showing similarity with known genes or proteins were grouped into 14 functional categories (Table 4) according to Gene Ontology [18] and other resources developed for gene functional annotation [19]. A list of all annotated transcripts is shown in Additional file 1. A large majority of ESTs (20.51%), displaying putative identity with ribosomal sequences and genes for the translation machinery, were grouped in the translation functional category, characterized by 8% of all known transcripts. We found genes with regulative functions in the translational initiation, like translation factor SUI1,initiation factor 4A and 3, in the translational elongation like elongation factor 1α, 1β, 2 and a specific tail muscle elongation factor 1γ. Other abundant E. superba sequences fall into gene categories related to cell structure, cell motility and functional homeostasis. For instance, genes involved in the mechanisms of DNA transcription (1.47%), transport (1.97%), skeletal muscle contraction (3.4%), and in amino acid, fatty acid and carbohydrate metabolism (4.03 %) are comprised in this category. This class includes also SEC61 β-subunit, an important transport protein that plays a crucial role in the insertion of secretory and membrane polypeptides into the endoplasmic reticulum and cellular retinoic acid/retinol binding protein (RBP1), involved in the transport of retinol from the digestive gland to peripheral tissues [20]. A transcript included in this class (ID: KRC00589) shows a good similarity with hemocyanin, the main oxygen carrier molecule in arthropods and molluscs [21].

Table 4: Classification of the 1,017 krill unique consensus sequences in functional categories.

#	Functional categories	Biological process	# Clusters	% Cluster	# ESTs	% ESTs
1	**DNA replication & binding**	DNA replication	4	**0.79**	**8**	**0.45**
		DNA binding	4			
2	**Transcription**	Transcription process	2	**1.47**	**16**	**0.90**
		Regulation of transcription	8			
		Chromatine assembly and remodelling	5			
3	**Translation**	Translation	70	**8.06**	**363**	**20.51**
		Translation initiation	4			
		Translation elongation	7			
		Translation termination	1			
4	**Transport**	Electron transport	8	**1.97**	**32**	**1.81**
		Calcium ion transport	2			
		Intracellular protein transport	4			
		Transport	6			
5	**Metabolic processes**	Amino acid metabolic process	2	**4.03**	**73**	**4.12**
		Carbohydrate metabolic process	13			
		Fatty acid biosynthetic process	3			
		Various metabolic process	23			
6	**Proteolysis, protein folding and modification**	Proteolysis	16	**3.24**	**48**	**2.71**
		Protein modification and binding	9			
		Protein folding	8			

7	**Striated muscle contraction**	Contraction	26	**3.54**	**95**	**5.37**
		Calcium binding	10			
8	**Signal transduction**		8	**0.79**	**9**	**0.51**
9	**Structural constituent of cuticle**		14	**1.38**	**30**	**1.69**
10	**Cell growth, proliferation and adhesion**		9	**0.88**	**12**	**0.68**
11	**Ion Binding**		5	**0.49**	**5**	**0.28**
12	**Hypotetical protein**		10	**0.98**	**16**	**0.90**
13	**Mitochondrial genes**		14	**1.38**	**174**	**9.83**
14	**Ribosomal RNA gene (18S and 28S)**		14	**1.38**	**29**	**1.64**
15	**Without similarity**		708	**69.62**	**860**	**48.59**
TO-TAL			**1017**	**100.00**	**1770**	**100.00**

The last four columns of the Table show respectively number and percentage of clusters and ESTs belonging to each functional category.

Other interesting krill transcripts that we were able to annotate are those involved in stress responses, proteolysis and immunoresponse (3.24%) like Hsp90, chaperones, cathepsine L-like cysteine protease, a lysosomal cysteine proteinase [22], peptidyl-prolyl cis-trans isomerase A1 (cyclophilin A1) and peptidylprolyl isomerase B (cyclophilinB). Cyclophilines are members of the immunophilin protein family, which play a role in immunoregulation and basic cellular processes involving protein folding and trafficking. ESTs with good similarity to hemocyanin are present in our collection: this protein has been recently reported to have antifungal and antiviral activities [23, 24].

Some krill ESTs identify histone 2A (KRC00431) and histone 3.3A (KRC00024) indicating the presence of unexpected polyadenylated histone transcripts displaying the polyadenylation signal and tail. In vertebrates, these evolutionary conserved housekeeping mRNAs are not polyadenylated, and this has been related to the high turnover of these transcripts in the dividing cells. Interestingly, polyadenylated H2A and H3 histone sequences were detected also in the systematic sequencing of 3'-end cDNA libraries obtained from brain and kidney of channel catfish Ictarulus punctatus [25, 26] and from various tissues (haemolymph, gills, digestive glands, mantles and adductor muscles) of the mussel Mytilus galloprovincialis [27]. The presence of polyadenylation signals in E. superba histone transcripts deserves a more detailed analysis. In fact, Eirin-Lopez et al. [28] have recently shown that all histone genes in the repetitive unit are characterized by two different mRNA termination signals in their 3' UTR: the typical stem-loop or hairpin-loop signal followed by a purine-rich element and a polyadenylation signal AATAAA located downstream to this last element. The presence of a double mRNA termination signal is unique to histone genes and common for other invertebrates such as Chironomus thummi [29], D. melanogaster [30], Chaetopterus variopedatus [31], M. galloprovincialis [32] and Crustacea [33]. Although in some invertebrates core histone transcripts (H2A, H2B, H3 and H4) include polyA tails, these sequences are among the most evolutionary conserved eukaryotic proteins [34].

Transcriptional signature of *E. superba* tissues

Gene expression profiling depends on the functional specificity of cells composing different tissues. So, the systematic sequencing of EST from unbiased cDNA libraries is a suitable approach for analyzing the gene expression profile of a given tissue [35]. In fact, the frequency of a given EST in the cDNA library can be related to the relative abundance of the corresponding mRNA in the source tissue.

To define tissue transcriptional signatures of E. superba, annotated ESTs obtained from the four tissue-specific cDNA libraries (head, abdomen, photophores, thoracopods) were separately grouped in 13 functional categories (Table 5) a further abundant category was created for those ESTs to which no function may be yet associated. Fig. 2 shows four different diagrams standing for ESTs distribution among functional categories in each cDNA library. The presence of highly represented functional categories is peculiar of strictly committed tissues such as abdomen and thoracopods in which transcripts involved in striated muscle contraction are very abundant (about 26% in abdomen and 7% in thoracopods). In the abdomen library, we were able to recognize the principal structural components of the sarcomeric contractile machinery (myosin heavy chain, myosin light chain 1, myosin-2, actin, alpha-tubulin, tropomyosin) and two subunits of the troponin complex (troponin T, troponin I), a key regulator of muscle contraction. About 10% of sequences produced from head and thoracopods libraries fall in functional categories related to metabolic processes (amino acid, fatty acid and carbohydrate metabolism). Interestingly, about 6% and 4% of ESTs respectively sequenced from head and thoracopods libraries identified structural constituents of cuticle (arthrodial cuticle protein AMP16.3, arthrodial cuticle protein AMP1A, calcification-associated peptide-1 precursor). This reflects the presence of cuticle traces in the head and thoracopods samples. In photophores and thoracopods transcripts displaying putative identity with ribosomal sequences are more abundant compared to other tissues (55% and 46%, respectively), indicating a relevant activity of the translation machinery.

Table 5: Classification of the annotated ESTs for each library.

#	Process	Head		Abdomen		Photophores		Thoracopods	
		# ESTs	%	# ESTs	%	# ESTs	%	# ESTs	%
1	**DNA replication**	3	1.43	2	0.90	2	0.88	1	0.45
2	**Transcription**	3	1.43	6	2.71	5	2.19	2	0.90
3	**Translation**	69	32.86	65	29.41	126	55.26	103	46.40
4	**Transport**	5	2.38	12	5.43	7	3.07	8	3.60
5	**Metabolic process**	22	10.48	16	7.24	14	6.14	21	9.46
6	**Proteolysis, protein folding and modification**	12	5.71	6	2.71	14	6.14	16	7.21
7	**Striated muscle contraction**	11	5.24	57	25.79	12	5.26	15	6.76
8	**Signal transduction**	4	1.90	1	0.45	3	1.32	1	0.45
9	**Structural constituent of cuticle**	13	6.19	1	0.45	7	3.07	9	4.05
10	**Cell growth, proliferation and adhesion**	2	0.95	3	1.36	2	0.88	5	2.25
11	**Ion Binding**	4	1.90	0	0.00	1	0.44	0	0.00
12	**Hypotetical protein**	4	1.90	6	2.71	6	2.63	0	0.00
13	**Mitochondrial genes**	58	27.62	46	20.81	29	12.72	41	18.47
	Total	**210**	100.00	**221**	100.00	**228**	100.00	**222**	100.0

Table showing the number and the percentage of ESTs in each functional category for each tissue specific library

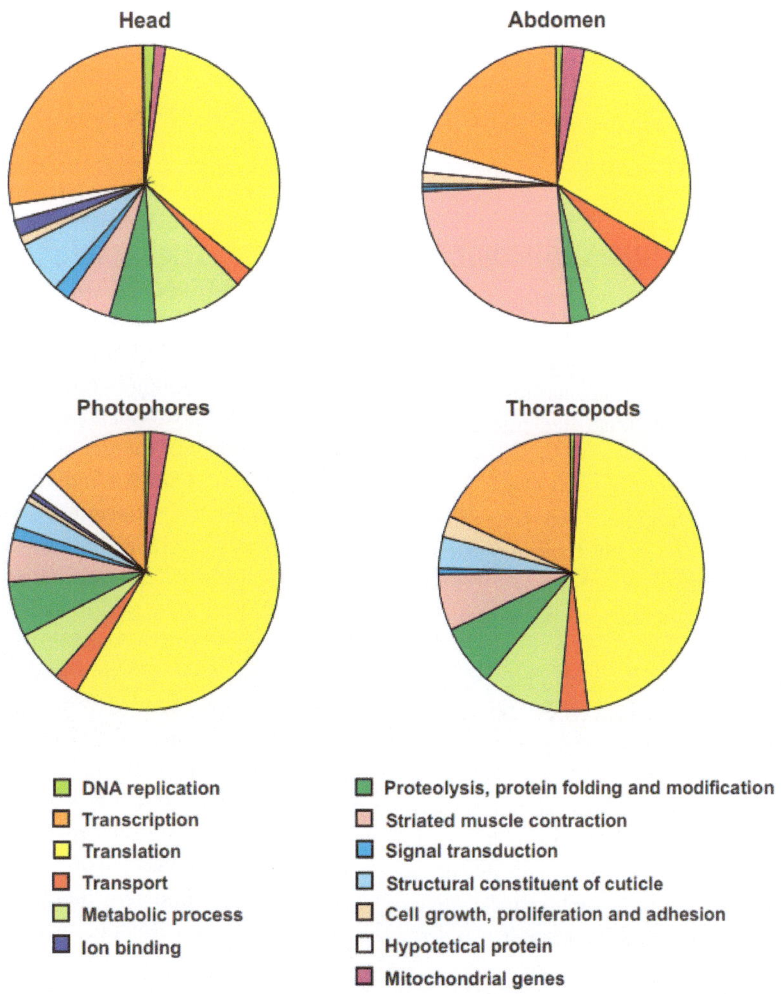

Figure 2: Classification of the annotated ESTs for each library into different 13 functional categories. Diagrams showing the proportion of each functional category in all four tissues. See Table 5 for more details.

We have also identified from the head cDNA library a novel opsin sequence (ID ESTs: KRC00735, KRC00802), a light-sensitive membrane-bound G protein-coupled receptors mediating the conversion of a photon of light into an electrochemical signal in the visual transduction cascade. In insects there are at least four main spectral classes: long-wavelenght-sensitive (LWS), middle-wavelenght-sensitive (MWS) and two short-wavelenght-sensitive (SWS) groups. The opsin sequences available for E. superba (GenBank accession no. DQ852576–DQ852580) show a spectral sensitivity with short wavelength (496–501 nm, λ_{max} = 487) and cannot be aligned with our consensus [36].

Quantitative RT-PCR analysis was performed to quantify and validate the expression level of some genes presenting different EST countings in krill tissues. We selected ten genes (compound eye opsin BCRH1, myosin light chain, myosin heavy chain, arthrodial cuticle protein AMP16.3, tail muscle elongation factor 1 gamma, cellular retinoic acid/retinol binding protein, eukaryotic initiation factor 4A, transport protein SEC61 subunit gamma, chromodomain helicase DNA binding protein and voltage-dependent calcium channel) representative of different levels of transcript abundance.

The housekeeping gene 18S rRNA was used as endogenous control. As reported in the Additional file 2, the expression values obtained with the quantitative RT-PCR for the tested transcripts were in agreement with the EST counting in the four libraries. In particular, we have demonstrated that the compound eye opsin is strongly expressed in head compared to other tissues and myosin light chain and myosin heavy chain are highly expressed in abdomen and thoracopods confirming their key role in the contractile machinery. Instead, the eukaryotic initiation factor 4A is expressed at about the same level in all tested tissues.

Identification of Microsatellite-Containing ESTs

Among the 1,017 non-redundant sequences examined in this study, 41 (4%) consensus sequences containing ESTs were identified by using MISA software. Twelve of these consensus present 2 distinct simple sequence repeats interrupted by more than 100 bp for a total of 69 identified microsatellites (SSR). The majority of these sequences (72%) fall into the 3 bp repeat type class with a preponderance of GAA and GAT. After a manual inspection of redundancy, raw sequence, data quality and the presence of sufficient flanking sequences we designed 9 pairs of specific PCR primers. We obtained successful amplifications for 6 of these 9 pairs of primers. Assessment of polymorphism information content (PIC), observed and expected heterozygosity and other population genetics analysis will be performed in the near future. These markers will increase the currently available Euphausiacea SSR markers. In fact, only five microsatellite loci isolated from the northern krill Meganyctiphanes norvegica have been reported so far [37]. Since our novel microsatellite markers were developed on the basis of expressed sequences and they are presumably conserved across other Euphausiacea species, they could also be useful for comparative mapping and for a molecular approach to Antarctic krill ecology.

CONCLUSION

Since genome sequencing and BAC libraries of Antarctic krill are not yet available, EST sequencing from randomly selected cDNA clones represents a powerful approach to identify large numbers of transcripts that could be used in gene expression and functional genomics studies [38]. The systematic sequencing of four cDNA libraries prepared from different E. superba tissues has allowed us to establish an EST database containing 1,017 unique sequences. Over 65% of the Antarctic krill sequences resulted in no BLAST matches with published sequences and they probably represent novel genes that could be functionally characterized. We have defined the transcriptional signatures of krill tissues and performed qRT-PCR to validate the level of expression of ten representative genes. All sequencing data have been deposited on the E. superba EST database available from our web site [16]. In addition, the EST collection is a potential source for the development of genetic markers including microsatellite and single nucleotide polymorphisms. Among the 1,017 unique sequences, 41 (4%) unique microsatellite containing ESTs were identified by using MISA software. Moreover, we have designed and successfully tested 6 pairs of specific PCR primers for microsatellite loci.

Our EST catalogue could provide a source for the design of microarray platform that will allow the study of the transcriptional responses of this abundant marine organism to environmental challenges [39].

METHODS

Tissues Samples, RNA Extraction and Quality Control

Antarctic krill (Euphausia superba) were fished from the Ross Sea (longitude: 167°28'81" E – 179°54'68" W, latitude 68°40'54" S – 77°01"81" S) in the January 2004 during the XIX Italian Antarctic Expedition. Specimens were collected at different time of the day (01:00, 06:00, 10:00, 15:00, 18:00), over a complete 24-hour cycle. Samples were frozen at -40°C in RNA stabilization solution (RNA-later, Ambion). For each fishing, selected tissues (head including compound eyes and brain, abdomen, thoracopods and photophores) from five animals were dissected individually in RNA later ice solution (Ambion). After dissection, tissues were rapidly rinsed in sterile water, weighed, frozen in Trizol reagent (Invitrogen) and stored at -80°C. A large excess of Trizol (15 ml for 0.5–1.5 g. of sample) was used in order to prevent RNA degradation by endogenous RNAse. Frozen tissues were minced and homogenized for 3–5 min using an ultra-turrax-T8.01 blender (IKA-Werke). Total RNA was isolated using the Trizol reagent (Invitrogen) following the manufacturer's instruction and further purified with LiCl in order to remove glucidic contaminants. All RNA samples were checked for quality by capillary electrophoresis (RNA 6000 Nano LabChip, Agilent Bioanalyzer 2100, Agilent Technologies). For each tissue (head, abdomen, thoracopods and photophores), equal amounts of total RNA (2 µg) extracted from every collection were pooled.

Construction of cDNA Libraries

Five independent cDNA libraries, named K01 and K05 (head), K06 (abdomen), K07 (photophores) and K09 (thoracopods), were constructed.

We have developed a new method using a combination of SMART protocol (Clontech), ensuring almost full-length cDNA, and Gateway technology (Invitrogen), allowing unidirectional cloning without enzymatic digestion. In this protocol, only fully-transcribed first strand cDNA (ss cDNA) is tagged with a short sequence complementary to a modified SMART oligo (Fig. 3). The SMART oligo sequence (SMART-16attB1-T3: 5'-TACAAAAAAGCAGGCTAATTAACCC TCACTAAAGGG-3') and the overhang of the oligo(dT) primer (5'-GGGGACCACTTTGTACAAGAAAGCTGGGCG GCCGC $[dT]_{20}$VN-3') used for first strand synthesis include an attB1 and attB2 recombination site respectively.

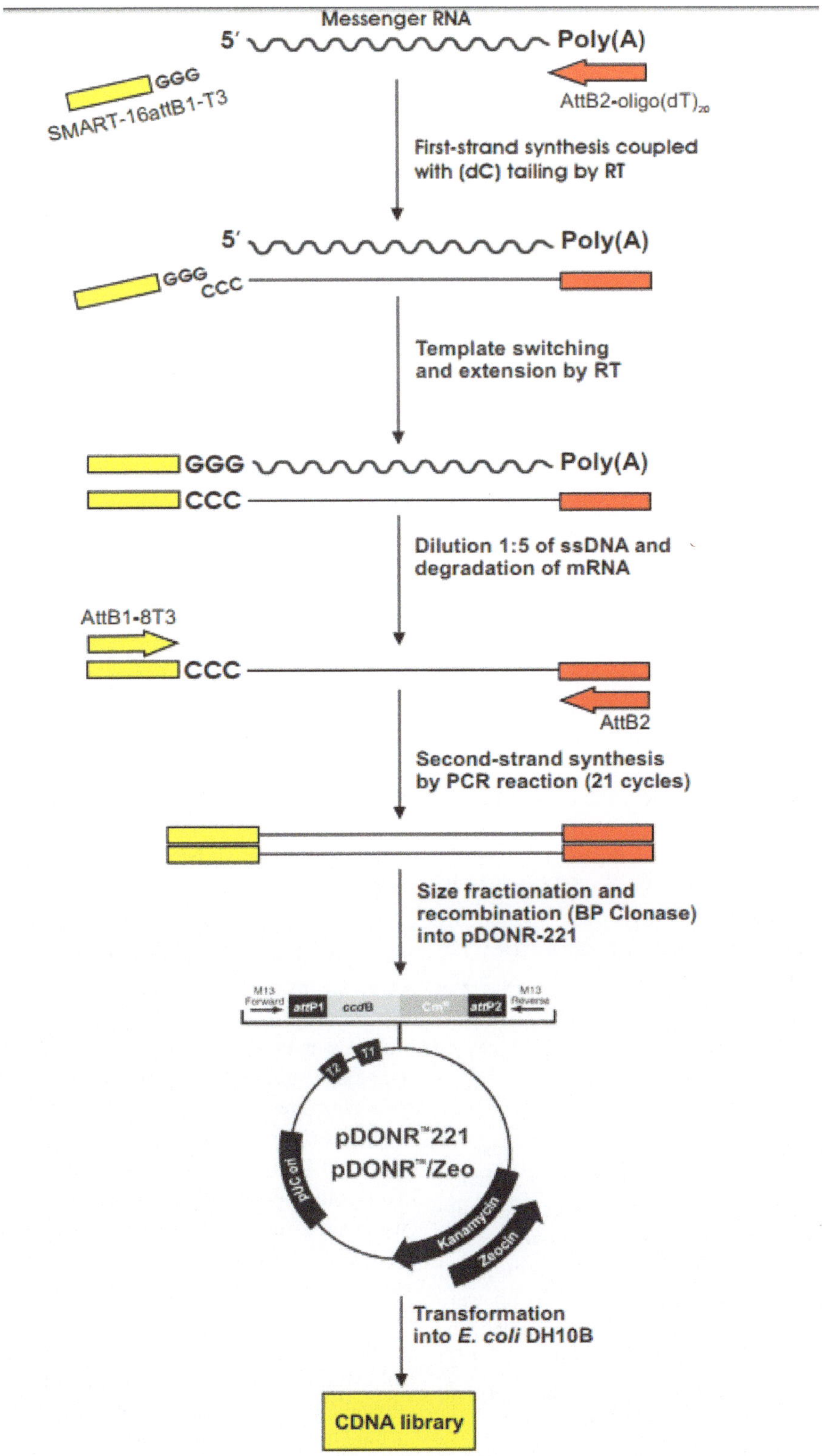

Figure 3: Schematic diagram of the method used for the construction of the cDNA libraries from different krill tissues. See the Methods for more details.

First strand cDNA synthesis was performed from 1.5 μg of total RNA in a 15 μl reaction. Then, the reaction was then diluted 1:5 ratio and incubated at 72°C for 2 min. Second strand reaction mix was added to 1 μl of diluted first strand cDNA to give a final concentration of 1× BD Advantage 2 PCR reaction buffer (Clontech), 0.2 mM dNTPs, 120 nM primers (attB1-8T3: 5›-GGGGACAAGTTTGTACAAAAAA GCAGGCTAATTAACC-3› and attB2: 5›-GGGGACCAC TTTGTACAAGAAAGCTGGG-3›) and 1× of Advantage2 DNA polymerase mix (Clontech) in a volume of 50 μl. This second strand reaction mixture was incubated for 21 cycles of 15 sec 95°C, 30 sec 66°C and 3 min 68°C. Only those ss cDNAs having a SMART anchor sequence at the 5› end were used as template and exponentially amplified. The second strand reaction was glass fibre column purified and cDNA was size selected by Sepharose CL-4B SPUN COLUMN (GE Healthcare). The cDNA was inserted in the cloning vector by a recombination reaction performed at 25°C for 18 h with about 35 ng of attB-cDNA, 150 ng of pDONR221 (Invitrogen) and 2 μl of BP Clonase II (Invitrogen) in 10 μl final volume. 1/5 of the purified reaction was used to transform electrocompetent DH10B E. coli cells. Recombinant colonies were selected on agar SOB medium plus kanamicin. Individual library colonies were arrayed by manual picking on 96 well plates in liquid selective SOB medium plus 7.5% of glycerol for independent growth [40].

DNA sequencing

After lysis of the bacterial colonies, cDNA inserts were directly amplified with universal primers M13 forward (5'-TGTAAAACGACGGCCAGTCTTA-3') and M13 reverse (5'-CAGGAAACAGCTATGACCATGT-3'). The polymerase chain reaction (PCR) profile consisted of: (1) initial denaturation for 5 min at 95°C, (2) 35 cycles of 40-s denaturation at 95°C, 40-s annealing at 60°C and 1-min elongation at 72°C and (3) final extension for 5 min at 72°C; samples with a size over 0.5 Kb were selected for sequencing. Single pass DNA sequencing from plasmids was performed by using the vector specific primer attB1_seq (5'-CTTTGTACAAAAAAGCAGGCT-3') and a modified Sanger dideoxy terminator cycle sequencing chemistry, the ABI BigDye kit version 3.1, on a ABI 3730 48-capillary sequencer and 36 cm capillaries (Sequencing Service of Max-Plank Institute for Molecular Genetics, Berlin, Germany).

Computer Management of Data

Trace2dbest and Partigene [41] were used to process chromatograms, clusterize sequences, and build an annotation database. Trace2dbest extracts sequences and quality information from traces (Phred algorithm), removes vector contamination and poly(A), and performs the trimming of low quality sequences. Sequences shorter than 100 bp were discarded. Partigene reads all sequence files and performs an assembling process in two step: 1) CLOBB software [42] clusterizes sequences on the basis of BLAST similarity; 2) Phrap [43] makes a consensus from each cluster.

Each consensus, converted in FASTA format, was searched locally in nucleotides database, downloaded from NCBI [44] and UniProtKB database [45], using Blast-N and Blast-X, respectively. First 10 HSPs (High Scoring Pair) from each blast result were collected and stored in a local PostgreSQL table, as a collection of automatic annotations.

Each cluster annotation in our database was further manually examined to assign the best describing text to the correspondent cluster: matches with expectations values greater than e^{-10} for protein (Blast-X) and e^{-40} for nucleotide (Blast-N) were considered as poorly informative. Moreover, for each UniProt ID, taken from Blast-X description field, we associated specific Gene Ontology annotation, that integrates information about process, function, and component. Clusters, consensus and related similarity and gene ontology searches were electronically organized and stored in a dedicated PostgreSQL database.

Identification of microsatellite containing ESTs

The unique sequences were screened for microsatellites by using the MISA software [46]. Only di-, tri-, tetra-, penta- and esanucleotide repeats were targeted, since mononucleotide repeats are not useful for mapping or population genetics due to difficulties in their genotyping. Strings of oligonucleotide sequences were used to search for microsatellites: 6 repeats for dinucleotide; 5 repeats for trinucleotide; 5 repeats for tetranucleotide, pentanucleotide and esanucleotide. Primers were designed for the flanking regions of the SSR using a web-based software, "Primer3" [47], and based on the criteria of 50% GC content, a minimum melting temperature of 55°C, and absence of secondary structure. Primers ranged from 18–27 nucleotides in length and amplified products of 150–390 bp. The primers were synthesized with a 5'-KS-tail (KS sequence: 5'-cgaggtcgacggtatcg-3') allowing to amplify microsatellite alleles in combination with a 5'-fluorescent-labeled KS primer [48].

Quantitative RT-PCR

Quantitative RT-PCR was conducted for some genes using the same tissues tested (head, abdomen, thoracopods and photophores) to confirm the integrity and robustness of EST sequencing.

Three μg of total RNA from each tissue was used to perform three independent cDNA syntheses in a final volume of 10 μl, using random decamers and SuperScript II reverse transcriptase (Invitrogen). 1 μl aliquot of diluted first-strand cDNA was PCR amplified in 10 μl volume using SYBR Green chemistry, according to the manufacturer›s recommendations (Finnenzymes). Gene-specific primers were designed using Primer Express® Software (Applera) to amplify fragments of 120–180 bp in length, close to the 3›-end of the transcript. To avoid the amplification of contaminant genomic DNA, we treated total RNA samples with DNase I (Qiagen). The dissociation curve was used to confirm the specificity of the amplicon. PCR reactions were performed in a 7500 Real-Time PCR System (Applied Biosystems). Thermal cycling conditions were as follows: 15 min denaturation at 95°C; followed by 40 cycles of 30 sec denaturation step at 95°C, annealing and elongation steps for 1 min each at 60°C and a final 3 min elongation at 72°C. To evaluate differences in gene expression a relative quantification method was chosen where the expression of the target gene is standardized by a non-regulated reference gene; consequently, three replicates of each sample and endogenous control were amplified. 18S rRNA was used as an endogenous control because the level of rRNA remains essentially constant from sample to sample (QuantumRNA™ 18S Internal Standards, Ambion). To calculate the relative expression ratio, the $2^{-\Delta\Delta Ct}$ (RQ, relative quantification) method implemented in the 7500 Real Time PCR System software [49] was used. This method determines the change in expression of a nucleic acid sequence (target) in a test sample relative to the same sequence in a calibrator sample. In our experiments, the expression of ten targets was tested (compound eye opsin BCRH1, myosin light chain, myosin heavy chain, arthrodial cuticle protein AMP16.3, tail muscle elongation factor 1 gamma, cellular retinoic acid/retinol binding protein, eukaryotic initiation factor 4A, transport protein SEC61 subunit gamma, chromodomain helicase DNA binding protein and voltage-dependent calcium channel) which displayed differential expression in the head, thoracopods and photophores, compared with the abdomen.

DECLARATIONS

Acknowledgements

This work was supported by the Italian Programma Nazionale di Ricerche in Antartide – PNRA (grant 2003/1.3 and grant 2005/1.04 to RC and CB). RC also thanks the European Community (6th Framework Project EUCLOCK No. 018741) and the Italian Space Agency (DCMC grant). We are also grateful to Silvia Casara (C.R.I.B.I. Biotechnology Centre-University of Padova, Italy) for help in total RNA sample preparation, Patrizia Tornincasa (Dept. of Biology-University of Trieste) for technical support in identification of microsatellite and to Antonello Sala (National Research Council, Institute of Marine Science, Ancona, Italy) for help in collecting the samples.

Electronic Supplementary Material

12864_2007_1239_MOESM1_ESM.PDF Additional file 1: This table lists 309 non-redundant sequences identifying known E. superba genes or sequences showing significant similarity with genes from arthropods and other species. These transcripts have been grouped into 13 different functional categories. (PDF 94 KB)

12864_2007_1239_MOESM2_ESM.PDF Additional file 2: Quantitative RT-PCR validation. Forward and reverse primer pairs for ten tested transcripts are detailed at the top. Relative expression values resulting from the quantitative real-time PCR are in the middle and related data plots are also reported. For each transcriptional profiling we have associated the EST counting obtained from the cDNA libraries systematic sequencing. (PDF 147 KB)

REFERENCES

1. Baker ADC, Boden BP, Brinton E: A practical guide to Euphausiids of the world. 1990, London: Natural history Museum Pub
2. Jarman SN, Nicol S, Elliot NG, McMinn A: 28S rDNA evolution in the Eumalacostraca and the phylogenetic position of krill. Mol Phylogenet Evol. 2000, 17: 26-36. 10.1006/mpev.2000.0823.
3. Miller DGM, Hampton I: Biology and ecology of the Antarctic krill. Biomass Scientific Series. 1989, 9: 1-166.
4. Nicol S, Endo Y: Krill fisheries: Development, management and ecosystem implications. Aquat Living Resour. 1999, 12: 105-120. 10.1016/S0990-7440(99)80020-5.
5. Chen YC, Jaczynski J: Gelation of protein recovered from whole Antarctic krill (Euphausia superba) by isoelectric solubilisation/precipitation as affected by functional additives. J Agric Food Chem. 2007, 55: 1814-1822. 10.1021/jf0629944.
6. Hewitt RP, Linen Low EH: The fishery on Antarctic krill: defining an ecosystem approach to management. Rev Fish Sci. 2000, 8: 235-298. 10.1080/10641260091129224.
7. Taki K, Hayashi T, Naganobu M: Characteristics of seasonal variation in diurnal vertical migration and aggregation of Antarctic krill (Euphausia superba) in the Scotia Sea, using Japanese fishery data. CCAMLR Sci. 2005, 12:
8. Mauchline J: The biology of mysids and euphausiids. Adv Mar Biol. 1980, 18: 1-681.
9. Godlewska M: Vertical migrations of krill (Euphausia superba, Dana). Pol Arch Hydrobiol. 1996, 43: 9-63.
10. Machida RJ, Miya MU, Yamauchi MM, Nishida M, Nishida S: Organization of the mitochondrial genome of Antarctic krill Euphausia superba (Crustacea: Malacostraca). Mar Biotechnol. 2004, 6: 238-50. 10.1007/s10126-003-0016-6.
11. Velculescu VE, Zhang L, Vogelstein B, Kinzler KW: Serial analysis of gene expression. Science. 1995, 270: 484-487. 10.1126/science.270.5235.484.
12. Liang P, Pardee AB: Differential display of eukaryotic messenger RNA by means of the polymerase chain reaction. Science. 1992, 257: 967-971. 10.1126/science.1354393.
13. Adams MD, Kerlavage AR, Fields C, Venter JC: 3400 new expressed sequence tags identify diversity of transcripts in human brain. Nat Genet. 1993, 4: 256-267. 10.1038/ng0793-256.
14. Tassanakajon A, Klinbunga S, Paunglarp N, Rimphanitchayakit V, Udomkit A, Jitrapakdee S, Sritunyalucksana K, Phongdara A, Pongsomboon S, Supungul P, Tang S, Kuphanumart K, Pichyangkura R, Lursinsap C: Penaeus monodongene discovery project: the generation of an EST collection and establishment of a database. Gene. 2006, 384: 104-112. 10.1016/j.gene.2006.07.012.
15. Okubo K, Hori N, Matoba R, Nuyama T, Fukushima A, Kojima Y, Matsubara K: Large scale cDNA sequencing for analysis of quantitative and qualitative aspects of gene expression. Nat Genet. 1992, 2: 173-179. 10.1038/ng1192-173.
16. E. superba database. [http://krill.cribi.unipd.it]
17. Pacchioni B, Trevisan S, Gomirato S, Toppo S, Valle G, Lanfranchi G: Semi-multiplex PCR technique for screening of abundant transcripts during systematic sequencing of cDNA libraries. Biotechniques. 1996, 21: 644-646.

18. Bard JBL, Rhee SY: Ontologies in biology: design, applications and future challenges. Nat Rev Genet. 2004, 5: 213-222. 10.1038/nrg1295.
19. Rapp BA, Wheeler DL: Bioinformatics resources from the national center for biotechnology information: an integrated foundation for discovery. J Am Inform Sci Technol. 2005, 56: 538-550. 10.1002/asi.20142.
20. Toikkanen JH, Miller KJ, Soderlund H, Jantti J, Keranen S: The beta subunit of the Sec61p endoplasmic reticulum translocon interacts with the exocyst complex in Saccharomyces cerevisiae. J Biol Chem. 2003, 278: 20946-20953. 10.1074/jbc.M213111200.
21. Beltramini M, Colangelo N, Giomi F, Bubacco L, Di Muro P, Hellmann N, Jaenicke E, Decker H: Quaternary structure and functional properties of Penaeus monodon hemocyanin. FEBS J. 2005, 272: 2060-2075. 10.1111/j.1742-4658.2005.04634.x.
22. Golstein P, Kroemer G: Cell death by necrosis: towards a molecular definition. Trends Biochem Sci. 2007, 32: 37-43. 10.1016/j.tibs.2006.11.001.
23. Zhang X, Huang C, Qin Q: Antiviral properties of hemocyanin isolated from shrimp Penaeus monodon. Antivir Res. 2004, 61: 93-99. 10.1016/j.antiviral.2003.08.019.
24. Destoumieux-Garzon D, Saulnier D, Garnier J, Jouffrey C, Bulet P, Bachere E: Crustacean immunity. Antifungal peptides are generated from the C terminus of shrimp hemocyanin in response to microbial challenge. J Biol Chem. 2001, 276: 47070-47077. 10.1074/jbc.M103817200.
25. Ju Z, Karsi A, Kocabas A, Patterson A, Li P, Cao D, Dunham R, Liu Z: Transcriptome analysis of channel catfish (Ictalurus punctatus): genes and expression profile from the brain. Gene. 2000, 261: 373-382. 10.1016/S0378-1119(00)00491-1.
26. Cao D, Kocabas A, Ju Z, Karsi A, Li P, Patterson A, Liu Z: Transcriptome of channel catfish (Ictalurus punctatus): initial analysis of genes and expression profiles of the head kidney. Animal Genet. 2001, 32: 169-188. 10.1046/j.1365-2052.2001.00753.x.
27. Venier P, Pallavicini A, De Nardi B, Lanfranchi G: Towards a catalogue of genes transcribed in multiple tissues of Mytilus galloprovincialis. Gene. 2003, 314: 29-40. 10.1016/S0378-1119(03)00708-X.
28. Eirin-Lopez JM, Gonzalez-Tizon AM, Martinez A, Mendez J: Birth-and-death evolution with strong purifying selection in the histone H1 multigene family and the origin of orphon H1 genes. Mol Biol Evol. 2004, 21: 1992-2003. 10.1093/molbev/msh213.
29. Hankeln T, Schmidt ER: The organization, localization and nucleotide sequence of the histone genes of the midge Chironomus thummi. Chromosoma. 1991, 101: 25-31. 10.1007/BF00360683.
30. Akhmanova A, Miedema K, Kremer H, Henning W: Two types of polyadenilated mRNAs are synthesized from Drosophilareplication-dependent histones. Eur J Biochem. 1997, 244: 294-300. 10.1111/j.1432-1033.1997.00294.x.
31. del Gaudio R, Potenza N, Stefanoni P, Chiusano ML, Geraci G: Organization and nucleotide sequence of the cluster of five histone genes in the polychaete worm Chaetopterus variopedatus: First record of a H1 histone gene in the phylum Annelida. J Mol Evol. 1998, 46: 64-73. 10.1007/PL00006284.
32. Eirýn-Lopez JM, Gonzalez-Tizon AM, Martýnez A, Mendez J: Molecular and evolutionary analysis of mussel histone genes (Mytilus spp.): Possible evidence of an “orphon” origin for H1 histone genes. J Mol Evol. 2002, 55: 272-283. 10.1007/s00239-002-2325-1.
33. Barzotti R, Pelliccia F, Bucciarelli E, Rocchi A: Organization, nucleotide sequence, and chromosomal mapping of a tandemly repeated unit containing the four core histones genes and a 5S rRNA gene in an isopod crustacean species. Genome. 2000, 43: 341-345. 10.1139/gen-43-2-341.
34. Marino-Ramirez L, Jordan IK, Landsman D: Multiple independent evolutionary solutions to core histone gene regulation. Genome Biol. 2006, 7: R122-10.1186/gb-2006-7-12-r122.
35. Bortoluzzi S, Danieli GA: Towards an in silico analysis of transcription patterns. Trends Genet. 1999, 15: 118-119. 10.1016/S0168-9525(98)01682-5.
36. Porter ML, Cronin TW, McClellan DA, Crandall KA: Molecular characterization of crustacean visual pigments and the evolution of pancrustacean opsins. Mol Biol Evol. 2007, 24: 253-268. 10.1093/molbev/msl152.
37. Ostellari L, Zane L, Maccatrozzo L, Bargelloni L, Patarnello T: Novel microsatellite loci isolated from the northern krill, Meganyctiphanes norvegica (Crustacea, Euphausiacea). Mol Ecol. 2000, 9: 377-8. 10.1046/j.1365-294x.2000.00874-7.x.
38. Rudd S: Expressed sequence tags: alternative or complement to whole genome sequences?. Trends Plant Sci. 2003, 7: 321-329. 10.1016/S1360-1385(03)00131-6.
39. Jackson RB, Linder CR, Lynch M, Purugganan M, Somerville S, Thayer SS: Linking molecular insight and ecological research. Trends Ecol Evol. 2002, 17: 409-414. 10.1016/S0169-5347(02)02571-5.
40. Pallavicini A, Del Mar Costa M, Gestal C, Dreos R, Figueras A, Venier P, Novoa B: High sequence variability of myticin transcripts in hemocytes of immune-stimulated mussels suggests ancient host-pathogen interactions. Dev Comp Immunol. 2008, 32: 213-26. 10.1016/j.dci.2007.05.008.
41. Parkinson J, Anthony A, Wasmuth J, Schmid J, Hedley A, Blaxter M: PartiGene – constructing partial genomes. Bioinformatics. 2004, 20: 1398-1404. 10.1093/bioinformatics/bth101.
42. Parkinson J, Guiliano DB, Blaxter M: Making sense of EST sequences by CLOBBing them. BMC Bioinformatics. 2002, 3: 31-10.1186/1471-2105-3-31.
43. Phrap software. [http://www.phrap.org]
44. The National Center for Biotechnology Information. [ftp://ftp.ncbi.nih.gov/blast/db]

45. UniProtKB database. [http://www.pir.uniprot.org/database/download.shtml]

46. MISA software. [http://pgrc.ipk-gatersleben.de/misa/]

47. Rozen S, Skaletsky H: Primer3 on the WWW for general users and for biologist programmers. Methods Mol Biol. 2000, 132: 365-86.

48. Schuelke M: An economic method for the fluorescent labeling of PCR fragments. Nature Biotechnology. 2000, 18: 233-234. 10.1038/72708.

49. Livak KJ, Schmittgen TD: Analysis of relative gene expression data using Real-Time Quantitative PCR and the $2^{-\Delta\Delta Ct}$method. Methods. 2001, 25: 402-408. 10.1006/meth.2001.1262.

50. Cook CE, Yue Q, Akam M: Mitochondrial genomes suggest that hexapods and crustaceans are mutually paraphyletic. Proc Biol Sci. 2005, 272: 1295-1304. 10.1098/rspb.2004.3042.

51. Crease TJ: The complete sequence of the mitochondrial genome of Daphnia pulex (Cladocera: Crustacea). Gene. 1999, 233: 89-99. 10.1016/S0378-1119(99)00151-1.

52. Umetsu K, Iwabuchi N, Yuasa I, Saitou N, Clark PF, Boxshall G, Osawa M, Igarashi K: Complete mitochondrial DNA sequence of a tadpole shrimp (Triops cancriformis) and analysis of museum samples. Electrophoresis. 2002, 23: 4080-4084. 10.1002/elps.200290024.

53. Lavrov DV, Brown WM, Boore JL: Phylogenetic position of the Pentastomida and (pan)crustacean relationships. Proc Biol Sci. 2004, 271: 537-544. 10.1098/rspb.2003.2631.

54. Ivey JL, Santos SR: The complete mitochondrial genome of the Hawaiian anchialine shrimp Halocaridina rubra Holthuis, 1963 (Crustacea: Decapoda: Atyidae). Gene. 2007, 394: 35-44. 10.1016/j.gene.2007.01.009.

55. Segawa RD, Aotsuka T: The mitochondrial genome of the Japanese freshwater crab, Geothelphusa dehaani (Crustacea: Brachyura): evidence for its evolution via gene duplication. Gene. 2005, 355: 28-39. 10.1016/j.gene.2005.05.020.

56. Kilpert F, Podsiadlowski L: The complete mitochondrial genome of the common sea slater, Ligia oceanica (Crustacea, Isopoda) bears a novel gene order and unusual control region features. BMC Genomics. 2006, 7: 241-10.1186/1471-2164-7-241.

57. Miller AD, Austin CM: The complete mitochondrial genome of the mantid shrimp Harpiosquilla harpax, and a phylogenetic investigation of the Decapoda using mitochondrial sequences. Mol Phylogenet Evol. 2006, 38: 565-574. 10.1016/j.ympev.2005.10.001.

58. Miller AD, Murphy NP, Burridge CP, Austin CM: Complete mitochondrial DNA sequences of the decapod crustaceans Pseudocarcinus gigas (Menippidae) and Macrobrachium rosenbergii (Palaemonidae). Mar Biotechnol (NY). 2005, 7: 339-349. 10.1007/s10126-004-4077-8.

59. Yamauchi MM, Miya MU, Machida RJ, Nishida M: PCR-based approach for sequencing mitochondrial genomes of decapod crustaceans, with a practical example from kuruma prawn (Marsupenaeus japonicus). Mar Biotechnol (NY). 2004, 6: 419-429. 10.1007/s10126-003-0036-2.

60. Sun H, Zhou K, Song D: Mitochondrial genome of the Chinese mitten crab Eriocheir japonica sinenesis (Brachyura: Thoracotremata: Grapsoidea) reveals a novel gene order and two target regions of gene rearrangements. Gene. 2005, 349: 207-217. 10.1016/j.gene.2004.12.036.

61. Miller AD, Nguyen TT, Burridge CP, Austin CM: Complete mitochondrial DNA sequence of the Australian freshwater crayfish, Cherax destructor (Crustacea: Decapoda: Parastacidae): a novel gene order revealed. Gene. 2004, 331: 65-72. 10.1016/j.gene.2004.01.022.

62. Yamauchi M, Miya M, Nishida M: Complete mitochondrial DNA sequence of the Japanese spiny lobster, Panulirus japonicus(Crustacea: Decapoda). Gene. 2002, 295: 89-96. 10.1016/S0378-1119(02)00824-7.

63. Yamauchi M, Miya M, Nishida M: Complete mitochondrial DNA sequence of the swimming crab, Portunus trituberculatus(Crustacea: Decapoda: Brachyura). Gene. 2003, 311: 129-135. 10.1016/S0378-1119(03)00582-1.

64. Hickerson MJ, Cunningham CW: Dramatic mitochondrial gene rearrangements in the hermit crab Pagurus longicarpus(Crustacea, Anomura). Mol Biol Evol. 2000, 17: 639-644.

65. Machida RJ, Miya MU, Nishida M, Nishida S: Complete mitochondrial DNA sequence of Tigriopus japonicus (Crustacea: Copepoda). Mar Biotechnol (NY). 2002, 4: 406-417. 10.1007/s10126-002-0033-x.

66. Tjensvoll K, Hodneland K, Nilsen F, Nylund A: Genetic characterization of the mitochondrial DNA from Lepeophtheirus salmonis (Crustacea; Copepoda). A new gene organization revealed. Gene. 2005, 353: 218-230. 10.1016/j.gene.2005.04.033.

67. Ogoh K, Ohmiya Y: Complete mitochondrial DNA sequence of the sea-firefly, Vargula hilgendorfii (Crustacea, Ostracoda) with duplicate control regions. Gene. 2004, 327: 131-139. 10.1016/j.gene.2003.11.011

CHAPTER 2

TRANSCRIPTOME ANALYSIS OF THE ORIENTAL RIVER PRAWN, MACROBRACHIUM NIPPONENSE USING 454 PYROSEQUENCING FOR DISCOVERY OF GENES AND MARKERS

Keyi Ma, Gaofeng Qiu, Jianbin Feng and Jiale Li

INTRODUCTION

The oriental river prawn *Macrobrachium nipponense*, a member of the Palaemonidae family of decapod crustaceans, is widely distributed in freshwater and low-salinity regions of estuaries [1]. This prawn is a useful model organism for studying reproduction and development in decapod species for multiple reasons. First, it produces large numbers of eggs due to its high fecundity. Each gravid female can lay thousands of eggs after mating. Second, *M. nipponense* is a relatively hardy species that adapts to a wide range of environmental conditions and is reared easily in a laboratory. Third, it can survive in freshwater for its entire life cycle. Hence, the costs are considerably lower than saltwater species. *M. nipponense* is also commercially important and is cultured extensively throughout China [2] and other Asian countries [3]–[7]. The basic production techniques for *M. nipponense* were developed in China approximately 40 years ago. In 2008, the culture yields of this prawn exceeded 205,000 tons, which accounts for the majority of its worldwide production. Owing to its high benefit and excellent adaptability, its culture production has gradually increased. However, due to the precocity of the farmed prawns, especially the females, the growth performance has decreased in recent years. On the other hand, there is a significant difference in growth performance between males and females in this species. Males grow much faster than females and reach a larger size at harvest, similarly to many other *Macrobrachium* species.

In monosex culture, energy from reproduction is diverted to growth, resulting in larger size of cultured individuals than that of mixed sex. Monosex culture is a common practice in aquaculture, and many attempts have been made to apply this technology to crustacean aquaculture [8]. The androgenic gland of crustaceans, an endocrine organ unique to males, plays a crucial role in the sexual differentiation to maleness [9]. Ablation or implantation of the androgenic gland at a certain stage of development could result in sex reversal to female or male, respectively [9]–[12]. Although all-male populations have been produced in *Macrobrachium rosenbergii* [8], corresponding research in *M. nipponense* has not been performed. Furthermore, the molecular mechanisms involved in sex determination/differentiation are poorly understood [13]. Due to a lack of genetic and genomic information, sex-determining genes have not been identified in decapod crustaceans; even genes related to sex determination/differentiation have rarely been reported [13]–[15]. Despite the aquacultural and biological importance of *M. nipponense*, previous studies have mostly focused on the isolation of microsatellites [16], [17], the investigation of genetic diversity [18], [19], Sanger-based sequencing of expressed sequence tags (ESTs) [20], the characterization of single functional genes [21]–[23] and sequencing of the mitochondrial genome of *M. nipponense* [1].

Citation: Ma K, Qiu G, Feng J, Li J (2012) Transcriptome Analysis of the Oriental River Prawn, Macrobrachium nipponense Using 454 Pyrosequencing for Discovery of Genes and Markers. PLoS ONE 7(6): e39727. doi:10.1371/journal.pone.0039727

It is well known that the generation of large-scale ESTs is a very useful approach to accelerating research in non-model species. ESTs represent a valuable resource for research, as they provide comprehensive information about the transcriptome [24]. ESTs have played significant roles in accelerating gene discovery [25], [26], developing SSRs and SNPs markers [27]–[30], improving genome annotation, facilitating large-scale expression analysis, elucidating phylogenetic relationships, etc. With these useful functions, the number of ESTs in public databases is rapidly increasing. To date, there are more than 72 million ESTs in the NCBI public collection (release 030112; March, 1 2012. http://www.ncbi.nlm.nih.gov/dbEST/dbEST_summary.html). However, less than 8,500 ESTs are available for *M. nipponense*. Transcriptome sequencing enables various functional genomics studies for an organism. Roche 454 Genome Sequencing (GS) FLX technology has generally been used for transcriptome analyses in a large number of plants [24], [27], [31]–[33], animals [29], [30], [34]–[36] and microorganisms [37]. It is particularly useful as a shotgun method for generating EST data, and its costs are relatively low [33], [38]. Moreover, by generating sufficiently long-sequence reads, the 454 GS FLX technology can compensate for the lack of a reference genome during *de novo* sequence assembly [39].

In this study, we prepared a cDNA library by pooling total RNA from various organs and tissues, including the eyestalk, gill, heart, ovary, testis, hepatopancreas, muscle, and embryos at the cleavage, gastrula, nauplius and zoea stages. We sequenced ESTs from this library to expand our knowledge of the transcriptome, facilitate future whole-genome sequence assembly in the oriental river prawn, and provide genes for future studies of the molecular mechanisms involved in sex determination/differentiation in *M. nipponense*. This transcriptome dataset provides an invaluable new resource for functional genomics and biological research in *M. nipponense*. The molecular markers identified in this study provide a material basis for future genetic linkage and quantitative trait loci (QTL) analysis and will be essential for accelerating aquaculture breeding programs to increase culture production in the future.

RESULTS AND DISCUSSION

Roche 454 GS FLX sequencing and reads assembly

In order to achieve a whole-body *M. nipponense* transcriptome, total RNA was extracted from a variety of adult organs and tissues, including the eyestalk, gill, heart, ovary, testis, hepatopancreas, muscle, and embryos at the cleavage, gastrula, nauplius and zoea stages. Equal quantities of RNA were mixed together to construct a cDNA library. The library was sequenced by a Roche 454 GS FLX.

The reads produced by the Roche 454 GS FLX were used for clustering and *de novo* assembly. After eliminating adapter sequences and filtering out the low-quality reads, Roche 454 sequencing yielded a total of 984,204 high-quality transcriptomic reads with a total size of 338,592,850 bp, resulting in an average of 344 bp per read (**Figure 1A**). Clustering and assembly of these reads yielded a non-redundant set of 81,411 ESTs, comprising 42,551 contigs and 38,860 singletons with average lengths of 939 bp and 345 bp, respectively (**Table 1**). Most of these contigs (57.68%) were distributed in the 201~1000 bp region (**Figure 1B**). And most of these singletons (99.43%) fell between 101 and 800 bp in length (**Figure 1C**). A total of 3,058 contigs were greater than 2.5 kb, with the largest contig at 14,768 bp. The average length of our assembled contigs was longer than that previously reported for *M. rosenbergii* (average of 845 bp) and *Euphausia superba* (average of 492 bp), whose transcriptomes were also obtained with a Roche 454 GS FLX in non-model species [30], [36]. Long sequences of good quality could enable us to obtain more information about genes. Therefore, this transcriptome dataset provides a useful resource for future analyses of genes related to sex determination and sex differentiation. To the best of our knowledge, this is the first comprehensive study of the transcriptome of *M. nipponense*.

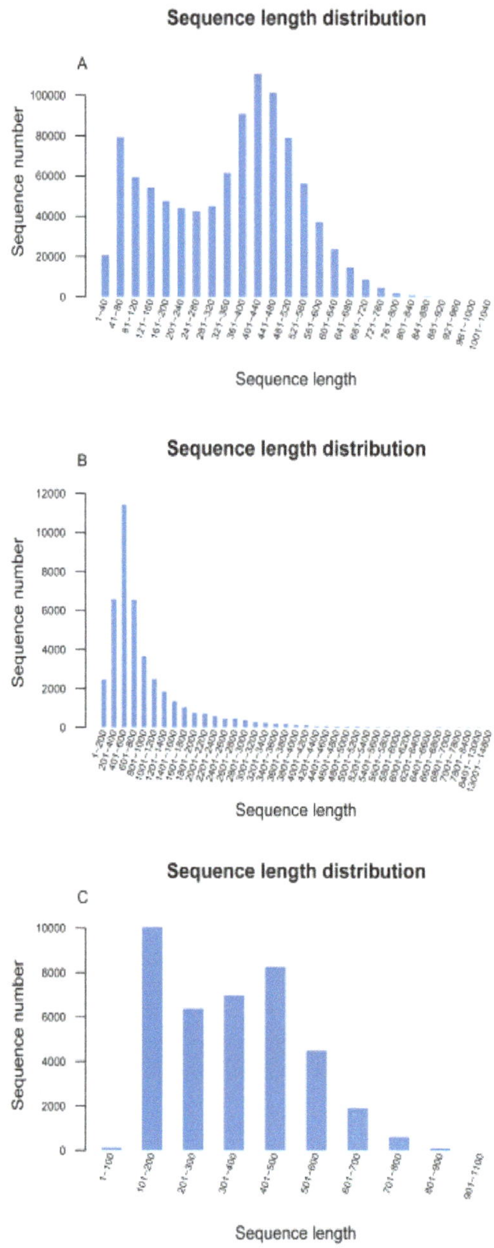

Figure 1: Length distribution of *M. nipponense* transcriptomic ESTs.

Table 1: Summary of Roche 454 GS FLX assembly and analysis of *M. nipponense* transcriptomic sequences.

Dataset name	All
transcriptomic reads	984,204
total number of bases	338,592,850 bp
average read length	344 bp
No. of contigs	42,551
longest contig	14,768 bp
No. of singletons	38,860

doi:10.1371/journal.pone.0039727.t001

GC content and alternative splicing

The GC content (ratio of guanine and cytosine) of all unique sequences of *M. nipponense* was determined. The concentration of GC was 39.45% and 39.28% in contigs and singletons, respectively, giving rise to an overall GC content of 39.40%, indicating a low GC content in the cDNA of *M. nipponense*.

The *M. nipponense* contigs were further assembled into 38,339 isogroups with an average length of 1,042 bp. More contigs than isogroups were found because some contigs (called isotigs) are attributed to the same isogroups due to alternative splicing. A total of 14.75% of isotigs (6,261 of 42,440) have no less than two methods of alternative splicing. A large number of alternative splicing could improve the utilization rate of the encoding genes. Alternative splicing is an important mechanism for regulating gene expression in eukaryotic cells, and it contributes to protein diversity.

Comparative analyses of unique sequences

We noted that only 143 (0.18%) unique sequences matched the registered sequences of *M. nipponense* in the GenBank non-redundant database, and 81,298 unique sequences were newly discovered in our present result. This result suggests that we have made a meaningful contribution to the knowledge of *M. nipponense* by characterizing these unique sequences.

Amongst the 81,441 unique sequences, a BLASTx sequencing similarity search of the GenBank nr database matched against 17,342 (40.76%) contigs and 6,882 (17.71%) singletons (with an e-value of 10^{-6}). Similarly, low matched number of contigs and singletons has been reported in *M. rosenbergii*transcriptomic dataset (46% of contigs and 19% of singletons) using the 454 GS FLX platform [36]. This is due to the lack of genomic information in non-model species. Because *M. nipponense* and *M. rosenbergii*both belong to the genus *Macrobrachium*, they are closely related to one another in taxology. Thus, we found a great number of unique sequences matches (80% sequence similarity) using a BLAST search against the transcriptome of *M. rosenbergii* (data not shown).

A previous report that Roche 454 sequencing of the non-model species, *Mytilus galloprovincialis* produced enough mitochondrial reads to enable alignment of these new data against the relevant mitochondrial sequence [40]. However, this alignment could not be performed with the *M. nipponense* transcriptomic data, as only 10 of the mitochondrial protein-coding genes (PCGs) (cytochrome c oxidase subunit 1, 2 and 3; ATP synthase F0 subunit 6; cytochrome b; and NADH dehydrogenase subunit 1, 2, 3, 4 and 5) were present in our contig dataset. Other PCGs in the *M. nipponense* mitochondrion (ATP synthase F0 subunit 8; and NADH dehydrogenase subunit 4L and 6) were not present in the contig dataset [1]. Singletons potentially contain useful lowly expressed sequences [41]. After searching the singleton dataset, however, we did not find other mitochondrial protein-coding genes. This phenomenon may be because some transcripts with low copies may be lost when constructing non-normalized libraries. The *M. nipponense* transcriptomic library was subjected to many treatments during its construction, and these programmes would contain a substantial proportion of artifacts derived from cDNA synthesis and sequencing to the subsequent reads assembly.

Amongst the unique sequences derived from contigs and singletons, coding sequences with homology to 'ubiquitin-protein ligase', 'ATP synthase', 'cell division cycle protein', 'cytochrome c oxidase', 'cytochrome P450', 'zinc finger protein', 'NADH dehydrogenase', 'pre-mRNA-splicing factor', 'serine/arginine-rich splicing factor' and 'splicing factor' were the most abundant. Similarly, most of the transcripts determined in this initial survey were also obtained in the *M. rosenbergii* transcriptomic dataset [42]. Although our research mainly focused on finding putative genes related to sex determination and sex differentiation, other putative functional transcripts identified here could provide a foundation for future investigations of the roles of stress response, reproduction and defense reaction. The transcriptomic findings could also be the best source for deciphering the putative functions of novel genes, but further studies would need to be conducted to understand their molecular functions.

Gene Ontology assignments

Gene Ontology (GO) could provide a structured and controlled vocabulary for describing gene products in three categories: molecular function, cellular component and biological process [43]. We added GO terms using Blast2GO [44], which is based on the automated annotation of all unique sequences in the GenBank nr protein database from NCBI. According to the database, a total of 19,092 unique *M. nipponense*sequences were assigned to one or more ontologies based on their similarity to sequences with previously known functions, including 30,425 sequences assigned to the molecular function category, 44,112 to the cellular component category and 67,679 to the biological process category. The assigned sequences were divided into 58 functional terms. Because several of the sequences were assigned to more than one GO term, the total number of GO terms obtained in our dataset was much bigger than the total number of the unique sequences.

In the molecular function category, 'binding', 'catalytic activity', 'molecular transducer activity', 'transporter activity' and 'enzyme regulator activity' comprised the largest proportion, accounting for 89.47% of the total (**Figure 2**). Whilst the cellular component category showed that many unique sequences were to likely possess 'cell' (27.46%), 'cell part' (27.46%) and 'organelle' (17.19%) functions. Moreover, the top five biological functions assigned were involved in 'cellular process' (18.33%), 'metabolic processes' (16.74%), 'biological regulation' (9.07%), 'regulation of biological process' (8.57%) and 'response to stimulus' (6.17%). In summary, these terms account for a large fraction of the overall assignments in *M. nipponense* transcriptomic dataset. Understandably, genes encoding these functions may be more conserved across different species and are thus easier to annotate in the database.

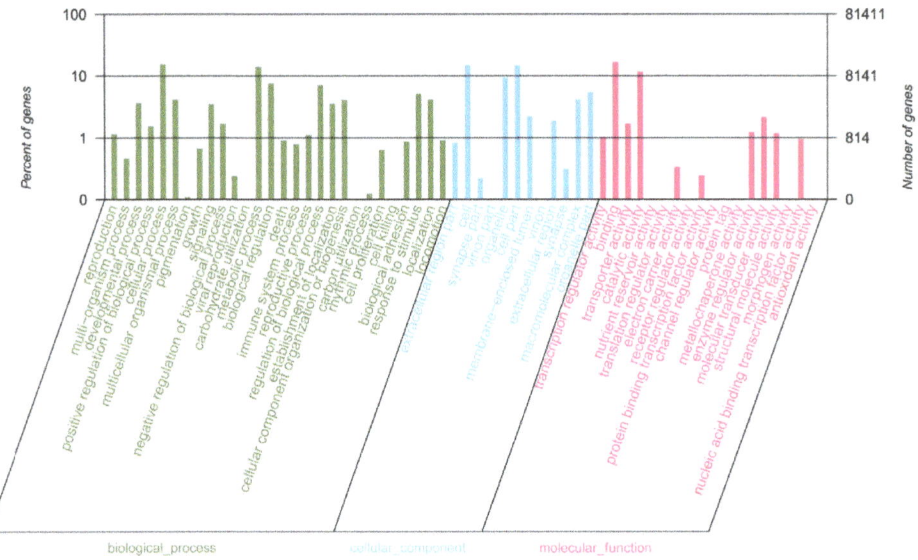

Figure 2: Gene Ontology (GO) terms for the transcriptomic sequences of *M. nipponense*.

KEGG analysis and COG analysis

The unique sequences were mapped to the KEGG database to define metabolic pathways. According to the KEGG results, 42,126 unique sequences, comprising 39,302 contigs and 2,824 singletons, were mapped onto 123 and 90 predicted KEGG metabolic pathways, respectively. In order to perform a comprehensive KEGG study, we added the unique sequences derived from singletons to our analysis. Ultimately, we found one type of metabolic pathway, 'bisphenol degradation', which did not include in the contigs.

The numbers of unique sequences mapped to various pathways ranged from 1 to 11,667. The main metabolic pathways of unique sequences in the *M. nipponense* transcriptome are 'purine and pyrimidine metabolism', 'amino acid metabolism', 'fatty acid metabolism' and 'carbohydrate metabolism'. As might be expected, some putative enzymes belonging to arthropod were found in the KEGG analysis of amino sugar and nucleotide sugar metabolism. The enzymes 'chitin synthase' and 'chitinase' play an important role in carapace formation and exuviation during the molting cycle [45], [46]. Considering the growth process habit of *M. nipponense*, finding these pathways in the dataset was not surprising.

Assignments of COG were used to predict and classify possible functions of the unique sequences. Based on sequence homology, 3,114 unique sequences had a COG functional classification. These sequences were classified into 24 COG categories (**Figure 3**). The most common category was 'general function prediction only' with 938 unique sequences. This category was followed by 'posttranslational modification, protein turnover, chaperones' (264) and 'function unknown' (248). 'Nuclear structure' (2), 'defense mechanisms' (3) and 'cell motility' (7) were the smallest COG categories.

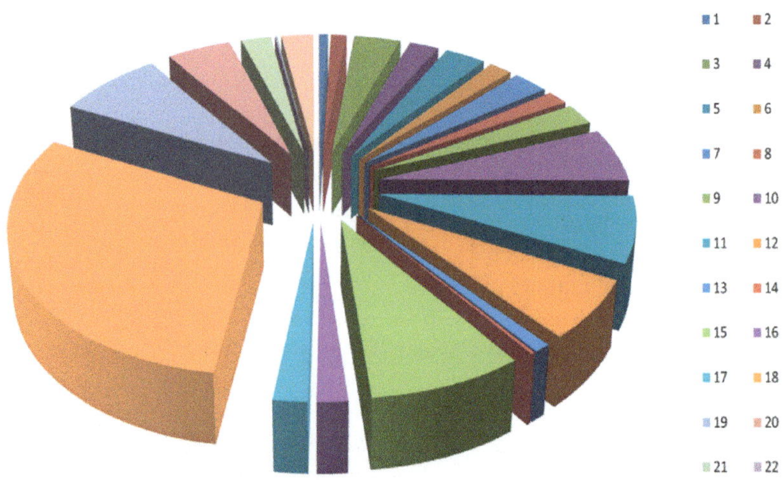

Figure 3: Clusters of orthologous groups (COG) classifications of unique *M. nipponense* sequences.

KEGG pathway analysis and COG analysis are helpful for predicting potential genes and their functions at a whole-transcriptome level. The predicted metabolic pathways, together with the COG analysis, are useful for further investigations of gene function in future studies.

Genes of interest related to sex determination

The transcriptome of *M. nipponense* was primarily examined to identify a wide range of candidate genes that might be functionally associated with sex determination and sex differentiation. Traditionally, such gene discovery in non-model organisms has required degenerate PCR, which is labor-intensive and prone to failure [47]. The annotated transcriptome reported here allows researchers to identify genes of interest more easily than that of using degenerate PCR. According to our sequence analysis and published literature, many genes involved in sex determination and sex differentiation, including sex-determining region Y-chromosome (*SRY*), doublesex- and mab-3-Related transcription factor 1 (*DMRT1*), fushi tarazu factor 1 (*FTZ-F1*), forkhead box L2 (*FOXL2*), feminization 1 (*FEM1*), *VASA* and other potential candidates, were identified (**Table 2**).

Table 2: Selected genes of interest for sex determination/differentiation in the *M. nipponense* transcriptome, including the contigs and singletons.

Candidate genes	Hit(s)	Similarity (%)	E-value	Length (bp)
Dmrt1	1	45	3.23E-08	377
SRY related	1	33	3.00E-07	807
FTZ-F1	4	45	1.84E-08~4.01E-08	799~1330
SF	2	60~88	1.00E-14~1.77E-06	318~404
P450	31	56~90	0~6.04E-06	157~1672
SOX/HMG-box	9	50~92	4.01E-99~4.90E-09	136~1521
FOXL2	1	50	8.03E-07	619
ECM	1	58	0	6049
FEM1	8	59~88	0~1.12E-08	109~3831
STAR	1	50	5.87E-14	751
WT1	1	82	8.80E-134	1164
GATA	4	58~78	5.94E-81~2.03E-10	160~1976
ZFY1	4	43~64	1.00E-16~1.00E-11	288~1127
WNT	6	65~87	4.38E-61~5.80E-10	179~560
GCL	2	53~86	2.16E-08~6.63E-06	136~329
Argonaute	6	60~98	2.65E-156~3.70E-13	266~1968
Piwi	2	58~60	0~4.18E-133	2625~3790
Nanos	1	74	9.21E-16	391
Tudor	11	41~89	0~2.70E-07	576~5692
Pumilio	6	52~83	0~2.29E-06	127~6975
VASA	1	100	0	2357
PL10	1	98	0	5421
testis-specific	11	40~77	0~6.48E-07	369~2524

doi:10.1371/journal.pone.0039727.t002

In mammals, the male sex determination switch is controlled by *SRY* [48]. One unique sequence similar to *SRY* was identified in the *M. nipponense* transcriptome, displaying 33% similarity with homologues in mouse. *SRY* has DNA-binding properties that are mediated by a high mobility group (HMG-box) motif [49], named *SOX* [50]. After searching the transcriptomic dataset, three unique sequences were annotated as *SOX2*, *SOX14* and *SOX19*. The sequences exhibited significant BLAST e-values (8.92E-67, 6.74E-61 and 3.61E-64) with those of *Pediculus humanus corporis*, *Scylla paramamosain* and *Tribolium castaneum*, respectively. Whilst two unique sequences were identified as HMG-box a and HMG-box b and displayed significant similarities (83% and 92%) with the corresponding sequences in *Litopenaeus vannamei*.

DMRT1, a gene related to worm and fly sexual regulators, is required for mammalian testis differentiation [51]. In all birds, *DMRT1* is located on the Z chromosome but is absent from the W. In chicken embryos, it is expressed in the early bipotential gonad and shows higher expression in males than in females [52]. Knockdown of *DMRT1* in early chicken embryos can result in the feminization of male gonads [53]. One unique sequence was aligned as *DMRT1*, displaying a BLAST e-value of 3.23E-08 and 45% similarity to the encoded amino acid sequence of homologues in *Pan troglodytes*. *FOXL2* is a member of the forkhead/HNF-3-related family of transcription factors involved in ovarian differentiation [54]. Recent research has shown that *FOXL2* suppresses testicular differentiation mainly through repression of the *SOX9*regulatory element, which promotes testis-specific expression of *SOX9* and leads to gonadal sex reversal in mice [55]. Likewise, one unique sequence, which was identified as *FOXL2*, exhibited 50% similarity in its encoded amino acid sequence with homologues in *Homo sapiens* as well as that of *DMRT1*. Lower sequence similarity and less quantity of the transcript were interpreted as a consequence of earlier genetic differentiation between phyla. Meanwhile, *DMRT1* and *FOXL2* might not be the dominant sex determination/differentiation genes in *M. nipponense*.

FTZ-F1 is a member of the nuclear receptor superfamily and was originally found as a regulator of the *Drosophila* homeobox segmentation gene *FTZ* [56]. In mammals, *FTZ-F1* is an essential factor in sex differentiation. To date, its homologues have been identified in human, mouse and a number of teleost species [57]–[60]. Four unique sequences were annotated as *FTZ-F1* with *Apis mellifera FTZ-F1*. However, no studies about *FTZ-F1* have been published for *Macrobrachium*.

FEM1, which is considered as a signal-transducing regulator, was first reported in the *Caenorhabditis elegans* sex determination pathway. *FEM1* of *C. elegans* plays a global role in sex determination. Nevertheless, a single *FEM1* transcript and protein are expressed at equivalent levels in both sexes, suggesting primarily posttranscriptional and posttranslational regulation of its activity [61]. Its homologs, including *FEM1A*, *FEM1B* and *FEM1C* of sex determination for candidate genes have been found in human and house mouse (*Mus musculus*). In *M. nipponense*, 11 unique sequences were identified as *FEM1A*, *FEM1B* and *FEM1C*, with similarities ranging from 49%~93%, respectively. To the best of our knowledge, these putative *FEM1A*, *FEM1B* and *FEM1C* genes are first reported in crustacean.

VASA, belonging to the DEAD-box family of ATP-dependent RNA helicase, plays a crucial role in germ cells development [62]. Its product is required maternally for germ plasm assembly and completion of oogenesis. Since *VASA* is expressed specifically in germ cells of most animals, it was used as a reliable molecular marker to trace the origin, migration, and differentiation of primordial germ cells [63], [64]. In the *M. nipponense* transcriptome, one unique sequence was aligned with *VASA*, displaying 100% similarity in encoded amino acid sequence with homologues in *M. nipponense* which amino acid sequence was previously submitted to NCBI database by our laboratory. This similarity suggests that the *M. nipponense*transcriptome presented here is accurate and can be trusted.

In this study, we also identified other putative genes related to sex determination/differentiation in *M. nipponense* for the first time, such as cytochrome P450 aromatase *(P450)* [65], steroidogenic acute regulatory protein (*STAR*) [66], Zinc finger Y-chromosomal protein 1 (*ZFY1*) [67], Wilms' tumor protein 1 (*WT1*) [68] and *Tudor* [69]. The research described herein provides no evidence for a master gene whose presence or absence determines *M. nipponense* sex or sex differentiation. Some studies showed that in *M. rosenbergii*, as in other macruran species, the female prawn is the heterogametic sex, bearing WZ sex chromosomes, whereas the male prawn is the homogametic sex, bearing ZZ sex chromosomes [8], [12], [70]. While *M. nipponense* belongs to the same genus as *M. rosenbergii*, it is uncertain whether *M. nipponense* possesses the same sex determination mechanism. Further studies are needed to understand the molecular functions of these putative genes. Sex determination and sex differentiation are very complicated processes. Understanding gene regulatory mechanisms has not only theoretical value, but also practical value and meaning for aquaculture.

Putative Molecular Markers

SSRs, or microsatellites, are polymorphic loci present in genomic DNA. They consist of repeated core sequences of 2~6 base pairs in length. Among the various molecular markers, SSRs have been proven to be an efficient tool for constructing genetic linkage, performing QTL analysis and evaluating the level of genetic variation in a species because of the high variability, abundance, neutrality and co-dominance of microsatellite DNA [71].

We obtained a total 6,689 SSRs in the transcriptomic dataset. Of these, 67.99% were di-nucleotide repeats, followed by 27.69% tri-nucleotide repeats and 4.32% tetra/penta/hexa-nucleotide repeats (**Figure 4**). Among the di-nucleotide repeats motifs, $(GA/AG)_n$, $(TA/AT)_n$ and $(CT/TC)_n$ were the three predominant types with frequencies of 29.73%, 27.88% and 23.04%, respectively. There was a bias towards tri-nucleotide repeats motifs composed of A and T. In the 20 types of tri-nucleotide repeats, $(TAT/ATT/TTA)_n$ and $(ATA/TAA/AAT)_n$ were the most common types with a combined frequency of 33.37%. To date, only a few microsatellites have been available for *M. nipponense* from NCBI. Thus, the development of SSRs for this species is highly desirable.

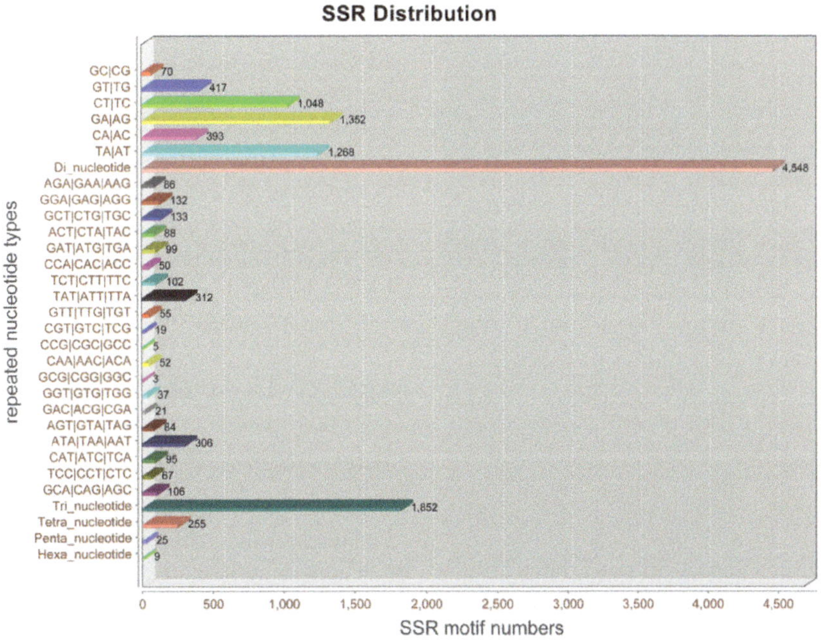

Figure 4: Distribution of simple sequence repeat (SSR) nucleotide classes among different nucleotide types found in the transcriptome of *M. nipponense.*

SNPs were identified from alignments of multiple sequences used for contig assembly. By excluding those that had mutation frequency of bases lower than 1%, we obtained a total of 24,786 SNPs, of which 6,679 were putative indels (In), 10,261 were putative transitions (Ts) and 7,846 were putative transversions (Tv), giving a mean In: Ts: Tv ratio of 1:1.54:1.17 across the transcriptome of *M. nipponense* (**Figure 5**). The AG/GA, CT/TC and AT/TA SNP types were the most common. In contrast, GC/CG types were the smallest SNP types because of the differences in the base structure and the number of hydrogen bonds between different bases. Multiple sequence alignment also identified a total of 6,679 indels across the transcriptome. It should be treated with caution because of technical problems associated with Roche 454 GS FLX pyrosequencing [32]. To verify the potential SNPs, a subset of 20 ESTs containing 39 SNP loci was selected randomly. A pooled cDNA sample of eight wild *M. nipponense* was amplified by PCR. Subsequently, PCR products were sequenced bidirectionally with forward or reverse primers (**Table 3**). Of the 39 SNP loci predicted to reside in the amplified sequences, 26 (66.67%) showed polymorphisms in the sample and were validated (**Table 3**). The rate of polymorphic SNPs was probably an underestimate because only eight individuals were used. Consequently, using more *M. nipponense* samples, the polymorphic rate of potential EST-SNPs should be higher than that found in our validation.

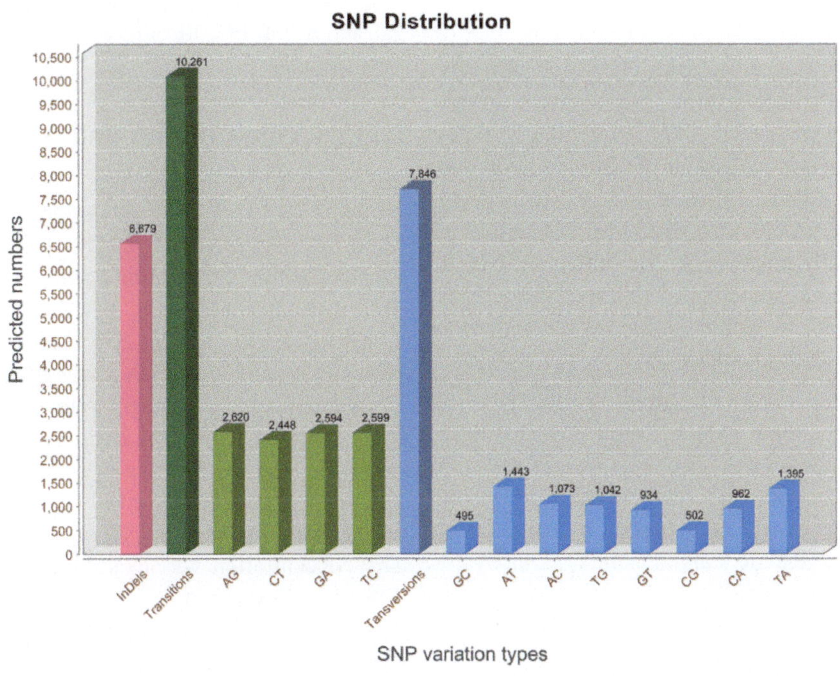

Figure 5: Distribution of putative single nucleotide polymorphisms (SNP) in the transcriptome of *M. nipponense*.

Table 3: Primers used and verified SNPs in the ESTs of *M. nipponense*.

Contig ID	Primers sequence (5′–3′)	Products length (bp)	No. of SNPs
06300	AACTTAGGGCTTGCGATTC CTCTTCCGAGTGCGTGAT	262	1/1/0
06559	GTTCTTTGGATGTTAGGTG AAGTTTGTTGATGTGGTTT	221	1/1/0
06567	GACATGACCATGCTCCCA AGGAACTGTCTCGCCAAC	551	0/1/0
07723	ACACTATTTAGGCATTTAT GGAAGTTCAGTAGGAGGT	208	2/2/0
01235	AGCAGACGCAGGTAGAAC AGTGAAGGCTTGAGGAAC	432	1/2/0
01295	GAGCCTCAAATAAGAAAG TATGAAGCCCATCAACCC	431	4/4/0
01191	ACCAGGAATGTTGGAGCA GCCTAGCATACCCGATGA	351	0/5/2
01399	TATCGCATACACTAAGGC ATTTAATGGTCGCTACTC	396	4/5/0
06092	GGAGGAGTGGGTTAGGTC TTACTGTCGCTGTGGTGA	319	2/2/1
08341	ACAGCCATGAAGTCATAAG AGAAGGTCCAGCGATACG	454	3/3/3
08357	GTTATTGCCGTGTTCTTA ATGGTGCTCCCTTATCTT	373	2/3/0
	CAGAAGCACATAGGCACA CAGGTCATCAAGAGGAGG	316	0/2/0
08359	ACTTGCTCCTGATAACCC CCACTTCTTTGACACACC	302	1/1/0
08391	CTCTTCCGCATTTCTACG TAAACCCAGGGATTCAGT	200	1/1/0
08461	TGAATGGGTCTGGTTGGT CATTTGTGGCGGAGGAAG	370	1/1/0
08476	AGTTCACAGCAGTGGGATT TGGTGAGGGAGACGAAGA	200	1/1/0
08491	GGAGAAACGCAAGAACGC TCCTGATGGTGGAGACCC	345	1/1/0
15044	TATTGAAGCTGCACAACC TAGAGCCATAAGCGAGGA	n/a	
27660	TCCGTATCTTTCGAGTTT GTCCCGCATTGTATTTTT	166	1/2/0
08344	GGAAGACAACAACGACAC CCAGCTAACACAAACAAT	496	0/1/0

For each contig, the following are indicated: the number of SNPs verified/the number of SNPs predicted in the fragment/the number of additional SNPs using Sanger technology; n/a indicates that sequencing was not possible because the sequencing signal was interrupted.
doi:10.1371/journal.pone.0039727.t003

Six additional SNPs, which were not detected in the transcriptomic dataset, were identified by genotyping 20 ESTs (**Table 3**). This phenomenon of missing SNPs may be due to the experimental methodology. Because 454 pyrosequencing was used to randomly sequence cDNAs, the detection of polymorphisms was likely hindered by the requirement of at least two reads with the variant alleles. Moreover, these reads had to contain at least 20 nucleotides of conserved sequence upstream and downstream of the locus.

SSRs and SNPs detected in this study are likely to be highly transferable to other closely related species, as has been the case in other crustacean species [72], [73]. We also envision that these potential markers identified within the ESTs will be valuable for studying the evolution and molecular ecology of *Macrobrachium* species, genome mapping, and QTL analysis.

CONCLUSIONS

In this study, *de novo* transcriptome sequencing for *M. nipponense* using the 454 GS FLX was performed for the first time. A total of 984,204 high-quality transcriptomic reads were obtained, giving rise to an average of 344 bp per read. A significant number of putative metabolic pathways and functions associated with the unique sequences were identified. Moreover, a large number of SNPs and SSRs were predicted and can be used for subsequent marker development, genetic linkage and QTL analysis. Many candidate genes that are potentially involved in sex determination and sex differentiation were identified for the first time and are worthy of further investigation. Our study provides the largest number of ESTs to date and lays the initial groundwork for in-depth, functional transcriptomic profiling of *M. nipponense*.

MATERIALS AND METHODS

Ethics statement

Collecting the wild oriental prawns from Chongming Island (Shanghai, China) was permitted by the Department of Fishery of Chongming County (permit number 2011HY000108). All handling of prawns were conducted in accordance with guidelines on the care and use of animals for scientific purposes set up by the Institutional Animal Care and Use Committee (IACUC) of Shanghai Ocean University, Shanghai, China.

Tissue material, RNA extraction and quality controls

Adult *M. nipponense* prawns (nine males and six females) used in this study were collected from Chongming Island (Shanghai, China) in 2011. The prawns were all anaesthetized on ice and dissected to collect samples, including the eyestalk, gill, heart, ovary, testis, hepatopancreas, muscle, and embryos at the cleavage, gastrula, nauplius and zoea stages. All of the samples were immediately frozen in liquid nitrogen and stored at −80°C until use. Total RNA was extracted from these materials using TRIzol Reagent (Invitrogen, USA) according to the manufacturer's instructions. The quality of total RNA was determined using a Nanodrop spectrophotometer (Thermo, USA). Only RNA samples with a 260 of 280 ratio from 1.9 to 2.1 and a 260 of 230 ratio from 2.0 to 2.5 were used for the next analysis.

Library construction and 454 pyrosequencing

We combined equivalent amounts of total RNA from each sample (~0.20 mg total RNA) and delivered it to Shanghai Majorbio Bio-pharm Biotechnology Co., Ltd. (Shanghai, China) for the construction of the cDNA library. The cDNA library was constructed by the SMART cDNA library construction kit (Clontech, USA) following the manufacturer's protocol step-by-step. cDNA was sheared by nebulization and DNA bands (500~800 bp) were extracted after agarose gel electrophoresis. Then, the obtained cDNA was purified, blunt ended, ligated to adapters and finally small fragments were removed. The quality control of a double DNA library was performed using High Sensitivity Chip (Agilent Technologies). The concentration was examined by TBS 380 Fluorometer. After the examination, one-plate, whole-run sequencing was performed on Roche 454 GS FLX Titanium chemistry (Roche Diagnostics, Indianapolis, IN, USA) by Shanghai Majorbio Bio-pharm Biotechnology Co., Ltd. following the manufacturer's protocol.

Sequence cleaning and assembly

For each sequence, low-quality bases and the sequencing adapter were trimmed using LUCY (http://lucy.sourceforge.net/) and SeqClean (http://compbio.dfci.harvard.edu). The remained sequencing reads were first assembled using the Newbler software with the "extend low depth overlaps" parameter. In order to improve the assembler quality, we collected all available ESTs of *Macrobrachium* from NCBI-EST database (http://www.ncbi.nlm.nih.gov/nucest?term=macrobrachium). All of the ESTs from the Roche 454 and NCBI-EST databases were used to run the final assembly of *M. nipponense.*

Annotation of mRNAs

BLASTx searches [74] of the GenBank nr database hosted by NCBI (http://www.ncbi.nlm.nih.gov/) were performed on all unique sequences to identify the putative mRNA functions (e-value threshold $\leq 1e^{-6}$, e values less than 1.0×10^{-6} were considered significant). Additionally, GO terms (http://www.geneontology.org) were extracted from the best hits obtained from the BLASTx against the nr database using Blast2GO. These results were then sorted by GO categories using in-house Perl scripts. BLASTx was also used to align unique sequences to the Swiss-Prot database (http://web.expasy.org/docs/swiss-prot_guideline.html), Kyoto Encyclopedia of Genes and Genomes (KEGG) and Clusters of Orthologous Groups (COG) (http://www.ncbi.nlm.nih.gov/COG/) (with the e-value of 10^{-6}) to predict possible functional classifications and molecular pathways [75], [76].

Identification of EST-SSR motifs and EST-SNPs

The unique sequences were screened for microsatellites using the Mreps software (http://bioinfo.lifl.fr/mreps/) with default parameters. Perfect di-, tri-, tetra-, penta-, and hexa-nucleotide motifs were detected, and all SSR types required a minimum of 6 repeats.

The SNPs were extracted using VarScan (http://varscan.sourceforge.net) with the default parameter only when both alleles were detected from the contigs. Since no reference sequences were available, SNPs were identified as superimposed nucleotide peaks where two or more reads contained polymorphisms at the variant allele. To validate the putative SNPs identified in ESTs, a cDNA pool of eight wild *M. nipponense*were used. Twenty ESTs containing 39 potential SNP loci and sufficient flanking regions were randomly selected for primer design to amplify 150~600 bp of fragments. PCR products were sequenced directly in both directions with forward and reverse primers using Sanger technology on the ABI3730 platform (Applied Biosystems). Sequencing chromatograms were visually analyzed with Chromas 2.32 (Technelysium Pty. Ltd.), and SNPs were identified as overlapping nucleotide peaks.

Data Deposition

The Roche 454 reads of *M. nipponense* were submitted to NCBI Sequence Read Archive under the accession number of SRA051767.2.

Acknowledgments

We acknowledge Xianglin Zhang and Yi Ren (Shanghai Majorbio Bio-pharm Biotechnology Co., Ltd.) for their kindly help in sequencing and bioinformatics analysis. We thank the lab members, Jingyun Lin and Ying Chen (Key laboratory of Freshwater Aquatic Genetic Resources Certificated by Ministry of Agriculture, College of Fisheries and Life Science, Shanghai Ocean University) for their help in RNA extraction and distinction to the different stages of embryo development.

Footnotes

Competing Interests: The authors have declared that no competing interests exist.

Funding: This research was supported by the Innovation Program of Shanghai Municipal Education Commission (project number 08ZZ80), the Marine Biology Program of Shanghai Leading Academic Discipline (project number J50701), the Natural Science Foundation of China (project number 31001111), and the Aquaculture Engineering and Technology Research Center of Shanghai (project number 09DZ2251300). The funders had no role in study design, data collection and analysis, decision to publish, or preparation of the manuscript.

REFERENCES

1. Ma KY, Feng JB, Lin JY, Li JL. The complete mitochondrial genome of *Macrobrachium nipponense.* Gene. 2011;487:160–165. [PubMed]
2. Feng JB, Li JL, Cheng X. Research progress on germplasm resource exploitation and protection of *Macrobrachium nipponense.* J Shanghai Fish Univ. 2008;17:371–376.
3. Yu H, Miyake S. Five species of the genus *Macrobrachium* (Crustacea, Decapoda, Palaemonidae) from Taiwan. Ohmu. 1972;3:45–55.
4. Cai Y, Shokita S. Report on a collection of freshwater shrimps (Crustacea: Decapoda: Caridea) from the Philippines, with descriptions of four new species. Raffles B Zool. 2006;54:245–270.
5. De Grave S, Ghane A. The establishment of the oriental river prawn, *Macrobrachium nipponense* (de Haan, 1849) in Anzali Lagoon, Iran. Aquat Invasions. 2006;1:204–208.
6. Mirabdullaev IM, Niyazov DS. Alien decapods (Crustacea) in Uzbekistan. Abstracts of the II International Symposium Invasion of alien species in Holarctic, Borok, Russia. 2005;2:113–114.
7. Salman SD, Page TJ, Naser MD, Yasser AG. The invasion of *Macrobrachium nipponense* (De Haan, 1849) (Caridea: Palaemonidae) into the southern Iraqi marshes. Aquat Invasions. 2006;1:109–115.
8. Aflalo ED, Hoang TTT, Nguyen VH, Lam Q, Nguyen DM, et al. A novel two-step procedure for mass production of all-male populations of the giant freshwater prawn *Macrobrachium rosenbergii.* Aquaculture.2006;256:468–478.

9. Sagi A, Cohen D, Milner Y. Effect of androgenic gland ablation on morphotypic differentiation and sexual characteristics of male freshwater prawns, *Macrobrachium rosenbergii*. Gen Comp Endocr.1990;77:15–22. [PubMed]

10. Nagamine C, Knight AW, Maggenti A, Paxman G. Effects of androgenic gland ablation on male primary and secondary sexual characteristics in the Malaysian prawn, *Macrobrachium rosenbergii* (de Man) (Decapoda, Palaemonidae), with first evidence of induced feminization in a nonhermaphroditic decapod. Gen Comp Endocr. 1980;41:423–441. [PubMed]

11. Manor R, Aflalo ED, Segall C, Weil S, Azulay D, et al. Androgenic gland implantation promotes growth and inhibits vitellogenesis in *Cherax quadricarinatus* females held in individual compartments. Invertebr Reprod Dev. 2004;45:151–159.

12. Malecha SR, Nevin PA, Ha P, Barck LE, Lamadrid-Rose Y, et al. Sex-ratios and sex-determination in progeny from crosses of surgically sex-reversed freshwater prawns, *Macrobrachium rosenbergii*. Aquaculture. 1992;105:201–218.

13. Zhang EF, Qiu GF. A novel Dmrt gene is specifically expressed in the testis of Chinese mitten crab, *Eriocheir sinensis*. Dev Genes Evol. 2010;220:151–159. [PubMed]

14. Nakkrasae L-I, Damrongphol P. A vasa-like gene in the giant freshwater prawn, *Macrobrachium rosenbergii*. Mol Reprod Dev. 2007;74:835–842. [PubMed]

15. Aflalo ED, Bakhrat A, Raviv S, Harari D, Sagi A, et al. Characterization of a vasa-like gene from the Pacific white shrimp *Litopenaeus vannamei* and its expression during oogenesis. Mol Reprod Dev.2007;74:172–177. [PubMed]

16. Feng JB, Li JL. Twelve polymorphic microsatellites in oriental river prawn, *Macrobrachium nipponense*. Mol Ecol Resour. 2008;8:986–988. [PubMed]

17. Ma KY, Feng JB, Li JL, Ding HY. Twenty-four novel polymorphic microsatellite markers from oriental river prawn (*Macrobrachium nipponense*). Conserv Genet Resour. 2010;2:125–128.

18. Feng JB, Wu CL, Ding HY, Hua XM, Li YS, et al. Microsatellite analysis of genetic diversity in 9 wild stocks of oriental river prawn *Macrobrachium nipponense* in Hongze Lake. J Fish Sci Chn.2010;17:218–227.

19. Ma KY, Feng JB, Xie N, Feng XY, Li JL. Microsatellite analysis of genetic variation of the oriental river prawn *Macrobrachium nipponense* in Qiantang River. Zool Res. 2011;32:363–370. [PubMed]

20. Wu P, Qi D, Chen LQ, Zhang H, Zhang XW, et al. Gene discovery from an ovary cDNA library of oriental river prawn *Macrobrachium nipponense* by ESTs annotation. Comp Biochem Phys D. 2009;4:111–120. [PubMed]

21. Wang YT, Mao H, Hou CC, Sun X, Wang DH, et al. Characterization and expression pattern of KIFC1-like kinesin gene in the testis of the *Macrobrachium nipponense* with discussion of its relationship with structure lamellar complex (LCx) and acroframosome (AFS). Mol Biol Rep. 2012. pp. 1–8. [PubMed]

22. Zhang FY, Chen LQ, Wu P, Zhao WH, Li EC, et al. cDNA cloning and expression of Ubc9 in the developing embryo and ovary of oriental river prawn, *Macrobrachium nipponense*. Comp Biochem Phys B.2010;155:288–293. [PubMed]

23. Zhang FY, Chen LQ, Qin JG, Zhao WH, Wu P, et al. cDNA cloning and expression analysis of gustavus gene in the oriental river prawn *Macrobrachium nipponense*. PLoS One. 2011;6:e17170.[PMC free article] [PubMed]

24. Blanca J, Cañizares J, Roig C, Ziarsolo P, Nuez F, et al. Transcriptome characterization and high throughput SSRs and SNPs discovery in *Cucurbita pepo* (Cucurbitaceae). BMC Genomics. 2011;12:104.[PMC free article] [PubMed]

25. Bouck A, Vision T. The molecular ecologist's guide to expressed sequence tags. Mol Ecol.2007;16:907–924. [PubMed]

26. Kim JM, Lee KH, Jeon YJ, Oh JH, Jeong SY, et al. Identification of genes related to Parkinson's disease using expressed sequence tags. DNA Res. 2007;13:275–286. [PubMed]

27. Garg R, Patel RK, Tyagi AK, Jain M. *De novo* assembly of chickpea transcriptome using short reads for gene discovery and marker identification. DNA Res. 2011;18:53–63. [PMC free article] [PubMed]

28. Manickavelu A, Kawaura K, Oishi K, Shin T, Kohara Y, et al. Comparative gene expression analysis of susceptible and resistant near-isogenic lines in common wheat infected by *Puccinia triticina*. DNA Res.2010;17:211–222. [PMC free article] [PubMed]

29. Xia JH, He XP, Bai ZY, Lin G, Yue GH. Analysis of the Asian seabass transcriptome based on expressed sequence tags. DNA Res. 2011;18:513–522. [PMC free article] [PubMed]

30. Clark MS, Thorne MAS, Toullec JY, Meng Y, Peck LS, et al. Antarctic krill 454 pyrosequencing reveals chaperone and stress transcriptome. PLoS One. 2011;6:e15919. [PMC free article] [PubMed]

31. Bai X, Rivera-Vega L, Mamidala P, Bonello P, Herms DA, et al. Transcriptomic signatures of ash (*Fraxinus* spp.) phloem. PLoS One. 2011;6:e16368. [PMC free article] [PubMed]

32. Parchman TL, Geist KS, Grahnen JA, Benkman CW, Buerkle CA. Transcriptome sequencing in an ecologically important tree species: assembly, annotation, and marker discovery. BMC Genomics.2010;11:180. [PMC free article] [PubMed]

33. Srivastava A, Rogers WL, Breton CM, Cai L, Malmberg RL. Transcriptome analysis of *Sarracenia*, an insectivorous plant. DNA Res. 2011;18:253–261. [PMC free article] [PubMed]

34. Bai X, Mamidala P, Rajarapu SP, Jones SC, Mittapalli O. Transcriptomics of the bed bug (*Cimex lectularius*). PLoS One. 2011;6:e16336. [PMC free article] [PubMed]

35. Salem M, Rexroad CE, Wang JN, Thorgaard GH, Yao JB. Characterization of the rainbow trout transcriptome using Sanger and 454-pyrosequencing approaches. BMC Genomics. 2010;11:564.[PMC free article] [PubMed]

36. Jung H, Lyons RE, Dinh H, Hurwood DA, McWilliam S, et al. Transcriptomics of a giant freshwater prawn (*Macrobrachium rosenbergii*): *de novo* assembly, annotation and marker discovery. PLoS One.2011;6:e27938. [PMC free article] [PubMed]

37. Poroyko V, White JR, Wang M, Donovan S, Alverdy J, et al. Gut microbial gene expression in mother-fed and formula-fed piglets. PLoS One. 2010;5:e12459. [PMC free article] [PubMed]

38. Vera JC, Wheat CW, Fescemyer HW, Frilander MJ, Crawford DL, et al. Rapid transcriptome characterization for a nonmodel organism using 454 pyrosequencing. Mol Ecol. 2008;17:1636–1647.[PubMed]

39. Rokas A, Abbot P. Harnessing genomics for evolutionary insights. Trends Ecol Evol. 2009;24:192–200. [PubMed]

40. Craft JA, Gilbert JA, Temperton B, Dempsey KE, Ashelford K, et al. Pyrosequencing of *Mytilus galloprovincialis* cDNAs: tissue-specific expression patterns. PLoS One. 2010;5:e8875. [PMC free article] [PubMed]

41. Meyer E, Aglyamova GV, Wang S, Buchanan-Carter J, Abrego D, et al. Sequencing and *de novo*analysis of a coral larval transcriptome using 454 GSFlx. BMC Genomics. 2009;10:219. [PMC free article][PubMed]

42. Chareontawee K, Poompuang S, Na-Nakorn U, Kamonrat W. Genetic diversity of hatchery stocks of giant freshwater prawn (*Macrobrachium rosenbergii*) in Thailand. Aquaculture. 2007;271:121–129.

43. Ashburner M, Ball CA, Blake JA, Botstein D, Butler H, et al. Gene Ontology: tool for the unification of biology. Nat Genet. 2000;25:25–29. [PMC free article] [PubMed]

44. Conesa A, Götz S. Blast2GO: A comprehensive suite for functional analysis in plant genomics. Int J Plant Genomics. 2008;2008:619832. [PMC free article] [PubMed]

45. Rocha J, Garcia-Carreno FL, Muhlia-Almazan A, Peregrino-Uriarte AB, Yepiz-Plascencia G, et al. Cuticular chitin synthase and chitinase mRNA of whiteleg shrimp *Litopenaeus vannamei* during the molting cycle. Aquaculture. 2012;330:111–115.

46. Proespraiwong P, Tassanakajon A, Rimphanitchayakit V. Chitinases from the black tiger shrimp *Penaeus monodon*: phylogenetics, expression and activities. Comp Biochem Phys B. 2010;156:86–96.[PubMed]

47. Ewen-Campen B, Shaner N, Panfilio KA, Suzuki Y, Roth S, et al. The maternal and early embryonic transcriptome of the milkweed bug *Oncopeltus fasciatus*. BMC Genomics. 2011;12:61. [PMC free article][PubMed]

48. Sinclair AH, Berta P, Palmer MS, Hawkins JR, Griffiths BL, et al. A gene from the human sex-determining region encodes a protein with homology to a conserved DNA-binding motif. Nature.1990;346:240–244. [PubMed]

49. Ner SS. HMGs everywhere. Curr Biol. 1992;2:208–210. [PubMed]

50. Vilain E, McCabe ERB. Mammalian sex determination: from gonads to brain. Mol Genet Metab.1998;65:74–84. [PubMed]

51. Raymond CS, Murphy MW, O'Sullivan MG, Bardwell VJ, Zarkower D. Dmrt1, a gene related to worm and fly sexual regulators, is required for mammalian testis differentiation. Gene Dev. 2000;14:2587–2595.[PMC free article] [PubMed]

52. Raymond CS, Kettlewell JR, Hirsch B, Bardwell VJ, Zarkower D. Expression of Dmrt1 in the genital ridge of mouse and chicken embryos suggests a role in vertebrate sexual development. Dev Biol.1999;215:208–220. [PubMed]

53. Smith CA, Roeszler KN, Ohnesorg T, Cummins DM, Farlie PG, et al. The avian Z-linked gene DMRT1 is required for male sex determination in the chicken. Nature. 2009;461:267–271. [PubMed]

54. Carlsson P, Mahlapuu M. Forkhead transcription factors: key players in development and metabolism. Dev Biol. 2002;250:1–23. [PubMed]

55. Uhlenhaut NH, Jakob S, Anlag K, Eisenberger T, Sekido R, et al. Somatic sex reprogramming of adult ovaries to testes by FOXL2 ablation. Cell. 2009;139:1130–1142. [PubMed]

56. Lavorgna G, Ueda H, Clos J, Wu C. FTZ-F1, a steroid hormone receptor-like protein implicated in the activation of fushi tarazu. Science. 1991;252:848–851. [PubMed]

57. Oba K, Yanase T, Nomura M, Morohashi K, Takayanagi R, et al. Structural characterization of human Ad4bp (SF-1) gene. Biochem Bioph Res Co. 1996; 226:261–267. [PubMed]

58. Lala DS, Rice DA, Parker KL. Steroidogenic factor I, a key regulator of steroidogenic enzyme expression, is the mouse homolog of fushi tarazu-factor I. Mol Endocrinol. 1992; 6:1249–1258. [PubMed]

59. Von Hofsten J, Jones I, Karlsson J, Olsson P-E. Developmental expression patterns of FTZ-F1 homologues in zebrafish (*Danio rerio*). Gen Comp Endocr. 2001; 121:146–155. [PubMed]

60. Zhang WM, Zhang YY, Zhang LH, Zhao HH, Li X, et al. The mRNA expression of P450 aromatase, gonadotropin beta-subunits and FTZ-F1 in the orange-spotted grouper (*Epinephelus coioides*) during 17 alpha-methyltestosterone-induced precocious sex change. Mol Reprod Dev. 2007; 74:665–673. [PubMed]

61. Gaudet J, Vanderlst I, Spence AM. Post-transcriptional regulation of sex determination in *Caenorhabditis elegans*: widespread expression of the sex-determining gene fem-1 in both sexes. Mol Biol Cell. 1996; 7:1107–1121. [PMC free article] [PubMed]

62. Lasko PF, Ashburner M. The product of the *Drosophila* gene vasa is very similar to eukaryotic initiation factor-4A. Nature. 1988; 335:611–617. [PubMed]

63. Chang CC, Dearden P, Akam M. Germ line development in the grasshopper *Schistocerca gregaria*: vasa as a marker. Dev Biol. 2002; 252:100–118. [PubMed]

64. Shinomiya A, Tanaka M, Kobayashi T, Nagahama Y, Hamaguchi S. The vasa-like gene, olvas, identifies the migration path of primordial germ cells during embryonic body formation stage in the medaka, *Oryzias latipes*. Dev Growth Differ. 2000;42:317–326. [PubMed]

65. Conley A, Hinshelwood M. Mammalian aromatases. Reproduction. 2001;121:685–695. [PubMed]

66. Jan E, Motzny CK, Graves LE, Goodwin EB. The STAR protein, GLD-1, is a translational regulator of sexual identity in *Caenorhabditis elegans*. EMBO J. 1999;18:258–269. [PMC free article] [PubMed]

67. Nagamine CM, Carlisle C. Zfy2/1 fusion gene fails to replicate Zfy1 expression pattern in fetal gonads. Genomics. 1997;43:397–398. [PubMed]

68. Nachtigal MW, Hirokawa Y, Enyeart-Vanhouten DL, Flanagan JN, Hammer GD, et al. Wilms' tumor 1 and Dax-1 modulate the orphan nuclear receptor SF-1 in sex-specific gene expression. Cell. 1998;93:445–454. [PubMed]

69. Boswell RE, Mahowald AP. Tudor, a gene required for assembly of the germ plasm in *Drosophila melanogaster*. Cell. 1985;43:97–104. [PubMed]

70. Ventura T, Aflalo ED, Weil S, Kashkush K, Sagi A. Isolation and characterization of a female-specific DNA marker in the giant freshwater prawn *Macrobrachium rosenbergii*. Heredity. 2011;107:456–461.[PMC free article] [PubMed]

71. Liu ZJ, Cordes JF. DNA marker technologies and their applications in aquaculture genetics. Aquaculture. 2004;238:1–37.

72. Kim WJ, Jung H, Gaffney PM. Development of type I genetic markers from expressed sequence tags in highly polymorphic species. Mar Biotechnol. 2011;13:127–132. [PubMed]

73. Ellis JR, Burke JM. EST-SSRs as a resource for population genetic analyses. Heredity. 2007;99:125–132. [PubMed]

74. Altschul SF, Madden TL, Schäffer AA, Zhang J, Zhang Z, et al. Gapped BLAST and PSI-BLAST: a new generation of protein database search programs. Nucleic Acids Res. 1997;25:3389–3402.[PMC free article] [PubMed]

75. Kanehisa M, Goto S, Hattori M, Aoki-Kinoshita KF, Itoh M, et al. From genomics to chemical genomics: new developments in KEGG. Nucleic Acids Res. 2006;34:D354–D357. [PMC free article][PubMed]

76. Kanehisa M, Araki M, Goto S, Hattori M, Hirakawa M, et al. KEGG for linking genomes to life and the environment. Nucleic Acids Res. 2008;36:D480–D484

CHAPTER 3

INSIGHTS INTO SEXUAL PRECOCITY OF FEMALE ORIENTAL RIVER PRAWN MACROBRACHIUM NIPPONENSE THROUGH TRANSCRIPTOME ANALYSIS

Jiang H, Li X, Sun Y, Hou F, Zhang Y, Li F, Gu Z, Liu X.

INTRODUCTION

The oriental river prawn *Macrobrachium nipponense*, a member of the Palaemonidae family of decapod crustaceans, is a commercial freshwater prawn species that naturally distributed throughout most freshwaters and low-salinity estuarine regions in Japan, Korea, Vietnam, Myanmar, and China [1,2]. It is considered as an important fishery resource and widely farmed in China due to its flavor, high nutritive value and excellent adaptability, with an average annual production value of over 23 million USD [3]. However, sexual precocity has become prevalent with the expansion of production scale. Sexual precocity refers to early male and female gonad development, which has been observed in many crustaceans. Sexually precocious *M. nipponense* has a low value due to low growth rate, poor survival, and short life span [4–6], which seriously restricts the sustainable development of this species. The adverse effect of sexual precocity on female *M. nipponense* is particularly prominent. Sexually precocious female *M. nipponense*is far less than normal sexually mature female prawn in both weight and size. Some female prawns even become sexually mature at 0.5 grams, but fecundity is only 200–500 eggs. Eggs obtained from these females often result in poor-quality offspring. Thus, understanding the mechanisms that regulate ovarian maturation in female *M. nipponense* and controlling sexual precocity of this prawn is crucial to improving the production of this species.

The ovary is a multifunctional organ that plays a key role in reproduction and secretion of hormones for regulation of growth and development in female prawns [7]. Ovarian maturation in prawn is a complex process controlled by several factors, such as endocrine control, nutrition and environmental factors [8–11]. However, the molecular mechanisms involved in stimulating ovarian development in prawn are still unclear. Till now, some reproduction- and ovary development-related genes have been identified from ovaries in *M. nipponense*, such as *SUMO-conjugating enzyme* (*Ubc9*) gene [12], *heat shock protein 90* (*HSP90*) gene [13] and *Gonadotropin-releasing hormone receptor* (*GnRHR*) gene [14], etc. However, the genomic or transcriptomic level resources that come from *M. nipponense* ovary remain limited. So far, only one study has reported sequenced transcriptome from ovary of *M. nipponense*, which sequenced non-normalized EST library using traditional Sanger sequencing [15]. This recorded cDNA library contains 3256 ESTs that can greatly accelerate the research and discovery of new genes and molecular markers on *M. nipponense* ovary. However, the underlying mechanism of sexual precocity of this female prawn has not been fully revealed, especially at the molecular level, including genes and pathways. In a word, the lack of genomic and transcriptomic information of *M. nipponense* ovary poses an obstacle to identify genes and construct regulatory networks associated with sexual precocity of this prawn.

Citation: Jiang H, Li X, Sun Y, Hou F, Zhang Y, Li F, et al. (2016) Insights into Sexual Precocity of Female Oriental River Prawn Macrobrachium nipponense through Transcriptome Analysis. PLoS ONE 11(6): e0157173. doi:10.1371/journal.pone.0157173

Recently, the development of next-generation sequencing (NGS) technologies such as Illumina HiSeq 2000 [16], ABI SOLiD and 454 of Roche [17], and the newly developed deep sequencing methods such as Solexa/Illumina RNA-seq and Digital gene expression (DGE) [18], have opened a new avenue into transcriptome characterization and gene-expression profiling for various species, and rapidly dominated transcriptome studies because the higher-accuracy, higher-speed and lower-cost than the first-generation sequencing technology (Sanger sequencing). The RNA-Seq, a technique based on sequencing the poly-A RNA fraction, is a powerful tool to study complex transcriptomes because it allows for not only characterizing isoforms from known genes but also discovering novel or predicted coding genes [19]. It gives a general view of gene expression especially in these species lack of a fully sequenced and assembled genome such as *M. nipponense*. Some cDNA libraries and transcriptome-level datasets have been obtained from *M. nipponense* by RNA-Seq to lay a foundation for functional genomics approaches used for improving the aquaculture performance of this species [2,20,21]. Based on these transcriptome studies, there are about 81,411 expressed sequence tags (ESTs) from *M. nipponense* in the public databases up to date. However, there have been no transcriptome studies regarding the ovary of sexually precocious *M. nipponense* was reported until now.

In the present study, we performed high-throughput sequencing of the ovaries of sexually precocious and normal sexually mature *M. nipponense* using Illumina RNA-Seq to generate a transcriptome database that will enlarge the public EST database for this species and help support future studies. The identification of differentially expressed genes and pathways in the ovary of these two types of prawn will help build a more complete understanding of the regulatory mechanisms associated with sexual precocity. In addition, the simple sequence repeats (SSRs) and single nucleotide polymorphisms (SNPs) reported in this transcriptome study are also potentially useful for population genetics and functional genomics studies in this species.

MATERIALS AND METHODS

Sample preparation and RNA extraction

There were two groups of female experimental prawn, one group was sexually precocious *M. nipponense* (MNOP) (2.5–3.5 cm, 0.5–1.5 g) which has grown about 90 days from hatching to sexual maturity, another group was normal sexually mature *M. nipponense* (MNON) (4.5–5.5 cm, 2.5–4.5 g) which took about one year to reach sexual maturity after hatching. Each group had ten samples which were all obtained from aquaculture base in Zhejiang institute of freshwater fisheries, Huzhou, China, and their ovaries were all at the stage V (maturation stage) based on the external morphology, color, gonado-somatic index (GSI), and histological features [13,15]. After prawns were anesthetized and finally sacrificed by placing them in an ice bath, ovaries were dissected and snap frozen in liquid nitrogen and stored at– 80°C until use. The study was approved by the Institutional Animal Care and Use Ethics Committee of Agriculture Ministry Key Laboratory of Healthy Freshwater Aquaculture, Zhejiang Institute of Freshwater Fisheries (Huzhou, China). Total RNA was extracted using Trizol Reagent (Invitrogen, Carlsbard, CA, USA). The integrity and purity of the total RNA were detected by an Agilent-2100 bioanalyzer (Agilent Technologies, Inc., Santa Clara CA, USA) and its concentration was determined using the NanoDrop 2000 spectrophotometer (Nano-Drop, Thermo Scientific, Wilmington, DE, USA). Ten RNA samples from each group were pooled in equal amounts, generating two mixed RNA samples from MNOP and MNON respectively, which were then used to prepare two separate RNA-seq transcriptome libraries.

Transcriptome library preparation and Illumina sequencing

RNA purification, reverse transcription, library construction and sequencing were performed at Shanghai Majorbio Biopharm Biotechnology Co., Ltd. (Shanghai, China) according to the manufacturer's instructions (Illumina, San Diego, CA). The two RNA-seq transcriptome libraries were prepared using Illumina TruSeqTM RNA sample preparation Kit (San Diego, CA). Poly(A) mRNA was purified from total RNA using oligo-dT-attached magnetic beads and then fragmented by fragmentation buffer. Taking these short fragments as templates, double-stranded cDNA was synthesized using a SuperScript double-stranded cDNA synthesis kit (Invitrogen, CA) with random hexamer primers (Illumina). Then the synthesized cDNA was subjected to end-repair, phosphorylation and 'A' base addition according to Illumina's library construction protocol. Libraries were size selected for cDNA target fragments of 200–300 bp on 2% Low Range Ultra Agarose followed by PCR amplified using Phusion DNA polymerase (New England Biolabs, Boston, MA) for 15 PCR cycles. After quantified by TBS380, two RNA-seq libraries were sequenced in single lane on an Illumina Hiseq 2500 sequencer (Illumina, San Diego, CA) for 2×100bp paired-end reads [22,23].

Sequencing data assembly and functional annotation

Image data output from the sequencer was transformed into raw reads by base calling, and stored in FASTQ format. The raw reads were filtered into the high quality clean reads after removing adaptor sequences, empty reads, and low quality sequences (reads with unknown sequences 'N' or less than 20bp) using SeqPrep (https://github.com/jstjohn/SeqPrep) and Sickle(https://github.com/najoshi/sickle). The clean reads were then assembled *de novo* into transcripts using the Trinity program (http://trinityrnaseq.sourceforge.net/, version number: trinityrnaseq-r2013-02-25) [24]. To eliminate redundant sequences, transcripts were clustered based on sequence similarities and the longest transcript in each cluster represented the final unique sequence (unigene). All the unigenes were searched against the protein databases such as non-redundant (NR) protein sequence database, Kyoto Encyclopedia of Genes and Genomes (KEGG) database, Search Tool for the Retrieval of Interacting Genes (String) database, Swissprot and Pfam databases to identify the proteins with the highly sequence similarity to the given unigenes and retrieve their function annotations, using BLASTX with a typical cut-off *E*-value $<1e^{-5}$. Biological processes, molecular functions and cellular components were described by GO annotations of unigenes using BLAST2GO (http://www.blast2go.com/b2ghome) [25].

Clusters of Orthologous Groups (COG) (http://www.ncbi.nlm.nih.gov/COG/) and KEGG (http://www.genome.jp/kegg/) were used to predict possible functional classifications and metabolic pathways, respectively [26,27].

Molecular Marker Detection

MSATCOMMANDER V. 0.8.2 (http://code.google.com/p/msatcommander/) [28] was used to search for microsatellites (simple sequence repeats, SSRs). The minimum repeat lengths considered were as follows: ten repeats for mono-SSRs; six repeats for di-SSRs; and four repeats for tri-, tetra-, penta-, and hexa-SSRs. SSR-containing unigenes were annotated based on BLAST similarity searches. The location of SSRs was estimated based on open reading frames (ORFs) and untranslated regions (UTRs) within the unigenes that were identified using Trinity [24]. All the consensus assembly sequences generated from the two transcriptome libraries were employed as reference sequences to detect potential single nucleotide polymorphisms (SNPs) in MNON and MNOP using Samtools V. 0.1.18(http://samtools.sourceforge.net) [29] and VarScan v.2.2.7(http://varscan.sourceforge.net/) [30]. In addition, to identify SNPs between MNON and MNOP groups, all sequences generated from the MNON transcriptome library were employed as reference sequences and SNPs in MNOP were detected using Samtools V. 0.1.18 and VarScan v.2.2.7.

Identification and annotation of differentially expressed genes

We used the RNASeq by Expectation Maximization (RSEM) (http://deweylab.biostat.wisc.edu/rsem/) [31] to calculate unigene abundances from RNA-Seq data. FPKM (fragments per kilobase of exon model per million mapped reads) was used to quantify the gene expression levels. FPKM value can be directly used for the comparison of gene expression level between samples. FPKM is calculated as follows:

$$\text{FPKM} = \frac{\text{total exon Fragments}}{\text{mapped reads (Millions)} \times \text{exon length (kb)}}$$

In this formula, total exon Fragments is the number of reads that aligned to a specific unigene, mapped reads (Millions) is the total number of reads that aligned to all unigenes, exon Length (kb) is the length of the unigene.

The detection of differentially expressed genes (DEGs) between ovaries of MNON and MNOP was performed using edgeR package (Empirical analysis of Digital Gene Expression in R) (http://www.bioconductor.org/packages/2.12/bioc/html/edgeR.html) [32] with the rigorous algorithm method. Following the procedures described in the edgeR documentation, read count tables were loaded into R, normalized using the default method for edgeR (trimmed mean of M values, or TMM), and then tested for differential expression using the Fisher exact Test method. 0.1, a typical value for the common biological coefficient of variation (square root-dispersion) for datasets arising from well-controlled experiments, was chosen as a nominal dispersion value for the data without biological replicates. The *P*-values reported by edgeR were adjusted for multiple hypothesis testing using the False Discovery Rate (FDR) control method [33], and the ratio of FPKMs was converted to the fold-change in the expression of each gene in two samples simultaneously. FDR < 0.05 and the absolute value of logFC (Fold change) ≥ 1 were set as the threshold for significantly differential expression. The function of DEGs was annotated by the GO and KEGG pathway enrichment analyses at a $P \leq 0.05$.

Quantitative real-time PCR (QPCR) validation

Twenty candidate DEGs (Thirteen annotated DEGs and seven unannotated DEGs) were randomly selected for validation by QPCR. Total RNA was extracted from ovaries of MNON and MNOP using Trizol Reagent (Invitrogen, Carlsbard, CA, USA), and the RNA samples were reverse-transcribed using PrimeScript®RT reagent Kit with gDNA Eraser (Perfect Real Time) (TaKaRa, Dalian, China). Oligonucleotide primers were designed with the Primer Premier 5.0 software and synthesized by Invitrogen (Shanghai, China). QPCR was performed using SYBR® Premix Ex Taq™ II (Tli RNaseH Plus) (TaKaRa, Dalian, China) on a CFX96™ Real-Time PCR Detection System (Bio-Rad, USA). Each PCR reaction (25 μL) contained 12.5 μL SYBR® Premix Ex Taq II, 0.4 μM of each primer and appropriately diluted cDNA. The thermal cycler program was 95°C for 30 s, followed by 40 cycles of 95°C for 5 s and 60°C for 30 s. Six biological and three technical replicates were performed respectively. Beta-actin (-actin) was amplified in parallel as an internal control for normalization of QPCR data due to its stable expression in *M. nipponense* after the expressions of three reference genes including -actin, elongation factor 1-alpha (Ef1α) and ribosomal protein L8 (RpL8) were compared in different tissues of MNOP and MNON by the method of Qian *et al* (2014) [34]. The relative changes in gene expression levels were calculated using the $2^{-\Delta\Delta CT}$ method [35,36]. Statistical analyses were performed with SPSS 20.0 software (SPSS Inc., Chicago, IL, USA). Results were presented as the means±S.D. (n = 6) for each group. One-way analysis of variance (ANOVA) followed by the post hoc test were carried out to determine whether the differences between groups were significant ($P < 0.05$).

RESULTS

Transcriptome sequencing and assembly

Two cDNA libraries were constructed from ovaries of MNOP and MNON. The Illumina Hiseq 2500 sequencing generated 47,254,012 raw reads from the MNOP and 56,793,358 raw reads from the MNON. The raw sequencing data were deposited in the Short Read Archive (SRA) of the NCBI with accession numbers of SRP063589. After filtering out the adaptor primers and empty and low-quality reads, 42,115,012 and 50,789,578 clean reads were generated from MNOP and MNON libraries, respectively (Table 1). After *de novo* assembly of clean reads using Trinity, a total 63,336 unigenes were identified. The lengths of all unigenes were distributed as 893.88 bp, 28,661 bp, 201 bp and 1846 bp from average, longest, shortest and N50 length, respectively (Fig 1A). Of these, 31,441 (49.64%) were 1–400 bp; 9,273 (14.64%) were 401–600 bp; 4,634 (7.32%) were 601–800 bp; 3,021 (4.77%) were 801–1000 bp and 2,187(3.45%) were 1001–1200 bp (Fig 1B).

Figure 1: Length distribution of 63,336 assembled unigenes.

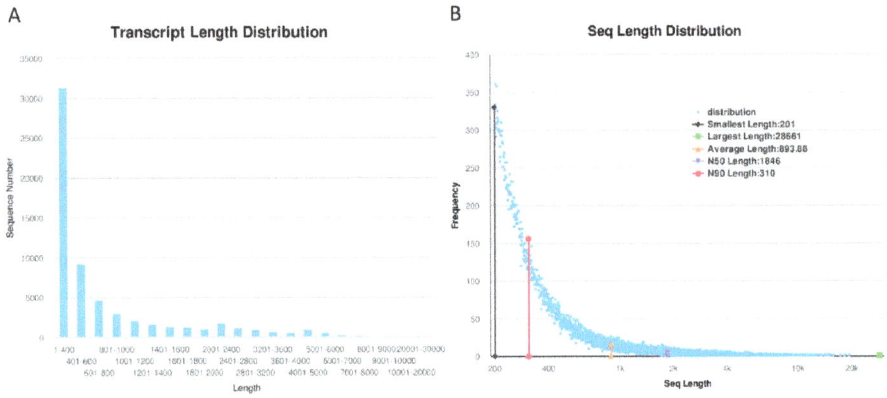

Table 1: Statistics of the sequencing results from two samples.

The sample name	Number of reads	Number of bases (bp)	Error%	Q20%	Q30%	GC%
Raw Data						
MNOP	47,254,012	4,772,655,212	0.0532	90.91	80.99	41.01
MNON	56,793,358	5,736,129,158	0.0503	91.41	81.87	41.04
Clean Data						
MNOP	42,115,012	4,036,709,557	0.0296	97.74	89.37	40.65
MNON	50,789,578	4,886,896,178	0.0288	97.86	89.85	40.71

doi:10.1371/journal.pone.0157173.t001

Functional annotation and classifictiaon

All the 63,336 assembled unigenes were searched against the public databases including NR, KEGG, String, Swissprot and Pfam databases, and a total of 15,134 unigenes were annotated, accounting for 23.89%. Unigenes that were annotated as unique in public databases are as follows: 10,692 unigenes in the Pfam database, 7,391 unigenes in the KEGG database, 3,865 unigenes in the String database, 10,364 unigenes in the Swissprot database, and 14,335 unigenes in the NR database. Besides, about 76.11% of unigenes (48,202) did not show any matches to known genes (Fig 2, Table 2). Of 14,335 unigenes that had BLAST matches against the NR database, 1703 matched unigenes (11.88%) showed significant homology with gene sequences from *Zootermopsis nevadensis*, followed by *Daphnia pulex* (1027, 7.16%), *Tribolium castaneum* (543, 3.79%), *Branchiostoma floridae* (365, 2.55%), *Pediculus humanus corporis* (341, 2.38%), and *Saccoglossus kowalevskii* (324, 2.26%). *Litopenaeus vannamei* (169, 1.18%) was the only decapod crustacean of the suborder Natantia in the top twenty species that match to the sequences of *M. nipponense* (Fig 3).

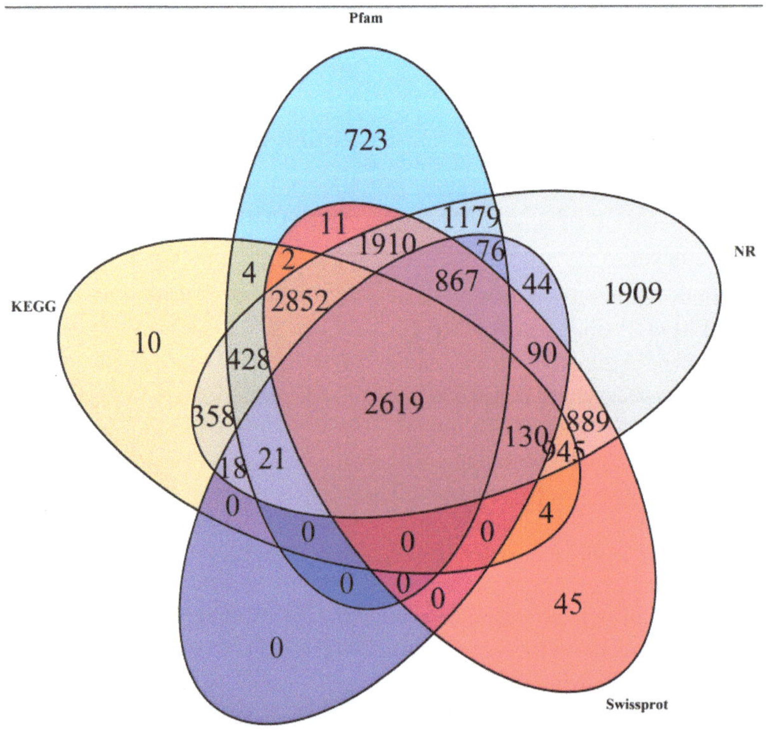

Figure 2: The Venn diagram shows the distribution of unigenes in different databases.

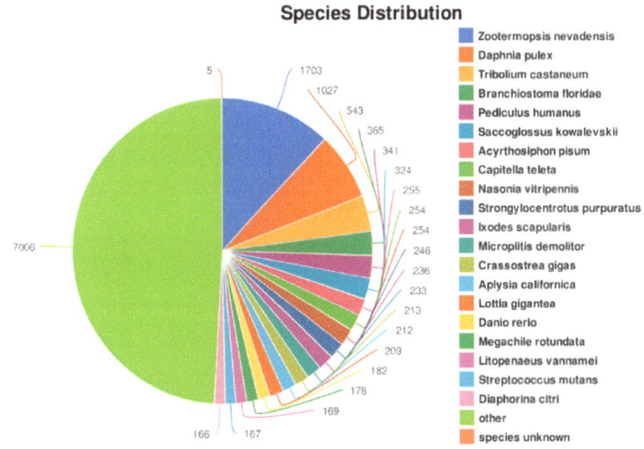

Figure 3: The top twenty species that match to the sequences of *M. nipponense*.

Table 2: Summary of the annotation percentage of *M. nipponense* as compared to public database.

Database	Number of unigenes	Annotation percentage (%)
Pfam	10,692	16.88%
KEGG	7,391	11.67%
String	3,865	6.102%
Swissprot	10,364	16.36%
NR	14,335	22.63%
All annotated unigenes	15,134	23.89%
Total unigenes	63,336	

doi:10.1371/journal.pone.0157173.t002

Gene Ontology (GO) and Cluster of Orthologous Groups of proteins (COG) analyses were used to classify the functions of unigenes. A total of 5,195 annotated unigenes were classified into 3 functional categories (level 2 terms): molecular function, cellular component and biological process (Fig 4). In the category of molecular function, unigenes were clustered into 13 classifications. The largest subcategory of the molecular function was 'catalytic activity' (2585, 49.76%) and the second was 'binding' (2536, 48.82%). In the category of cellular components, unigenes were divided into 20 classifications: the most represented cellular components were "cell" (1517, 29.20%), and 'cell part' (1516, 29.18%). In the category of biological processes, unigenes were grouped into 24 classifications: the most represented biological processes were 'metabolic process' (2780, 53.52%) and 'cellular process' (2710, 52.17%). In addition, a total of 3,227 unigenes were clustered into 25 categories of COG (Fig 5). The largest category was 'general function prediction only' (684, 20.08%), the second category was 'Transcription' (316, 9.79%), and the third category was 'Replication, recombination and repair' (315, 9.76%). To systematically analyze the active biochemical pathways in ovary of *M. nipponense*, unigenes were compared against the KEGG database using BLASTX and the corresponding pathways were established. A total of 7,391 unigenes were assigned to 332 KEGG pathways. 'Purine metabolism' pathway had the largest number of unigenes (166), followed by 'Lysosome' (164 unigenes), 'HTLV-I infection' (159 unigenes), 'Spliceosome' (156 unigenes), 'Huntington's disease' (155 unigenes) and 'PI3K-Akt signaling pathway' (147 unigenes) (Fig 6).

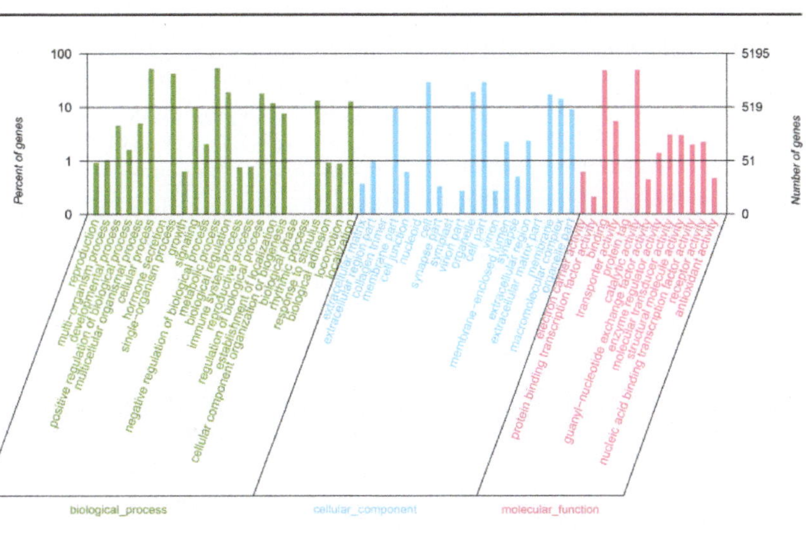

Figure 4: Gene ontology (GO) category of the unigenes.

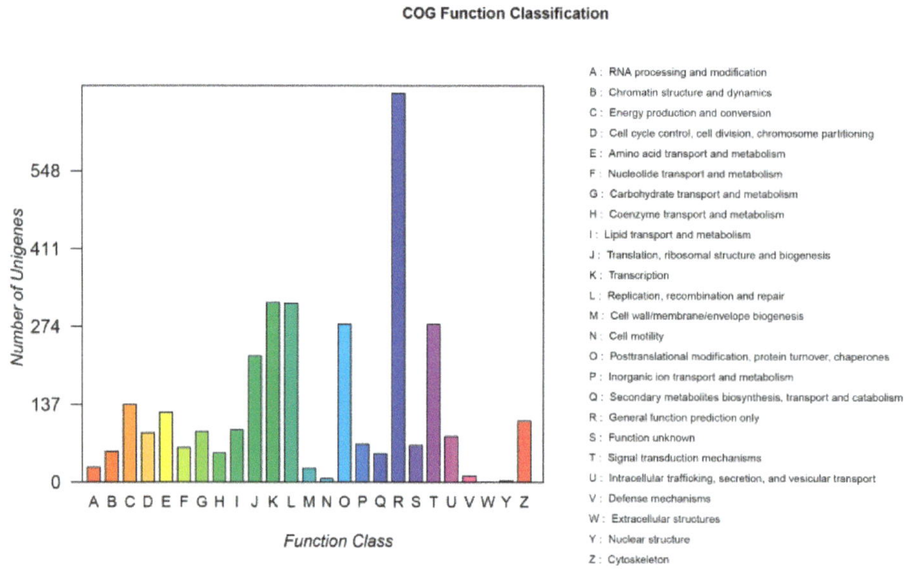

Figure 5: COG function classification of the unigenes.

Figure 6: The top twenty most abundant KEGG pathways.

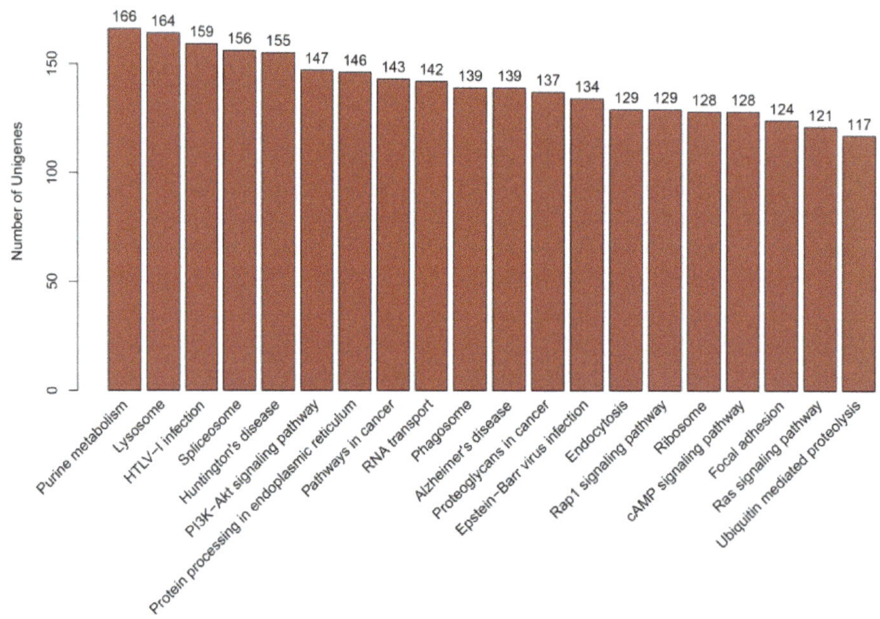

Identification of reproduction-related pathways and genes

Among 332 KEGG pathways, some reproduction-related pathways were found, such as 'Ovarian steroidogenesis', 'Progesterone-mediated oocyte maturation', 'MAPK signaling pathway', 'Wnt signaling pathway', 'Oocyte meiosis', 'GnRH signaling pathway', 'Estrogen signaling pathway', 'p53 signaling pathway', 'Steroid hormone biosynthesis' and 'Fatty acid biosynthesis'. As might be expected, a lot of reproduction-related genes were indentified from these pathways, for example, mitogen-activated protein kinase gene family (*MAPKs*), cell cycle dependent protein kinase gene family (*CDKs*), Wnt protein gene family (*WNTs*), adenylate cyclase gene family (*ADCYs*), anaphase-promoting complex gene family (*APCs*), heatshock protein 90 (*HSP90*), insulin-like receptor (*INSR*), insulin-like growth factor 1 receptor (*IGF1R*), fatty acid synthase (*FASN*), cyclooxygenase (*COX2*), protein kinase A (*PKA*), membrane-associated tyrosine- and threonine-specific cdc2-inhibitory kinase-like protein (*PKMYT*), cytoplasmic polyadenylation element-binding protein (*ORB*), p38 MAP kinase (*P38*) and low-density lipoprotein receptor-related protein 5/6 (*LRP5_6*) genes, *etc* . In addition, some other reproduction-related genes were also indentified from our KEGG database based on comparisons with published data for crustaceans, for instances, cathepsin L (*CTSL*), ribosomal protein L24 (*RPL24*), Broad-complex (*BR-C*), valosin-containing protein (*VCP*), protein disulfide isomerase A6 (*PDIA6*), adipose differentiation-related protein (*ADRP*), prostaglandin reductase 1 (*PTGR1*), nuclear autoantigenic sperm protein (*NASP*), estrogen-related receptor (*ERR*), G protein-coupled receptor family (*GPRs*), *VASA*, and progestin membrane receptor component (*PGRMC1-2*) genes, *etc* [37–42].

SSR and SNP discovery

A total of 26,008 SSRs were identified by analysis of the unigenes with MSATCOMMANDER V. 0.8.2 (Fig 7). The largest number of SSR motifs were mononucleotides (11,679), which accounted for 44.91% of all the predicted SSRs, followed by trinucleotides (8,238, 31.67%), dinucleotide (5,511, 21.19%), tetranucleotides (495, 1.90%), pentanucleotide (48, 0.18%) and hexanucleotides (37, 0.14%). In addition, 7,927 SSRs in the whole predicted SSRs were longer than 15bp, which comprised 2,621 mononucleotide SSRs, 3,022 dinucleotide SSRs, 1,704 trinucleotide SSRs, 495 tetranucleotide SSRs, 48 pentanucleotide SSRs and 37 hexanucleotide SSRs.

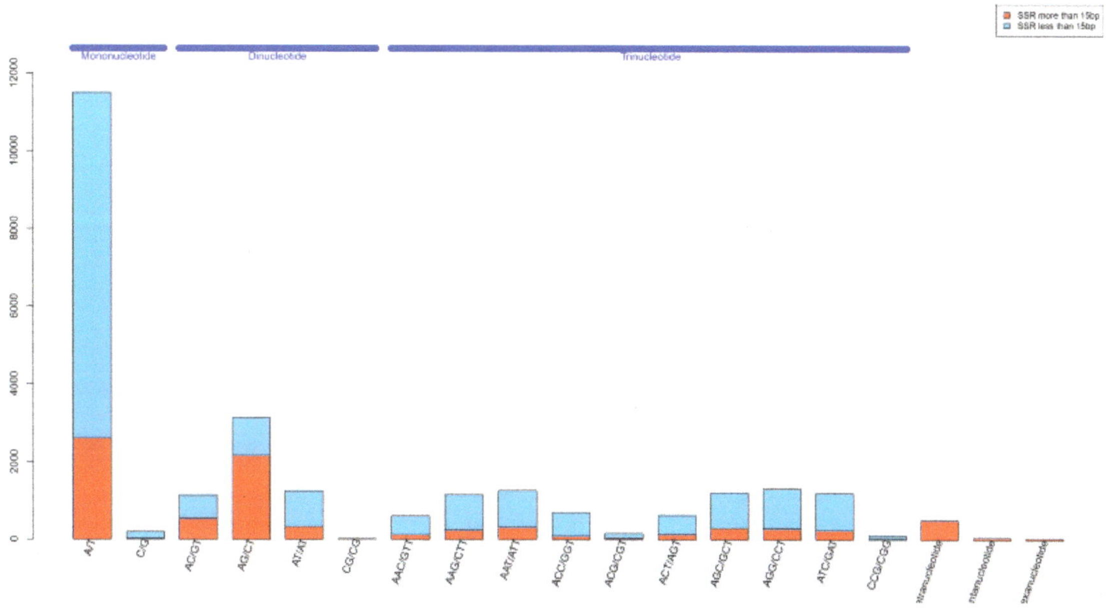

Figure 7: Distribution of simple sequence repeat (SSRs) nucleotide classes among different nucleotide types found in *M. nipponense* sequences.

Using the Samtools and VarScan v.2.2.7 software, we predicted 80,529 potential SNPs in MNOP and 80,516 potential SNPs in MNON when all the consensus assembly sequences generated from the two transcriptome libraries were employed as reference sequences. The predicted SNPs in MNOP included 48,223 transitions (59.88%) and 32,306 transversions (40.12%), and the predicted SNPs in MNON included 48,051 transitions (59.68%) and 32,465 transversions (40.32%). The most frequent SNPs types in both MNOP and MNON were C/T, A/G and A/T, while C/G was the least common type (Table 3). Of all the SNPs in MNOP, 5,851 (7.27%), 4,307 (5.35%) and 18,971 (23.56%) SNPs were located at the first, the second and the third position of the codon, respectively, 6,645 (8.25%) and 19,792 (24.58%) SNPs were located at 5'UTR and 3'UTR, respectively (Fig 8A). Of all the SNPs in MNON, 5,808 (7.21%), 4,229 (5.25%), 18,695 (23.22%), 6,487 (8.06%) and 20,234 (25.13%) SNPs were located at the first, the second and the third position of the codon, 5'UTR and 3'UTR, respectively (Fig 8B). In addition, we also predicted 29,851 potential SNPs between MNON and MNOP groups (Table 4). These predicted SNPs included 17,570 transitions (58.86%) and 12,281 transversions (41.14%). 2,019 (6.76%), 1,495 (5.01%) and 6,471 (21.68%) SNPs were located at the first, the second and the third position of the codon, respectively, 2,181 (7.31%) and 7,454 (24.97%) SNPs were located at 5'UTR and 3'UTR, respectively.

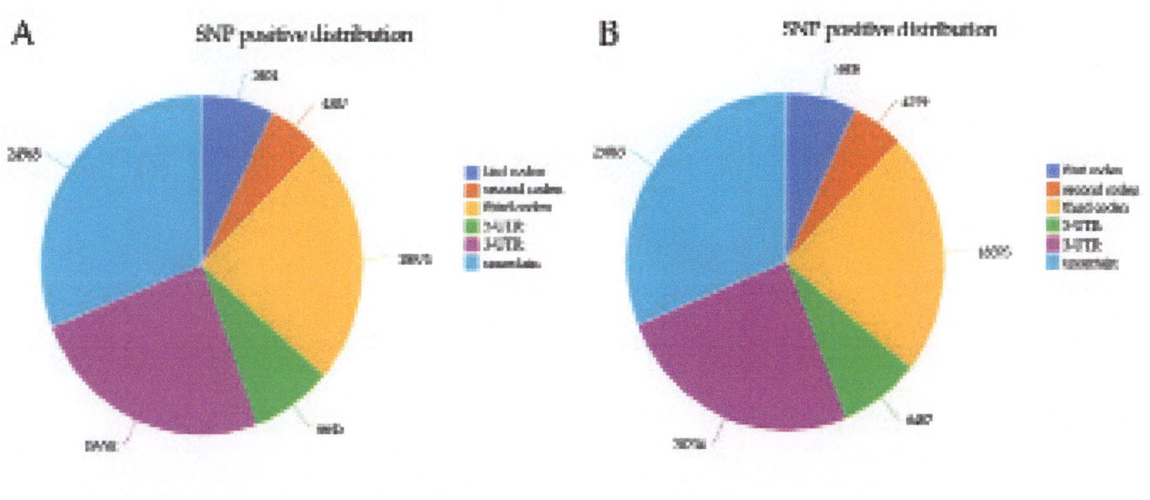

Figure 8: Positions of single nucleotide polymorphisms (SNPs).

Table 3: Statistics of SNP types in the transcriptomes of MNOP and MNON when all the consensus assembly sequences generated from the two transcriptome libraries were employed as reference sequences.

	MNOP			MNON	
SNP types	Count	Frequence per kb	SNP types	Count	Frequence per kb
Transition	48,223		Transition	48,051	
C/T	23,967	0.30	C/T	23,897	0.29
A/G	24,256	0.30	A/G	24,154	0.30
Transversion	32,306		Transversion	32,465	
A/T	11,853	0.14	A/T	11,977	0.15
A/C	7,906	0.09	A/C	7,967	0.09
T/G	8,317	0.10	T/G	8,259	0.10
C/G	4,230	0.05	C/G	4,262	0.05
Total	80,529	1.01	Total	80,516	1.01

doi:10.1371/journal.pone.0157173.t003

Table 4: Statistics of SNP types and positions in the transcriptomes of MNOP when all sequences generated from the MNON transcriptome library were employed as reference sequences.

	SNP types in MNOP		SNP positions in MNOP	
SNP types	Count	Frequence per kb	SNP positions	Count
Transition	17,570		The first position of thecodon	2,019
C/T	8,735	0.10	The second position of thecodon	1,495
A/G	8,835	0.11	The third position of thecodon	6,471
Transversion	12,281		5'UTR	2,181
A/T	4,492	0.05	3'UTR	7,454
A/C	3,069	0.03	Uncertain	10,231
T/G	3,005	0.03		
C/G	1,715	0.02		
Total	29,851	0.37	Total	29,851

doi:10.1371/journal.pone.0157173.t004

Identification and validation of differentially expressed genes

Through differential expression analysis (FDR ≤ 0.05, logFC ≥ 1) using edgeR, 549 unigenes (0.87% of all unigenes) were identified as DEGs between MNOP and MNON, which comprised 212 up-regulated unigenes (38.62% of all DEGs) and 337 down-regulated unigenes (61.38% of all DEGs) in MNOP compared with MNON. However, 62,787 unigenes (99.13% of all unigenes) had no significant difference between MNOP and MNON (FDR > 0.05) (Fig 9, Table 5). Moreover, 187 MNON-specifically expressed unigenes and 90 MNOP-specifically expressed unigenes were identified from the 549 DEGs, and the remaining 272 DEGs were common unigenes expressed both in the MNOP and MNON (Fig 10). Of 549 DEGs, 186 (33.88%) unigenes were annotated successfully, and the remaining 363 unigenes had nearly no information. Based on the published literature and sequence annotations, we identified nine reproduction-related genes that may be critical for sexual precocity of *M. nipponense* from those annotated DEGs (Table 6). They include *vitellogenin* (*VTG*), *cathepsin L* (*CTSL*), *cystatin-M* (*CST6*), *insulin-like receptor* (*INSR*), *insulin-like growth factor 1 receptor* (*IGF1R*), *cyclooxygenase* (*COX2*), *glutathione peroxidase* (*GPX*), *copper zinc superoxide dismutase* (*SOD1*), and *fatty acid synthase* (*FASN*) genes which involved in 'ovary development', 'ovarian steroidogenesis', 'Peroxisome', 'Glutathione metabolism' and 'Fatty acid biosynthesis' pathways. *COX2 and SOD1* genes were down-regulated, while *VTG*, *CTSL*, *CST6*, *INSR*, *IGF1R*, *GPX* and *FASN* genes were up-regulated in MNOP compared with MNON.

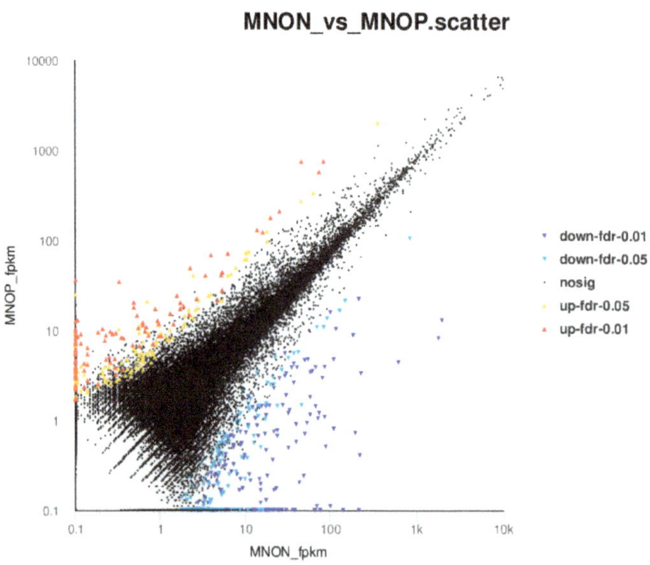

Figure 9: Scatter plot of differentially expressed unigenes (DEGs) between MNOP and MNON.

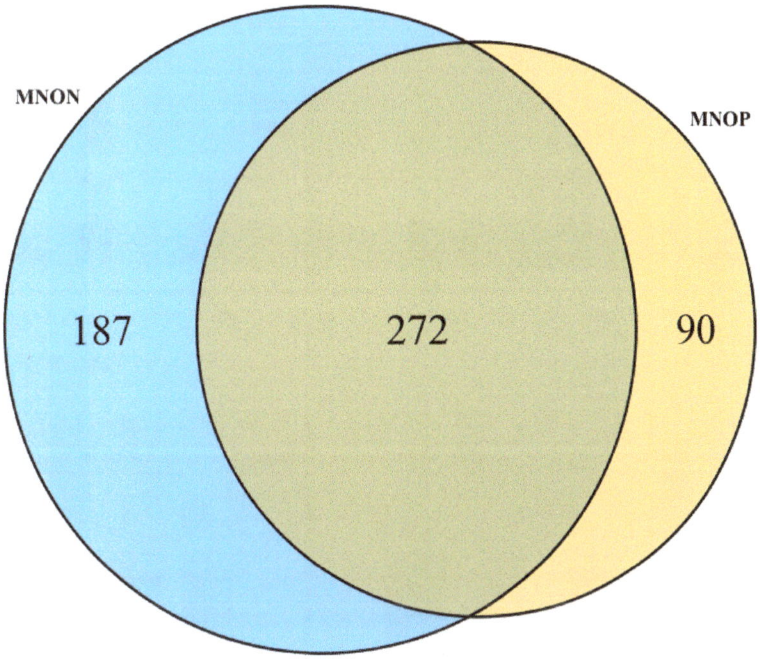

Figure 10: Venn diagram for exclusive DEGs found in MNON and MNOP.

Table 5: The differential expression unigenes (DEGs) between MNOP and MNON.

Items	MNOP versus MNON transcriptome		
	All[a]	Up[b]	Down[c]
Differential expression unigenes (DEGs)	549	212	337
DEGs: fdr < 0.01	316	104	212
DEGs: 0.01 < fdr < 0.05	233	108	125
No Significant difference unigenes: fdr > 0.05	62,787		

[a] Means summary of unigenes expressed up and down between MNOP and MNON transcriptome
[b] Indicates number of up-regulated unigenes in MNOP compared with MNON
[c] Means number of down-regulated unigenes in MNOP compared with MNON

doi:10.1371/journal.pone.0157173.t005

Table 6: Nine key reproduction-related DEGs that affect sexual maturity of *M. nipponense*.

Pathway	Unigene ID	Gene name	log2FC(MNOP/MNON)	FDR	Regulate
Ovary development	c33992_g1	*Vitellogenin (VTG)*	5.45	1.58E-09	Up
	c25097_g1	*Cathepsin L (CTSL)*	4.01	2.21E-04	Up
	c31957_g1	*Cystatin-M (CST6)*	2.48	3.49 E-02	Up
Ovarian steroidogenesis	c23275_g1	*Insulin-like receptor (INSR)*	2.65	3.31 E-02	Up
	c33061_g1	*Insulin-like growth factor 1 receptor (IGF1R)*	2.44	3.33 E-02	Up
	c13375_g1	*Cyclooxygenase (COX2)*	-3.2	4.72 E-03	Down
Glutathione metabolism	c25682_g1	*Glutathione peroxidase (GPX)*	3.00	6.28 E-03	Up
Peroxisome	c8193_g1	*Copper zincsuperoxide dismutase (SOD1)*	-2.69	3.31 E-02	Down
Fatty acid biosynthesis	c32305_g1	*Fatty acid synthase (FASN)*	2.61	2.13 E-02	Up

Regulate: Up-regulated or down-regulated of genes in MNOP compared with MNON

Here, we used Quantitative real-time PCR (QPCR) to confirm the differential expression of 20 DEGs (include those nine genes that may be critical for sexual precocity of *M. nipponense*). As shown in Fig 11, the relative expression levels of five unannotated unigenes (unigene ID: c28611_g1, c33575_g1, c24064_g1, c28116_g1, c32848_g1) and two annotated genes (*SERPINB* and *COX2*) showed significant down-regulation in MNOP compared with MNON, and the relative expression levels of one unannotated unigene (unigene ID: c33909_g1) and eleven annotated genes (*HC*, *MYH*, *PPAP2*, *VTG*, *CTSL*, *CST6*, *FASN*, *INSR*, *IGF1R*, *GPX* and *SOD1*) showed significant up-regulation in MNOP compared with MNON. These QPCR results all agreed with the results of RNAseq-based differential expression analysis. Only one unannotated unigene (unigene ID: c31008_g1) did not show consistent expression between QPCR and the RNAseq-based differential expression analysis, its relative expression levels showed no significant difference between MNON and MNOP.

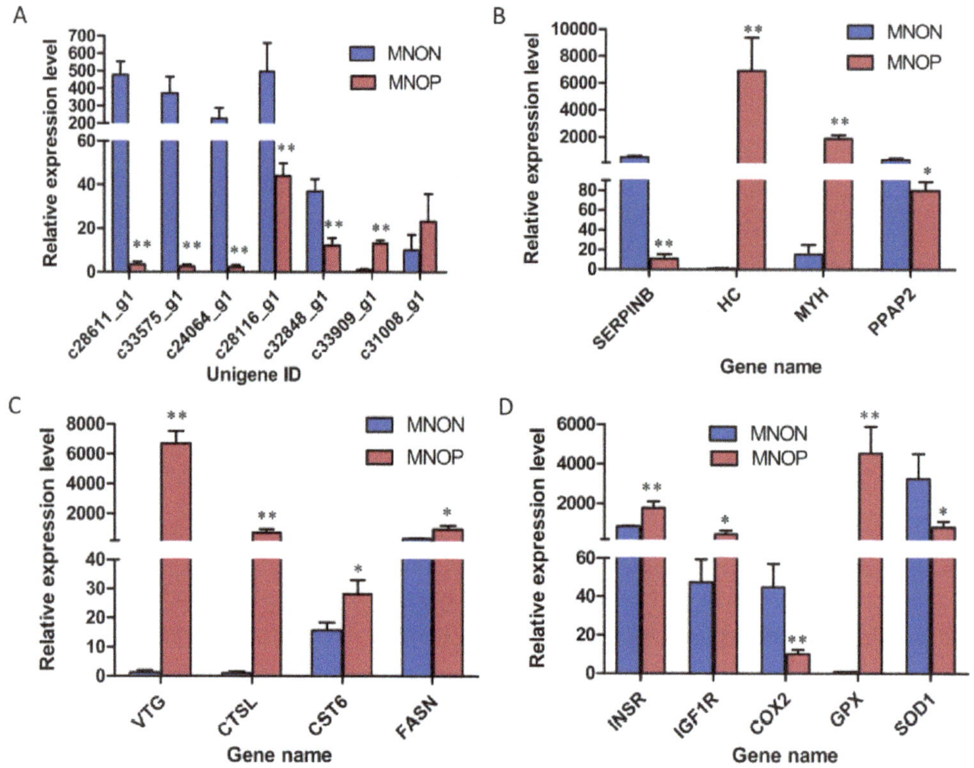

Figure 11: QPCR analysis of twenty selected DEGs.

DISCUSSION

Little is known about the genes and biological mechanisms controlling sexual precocity in crustaceans. In this study, by profiling the ovary transcriptomes of sexually precocious and normal sexually mature *M. nipponense*, we expand the limited amount of sequence data that are available for *M. nipponense* in the public EST database and identified differentially expressed genes potentially influencing the sexual precocity of this prawn.

In this study, the RNA sequencing produced total 63,336 unigenes with average length was 893.88 bp from the ovaries of MNOP and MNON. This is considerably higher than the sum of unigenes (1,514) and average length of unigenes (734) obtained from ovary of *M. nipponense* using traditional method of expressed sequence tag (EST) reported by Wu et al. (2009). For one thing, the results show that high-throughput RNA-Seq technology has great superiority in large-scale transcriptome studies. For another, this discrepancy probably due to the difference of ovarian development stage of *M. nipponense* used in these two studies. Wu et al. (2009) used ovary at previtellogenesis stage as their experimental material, while the ovaries used in this study were at maturation stage. Besides, we used two sequencing samples (ovaries of sexually precocious and normal sexually mature *M. nipponense*), giving rise to a huge amount of raw data as compared to the only one sequencing sample (ovary of healthy adult prawn) that was used for transcriptome analysis in study of Wu et al. (2009). To the best of our knowledge, this is the first comprehensive ovary transcriptome study of not only normal sexually mature but also sexually precocious *M. nipponense*. Therefore, this transcriptome dataset provides a useful resource for future analyses of genes related to reproduction, gonadal development and sexual precocity in *M. nipponense*.

A total of 15,134 (23.89%) unigenes were annotated successfully in the present study, whereas the rest 48,202 (76.11%) unigenes had no annotations which will make a meaningful contribution to the knowledge of *M. nipponense* by further characterizing their functions. BLAST results against the Nr database showed that the *M. nipponense* sequences had the highest number of matches with the gene sequences from *Z. nevadensis*, followed by *D. pulex*. *Z. nevadensis* was the latest arthropod of the class insecta and *D. pulex* was the first arthropod of the class crustacea to have their whole genome sequenced [43,44]. However, only about 11.92% and 7.19% of all the annotated sequences matched *Z. nevadensis* and *D. pulex* respectively, probably because *Z. nevadensis* belongs to pterygota insect, *D. pulex* belongs to branchiopoda crustacean, while *M. nipponense* belongs to malacostraca crustacean. In addition, the transcriptome research in *L. vannamei* was the most extensive of all decapod crustaceans of the suborder Natantia in recent years [45–50], therefore, it is not surprising that *L. vannamei* was the only decapod crustaceans of the suborder Natantia in the top twenty species that match to the sequences of *M. nipponense*.

The results of GO and KEGG analysis extend our knowledge on the gene function classification and potential involvement of genes in cellular metabolic pathways [51,52]. In this study, the *M. nipponense* ovarian sequences fall into GO categories with a roughly similar distribution to that of some other arthropods, such as *Drosophila melanogaster*, *Parhyale hawaiensis*, *D. pulex* [53] and *Oncopeltus fasciatus* [54], in which most ovarian sequences were enriched in functional category of "catalytic activity", suggesting that arthropod ovaries contain a large diversity of enzyme genes involved in various molecular function. As might be expected, many reproduction-related pathways and genes were found in the KEGG analysis, which maybe play an important role in development of oocytes and ovaries in crustaceans. However, further studies are also needed to understand the molecular functions of some putative reproduction-related genes in these pathways in *M. nipponense*.

The focus of our project was to mine the differentially expressed genes (DEGs) between MNOP and MNON. In the present study, we identified 549 DEGs which would give clues on the molecular mechanism of sexual precocity determination. In these DEGs, the number of down-regulated unigenes (337) was more than the number of up-regulated unigenes (212), as well as the number of MNON-specifically expressed unigenes (187) was more than the number of MNOP-specifically expressed unigenes (90), indicating that the lack or low expression of quite a few genes in the ovarian development of *M. nipponense* is one of the important causes of sexual precocity of this species. Besides, in these DEGs, the number of unannotated sequences (363) was much larger than the annotated ones (186), which made it impossible to exploit all DEGs, some potentially useful genetic information relate to sexual precocity of *M. nipponense* may be missed. From these 186 annotated DEGs, we identified nine genes of interest that may be play an important role in sexual precocity of *M. nipponense*. They were *VTG*, *CTSL*, *CST6*, *INSR*, *IGF1R*, *COX2*, *GPX*, *SOD1* and *FASN* genes.

VTG, *CTSL*, *CST6* are 'ovary development'- related genes. The expressions of these three genes were significantly up-regulated in MNOP compared with MNON in the present study. VTG is the precursor of vitellin (VN), the major yolk protein that provides nutrition during embryonic development. VTG is synthesized extraovary or intraovary, transported by the hemolymph and finally sequestered into the growing oocytes through vitellogenin receptor (VTGR) mediated endocytosis [55,56]. In oviparous animals, oocyte development and ovarian maturation depend on massive production and accumulation of VTG. In vertebrates, estrogen hormones play a key role in regulating the synthesis of VTG through the enhancement of the transcription of the VTG gene [57,58]. This process is mediated by an E-ER-HSP90-VTG regulation pathway which widely existed in vertebrates [56–61]. During vitellogenesis, estrogen (E) binds to its receptor (ER) form a hormone-receptor complex and then acts on estrogen responsive elements (ERE) located at the upstream of the VTG DNA, which leads to the activation or enhancement of the VTG gene transcription [61]. However, HSP90 binds to estrogen receptor (ER) in the absence of estrogen (E) to increase the activity of estrogen hormone receptor complex to transcribe target genes [59,60,62,63]. In our transcriptomic data, *HSP90* genes were identified. Although an *ER* gene has not been found yet, we identified estrogen related receptor (*ERR*) gene in our transcriptome database. However, *HSP90* and *ERR* genes had no significantly differences between MNOP and MNON. So, whether estrogen (E) caused the significant up-regulation of *VTG* gene expression in MNOP pass through the E-ER-HSP90-VTG regulation pathway need to be confirmed by more studies. After all, the regulatory mechanism of E-ER-HSP90-VTG pathway for VTG synthesis in crustaceans is still controversial [13,37]. Cathepsins are lysosomal cysteine proteases, but cystatins are natural tight-binding reversible inhibitors of cysteine proteases. Based on sequence homology and conserved amino acid motifs, the group of cathepsins can be divided into three subgroups: cathepsin L (CTSL), cathepsin B (CTSB) and cathepsin F (CTSF) [64]. CTSL plays an important role during yolk formation and processing in vertebrates, for example, CTSL can degrade VTG into smaller yolk proteins which are stored in oocyte and will be the primary source of nutrients and energy for the developing embryo [65]. Current findings in *Metapenaeus ensis* [66] and *M. nipponense* [67] suggest that the ovary is the unique non-digestive organ showing *CTSL* expression and *CTSL* in the ovary in crustacean is also functionally important in reproductive physiology. On the contrary, as a cysteine protease inhibitor, CST6 plays an important role in regulating and protecting against uncontrolled proteolysis by cathepsins L (CTSL) and V (CTSV) and the asparaginyl endopeptidase legumain (LGMN) [68]. In the present study, the significant up-regulation of *CST6* gene expression in MNOP ovary means that the large amounts of CST6 produced and inhibited the hydrolysis of VTG by CTSL. Although CTSL also produced abundantly because of the significant up-regulation of *CTSL* gene expression, the inhibition of CST6 probably overwhelmed the hydrolysis of CTSL. In sum, the magnificent VTG production due to the significant up-regulation of *VTG* gene expression, together with the inhibition of the hydrolysis of VTG by CST6, led to the accumulation of large amounts of VTG in the ovary, and finally resulted in the sexual precocity of *M. nipponense*.

INSR, *IGF1R*, *COX2* genes were three DEGs in 'ovarian steroidogenesis' pathways (Table 3). Compared with MNON, *INSR* and *IGF1R* gene expressions were significantly up-regulated, while *COX2* gene expression was significantly down-regulated in MNOP. INSR and IGF1R are transmembrane receptors that are activated by hormones called insulin and insulin-like growth factor 1 (IGF-1) respectively. Insulin is a key peptide hormone that regulates growth, metabolism and reproduction in vertebrates [69], However, insulin-like peptides (ILPs), the counterpart of vertebrate insulin, belong to the insulin-like superfamily and were widely found in invertebrate species [70–72]. The insulin-like superfamily includes two major subgroups, ILPs and the insulin-like growth factors (IGFs) [72]. IGF-I is a member of this superfamily, its function overlaps with that of insulin/ILPs [73]. In vertebrates, insulin and IGF-1 banded to the INSR and IGF1R respectively, and then they can stimulate DNA synthesis and promote the proliferation of ovarian theca-interstitial cell through insulin signaling pathway [74]. Insulin signaling pathway is also present in invertebrates, and is closely related to their reproduction. For instance, insulin-like growth factor signaling pathway in *Drosophila*was considered important for the regulation of vitellogenesis, oogenesis, and reproductive diapause [75]. The interaction and crosstalk between ILP signaling networks in *Aedes aegypti* was thought to be relevant to the regulation of yolk protein precursor, juvenile hormone, ecdysteroids, and nutrients [76]. In the present study, the significantly up-regulation of *INSR* and *IGF1R* gene expressions in MNOP means that a great deal of ILPs and IGF-1 generated in the sexually precocious prawn, which promoted vitellogenesis, oogenesis and proliferation of ovarian theca-interstitial cell, leading to the rapid ovarian maturation of *M. nipponense*. Cyclooxygenase-2 (COX2) is an inducible enzyme that catalyzes the formation of prostaglandins (PGs) from arachidonic acid (AA). In crustaceans, PGs have important significance in the regulation of female reproductive maturation. In *Procambarus paeninsulanus* [77] and *Oziothelphusa senex senex* [78], there has a positive correlation between the production of PGs (PGE_2 and $PGF_{2\alpha}$) and the ovarian maturation. Furthermore, the amount of PGs in ovaries of crustaceans varies along with developing ovary stages. For instance, the amounts of PGE_2 and $PGF_{2\alpha}$ were highest in ovaries at stage I and continued to decrease until stage IV in *Metanephrops japonicus* [79]. In this study, *COX2* gene expression was significantly down-regulated in ovary at stage V of MNOP compared with MNON, suggesting the decrease of PGs in ovary at stage V of sexually precocious *M. nipponense*. We believe that the expression of *COX2* gene in immature ovaries (stage I -VI ovaries) of sexually precocious *M. nipponense* was higher than that in mature ovary (stage V ovary) as *M. japonicus*, which promoted the synthesis of the PGs in large quantities in immature ovaries, and finally resulted in the rapid ovarian maturation. However, due to the reduced need for PGs in mature ovary, the expression of *COX2* gene decreased sharply in the stage V ovary of sexually precocious *M. nipponense*, even significantly lower than that in the stage V ovary of normal sexually mature prawn. But the real expression levels of *COX2* gene in ovaries at different stages of sexually precocious and normal sexually mature *M. nipponense* require further investigation.

Aerobic metabolism in animals is associated with the generation of free radicals or reactive oxygen species (ROS) that include the hydroxyl radicals, superoxide anion, hydrogen peroxide, and nitricoxide. ROS have been recently recognized as key signaling molecules involved in reproductive biology including oocyte maturation, fertilization, and embryo development [80–82]. However, because of the double-edged sword property of ROS, excessive ROS can also lead to serious damage to major cellular components including protein, lipid and DNA [83]. To minimize the oxidative damage caused by ROS, cells possess a range of enzymatic scavenger systems including superoxide dismutase (SOD), catalase (CAT) and glutathione peroxidase (GPX). In this study, SOD and GPX belong to 'peroxisome' and 'glutathione metabolism' pathways, respectively, while they are important antioxidant enzymes in antioxidant system. SOD first catalyzes the dismutation of superoxide radicals to H_2O_2, which is further metabolized to H_2O and O_2 by CAT and GPX [84]. GPX, a selenoprotein, plays an important role in detoxifying lipids and hydrogen peroxide, with the concomitant oxidation of glutathione [85]. In this study, compared with MNON, *SOD* and *GPX* gene expressions in MNOP were significantly down-regulated and up-regulated respectively, while *CAT* gene expression had no significant difference between these two groups. There must be a precise balance in ROS production and recycling under the regulation of these antioxidant enzymes for avoiding the oxidative damage to cells in ovary of *M. nipponense*. The Low expression of *SOD* gene and the over expression of *GPX* gene disrupted this balance and caused overproduction of ROS. The excess ROS promoted the mature of the ovary as signaling molecules, and finally leading to the sexual precocity of *M. nipponense*.

It was reported that lipids (including all kinds of fatty acids) are the essential nutrients which stored in the oocytes of female broodstocks to nourish subsequent embryonic development during ovarian maturation of prawn [86]. Moreover, fatty acids also have effects on growth and reproductive success in female prawns through stimulating ovarian maturation and oocyte differentiation [87,88]. For example, C-20 polyunsaturated fatty acids (PUFAs), include arachidonic acid (AA), eicosapentaenoic acid (EPA) and docosahexaenoic acid (DHA), serve as prostanoid precursors and convert into a variety of prostanoids via enzymatic catalysis. One of the more prominent roles of prostanoids in crustaceans is the regulation of female reproductive maturation [89]. Fatty acid synthase (FASN), a key enzyme in the process of the *de novo* fatty acid synthesis in 'fatty acid biosynthesis' pathway, catalyzes all of the reaction steps for synthesizing saturated fatty acids (SFA) from acetyl-CoA and malonyl-CoA in an NADPH-dependent manner [90]. Then, various unsaturated fatty acids (USFA) are synthesized from SFA under catalysis of the fatty acid desaturases. In this study, various fatty acids and prostanoids in ovary of MNOP should be richer than those in ovary of MNON because of the up-regulation of *FASN* gene expression in ovary of MNOP compared with MNON, which may be one of the key factors of sexual precocity of *M. nipponense*.

Conventional Quantitative real-time PCR was frequently used to confirm the data obtained from high-throughput sequencing. Nineteen out of twenty selected DEGs from RNA-Seq were confirmed by QPCR, which indicates the high reliability of RNA-Seq for detecting DEGs. The differential expressions of above nine key genes that may be related to sexual precocity of *M. nipponense* were also verified with QPCR (Fig 11). Although these nine genes are highly related to reproduction of crustaceans, their involvement in sexual precocity of crustaceans has not yet been reported. Our findings in this study firstly revealed the potential relationship between these nine genes and sexual precocity of *M. nipponense*. Nevertheless, conclusion on this point cannot be drawn completely until the regulation pathways of these nine genes in sexual precocity of *M. nipponense* are fully understood.

SSRs and SNPs are useful molecular markers that have been widely applied into genetic mapping, population genetic analysis and breeding [35]. In the present study, we identified large volumes of SSRs and SNPs which were a lot higher than those in previous reports on *M. nipponense* [2,20,21]. This is a powerful supplement to SSR and SNP databases available for *M. nipponense*. Moreover, further in-depth research of the potential SNPs between MNON and MNOP groups predicted in this study will also provide certain help for the understanding of the sexually precocious mechanism of this prawn.

CONCLUSION

This is the first comprehensive transcriptome dataset for the ovaries of sexually precocious and normal sexually mature *M. nipponense* to be reported. A total of 63,336 unigenes were obtained and a number of reproduction-related unigenes and metabolic pathways were identified. Many potential SSRs and SNPs were predicted and can be used for subsequent marker development, genetic linkage and gene localization analysis. Moreover, 549 DEGs and nine key DEGs that are potentially involved in sexual precocity of *M. nipponense* were identified for the first time and are worthy of further investigation. In sum, these findings will serve as the important foundation for insight into the regulatory mechanism of sexual precocity of *M. nipponense*, and may also provide some references for future molecular research in freshwater prawn.

FUNDING STATEMENT

Zhejiang Province Key S&T Special Projects of New Variety Breeding, China No.2012C12907 to Zhimin Gu. the open project of Agriculture Ministry Key Laboratory of Healthy Freshwater Aquaculture No.ZJK201506 to Hongxia Jiang.

The work was financially supported by the Zhejiang Province Key S&T Special Projects of New Variety Breeding, China (No.2012C12907), and the open project of Agriculture Ministry Key Laboratory of Healthy Freshwater Aquaculture (No.ZJK201506).

DATA AVAILABILITY

The raw sequencing data were deposited in the Short Read Archive (SRA) of the NCBI with accession numbers of SRP063589.

REFERENCES

1. Cai YX, Ng PKL (2002) The freshwater palaemonid prawns (Crustacea: Decapoda: Caridea) of Myanmar. Hydrobiologia 487: 59–83.
2. Ma KY, Qiu GF, Feng JB, Li JL (2012) Transcriptome analysis of the oriental river prawn, *Macrobrachium nipponense* using 454 pyrosequencing for discovery of genes and markers. PloS One 7: e39727 doi: 10.1371/journal.pone.0039727
3. Fu HT, Jiang SF, Xiong YW (2012) Current status and prospects of farming the giant river prawn (*Macrobrachium rosenbergii*) and the oriental river prawn (*Macrobrachium nipponense*) in China. Aquac Res 43: 993–998.
4. Li X, Li Z, Liu J, De Silva SS (2011) Advances in precocity research of the Chinese mitten crab *Eriocheir sinensis*. Aquacult Int 19: 251–267.
5. Qiao H, Fu HT, Jin SB, Wu Y, Jiang SF, et al. (2012) Constructing and random sequencing analysis of normalized cDNA library of testis tissue from oriental river prawn (*Macrobrachium nipponense*). Comp Biochem Phys D 7: 268–276.
6. Bai HK, Qiao H, Li FJ, Fu HT, Sun SM, et al. (2015) Molecular characterization and developmental expression of vitellogenin in the oriental river prawn *Macrobrachium nipponense* and the effects of RNA interference and eyestalk ablation on ovarian maturation. Gene 562: 22–31. doi: 10.1016/j.gene.2014.12.008
7. Chu SL, Weng CF, Hsiao CD, Hwang PP, Chen YC, et al. (2006) Profile analysis of expressed sequence tags derived from the ovary of tilapia, *Oreochromis mossambicus*. Aquaculture 251: 537–548.
8. Ibarra AM, Racotta IS, Arcos FG, Palacios E (2007) Progress on the genetics of reproductive performance in penaeid shrimp. Aquaculture 268: 23–43.
9. Le Moullac G, Goyard E, Saulnier D, Haffner P, Thouard E, et al. (2003) Recent improvements in broodstock management and larviculture in marine species in Polynesia and New Caledonia: genetic and health approaches. Aquaculture 227: 89–106.
10. Racotta IS, Palacios E, Ibarra AM (2003) Shrimp larval quality in relation to broodstock condition. Aquaculture 227: 107–130.
11. Makkapan W, Maikaeo L, Miyazaki T, Chotigeat W (2011) Molecular mechanism of serotonin via methyl farnesoate in ovarian development of white shrimp: *Fenneropenaeus merguiensis* de Man. Aquaculture321: 101–107.
12. Zhang FY, Chen LQ, Wu P, Zhao WH, Li EC, et al. (2010) cDNA cloning and expression of Ubc9 in the developing embryo and ovary of oriental river prawn, *Macrobrachium nipponense*. Comp Biochem Phys B155: 288–293.
13. Zhao WH, Chen LQ, Qin JG, Wu P, Zhang FY, et al. (2011) MnHSP90 cDNA characterization and its expression during the ovary development in oriental river prawn, *Macrobrachium nipponense*. Mol Biol Rep38: 1399–1406. doi: 10.1007/s11033-010-0243-7
14. Du YX, Ma KY, Qiu GF (2015) Discovery of the genes in putative GnRH signaling pathway with focus on characterization of GnRH- like receptor transcripts in the brain and ovary of the oriental river prawn *Macrobrachium nipponense*. Aquaculture 442: 1–11.
15. Wu P, Qi D, Chen LQ, Zhang H, Zhang XW, et al. (2009) Gene discovery from an ovary cDNA library of oriental river prawn *Macrobrachium nipponense* by ESTs annotation. Comp Biochem Phys D 4: 111–120.
16. Quail MA, Kozarewa I, Smith F, Scally A, Stephens PJ, et al. (2008) A large genome center's improvements to the Illumina sequencing system. Nat Methods 5: 1005–1010. doi: 10.1038/nmeth.1270
17. Meyer M, Stenzel U, Hofreiter M (2008) Parallel tagged sequencing on the 454 platform. Nat Protoc 3: 267–278. doi: 10.1038/nprot.2007.520
18. Shendure J (2008) The beginning of the end for microarrays? Nat Methods 5: 585–587. doi: 10.1038/nmeth0708-585
19. Mortazavi A, Williams BA, Mccue K, Schaeffer L, Wold B (2008) Mapping and quantifying mammalian transcriptomes by RNA-Seq. Nat Methods 5: 621–628. doi: 10.1038/nmeth.1226
20. Jung H, Lyons RE, Dinh H, Hurwood DA, McWilliam S, et al. (2011) Transcriptomics of a giant freshwater orawn (*Macrobrachium rosenbergii*): de novo assembly, annotation and marker discovery. Plos One 6: e27938 doi: 10.1371/journal.pone.0027938
21. Jin SB, Fu HT, Zhou Q, Sun SM, Jiang SF, et al. (2013) Transcriptome analysis of androgenic gland for discovery of novel genes from the oriental river orawn, *Macrobrachium nipponense*, using Illumina Hiseq 2000. Plos One 8: e76840 doi: 10.1371/journal.pone.0076840
22. Cock PJA, Fields CJ, Goto N, Heuer ML, Rice PM (2010) The Sanger FASTQ file format for sequences with quality scores, and the Solexa/Illumina FASTQ variants. Nucleic Acids Res 38: 1767–1771. doi: 10.1093/nar/gkp1137
23. Erlich Y, Mitra PP, delaBastide M, McCombie WR, Hannon GJ (2008) Alta-Cyclic: a selfoptimizing base

caller for next-generation sequencing. Nat Methods 5: 679–682. doi: 10.1038/nmeth.1230

24. Grabherr MG, Haas BJ, Yassour M, Levin JZ, Thompson DA, et al. (2011) Full-length transcriptome assembly from RNA-Seq data without a reference genome. Nat Biotechnol 29: 644–U130. doi: 10.1038/nbt.1883

25. Conesa A, Gotz S, Garcia-Gomez JM, Terol J, Talon M, et al. (2005) Blast2GO: a universal tool for annotation, visualization and analysis in functional genomics research. Bioinformatics 21: 3674–3676.

26. Kanehisa M, Goto S, Sato Y, Furumichi M, Tanabe M (2012) KEGG for integration and interpretation of large-scale molecular data sets. Nucleic Acids Res 40: D109–D114. doi: 10.1093/nar/gkr988

27. Kanehisa M, Goto S (2000) KEGG: Kyoto encyclopedia of genes and genomes. Nucleic Acids Res 28: 27–30.

28. Faircloth BC (2008) MSATCOMMANDER: detection of microsatellite repeat arrays and automated, locus-specific primer design. Mol Ecol Resour 8: 92–94. doi: 10.1111/j.1471-8286.2007.01884.x

29. Li H, Handsaker B, Wysoker A, Fennell T, Ruan J, et al. (2009) The Sequence Alignment/Map format and SAMtools. Bioinformatics 25: 2078–2079. doi: 10.1093/bioinformatics/btp352

30. Koboldt DC, Chen K, Wylie T, Larson DE, McLellan MD, et al. (2009) VarScan: variant detection in massively parallel sequencing of individual and pooled samples. Bioinformatics 25: 2283–2285. doi: 10.1093/bioinformatics/btp373

31. Li B, Dewey CN (2011) RSEM: accurate transcript quantification from RNA-Seq data with or without a reference genome. Bmc Bioinformatics 12: 323 doi: 10.1186/1471-2105-12-323

32. Robinson MD, McCarthy DJ, Smyth GK (2010) edgeR: a Bioconductor package for differential expression analysis of digital gene expression data. Bioinformatics 26: 139–140. doi: 10.1093/bioinformatics/btp616

33. Reiner A, Yekutieli D, Benjamini Y (2003) Identifying differentially expressed genes using false discovery rate controlling procedures. Bioinformatics 19: 368–375.

34. Qian Z, Liu TY, Liu Q, He SL, Liu YJ, et al. (2014) p53 is involved in shrimp survival via its regulation roles on MnSOD and GPx in response to acute environmental stresses. Comp Biochem Phys C 159: 38–51.

35. Schmittgen TD, Livak KJ (2008) Analyzing real-time PCR data by the comparative C-T method. Nat Protoc 3: 1101–1108.

36. Livak KJ, Schmittgen TD (2001) Analysis of relative gene expression data using real-time quantitative PCR and the $2^{-\Delta\Delta CT}$ method. Methods 25: 402–408.

37. Gao J, Wang XW, Zou ZH, Jia XW, Wang YL, et al. (2014) Transcriptome analysis of the differences in gene expression between testis and ovary in green mud crab (*Scylla paramamosain*). Bmc Genomics 15: 585 doi: 10.1186/1471-2164-15-585

38. Zhang ZP, Wang YL, Jiang YH, Lin P, Jia XW, et al. (2007) Ribosomal protein L24 is differentially expressed in ovary and testis of the marine shrimp *Marsupenaeus japonicus*. Comp Biochem Phys B 147: 466–474.

39. Buaklin A, Sittikankaew K, Khamnamtong B, Menasveta P, Klinbunga S (2013) Characterization and expression analysis of the *Broad-complex* (*Br-c*) gene of the giant tiger shrimp *Penaeus monodon*. Comp Biochem Phys B 164: 280–289.

40. Talakhun W, Roytrakul S, Phaonakrop N, Kittisenachai S, Khamnamtong B, et al. (2012) Identification of reproduction-related proteins and characterization of the *protein disulfide isomerase A6* cDNA in ovaries of the giant tiger shrimp *Penaeus monodon*. Comp Biochem Phys D 7: 180–190.

41. Sittikankaew K, Hiransuchalert R, Yocawibun P, Yamano K, Klinbunga S (2010) Identification, characterization and expression of *adipose differentiation-related protein* (*ADRP*) gene and protein in ovaries of the giant tiger shrimp *Penaeus monodon*. Aquaculture 308: S91–S99.

42. Karoonuthaisiri N, Sittikankeaw K, Preechaphol R, Kalachikov S, Wongsurawat T, et al. (2009) ReproArray^{GTS}: A cDNA microarray for identification of reproduction-related genes in the giant tiger shrimp *Penaeus monodon* and characterization of a novel *nuclear autoantigenic sperm protein* (*NASP*) gene. Comp Biochem Phys D 4: 90–99.

43. Terrapon N, Li C, Robertson HM, Ji L, Meng XH, et al. (2014) Molecular traces of alternative social organization in a termite genome. Nat Commun 5: 3636 doi: 10.1038/ncomms4636

44. Colbourne JK, Pfrender ME, Gilbert D, Thomas WK, Tucker A, et al. (2011) The ecoresponsive genome of *Daphnia pulex*. Science 331: 555–561. doi: 10.1126/science.1197761

45. Ghaffari N, Sanchez-Flores A, Doan R, Garcia-Orozco KD, Chen PL, et al. (2014) Novel transcriptome assembly and improved annotation of the whiteleg shrimp (*Litopenaeus vannamei*), a dominant crustacean in global seafood mariculture. Sci Rep-Uk 4: 7081.

46. Christie AE (2014) Expansion of the *Litopenaeus vannamei* and *Penaeus monodon* peptidomes using transcriptome shotgun assembly sequence data. Gen Comp Endocr 206: 235–254. doi: 10.1016/j.ygcen.2014.04.015

47. Yu Y, Wei JK, Zhang XJ, Liu JW, Liu CZ, et al. (2014) SNP discovery in the transcriptome of white Pacific shrimp *Litopenaeus vannamei* by next generation sequencing. Plos One 9: e87218 doi: 10.1371/journal.pone.0087218

48. Xue SX, Liu YC, Zhang YC, Sun Y, Geng XY, et al. (2013) Sequencing and de novo analysis of the hemocytes transcriptome in *Litopenaeus vannamei* response to white spot syndrome virus infection. Plos One 8: e76718 doi: 10.1371/journal.pone.0076718

49. Chen XH, Zeng DG, Chen XL, Xie DX, Zhao YZ, et al. (2013) Transcriptome analysis of *Litopenaeus vannamei* in response to White spot syndrome virus infection. Plos One 8: e73218 doi: 10.1371/journal.pone.0073218

50. Zeng DG, Chen XL, Xie DX, Zhao YZ, Yang CL, et al. (2013) Transcriptome analysis of Pacific white shrimp

(*Litopenaeus vannamei*) hepatopancreas in response to Taura syndrome virus (TSV) experimental infection. Plos One 8: e57515 doi: 10.1371/journal.pone.0057515

51. Ashburner M, Ball CA, Blake JA, Botstein D, Butler H, et al. (2000) Gene Ontology: tool for the unification of biology. Nat Genet 25: 25–29.
52. Lin CM, Feng W (2012) Microarray and synchronization of neuronal differentiation with pathway changes in the Kyoto Encyclopedia of Genes and Genomes (KEGG) databank in nerve growth factor-treated PC12 cells (Retracted article. See vol. 10, pg. 90, 2013). Curr Neurovasc Res 9: 222–229.
53. Zeng V, Villanueva KE, Ewen-Campen BS, Alwes F, Browne WE, et al. (2011) De novo assembly and characterization of a maternal and developmental transcriptome for the emerging model crustacean *Parhyale hawaiensis*. BMC Genomics 12: 581 doi: 10.1186/1471-2164-12-581
54. Ewen-Campen B, Shaner N, Panfilio KA, Suzuki Y, Roth S, et al. (2011) The maternal and early embryonic transcriptome of the milkweed bug *Oncopeltus fasciatus*. BMC Genomics 12: 61 doi: 10.1186/1471-2164-12-61
55. Warrier S, Subramoniam T (2002) Receptor mediated yolk protein uptake in the crab *Scylla serrata*: Crustacean vitellogenin receptor recognizes related mammalian serum lipoproteins. Mol Reprod Dev 61: 536–548.
56. Shi XD, Zhang SC, Pang QX (2006) Vitellogenin is a novel player in defense reactions. Fish Shellfish Immun 20: 769–772.
57. Soverchia L, Ruggeri B, Palermo F, Mosconi G, Cardinaletti G, et al. (2005) Modulation of vitellogenin synthesis through estrogen receptor beta-1 in goldfish (*Carassius auratus*) juveniles exposed to 17-β estradiol and nonylphenol. Toxicol Appl Pharm 209: 236–243.
58. Orn S, Yamani S, Norrgren L (2006) Comparison of vitellogenin induction, sex ratio, and gonad morphology between zebrafish and Japanese medaka after exposure to 17 alpha-ethinylestradiol and 17 beta-trenbolone. Arch Environ Con Tox 51: 237–243.
59. Caruso JA, Laird DW, Batist G (1999) Role of HSP90 in mediating cross-talk between the estrogen receptor and the Ah receptor signal transduction pathways. Biochem Pharmacol 58: 1395–1403.
60. Fliss AE, Benzeno S, Rao J, Caplan AJ (2000) Control of estrogen receptor ligand binding by Hsp90. J Steroid Biochem 72: 223–230.
61. Arukwe A, Goksøyr A. (2003) Eggshell and egg yolk proteins in fish: hepatic proteins for the next generation: oogenetic, population, and evolutionary implications of endocrine disruption. Comp Hepatol 2: 1–21.
62. Beato M, Klug J (2000) Steroid hormone receptors: an update. Hum Reprod Update 6: 225–236.
63. Brown MA, Zhu L, Schmidt C, Tucker PW (2007) Hsp90—From signal transduction to cell transformation. Biochem Bioph Res Co 363: 241–246.
64. Lecaille F, Kaleta J, Bromme D (2002) Human and parasitic papain-like cysteine proteases: Their role in physiology and pathology and recent developments in inhibitor design. Chem Rev 102: 4459–4488.
65. Tingaud-Sequeira A, Cerda J (2007) Phylogenetic relationships and gene expression pattern of three different cathepsin L (Ctsl) isoforms in zebrafish: Ctsla is the putative yolk processing enzyme. Gene 386: 98–106.
66. Hu KJ, Leung PC (2004) Shrimp cathepsin L encoded by an intronless gene has predominant expression in hepatopancreas, and occurs in the nucleus of oocyte. Comp Biochem Phys B 137: 21–33.
67. Zhao W, Chen L, Zhang F, Wu P, Li E, et al. (2013) Molecular characterization of cathepsin L cDNA and its expression during oogenesis and embryogenesis in the oriental river prawn *Macrobrachium nipponense*(Palaemonidae). Genet Mol Res 12: 5215–5225. doi: 10.4238/2013.October.30.6
68. Jansen PAM, van den Bogaard EH, Kersten FFJ, Oostendorp C, van Vlijmen-Willems IMJJ, et al. (2012) Cystatin M/E knockdown by lentiviral delivery of shRNA impairs epidermal morphogenesis of human skin equivalents. Exp Dermatol 21: 889–891. doi: 10.1111/exd.12022
69. Chan SJ, Steiner DF (2000) Insulin through the ages: Phylogeny of a growth promoting and metabolic regulatory hormone. Am Zool 40: 213–222.
70. Wang S, Luo XN, Zhang SH, Yin C, Dou YX, et al. (2014) Identification of putative insulin-like peptides and components of insulin signaling pathways in parasitic platyhelminths by the use of genome-wide screening. Febs J 281: 877–893. doi: 10.1111/febs.12655
71. Dircksen H, Neupert S, Predel R, Verleyen P, Huybrechts J, et al. (2011) Genomics, transcriptomics, and peptidomics of *Daphnia pulex* neuropeptides and protein hormones. J Proteome Res 10: 4478–4504. doi: 10.1021/pr200284e
72. Chandler JC, Aizen J, Elizur A, Hollander-Cohen L, Battaglene SC, et al. (2015) Discovery of a novel insulin-like peptide and insulin binding proteins in the Eastern rock lobster *Sagmariasus verreauxi*. Gen Comp Endocr 215: 76–87. doi: 10.1016/j.ygcen.2014.08.018
73. Caruso MA, Sheridan MA (2011) New insights into the signaling system and function of insulin in fish. Gen Comp Endocr 173: 227–247. doi: 10.1016/j.ygcen.2011.06.014
74. Duleba AJ, Spaczynski RZ, Olive DL (1998) Insulin and insulin-like growth factor I stimulate the proliferation of human ovarian theca-interstitial cells. Fertil Steril 69: 335–340.
75. Brogiolo W, Stocker H, Ikeya T, Rintelen F, Fernandez R, et al. (2001) An evolutionarily conserved function of the *Drosophila* insulin receptor and insulin-like peptides in growth control. Curr Biol 11: 213–221.
76. Hansen IA, Attardo GM, Rodriguez SD, Drake LL (2014) Four-way regulation of mosquito yolk protein precursor genes by juvenile hormone-, ecdysone-, nutrient-, and insulin-like peptide signaling pathways. Front Physiol 5: 103 doi: 10.3389/fphys.2014.00103

77. Spaziani EP, Hinsch GW, Edwards SC (1993) Changes in prostaglandin E_2 and $F_{2\alpha}$ during vitellogenesis in the Florida crayfish *Procambarus Paeninsulanus*. J Comp Physiol B 163: 541–545.

78. Reddy PS, Reddy PR, Nagaraju GPC (2004) The synthesis and effects of prostaglandins on the ovary of the crab *Oziotelphusa senex senex*. Gen Comp Endocr 135: 35–41.

79. Tahara D, Yano I (2004) Maturation-related variations in prostaglandin and fatty acid content of ovary in the kuruma prawn (*Marsupenaeus japonicus*). Comp Biochem Phys A 137: 631–637.

80. Surai PF, Blesbois E, Grasseau I, Chalah T, Brillard JP, et al. (1998) Fatty acid composition, glutathione peroxidase and superoxide dismutase activity and total antioxidant activity of avian semen. Comp Biochem Phys B 120: 527–533.

81. Iwata H, Ohota M, Hashimoto S, Nagai Y (2003) Free oxygen radicals are generated at the time of aspiration of oocytes from ovaries that have been stored for a long time. Zygote 11: 1–5.

82. Arenas-Rios E, Leon-Galvan MA, Mercado PE, Lopez-Wilchis R, Cervantes DLMI, et al. (2007) Superoxide dismutase, catalase, and glutathione peroxidase in the testis of the Mexican big-eared bat (*Corynorhinus mexicanus*) during its annual reproductive cycle. Comp Biochem Phys A 148: 150–158.

83. Wu LT, Chu KH (2010) Characterization of an ovary-specific glutathione peroxidase from the shrimp *Metapenaeus ensis* and its role in crustacean reproduction. Comp Biochem Phys B 155: 26–33.

84. Atli G, Canli M (2007) Enzymatic responses to metal exposures in a freshwater fish *Oreochromis niloticus*. Comp Biochem Phys C 145: 282–287.

85. Thompson JL, Thomas PM, Schuller KA (2006) Purification and properties of a glutathione peroxidase from Southern bluefin tuna (*Thunnus maccoyii*) liver. Comp Biochem Phys B 144: 86–93.

86. Chansela P, Goto-Inoue N, Zaima N, Hayasaka T, Sroyraya M, et al. (2012) Composition and localization of lipids in *Penaeus merguiensis* ovaries during the ovarian maturation cycle as revealed by imaging mass spectrometry. Plos One 7: e33154 doi: 10.1371/journal.pone.0033154

87. Glencross BD, Smith DM (2001) A study of the arachidonic acid requirements of the giant tiger prawn, *Penaues monodon*. Aquacult Nutr 7: 59–69.

88. Meunpol O, Duangjai E, Yoonpun R, Piyatiratitivorakul S (2010) Detection of prostaglandin E2 in polychaete *Perinereis sp* and its effect on *Penaeus monodon* oocyte development *in vitro*. Fisheries Sci 76: 281–286.

89. Wimuttisuk W, Tobwor P, Deenarn P, Danwisetkanjana K, Pinkaew D, et al. (2013) Insights into the prostanoid pathway in the ovary development of the penaeid shrimp *Penaeus monodon*. Plos One 8: e76934 doi: 10.1371/journal.pone.0076934

90. Smith S, Witkowski A, Joshi AK (2003) Structural and functional organization of the animal fatty acid synthase. Prog Lipid Res 42: 289–317.

CHAPTER 4

TRANSCRIPTOME ANALYSIS OF THE PORTUNUS TRITUBERCULATUS: DE NOVO ASSEMBLY, GROWTH-RELATED GENE IDENTIFICATION AND MARKER DISCOVERY

Jianjian Lv, Ping Liu, Baoquan Gao, Yu Wang, Zheng Wang, Ping Chen, and Jian Li

INTRODUCTION

Portunus trituberculatus (Crustacea: Decapoda: Brachyura), commonly known as the swimming crab, is widely distributed in the coastal waters of Korea, Japan, China, and southeast Asia [1]. This species inhabits estuaries and coastal waters, which belong to typical euryhaline crab species. In China, it is a major edible crab species and one of the most important fishery resources [2] and the production has now reached 92,907 tons2011 [3]. At present, Commercial crab farming largely depended on wild seed stock and the commercial characteristics (growth rate, flesh quality and disease resistance) of the cultured stocks have also declined after many years of culturing [4], and wild populations of *Portunus trituberculatus* have dramatically declined for the last decades due to over-exploitation and the deterioration of environmental conditions in China [5]. To improve the germplasm of swimming crab, a selective breeding for fast growing *Portunus trituberculatus* has been carried out since 2004 at the Yellow Sea Fisheries Research Institute, Chinese Academy of Fishery Sciences (Qingdao, China). The selected fast-growing population of the crab had been examined and approved by China National Aquaculture Variety Approval Committee as a new variety for aquaculture and named "HuangXuan No. 1" (Authorization number: GS-01-002-2012) in 2012, of which growth rate increased by 20.12% compared to wild seed stock.

However, the molecular mechanisms involved in growth are poorly understood. Due to a lack of genetic and genomic information, growth-determining genes have not been identified in the crab, even genes related to growth have rarely been reported. Therefore, more genome-wide or transcriptome-wide datasets should be generated as a basis for functional genomics approaches aimed at improving the aquaculture performance of this species. Despite the aquacultural and biological importance of *Portunus trituberculatus*, previous studies have mostly focused on the isolation of microsatellites [6]–[8], the investigation of genetic diversity [9], Sanger-based sequencing of expressed sequence tags (ESTs) [10], [11], the characterization of single functional genes [12]–[14] and sequencing of the mitochondrial genome of *Portunus trituberculatus*[15]–[17].

When no genome sequence is available, transcriptome sequencing is an effective way to obtain large numbers of molecular makers and identify transcripts involved in specific biological processes [18]. Massively parallel sequencing of RNA (RNA-Seq) has offered the opportunity to characterize the transcriptome with unprecedented sensitivity and depth. It has already revolutionized the way we study the transcriptome. The latest paired-end sequencing of RNA-Seq techniques have further improved the efficiency of DNA sequencing and expanded short read lengths, permitting a deeper understanding of the transcriptome [19]. Because it is not restricted by the unavailability of a genome reference sequence, this approach

Citation: Lv J, Liu P, Gao B, Wang Y, Wang Z, Chen P, et al. (2014) Transcriptome Analysis of the Portunus trituberculatus: De Novo Assembly, Growth-Related Gene Identification and Marker Discovery. PLoS ONE 9(4): e94055. doi:10.1371/journal.pone.0094055

has been applied in decoding the genomes of several non-model organisms, providing valuable information in the understanding of gene function, cell responses and evolution [20]–[22]. Significant progress has aslo been made in understanding the transcript of various marine crustacea by RNA-seq over the last two years, such as *Litopenaeus vannamei*, *Fenneropenaeus chinensis*, *Eriocheir sinensis* and *Macrobrachium nipponense* [23]–[27], which is essential to better understand a species' biology and to devise strategies to improve productivity in culture. However, such investigations in *Portunus trituberculatus* have not been reported.

In this study, we present the first *Portunus trituberculatus* transcriptome using massively parallel mRNA sequencing. We perform Illumina sequencing of the eyestalk, gill, heart, hepatopancreas, and muscle tissues to characterize the *Portunus trituberculatus* transcriptome. The transcriptome provides an invaluable new data for a functional genomics resource and future biological research in *Portunus trituberculatus*. According to our sequence analysis, many genes involved in growth were identified. In addition, a variety of markers potentially useful for genomic population studies including simple sequence repeats (SSRs) located within coding regions and single nucleotide polymorphisms (SNPs) detected amongst deep coverage sequence regions reads are also reported.

MATERIALS AND METHODS

Ethics Statement

The crabs used in the present study were marine-cultured animals, and all of the experiments were conducted according to the regulations of the local and central government.

Sample Preparation

The swimming crabs, *Portunus trituberculatus* at 100 days age (20.62~64.19 g in body weight), were obtained from a local farm in Qingdao, China. All the samples were acclimated in the laboratory (33 ppt, 18°C) for one week before the experiment treatment. The crabs (nine males and nine females) were all anaesthetized on ice and dissected to collect samples, including the eyestalk, gill, heart, hepatopancreas, and muscle. All of the samples were immediately in RNAlater (Ambion) at 4 °C overnight and then at −20 °C until RNA extraction within 2 weeks.

RNA Isolation, cDNA Library Construction and Illumina Deep Sequencing

Total RNA was isolated from each sample by trizol (Invitrogen, CA,USA). RNA degradation and contamination was monitored on 1% agarose gels. RNA purity was checked using the NanoPhotometer spectrophotometer (IMPLEN, CA, USA). RNA concentration was measured using Qubit RNA Assay Kit in Qubit 2.0 Flurometer (Life Technologies, CA, USA). RNA integrity was assessed using the RNA Nano 6000 Assay Kit of the Bioanalyzer 2100 system (Agilent Technologies, CA, USA). A total amount of 5 ug RNA per sample was used as input material for the RNA sample preparations and all samples had RIN values above 8. Then, all samples were pooled in equal amounts to generate one mixed sample. The pooling samples were then used to prepare one separate Illumina sequencing libraries.

cDNA libraries were generated using Illumina TruSeq RNA Sample Preparation Kit (Illumia, San Diego, USA) following manufacturer's recommendations. Briefly, mRNA was purified from total RNA using poly-T oligo-attached magnetic beads. Fragmentation was carried out using divalent cations under elevated temperature in Illumina proprietary fragmentation buffer. First strand cDNA was synthesized using random oligonucleotides and SuperScript II. Second strand cDNA synthesis was subsequently performed using DNA Polymerase I and RNase H. Remaining overhangs were converted into blunt ends via exonuclease/polymerase activities and enzymes were removed. After adenylation of 3′ ends of DNA fragments, Illumina PE adapter oligonucleotides were ligated to prepare for hybridization. In order to select cDNA fragments of preferentially 200 bp in length, the library fragments were purified with AMPure XP system (Beckman Coulter, Beverly, USA). DNA fragments with ligated adaptor molecules on both ends were selectively enriched using Illumina PCR Primer Cocktail in a 10 cycle PCR reaction. Products were purified (AMPure XP system) and quantified using the Agilent high sensitivity DNA assay on the Agilent Bioanalyzer 2100 system. In the final step, the library preparations were sequenced on an Illumina Hiseq 2000 platform and 100 bp paired-end reads were generated.

Availability of supporting data

The data sets of Illumina sequencing are being submitted to the NCBI Short Read Archive (SRA) database.

Bioinformatic Analysis

Quality Control

Raw data (raw reads) of fastq format were firstly processed through our self-written perl scripts. In this step, clean data(clean reads) were obtained by removing reads containing adapter, reads containing ploy-N and low quality reads from raw data. At the same time, Q20, Q30, GC-content and sequence duplication level of the clean data were calculated. All the downstream analyses were based on clean data with high quality.

Transcriptome Assembly

Reads were assembled using Trinity [28], followed by TIGR Gene Indices clustering tools (TGICL) [29], with default parameters. The longest assembled sequences were referred to as contigs. The reads were then mapped back to contigs with paired-end reads to detect contigs from the same transcript and the distances between these contigs. Finally, sequences were obtained that lacked Ns and could not be extended on either end [30]. Such sequences were defined as unigenes.

Transcriptome Annotation

The unigenes were predicted to mapping to protein-coding sequences by GetORF of EMBOSS [31]. The predicted protein-coding sequences were compared with the NCBI non-redundant (Nr) protein database and UniProtKB database using BLASTx with E values less than 1.0×10^{-5}(E values less than 1.0×10^{-5} were considered as significant) [32], [33]. Based on Nr annotation, we used BLAST2GO program (http://www.BLAST2go.org/) to get GO annotation of unigenes [34]. GO functional classification for all unigenes was performed using WEGO software (http://wego.genomics.org.cn/cgi-bin/wego/index.pl) [35]. KEGG metabolic pathway annotation and COG classification of unigenes were determined by BLASTx searching against KEGG database and COG database, respectively [36], [37].

Makers Detection

SSR of the transcriptome were identified using MISA (http://pgrc.ipk-gatersleben.de/misa/misa.html), and primer for each SSR was designed using Primer 3 (http://primer3.sourceforge.net/releases.php). SNP were detected according to align clean reads to the reference transcriptome using SOAP2, then duplicated reads and multi-mapped reads were filtered from the alignment results in order to eliminate the PCR interference and ambiguous mapping. SOAPsnp was used to call SNP based on the sorted alignment results. SNPs qualified for the following standards were selected as the final SNP sets: quality score is not lower than 20, and distance between two SNPs are greater than 5.

Real-time PCR Assays

14 annotated unigenes that may relate to growth were selected to be analyzed using real-time PCR, and their specific primers were listed. The crab were siblings generated from a single pair of broodstock. The large and small sizes of crab were selected at 100 days age, respectively, from the >90 and <10 percentile regions of the growth distribution curve. Eyestalks were collected from nine healthy crabs of small size group (SG, 20.6±5.4 g in average body weight) and large size group (LG, 64.2±6.1 g in average body weight), respectively. All the samples were acclimated in the laboratory (33 ppt, 18°C) for one week before the experiment treatment. Total RNA was isolated according to the manufacturer's instructions of TRIZOL LS reagent (Invitrogen, Carlsbad, CA, USA). Then, RNA samples of three individuals were pooled within each group in equal amounts to generate three mixed sample, respectively. (three biological replicates of each group). RpL8 gene was selected as an internal control for qPCR analysis and the primers reference Xu's literature [38]. First strand cDNA was synthesized from 1 mg of RNA using M-MuLV reverse transcriptase (Qiagen). The qPCR reaction mixture (20 uL) consisted of 26 Power SYBR Green PCR Master mix, 0.9 M each of the forward and reverse primers, and 1 mL of template cDNA. PCR amplification was performed under the following conditions: 50°C for 2 min and 95°C for 30 s, followed by 40 cycles of 95°C for 15 s and 62°C for 1 min, and a final extension at 72°C for 5 min.

SSR validation and Polymorphism evaluation

Genomic DNA of crabs was extracted from muscle tissue using genomic DNA extraction kit (BioTeke, Beijing, China) following the protocols. Electrophoresis through a 1.5% agarose gel was used to check DNA integrity. The SSR markers were initially tested for amplification using a pool DNA sample of 10 crabs. PCR amplifications were carried out using Master-cycler gradient thermal cycler (Eppendorf) in a final volume of 10 ul. Each reaction tube contains 1.0 μl of 10×PCR buffer, 0.8 μl of dNTP (2.5 mM), 0.4 μl of each primer (10 umol), and 0.5 μl of genomic DNA (20 ng/ul), 0.05 μl of rTaq DNA polymerase (5 U/ul,Takara), 6.85 uL of ddH_2O. The PCR reaction program was: DNA denaturation at 94°C for 5 min; 35 cycles of 94°C for 30 s, 50–60°C for 30 s, 72°C for 30 s.; and 72°C for 7 min as a final extension. The primers that were not successful for amplification or produced multiple bands were reanalyzed using the touchdown PCR method with 1°C increments. The optimized SSR primers were used to amplify DNA from 30 wild individuals of *P. trituberculatus* collected from Jiaozhou, Shandong province, China for polymorphism evaluation. Amplification products were resolved via 8% denaturing polyacrylamide gel, and visualized by silver-staining. A 10-bp DNA ladder (Invitrogen Inc.) was used as a reference marker for allele size determination.

The number of alleles (*Na*), polymorphism information content (PIC), expected and observed heterozygosities (*He* and *Ho*, respectively) were calculated with the software CERVUS 3.0 [39].

SNP validation

To validate the putative SNPs identified in transcripts, the same cDNA samples as for the transcriptome profiling (pool of eighteen wild *Portunus trituberculatus*) were used. Twenty transcripts containing 56 potential SNPs and sufficient flanking regions were randomly selected for primer design. PCR products were sequenced directly in both directions with forward and reverse primers using Sanger technology on the ABI3730 platform (Applied Biosystems). Sequencing chromatograms were visually analyzed with Chromas2.32 (Technelysium Pty. Ltd.), and SNPs were identified as overlapping nucleotide peaks.

RESULTS AND DISCUSSION

Illumina Draft Reads and Sequence Assembly

In order to achieve a comprehensive *Portunus trituberculatus* transcriptome, total RNA was extracted from a variety of tissues, including the eyestalk, gill, heart, hepatopancreas, and muscle. Equal quantities of RNA were mixed together to construct a cDNA library and perform Illumina sequencing. This pooling strategy was widely used in some similar studies [23], [24], [40], [41]. The schematic of Illumina deep sequencing and analysis are shown in Figure 1. The overall Illumina read pairs and clean bases for all samples are 65,846,872 and 12.86G, respectively (**Table 1**). Files containing these data were deposited in the Short Read Archive of the National Center for Biotechnology Information (NCBI) with accession numbers of SRR1168416 and SRR1168417.

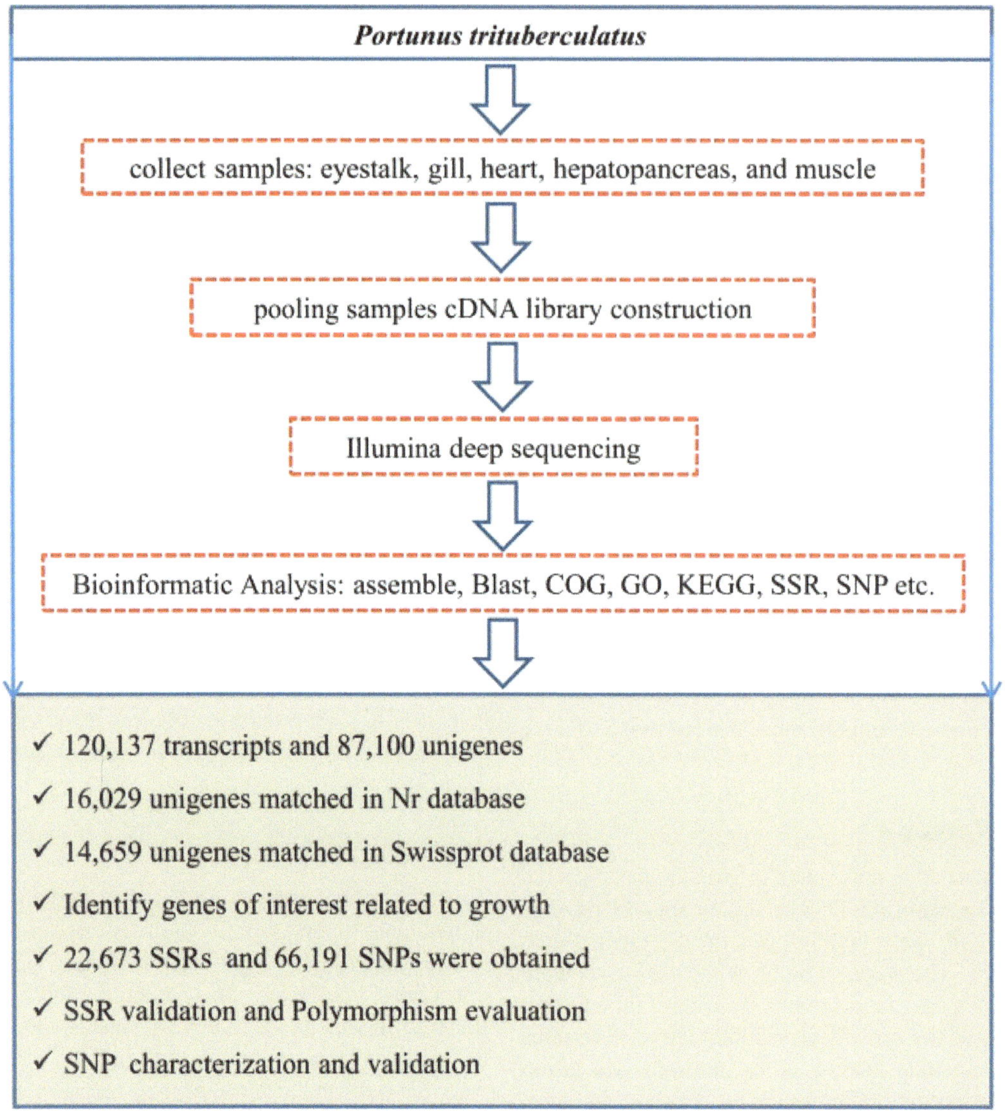

Figure 1: Schematic of Illumina deep sequencing and analysis.

Table 1: Summary of Illumina transcriptome sequencing, assembly and annotation for *Portunus trituberculatus*.

Raw results (after trimming)		Assembly results		Annotation results	
Clean bases (G)	12.86	Transcripts(bp)	120,137	Nr annotations	16,029
Read pairs	65,846,872	Average length of transcripts (bp)	1,037	UniProtKB annotations	14,659
Read length (bp)	100	Smallest transcripts (bp)	201	COG hits	14,263
		Largest transcripts (bp)	33,865	GO mapped	26,732
				KEGG hits	7,588

doi:10.1371/journal.pone.0094055.t001

After assembly analysis based on all Illumina reads, we identified 120,137 transcripts. The average length of all transcripts was 1,037 bp, with the smallest sequence of 201 bp and the largest one of 33,865 bp. The sequence length distribution of transcripts is indicated in **Figure 2** and **Table 1**. The average length of our assembled contigs was longer than that previously reported for *Litopenaeus vannamei* (average of 396 bp), *Eriocheir sinensis* (average of 385 bp) and *Fenneropenaeus chinensis* (average of 676 bp) [24], [25], [42]. Long sequences of good quality could enable us to obtain more information about genes. Therefore, this transcriptome dataset provides a useful resource for future analyses of genes related to economic traits. To the best of our knowledge, this is the first comprehensive study of the transcriptome in *Portunus trituberculatus*.

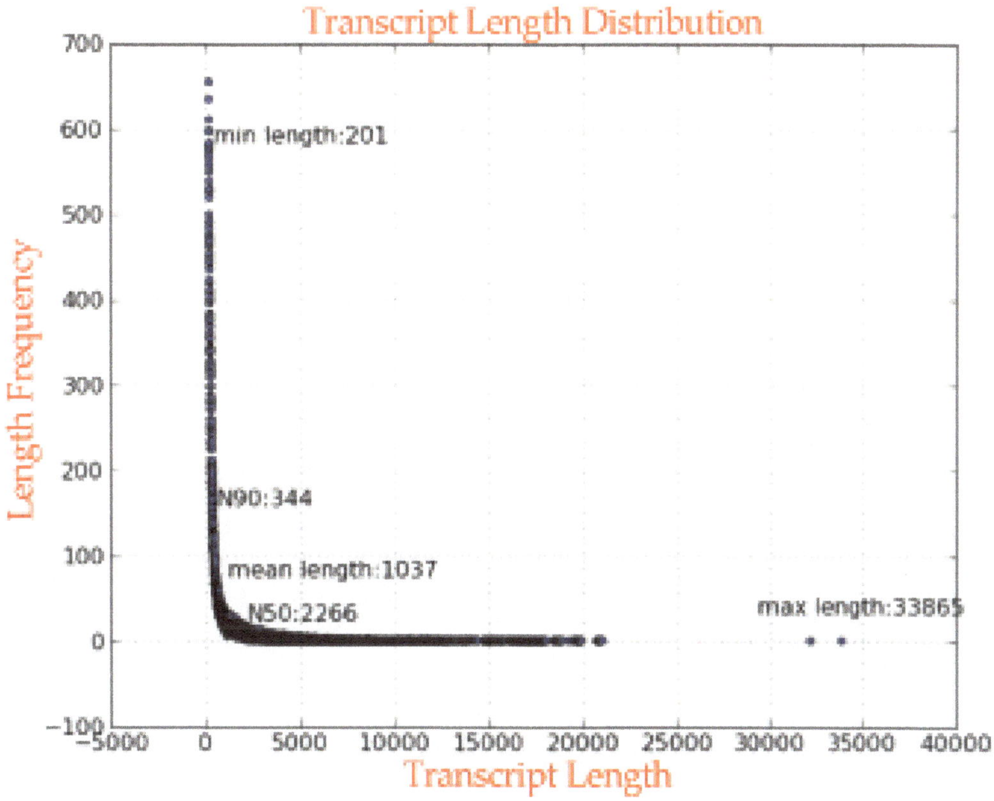

Figure 2: Sequence length distribution of transcripts assembled from Illumina reads.

To assess the abundance and coverage of the transcriptome data, we matched the assembled unigenes against the known EST library on Genbank. The 13,985 ESTs downloaded from NCBI were clustered and assembled, and 2,612 assembled EST-unigenes with mean length of 783 bp were generated. Comparisons between transcriptome unigenes and EST-unigenes were performed using BLASTn algorithm. All of the EST-unigenes can be matched in the transcriptome unigenes library, whereas only 3.0% of the transcriptome unigene sequences can be found in the ESTs library. It suggests the transcriptome data provide abundant information besides the now available ESTs sequences, and will vastly expand the number of genes identified in this species.

Annotation of Unigenes

After ruling out short-length and low-quality sequences, 87,100 unigenes were selected and subjected to annotation analysis by matching sequences against Nr and UniProtKB databases using BLASTx searching with an E value 1.0×10^{-5}. 16,029 unigenes (18.40% of the total) can be matched in Nr database, and 14,659 (16.83% of the total) matched in UniProtKB (**Table 1**). A significant number of *Portunus trituberculatus* unigenes did not matching any sequences in the GenBank nr database which is not surprising for crustacean transcriptome studies [26], [43]–[46]. Whilst most of these likely represent transcripts spanning only untranslated mRNA regions, chimeric transcript sequences derived from assembly errors or transcripts containing non-conserved protein regions, as reported in other transcriptome analyses [47]–[49], it is also possible that some may constitute novel genes unique to this species.

For main species distribution matched against Nr database, 9.58% of the matched unigenes showed similarities with *Daphnia pulex*, a microcrustacean, whose draft genome was recently published, followed by *Tribolium castaneum* (5.81%), *Pediculus humanu scorporis* (4.39%), *Branchiostoma floridae* (3.60%), *Strongylocentrotus purpuratus* (2.85%), *Nasonia vitripennis* (2.58%), *Ixodes scapularis* (2.50%), *Anopheles darlingi* (2.43%), *Megachile rotundata* (2.37%), *Camponotus floridanus* (2.00%), *Acyrthosiphon pisum* (2.00%)), and others (18.5%). As might be expected, unigenes of transcriptome matched well to crustacean and other arthropod proteins (**Figure 3**) which are in agreement with previous crustacean studies [26], [43]–[46].

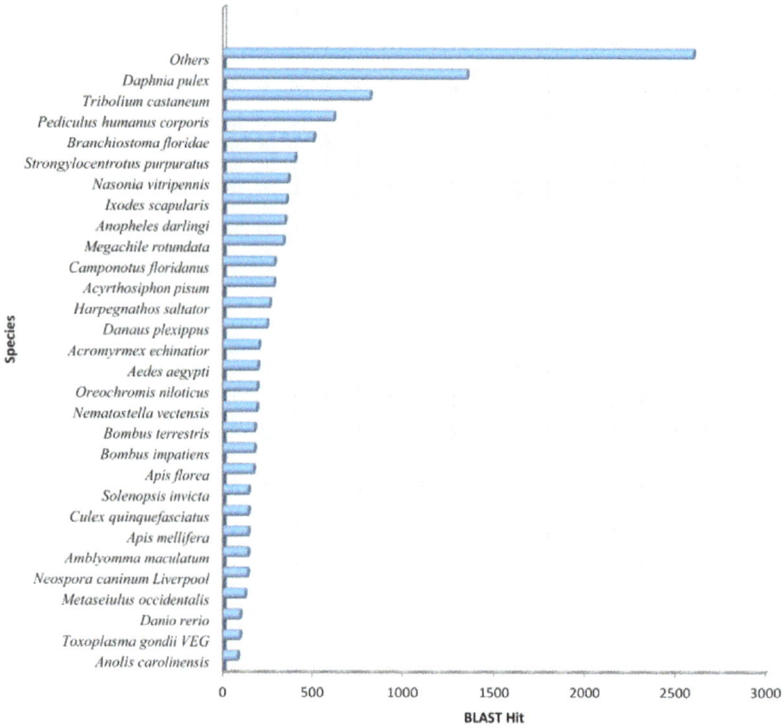

Figure 3: Top 30 hit species distribution based on BLASTx.

COG, GO and KEGG Classification

The assembled unigene sequences were subjected to BLAST searching against GO, COG and KEGG databases, and the summary statistics of BLAST assignment was shown in**Table 1**.

COG is a database where orthologous gene products were classified. Every protein in COG is assumed to be evolved from an ancestor protein, and the whole database is built on coding proteins with complete genome as well as system evolutionrelationships of bacteria, algae and eukaryotes [50]. Phylogenetic classifications of the predicted CDSs of unigenes were analyzed by searching against COG database to predict and classify possible functions of the unigenes. Possible functions of 14,263 unigenes were classified and subdivided into 26 COG categories, among which the cluster for'General function prediction only' represents the largest group (2652, 16.38% of the matched unigenes), followed by 'Signal Transduction' (2,279, 14.08%) and 'Posttranslational modification, protein turnover, chaperones' (1,483, 9.16%).

The Gene Ontology (GO) project provides structured, controlled vocabularies and classifications that cover several domains of molecular and cellular biology and are freely available for community use in the annotation of genes, gene products and sequences. Many model organism databases and genome annotation groups use the GO and contribute their annotation sets to the GO resource [51]. Among 87,100 assembled unigenes, 26,732 were successfully annotated by GO assignments, belonging to one or more of the three categories: biological process, cellular component, and molecular function. Among the annotated unigenes, 18,517 are involved in various biological process categories, cellular process (13,977 unigenes; 16.05%), metabolic process (12,657; 14.53%), biological regulation (4,824; 5.54%) and regulation of biological process(4,612; 5.30%) comprised the largest proportion. Further, 15,405 unigenes are involved in cellular component categories, among which, cell part (13,174; 15.12%), cell (13,174; 15.12%), organelle (5,712; 6.56%) and macromolecular complex (3,748; 4.30%) comprised the largest proportion. In addition, 21,988 unigenes are involved in molecular function categories, the top four categories were involved in binding (14,155; 16.25%), catalytic activity (10,779; 12.38%), transporter activity (2,467; 2.83%) and structural molecule activity (1,776; 2.04%). In summary, these terms account for a large fraction of the overall assignments in *Portunus trituberculatus* transcriptomic dataset. Understandably, genes encoding these functions may be more conserved across different species and are thus easier to annotate in the database.

The KEGG pathway database records networks of molecular interactions in the cells and variants of them specific to particular organisms. Pathway-based analysis helps us to further learn biological functions of genes [36], [52]. To systematically analyze their inner cell metabolic pathways and complicated biological behaviors, we classified the unigenes into biological pathways by mapping the annotated CDS sequences to the reference canonical pathways in the KEGG database. 7,588 unigenes were consequently assigned to 31 KEGG pathways, among which 858 members assigned to 'Translation', followed by 'Signal Transduction' (818 members), 'Folding' (750 members), 'Carbohydrate Metabolism' (699 members), 'Nervous System' (560 members), 'Transport and Catabolism' (542 members), Cell Growth and Death' (510 members), 'Replication and Repair' (493 members), Acid Metabolism' (475 members), 'Immune

System' (458 members), 'Nucleotide Metabolism' (431 members) and others.

GO, KEGG pathway and COG analysis are helpful for predicting potential genes and their functions at a whole-transcriptome level. The predicted GO categories, metabolic pathways, together with the COG analysis, are useful for further investigations of gene function in future studies.

Genes of interest related to growth

The transcriptome of *Portunus trituberculatus* was primarily examined to identify a wide range of candidate genes that might be functionally associated with growth. Traditionally, such gene discovery in non-model organisms has required degenerate PCR, which is labor-intensive and prone to failure [53]. The annotated transcriptome reported here allows researchers to identify genes of interest more easily than that of using degenerate PCR, especially in crustaceans whose genome information is relatively poor. According to our sequence analysis, many genes involved in growth were identified (**Table 2**) via three principal search strategies [54]: 1) associations between genes and growth reported in crustaceans, 2) growth-related genes involved with moulting, 3) muscle development and degradation genes involved in moulting.

Table 2: Genes of interest for growth and muscle development in *Portunus trituberculatus*.

Candidate genes	Transcript IDs Contig IDs
5-Hydroxytryptamine receptor	comp2095_c0;comp2005_c0;comp580344_c0;
Alpha-amylase	comp57976_c0;comp45223_c0;
Cathepsin L	comp459050_c0;comp14425_c0;comp31871_c0;comp31366_c1; comp11733_c0;comp14990_c0;comp38911_c0;comp722731_c0;comp27256_c0;
Cyclophilin	comp437375_c0;comp423352_c0;comp499501_c0;comp289233_c0;comp9973_c0;comp530052_c0; comp59112_c0;comp682946_c0;comp50416_c0;
Fatty acid-binding protein	comp31398_c0;
Fibrillarin	comp41046_c0;comp48390_c0;
Profilin	comp41014_c0;comp39876_c0;comp33022_c0;comp51876_c0;
Growth hormone and insulin-like growth factor	comp56179_c4;comp22818_c0;comp388372_c0;comp430291_c0;comp14260_c0;comp38363_c1; comp49319_c0;comp38363_c0;comp380665_c0;comp552_c0;comp46854_c0;comp514057_c0; comp55694_c0;comp58751_c0;
Myostatin and growth differentiation factor 8/11	comp623701_c0;comp317115_c0;
SPARC	comp45527_c0;comp49996_c0;
Ecdysteroid	comp31811_c0;comp30741_c0;comp56215_c0;
CHH	comp15032_c0;comp196993_c0;
Gonad/vitellogenesis-inhibiting hormone (G/VIH)	comp342264_c0;comp51889_c0;comp57850_c2;comp508364_c0;
Methyl farnesoate and farneosoic acid O-methyltransferase	comp17060_c0;comp55531_c1;comp48156_c0;
MIH	comp198837_c0;
Actin	comp31467_c0;comp58899_c0;comp48501_c1;comp54937_c0; comp57254_c0; comp87442_c0; comp378707_c0
Myosin	comp427727_c0;comp381601_c0;comp45465_c1;comp60783_c0;comp45934_c0; comp30670_c0;comp39835_c0;comp51426_c0;comp46623_c0;comp23652_c0; comp129272_c0;comp31352_c0;comp31357_c1;comp33150_c0;comp40995_c0;
Alpha skeletal muscle	comp50088_c0;comp324557_c0;comp45287_c0;
Calponin/calponin transgelin	comp41232_c0;comp52466_c0;comp57804_c0;
Tropomyosin	comp48547_c0;comp59654_c0;
Muscle lim protein	comp59750_c0;comp60108_c0;comp59060_c0;comp48583_c0;comp31515_c0;

doi:10.1371/journal.pone.0094055.t002

In this study, a total of 21 categories of growth-related genes were found, amongst these, ten were found in our transcriptome data which have been identified previously to have roles in growth in crustaceans. These genes include 1. 5-Hydroxytryptamine receptor, Alpha-amylase, Cathepsin L, Cyclophilin, Fatty acid-binding protein, Fibrillarin, Glyceradehyde-3-phosphate dehydrogenase, Growth hormone and insulin-like growth factor, Myostatin and growth differentiation factor 8/11, Signal transducer and activator of transcription, Secreted protein acidic and rich in cysteine(SPARC), and Translin-associated factor X (TRAX). Previous studies have shown that SNP in 5-Hydroxytryptamine receptor [55], Cathepsin L [56], Myostatin and growth differentiation factor 8/11 genes [57] show significant associations with growth traits in crustaceans. Besides, Cyclophilin, Fibrillarin and Secreted protein acidic and rich in cysteine(SPARC) gene expression showed a negative correlation with body weight in shrimp by Pearson's correlation analysis [58].

In crustaceans, periodic shedding of the exoskeleton is one of the most important physiological processes essential for crustacean growth and postembryonic development including moulting and regeneration [59]. Although the functions of many of the hormones and genes involved in this process are still not well defined, a number of studies have indicated that moulting and reproduction in crustaceans is regulated by the eyestalk derived CHH gene family which is one of the major groups of peptide hormones produced in the XO-SG [60], [61]. In addition, MIH is responsible for maintaining animals in the intermoult stage which is an important regulator of steroidogenesis in the YO [62]. In this research, a few genes which belong to crustacean hyperglycemic hormone neuropeptides (CHH) family were identified including MIH, CHH and ecdysteroids. Correlations between SNPs in the CHH and MIH gene with individual growth performance show that the CHH and MIH gene has high potential to impact body weight variation in crustaceans [63] and should, therefore, be considered as a primary gene of interest in growth studies.

Crustacean muscle growth is not continuous and is strongly influenced by the moulting cycle [59]. During the moult, muscles regenerate, and energy reserves including glycogen and lipids are accumulated in the hemolymph and the midgut for the next moult [64], [65]. Overall muscle protein synthesis is very important for growth, reproduction and other metabolic activities in crustaceans. Recent studies of invertebrates have highlighted the importance of muscle specific genes and proteins in crustaceans [43]. In this research, six genes related to muscle build-up or degradation during the moulting event were identified including Actin, Myosin, Alpha skeletal muscle, Calponin/calponin transgelin, Tropomyosin and Muscle lim protein. Among which, both actin and myosin proteins showed a high number of transcripts. It has been reported that actins are expressed in abundance as they are critical to formation of muscle filaments [66]. Different actin isoforms have been identified in various crustaceans [67], and are likely to be involved in playing important roles in cytoskeletal structure, cell division and mobility, and muscle contraction [68], [69]. As evidence for a role for actin in muscle build-up during the moult cycle in crustaceans, Cesar and Yang (2007) reported that muscle structural α-actin and cytoskeletal β-actin increased during the intermoult and premoult stages, a phase where high muscle growth occurred in the abdominal muscle of L. vannamei [70]. However, in a recent SNP association analysis study, four synonymous polymorphisms were identified in an actin fragment but SNP allele distributions were not related significantly to individual growth performance in the two studied groups of giant freshwater prawn *M.rosenbergii* [63], and further studies will be required to investigate. Myosins are a major component of the contractile apparatus and consist of two heavy (MHC) and four associated light chains (MLC) [71]. Previous study showed, myosin gene expression levels could provide a good molecular marker of individual growth potential in the Atlanticpink shrimp *Farfantepenaeus paulensis* that identified MHC as a possible growth candidate gene [72]. A high number of actin and myosin protein transcripts observed here may regulate muscle development and function in *Portunus trituberculatus*, and similar results have been found in *Macrobrachium rosenbergii* [43], however, further studies are needed to confirm these observations.

The current study identified a number of putative genes that are potentially involved with growth in *Portunus trituberculatus*. However, further studies are needed to understand the molecular functions of these putative genes with growth performance.

Real-time RT-PCR confirmation of growth-related genes

To futher confirm the growth-related genes obtained from the transcriptome data, 14 candidate genes were selected to be analyzed using real-time PCR. Since the XO-SG complex in the crab eyestalk produces a variety of neuropeptides/neurohormones [60], [61], it was selected, in this study, as the target tissue for differential gene expression analysis between large and small crab.

All of the 14 selected genes revealed significant differences in gene expression between small size group (SG) and large size group (LG) (**Figure 4**), which consistent with the fact that body weight is a complex trait regulated by the coordinate action of several genes. Most of gene (12) were significantly down-regulated in LG including cyclophilin A (comp59112_c0), fibrillarin (comp41046_c0) and SPARC (comp49996_c0). In *P. monodon*, Tangprasittipap et al. (2010) reported the index of relative cyclophilin, SPARC and fibrillarin-like expression was negatively correlated with body weight ($p<0.05$) [58], which were similar to our results, suggests that these genes may have some effect on individual growth performance, and warrants further study in crustacean species.

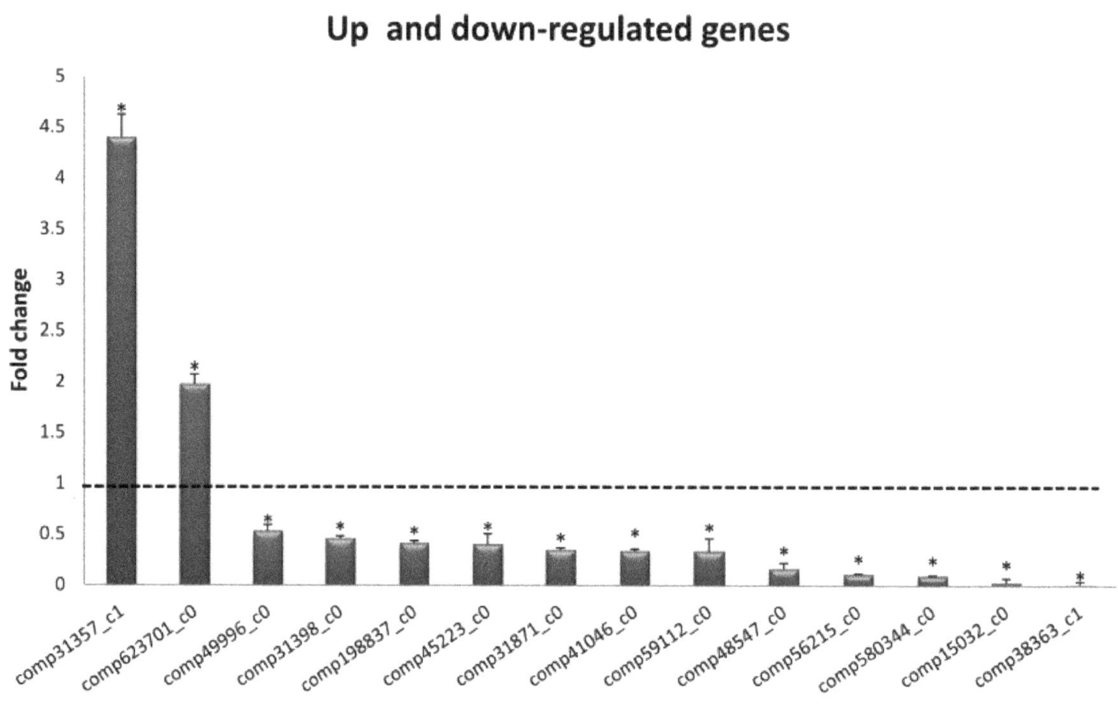

Figure 4: Relative expression of 14 growth-related genes (LG Vs SG).

Myostatin (comp623701_c0) and Myosin heavy chain (comp31357_c1) were up-regulated in LG. In vertebrates, Myostatin (MSTN), principally controls growth of muscle cells as a negative regulator of muscle development [73], however, MSTN show positive regulation of growth in the crab. Similar result was found in *P. monodon*, reduced levels of MSTN transcripts resulted in a dramatic slowing of growth rate compared with control groups [74], which suggests the genes may regulate growth positively in crustaceans. Myosins are a major component of the contractile apparatus and consist of two heavy and four associated light chains [71]. In south-western Atlantic pink shrimp *Farfantepenaeus paulensis*, higher MHC expression was observed in a high weight shrimp group [72], a result was similar to our research, which suggested MHC as a possible growth candidate gene in crustaceans.

The result here show that seeking the growth-related genes via high-throughput transcriptome sequencing and bioinformatics analysis is an effective way. Further studies should be carried out to elucidate the function of these genes in growth.

Putative molecular markers

SSR characterization and Polymorphism evaluation

SSRs, or microsatellites, are polymorphic loci present in genomic DNA. They consist of repeated core sequences of 2~6 base pairs in length. Among the various molecular markers, SSRs have been proven to be an efficient tool for constructing genetic linkage, performing QTL analysis and evaluating the level of genetic variation in a species because of the high variability, abundance, neutrality and co-dominance of microsatellite DNA [75].

We obtained a total 22,673 SSRs in the transcriptomic dataset. Of these, 67.03% were di-nucleotide repeats, followed by 30.06% tri-nucleotide repeats and 2.92% tetra/penta/hexa-nucleotiderepeats (**Figure 5**). Generally believed that SSR of animals are mainly di-nucleotide repeats [76], [77], and our findings support this conclusion. Among the di-nucleotide repeats motifs, (GT/TG)n, (GA/AG)n and (CA/AC)n were the three predominant types with frequencies of 36.46%, 27.03% and 22.80%, respectively. In the 20 types of tri-nucleotide repeats, (GGT/GTG/TGG)n, (GGA/GAG/AGG)n, and (CCA/CAC/ACC)n were the most common types with a combined frequency of 39.84%. The results of this study is differ from other studies which indicate that microsatellite repeats types have species-specific in crustacea [26], [43]. Primer for each SSR was designed using Primer 3 (http://primer3.sourceforge.net/releases.php), and 14.5% (3,858) SSRs can be designed priemers successfully (the data do not show). To date, only a few microsatellites have been available for *Portunus trituberculatus* from NCBI. Thus, the development of SSRs for this species is highly desirable.

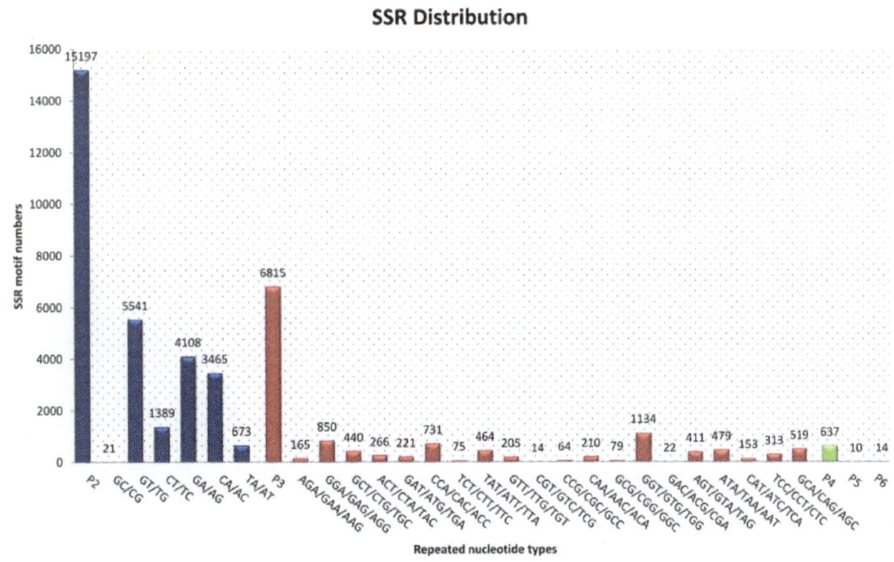

Figure 5: Distribution of simple sequence repeat (SSR) nucleotide classes among different nucleotide types found in *Portunus trituberculatus*.

Fifteen SSRs were randomly selected for primer synthesis and validation, among which, 9 were successful in PCR amplification using genomic DNA from *Portunus trituberculatus*. The remaining 6-pair primers failed to generate PCR products, even when the annealing temperature was reduced by 8°C. Of the 9 primers, 1 primers were monomorphic, the other 8 primers were polymorphic (Figure **6**), the proportion of polymorphic primers was 53.3%. From the 8 polymorphic loci, the number of alleles per locus ranged from 2 to 8 alleles. A total of 34 alleles were identified, with an average of 4.25 alleles per locus. Across 8 loci, the polymorphic information content (PIC) ranged from 0.15 to 0.78 (Table **3**), with an average of 0.58, suggesting that the developed EST-SSRs were highly polymorphic. The results obtained in this study indicated that these SSRs developed from EST in the swimming crab will be a useful tool for the genetic research such as population variation, parentage analysis, stock enhancement evaluation, and the establishment of effective conservation strategy of *Portunus trituberculatus*.

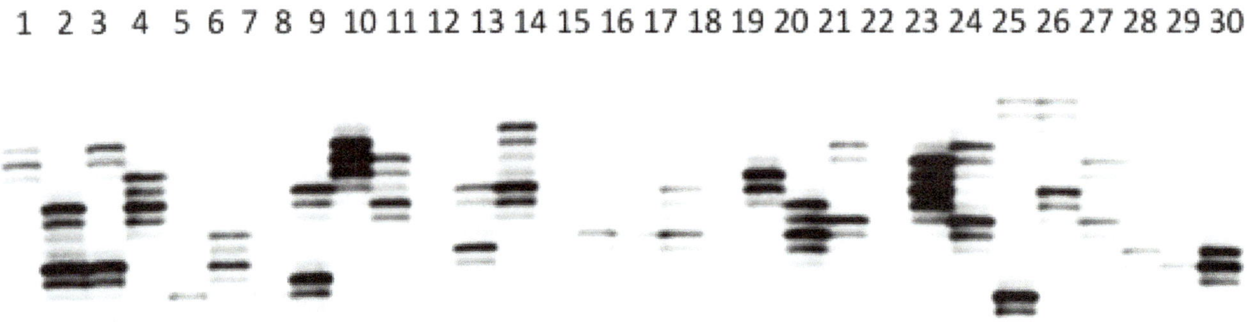

Figure 6: Polyacrylamide gel electrophoresis for one SSR markers (comp17478_c0) in the 30 individuals.

Table 3: Characterization of 15 polymorphic microsatellite loci in *Portunus trituberculatus*.

Transcript IDs	(Type) No.	Primer sequence (5′-3′)	Length	*Ta*	*Na*	*Ho*	*He*	PIC
comp57884_c0	(TG)9	F: TTGGAAGTTTTGTTGGCGGC	199	59	3	0.59	0.64	0.56
		R: ACGTTGTGGTCACTTGTTGC						
comp32503_c0	(TG)9	F: GCGGCGTGGTAATCTCTATGT	199	60	4	1	0.74	0.70
		R: CTAACAAGACATCCACGCACG						
comp333871_c0	(TG)8	F: GCAGGGGCTTCTTCACCTTA	235	60	5	0.96	0.79	0.74
		R: CCCCTCCTTCCAATTTGTACCT						
comp31578_c0	(CA)8	F: GACCAACAGGCGACCGAG	134	57	2	0.2	0.18	0.16
		R: AGCCGCTTCCCGAGATTC						
comp49316_c0	(TC)9	F: TGCGTTGCGTAACTGAGTCT	269	59	4	1	0.63	0.56
		R: CTTGTCAGTGCTTCCTTGCC						
comp55771_c0	(TG)10	F: AAGACCCCGAGGAAGAGC	199	55	5	1	0.75	0.69
		R: GCAAGGCATATCCAACAT						
comp54398_c4	(TC)9	F: CTTTGTGTGTTGGGTTGGGC	142	58	3	0.72	0.55	0.45
		R: AACAATGCCCACAACTCAGG						
comp17478_c0	(GT)10	F: GCATAGAACGAGTGATACTAGATGC	235	59	8	0.92	0.79	0.78
		R: CACACACACACACACACACA						
comp53876_c0	(GT)10	F: ACAGCGTCAGGAAGCAATCA	209	60	1	0	0	0
		R: ATTGCCTCCTTCCCAATGCA						
comp49119_c1	(CA)9	F: TGGTTTTGCCACTCCACACT	182	59	n/a			
		R: TGTCAGCCACGACACTACTT						
comp44386_c0	(CA)7	F: TTGCGTGTGGAGGAAGTGTT	277	60	n/a			
		R: GAGAGGTGGATGGGGAGACT						
comp16699_c0	(AG)7	F: TTCTCATTCTTGGCTCGCGT	205	60	n/a			
		R: TCACATGCTCCGAGGATTGG						
comp24638_c0	(AG)8	F: CGTGTGTGCGTGTTTGTCTT	279	58	n/a			
		R: AGTTTTCCTCCTTTACGCATCA						
comp58479_c1	(TG)6	F: AGGACCATAACAAGGCCACG	176	60	n/a			
		R: CTGCAACACAGCACTGACAG						
comp1580_c0	(TG)7	F:CACCCAAGCTCTCTTCCTGG	228	59	n/a			
		R: ACCAACAGACAGGGAGAGGA						

Ho observed heterozygosity, *He* expected heterozygosity, *Na* observed number of alleles, *Ta* annealing temperature, PIC polymorphic information content, n/a indicates that no PCR amplification.

doi:10.1371/journal.pone.0094055.t003

SNP characterization and validation

SNPs were identified from alignments of multiple sequences used for contig assembly. By excluding those that had mutation frequency of bases lower than 1%, we obtained a total of 66,191 SNPs, of which 23,734 were putative transitions (Ts) and 42,457 were putative transversions (Tv), giving a mean Ts: Tv ratio of 1:1.79 across the transcriptome of *Portunus trituberculatus* (**Figure** 7) which can help identify genes affected by selection [78]. Further analysis found the Ts: Tv ratio were species-specific in crustacean species, in Oriental River Prawn (*Macrobrachium nipponense*) and Giant Freshwater Prawn (*Macrobrachium rosenbergii*), the Ts: Tv ratios were 1.99:1.00 and 1.32:1.00, respectively. The AT/TA, AG/GA and CT/TC SNP types were the most common. In contrast, GC/CG types were the smallest SNP types because of the differences in the base structure and the number of hydrogen bonds between different bases [26].

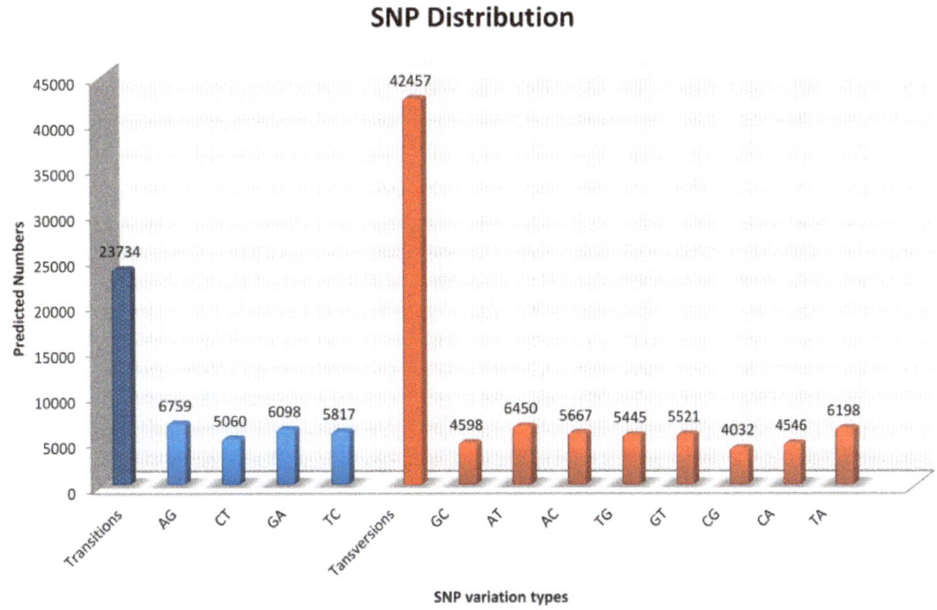

Figure 7: Distribution of putative single nucleotide polymorphisms (SNP) in *Portunus trituberculatus* sequences.

To verify the potential SNPs, a subset of 20 transcripts containing 56 SNPs were selected randomly. A pooled cDNA sample of eighteen wild *Portunus trituberculatus* was amplified by PCR. Subsequently, PCR products were sequenced bidirectionally with forward or reverse primers among which, sequencing of four transcripts failed. Of the 43 SNPs predicted to reside in the amplified sequences, 21 (48.8%) showed polymorphisms in the sample and were validated .The rate of polymorphic SNPs was probably an underestimate because only eighteen individuals were used. In addition, the way of SNP validation in this study has some limitations, SNPs with smaller variation frequency were difficult to identify via overlapping nucleotide peaks. Consequently, using more *Portunus trituberculatus* samples, the polymorphic rate of potential EST-SNPs should be higher than that found in our validation.

SSRs and SNPs detected in this study are likely to be highly transferable to other closely related species, as has been the case in other crustacean species [79], [80]. These potential markers identified within the ESTs will be valuable for studying the evolution and molecular ecology, genome mapping, and QTL analysis of *Portunus trituberculatus*.

CONCLUSION

Here we report the first comprehensive transcript dataset of the *Portunus trituberculatus*, a non-model species for which little molecular knowledge currently exists. The 120,137 transcripts identified and assembled will enable gene discovery in *Portunus trituberculatus*, and with the significant number of putative growth-related genes identified should facilitate genomics approaches to improving the growth performance of domesticated stocks used for aquaculture. In addition, the large number of SNPs and SSRs detected provide targets for identifying polymorphisms across *Portunus trituberculatus* populations useful for parentage assignment and for managing in breeding in cultured populations.

FUNDING STATEMENT

Financial support for this study was provided by the National High Technology Research and Development Program of China (Project 2012AA10A409), Agricultural science and Technology Achievements Transformation Fund Project of the Ministry of science and technology (Project 2013GB23260589), the National Natural Science Foundation of China (Grant No. 41306177), the National Science Foundation for Post-doctoral Scientists of China (Project 2013M531657), Special Scientific Research Funds for Central Non-profit Institutes, Yellow Sea Fisheries Research Institute (Project 20603022013039), Projects of independent innovation in Shandong Province (Project 2013CXC80202), Project development of science and technology in Shandong Province (Project 2011GHY11526). The funders had no role in study design, data collection and analysis, decision to publish, or preparation of the manuscript.

REFERENCES

1. Dai A, Yang S, Song Y (1986) Marine crabs in China Sea. Marine publishing company, Beijing.
2. Yu C, Song H, Yao G (2003) Geographical distribution and faunal analysis of crab resources in the East China Sea. J Zhejiang Ocean Univ 22: 108–113
3. Yearbook CFS (2012) China Agriculture Press. 1–108.
4. Chen P, Li JT, Gao BQ, Liu P, Wang QY, et al. (2011) cDNA cloning and characterization of peroxiredoxin gene from the swimming crab Portunus trituberculatus. Aquaculture 322: 10–15
5. Yu C, Song H, Yao G, Lu H (2006) Composition and distribution of economic crab species in the East China Sea. Oceanologia et Limnologia Sinica 37: 60
6. Lee HJ, Lee DH, Yoon SJ, Kim DH, Kim SG, et al. (2013) Characterization of 20 microsatellite loci by multiplex PCR in swimming crab, Portunus trituberculatus. Genes & Genomics 35: 77–85

7. Wei Q, Wang CL, Mu CK, Song WW, Li RH (2013) Development and characterization of EST-derived microsatellite makers for swimming crab, Portunus trituberculatus. Conservation Genetics Resources 5: 511–513
8. Wang HX, Cui ZX, Wu DH, Guo EM, Liu Y, et al. (2012) Application of microsatellite DNA parentage markers in the swimming crab Portunus trituberculatus. Aquaculture International 20: 649–656
9. Liu YG, Guo YH, Hao J, Liu LX (2012) Genetic diversity of swimming crab (Portunus trituberculatus) populations from Shandong peninsula as assessed by microsatellite markers. Biochemical Systematics and Ecology 41: 91–97
10. Cui ZX, Song CW, Liu Y, Wang SY, Li QQ, et al. (2012) Crustins from eyestalk cDNA library of swimming crab Portunus trituberculatus: Molecular characterization, genomic organization and expression analysis. Fish & Shellfish Immunology 33: 937–945
11. Liu YA, Cui ZX, Luan WS, Song CW, Nie Q, et al. (2011) Three isoforms of anti-lipopolysaccharide factor identified from eyestalk cDNA library of swimming crab Portunus trituberculatus. Fish & Shellfish Immunology 30: 583–591
12. Liu Y, Cui ZX, Li XH, Song CW, Shi GH, et al. (2013) Molecular cloning, genomic structure and antimicrobial activity of PtALF7, a unique isoform of anti-lipopolysaccharide factor from the swimming crab Portunus trituberculatus. Fish & Shellfish Immunology 34: 652–659
13. Xu QH, Qin Y (2012) Molecular cloning of heat shock protein 60 (PtHSP60) from Portunus trituberculatus and its expression response to salinity stress. Cell Stress & Chaperones 17: 589–601
14. Wang SY, Cui ZX, Liu Y, Li QQ, Song CW (2012) The first homolog of pacifastin-related precursor in the swimming crab (Portunus trituberculatus): Characterization and potential role in immune response to bacteria and fungi. Fish & Shellfish Immunology 32: 331–338
15. Guo EM, Liu Y, Cui ZX, Li XL, Cheng YX, et al. (2012) Genetic variation and population structure of swimming crab (Portunus trituberculatus) inferred from mitochondrial control region. Molecular Biology Reports 39: 1453–1463
16. Cho EM, Min GS, Kanwal S, Hyun YS, Park SW, et al. (2009) Phylogenetic Analysis of Mitochondrial DNA Control Region in the Swimming Crab, Portunus trituberculatus. Animal Cells and Systems 13: 305–314
17. Yamauchi MM, Miya MU, Nishida M (2003) Complete mitochondrial DNA sequence of the swimming crab, Portunus trituberculatus (Crustacea: Decapoda: Brachyura). Gene 311: 129–135
18. Fu BD, He SP (2012) Transcriptome Analysis of Silver Carp (Hypophthalmichthys molitrix) by Paired-End RNA Sequencing. DNA Research 19: 131–142
19. Fullwood MJ, Wei C-L, Liu ET, Ruan Y (2009) Next-generation DNA sequencing of paired-end tags (PET) for transcriptome and genome analyses. Genome research 19: 521–532
20. Ramayo-Caldas Y, Mach N, Esteve-Codina A, Corominas J, Castello A, et al. . (2012) Liver transcriptome profile in pigs with extreme phenotypes of intramuscular fatty acid composition. BMC Genomics 13.
21. Sanchez CC, Weber GM, Gao G, Cleveland BM, Yao J, et al. . (2011) Generation of a reference transcriptome for evaluating rainbow trout responses to various stressors. BMC Genomics 12.
22. Wang XW, Luan JB, Li JM, Su YL, Xia J, et al. . (2011) Transcriptome analysis and comparison reveal divergence between two invasive whitefly cryptic species. BMC Genomics 12.
23. Zeng D, Chen X, Xie D, Zhao Y, Yang C, et al. (2013) Transcriptome Analysis of Pacific White Shrimp (*Litopenaeus vannamei*) Hepatopancreas in Response to Taura Syndrome Virus (TSV) Experimental Infection. PLoS ONE 8: e57515.
24. Li S, Zhang X, Sun Z, Li F, Xiang J (2013) Transcriptome Analysis on Chinese Shrimp *Fenneropenaeus chinensis* during WSSV Acute Infection. PLoS ONE 8: e58627.
25. He L, Jiang H, Cao D, Liu L, Hu S, et al. (2013) Comparative Transcriptome Analysis of the Accessory Sex Gland and Testis from the Chinese Mitten Crab (*Eriocheir sinensis*). PLoS ONE 8: e53915.
26. Ma KY, Qiu GF, Feng JB, Li JL (2012) Transcriptome Analysis of the Oriental River Prawn, Macrobrachium nipponense Using 454 Pyrosequencing for Discovery of Genes and Markers. PLoS ONE 7.
27. He L, Wang Q, Jin XK, Wang Y, Chen LL, et al. . (2012) Transcriptome Profiling of Testis during Sexual Maturation Stages in Eriocheir sinensis Using Illumina Sequencing. PLoS ONE 7.
28. Grabherr MG, Haas BJ, Yassour M, Levin JZ, Thompson DA, et al. (2011) Full-length transcriptome assembly from RNA-Seq data without a reference genome. Nature Biotechnology 29: 644–652
29. Pertea G, Huang X, Liang F, Antonescu V, Sultana R, et al. (2003) TIGR Gene Indices clustering tools (TGICL): a software system for fast clustering of large EST datasets. Bioinformatics 19: 651–652
30. Cao C, Wang Z, Niu C, Desneux N, Gao X (2013) Transcriptome Profiling of *Chironomus kiinensis*under Phenol Stress Using Solexa Sequencing Technology. PLoS ONE 8: e58914.
31. Olson SA (2002) Emboss opens up sequence analysis. Briefings in bioinformatics 3: 87–91
32. Altschul SF, Madden TL, Schäffer AA, Zhang J, Zhang Z, et al. (1997) Gapped BLAST and PSI-BLAST: a new generation of protein database search programs. Nucleic Acids Research 25: 3389–3402
33. Zhao XL, Yu H, Kong LF, Li Q (2012) Transcriptomic Responses to Salinity Stress in the Pacific Oyster Crassostrea gigas. PLoS ONE 7.
34. Conesa A, Götz S (2008) Blast2GO: A comprehensive suite for functional analysis in plant genomics. International journal of plant genomics 2008.

35. Ye J, Fang L, Zheng H, Zhang Y, Chen J, et al. (2006) WEGO: a web tool for plotting GO annotations. Nucleic acids research 34: W293–W297

36. Kanehisa M, Goto S (2000) KEGG: kyoto encyclopedia of genes and genomes. Nucleic Acids Research28: 27–30

37. Tatusov RL, Fedorova ND, Jackson JD, Jacobs AR, Kiryutin B, et al. (2003) The COG database: an updated version includes eukaryotes. BMC bioinformatics 4: 41.

38. Xu QH, Liu Y, Liu RL (2010) Expressed sequence tags from cDNA library prepared from gills of the swimming crab, Portunus trituberculatus. Journal of Experimental Marine Biology and Ecology 394: 105–115

39. Kalinowski ST, Taper ML, Marshall TC (2007) Revising how the computer program CERVUS accommodates genotyping error increases success in paternity assignment. Molecular ecology 16: 1099–1106

40. Colaiacovo M, Subacchi A, Bagnaresi P, Lamontanara A, Cattivelli L, et al. (2010) A computational-based update on microRNAs and their targets in barley (Hordeum vulgare L.). BMC Genomics 11: 595.

41. Nie QH, Sandford EE, Zhang XQ, Nolan LK, Lamont SJ (2012) Deep Sequencing-Based Transcriptome Analysis of Chicken Spleen in Response to Avian Pathogenic Escherichia coli (APEC) Infection. PLoS ONE 7.

42. Li CZ, Weng SP, Chen YG, Yu XQ, Lu L, et al. . (2012) Analysis of Litopenaeus vannamei Transcriptome Using the Next-Generation DNA Sequencing Technique. PLoS ONE 7.

43. Jung H, Lyons RE, Dinh H, Hurwood DA, McWilliam S, et al. . (2011) Transcriptomics of a Giant Freshwater Prawn (Macrobrachium rosenbergii): De Novo Assembly, Annotation and Marker Discovery. PLoS ONE 6.

44. Zhang W, Wan H, Jiang H, Zhao Y, Zhang X, et al. (2011) A transcriptome analysis of mitten crab testes (Eriocheir sinensis). Genetics and Molecular Biology 34: 136–141

45. Leekitcharoenphon P, Taweemuang U, Palittapongarnpim P, Kotewong R, Supasiri T, et al. (2010) Predicted sub-populations in a marine shrimp proteome as revealed by combined EST and cDNA data from multiple Penaeus species. BMC research notes 3: 295.

46. Leu J-H, Chen S-H, Wang Y-B, Chen Y-C, Su S-Y, et al. (2011) A review of the major penaeid shrimp EST studies and the construction of a shrimp transcriptome database based on the ESTs from four penaeid shrimp. Marine Biotechnology 13: 608–621

47. Wang J-PZ, Lindsay BG, Leebens-Mack J, Cui L, Wall K, et al. (2004) EST clustering error evaluation and correction. Bioinformatics 20: 2973–2984

48. Liang H, Carlson JE, Leebens-Mack JH, Wall PK, Mueller LA, et al. (2008) An EST database for Liriodendron tulipifera L. floral buds: the first EST resource for functional and comparative genomics in Liriodendron. Tree Genetics & Genomes 4: 419–433

49. Mittapalli O, Bai X, Mamidala P, Rajarapu SP, Bonello P, et al. (2010) Tissue-specific transcriptomics of the exotic invasive insect pest emerald ash borer (Agrilus planipennis). PLoS One 5: e13708.

50. Tatusov RL, Natale DA, Garkavtsev IV, Tatusova TA, Shankavaram UT, et al. (2001) The COG database: new developments in phylogenetic classification of proteins from complete genomes. Nucleic Acids Research 29: 22–28

51. Harris M, Clark J, Ireland A, Lomax J, Ashburner M, et al. (2004) The Gene Ontology (GO) database and informatics resource. Nucleic Acids Research 32: D258–261

52. Ogata H, Goto S, Sato K, Fujibuchi W, Bono H, et al. (1999) KEGG: Kyoto encyclopedia of genes and genomes. Nucleic Acids Research 27: 29–34

53. Ewen-Campen B, Shaner N, Panfilio KA, Suzuki Y, Roth S, et al. (2011) The maternal and early embryonic transcriptome of the milkweed bug Oncopeltus fasciatus. BMC genomics 12: 61.

54. Jung H, Lyons RE, Hurwood DA, Mather PB (2013) Genes and growth performance in crustacean species: a review of relevant genomic studies in crustaceans and other taxa. Reviews in Aquaculture 5: 77–110

55. Martin Marti S, Onteru S, Du Z, Rothschild MF (2010) Short communication. SNP analyses of the 5HT1R and STAT genes in Pacific white shrimp, Litopenaeus vannamei. Spanish Journal of Agricultural Research, 2010, vol 8 , núm 1, p 53–55.

56. Glenn K, Grapes L, Suwanasopee T, Harris D, Li Y, et al. (2005) SNP analysis of AMY2 and CTSL genes in Litopenaeus vannamei and Penaeus monodon shrimp. Animal genetics 36: 235–236

57. Wang X, Meng X, Song B, Qiu X, Liu H (2010) SNPs in the myostatin gene of the mollusk *Chlamys farreri*: Association with growth traits. Comparative Biochemistry and Physiology Part B: Biochemistry and Molecular Biology 155: 327–330

58. Tangprasittipap A, Tiensuwan M, Withyachumnarnkul B (2010) Characterization of candidate genes involved in growth of black tiger shrimp Penaeus monodon. Aquaculture 307: 150–156

59. Jung H, Lyons RE, Hurwood DA, Mather PB (2013) Genes and growth performance in crustacean species: a review of relevant genomic studies in crustaceans and other taxa. Reviews in Aquaculture.

60. Chang ES, Mykles DL (2011) Regulation of crustacean molting: a review and our perspectives. General and Comparative Endocrinology 172: 323–330

61. Chung JS, Zmora N, Katayama H, Tsutsui N (2010) Crustacean hyperglycemic hormone (CHH) neuropeptides family: Functions, titer, and binding to target tissues. General and Comparative Endocrinology 166: 447–454

62. Nakatsuji T, Lee C-Y, Watson RD (2009) Crustacean molt-inhibiting hormone: structure, function, and cellular mode of action. Comparative Biochemistry and Physiology Part A: Molecular & Integrative Physiology 152: 139–148

63. Thanh NM, Barnes AC, Mather PB, Li Y, Lyons RE (2010) Single nucleotide polymorphisms in the actin and crustacean hyperglycemic hormone genes and

their correlation with individual growth performance in giant freshwater prawn *Macrobrachium rosenbergii*. Aquaculture 301: 7–15

64. Devaraj H, Natarajan A (2006) Molecular mechanisms regulating molting in a crustacean. FEBS Journal 273: 839–846
65. Kuballa A, Elizur A (2007) Novel molecular approach to study moulting in crustaceans. BULLETIN-FISHERIES RESEARCH AGENCY JAPAN 20: 53
66. Dominguez R, Holmes KC (2011) Actin structure and function. Annual review of biophysics 40: 169
67. Zhu X-J, Dai Z-M, Liu J, Yang W-J (2005) Actin gene in prawn, *Macrobrachium rosenbergii*: characteristics and differential tissue expression during embryonic development. Comparative Biochemistry and Physiology Part B: Biochemistry and Molecular Biology 140: 599–605
68. Kim BK, Kim KS, Oh C-W, Mykles DL, Lee SG, et al. (2009) Twelve actin-encoding cDNAs from the American lobster, *Homarus americanus*: Cloning and tissue expression of eight skeletal muscle, one heart, and three cytoplasmic isoforms. Comparative Biochemistry and Physiology Part B: Biochemistry and Molecular Biology 153: 178–184
69. Hooper SL, Thuma JB (2005) Invertebrate muscles: muscle specific genes and proteins. Physiological reviews 85: 1001–1060
70. Cesar JRO, Yang J (2007) Expression patterns of ubiquitin, heat shock protein 70, α-actin and β-actin over the molt cycle in the abdominal muscle of marine shrimp Litopenaeus vannamei. Molecular reproduction and development 74: 554–559
71. Harrington WF, Rodgers ME (1984) Myosin. Annual review of biochemistry 53: 35–73
72. Kamimura MT, Meier KM, Cavalli RO, Laurino J, Maggioni R, et al. (2008) Characterization of growth-related genes in the south-western Atlantic pink shrimp Farfantepenaeus paulensis (Pérez-Farfante 1967) through a modified DDRT-PCR protocol. Aquaculture Research 39: 200–204
73. McPherron AC, Lawler AM, Lee S-J (1997) Regulation of skeletal muscle mass in mice by a new TGF-p superfamily member.
74. De Santis C, Wade NM, Jerry DR, Preston NP, Glencross BD, et al. (2011) Growing backwards: an inverted role for the shrimp ortholog of vertebrate myostatin and GDF11. The Journal of experimental biology 214: 2671–2677
75. Liu Z, Cordes J (2004) DNA marker technologies and their applications in aquaculture genetics. Aquaculture 238: 1–37
76. Chen C, Zhou P, Choi YA, Huang S, Gmitter FG (2006) Mining and characterizing microsatellites from citrus ESTs. TAG theoretical and applied genetics 112: 1248–1257
77. Kantety RV, La Rota M, Matthews DE, Sorrells ME (2002) Data mining for simple sequence repeats in expressed sequence tags from barley, maize, rice, sorghum and wheat. Plant Molecular Biology 48: 501–510
78. Morton BR, Bi IV, McMullen MD, Gaut BS (2006) Variation in mutation dynamics across the maize genome as a function of regional and flanking base composition. Genetics 172: 569–577
79. Kim W-J, Jung H, Gaffney PM (2011) Development of type I genetic markers from expressed sequence tags in highly polymorphic species. Marine Biotechnology 13: 127–132
80. Ellis J, Burke J (2007) EST-SSRs as a resource for population genetic analyses. Heredity 99: 125–132

CHAPTER

5

COMPARATIVE ANALYSIS OF CATFISH BAC END SEQUENCES WITH THE ZEBRAFISH GENOME

Hong Liu, Yanliang Jiang, Shaolin Wang, Parichart Ninwichian, Benjaporn Somridhivej, Peng Xu, Jason Abernathy, Huseyin Kucuktas and Zhanjiang Liu

BACKGROUND

Comparative mapping is a powerful tool to transfer genomic information from sequenced genomes to closely related species for which whole genome sequence data are not yet available. Such an approach was initially demonstrated by Fujiyama *et al.* [1] for the construction of the human-chimpanzee comparative map. In these closely related primate species, approximately 98% of chimpanzee BAC end sequences (BES) had significant BLAST hits to the human genome sequence allowing putative orthologues to be identified [1]. A similar approach was used for the construction of the human-mouse comparative map [2]. Subsequently, this approach was extensively used in mammals including construction of the human-cattle, the human-horse, and the human-porcine comparative maps [3, 4, 5]. Most recently, this approach was utilized one step further for the construction of the comparative genome contig (CGC)-based physical map of the sheep genome [6], where CGC is established based on anchorage of the sheep BES onto the genome sequences of dog, cow, and human. These successes depended on high percentage of BLAST hits and/or high levels of genome collinearity.

Five teleost fish genomes have been fully sequenced http://www.ensembl.org/index.html including zebrafish (*Danio rerio*, from the order of Cypriniformes), Japanese pufferfish (*Fugu rubripes*, from the order of Tetraodontiformes), green spotted pufferfish (*Tetraodon nigroviridis*, from the order of Tetraodontiformes), medaka (*Oryzias latipes*, from the order of Beloniformes), and three-spined stickleback (*Gasterosteus aculeatus*, from the order of Gasterosteiformes), while whole genome sequencing is also underway for tilapia http://www.cichidgenome.org; http://www.broad.mit.edu/science/projects/mammals-models/vertebrates-invertebrates/tilapia/tilapia-genome-sequencing-project. The availability of these whole genome sequences lends great opportunities for comparative genome analysis. Recently, major genomic resources have been developed from a number of fish species such as Atlantic salmon (*Salmo salar*) [7, 8, 9], rainbow trout (*Oncorhynchus mykiss*) [10, 11], tilapia [12, 13], gilthead sea bream (*Sparus auratus*) [14, 15, 16, 17], European sea bass (*Dicentrarchus labrax*) [18, 19], and channel catfish (*Ictalurus punctatus*) (for a review, see [20, 21]).

Catfish is the major aquaculture species in the United States. It is one of the six species included in the U.S. National Animal Genome Project NRSP-8. A number of genome resources have been developed in catfish including a large number of molecular markers [22, 23, 24, 25], genetic linkage maps [26, 27, 28], several hundred thousands of ESTs ([29, 30, 31, 32, 33]; Z. Liu, unpublished data), microarray platforms [34, 35, 36, 37, 38], BAC libraries [39, 40], and BAC-based physical maps [41, 42]. To enable BAC end sequence-based comparative genome analysis, we previously reported generation of 20,366 BES in catfish [25]. In spite of the great value of those BES for the characterization of genome repeat structures [43] and for the identification of microsatellite markers, our previous comparative genome analysis using BES revealed very limited conservation between the catfish and zebrafish genomes. Of the 141 mate-paired BES with genes on both ends of the BAC inserts, only 34 (24.1%) were found in nearby genomic locations in the zebrafish genome,

Citation: Hong Liu, Yanliang Jiang, Shaolin Wang, Parichart Ninwichian, Benjaporn Somridhivej, Peng Xu, Jason Abernathy, Huseyin Kucuktas and Zhanjiang LiuEmail, Comparative analysis of catfish BAC end sequences with the zebrafish genome, doi: 10.1186/1471-2164-10-592

suggesting high levels of chromosomal rearrangements [40]. Such findings were in strong contrast to the situations found between medaka-sea bream, *Tetraodon*-sea bream, medaka-stickleback, *Tetraodon*-medaka, stickleback-sea bream, *Tetraodon*-stickleback genome comparisons where almost complete genome collinearities were found [44]. We speculated that our earlier inability to discover large extent of genome collinearity between catfish and zebrafish could be a result of the low numbers of BES and the lack of a physical map. Therefore, in this study, we extended our efforts in BAC end sequencing and generated additional 43,021 BES, bringing the total to 63,387 (25,676 mate-paired). Using these catfish BES and its BAC contig-based physical map [42], genetic linkage mapping of BAC end-anchored microsatellites, and the genome sequence of zebrafish, here we conducted extensive comparative genome analysis. We report the identification of conserved syntenies and demonstrate the construction of super scaffolds of contigs by genetic linkage mapping of BAC end-associated microsatellites.

RESULTS AND DISCUSSION

BAC end sequencing

As shown in Table 1, a total of 42,240 BAC inserts (6.13× clone-coverage of the channel catfish genome) were sequenced from both ends, resulting in 63,387 BES ≥ 200 bp in length (75% overall success rate), including 20,366 BES we previously reported [25]. Mate-paired BES were produced from 25,676 BAC clones, while only a single BES was obtained from 12,035 clones. The BES were of high quality as the Q20 length ranged from 200 to 810 bp, with an average Q20 read length of 596 bp. All these BES have been deposited into the GenBank GSS database with consecutive accession numbers of [Gen Bank: FI857756-FI900776]. A total of 37,784,877 bp of genomic sequences was generated from this study, representing approximately 4% of the catfish genome. Analysis using the 37,784,877 bp BES resulted in 11.91% of base pairs masked using the *Danio* repeat database, with the most abundant type of repeat being the DNA transposons. We previously reported the assessment of repetitive elements in the catfish genome and the additional 43,021 BES generated in this study confirmed our previous findings in general [25]. These BES [Gen Bank: DX083364-DX103729] were also used for comparative genome analysis in this study.

Table 1: A summary of BAC end sequences

Category	Numbers
BAC sequence reactions	84,480
Total clean sequences	63,387 (75% success)
T7 sequences	32,074
SP6 sequences	31,313
Pair BAC end sequences	25,676
Total length sequenced	37,784,877 bp
Average length	596 bp

In silico analysis of the BAC-associated catfish genes on the zebrafish genome

TBLASTX searches using the 63,387 catfish BES against the ENSEMBL zebrafish cDNA database with chromosome information resulted in 5,066 significant hits (Table 2). Of the 5,066 significant hits, 2,197 unique zebrafish genes were hit by a single BES while 1,024 unique zebrafish genes were hit by two or more catfish BES, making a total of 3,221 unique zebrafish genes with significant hits from the catfish BES. The 3,221 genes cover all 25 zebrafish chromosomes, with the largest number of gene hits being located on chromosome 5 (224 significant hits), followed by chromosome 7 (191 significant hits), chromosome 20 (171 significant hits), chromosome 6 (151 significant hits) and chromosome 19 (134 significant hits); and the smallest number of gene hits on chromosome 24 with 78 hits (Table 2). The number of gene hits on various chromosomes was approximately proportional to the sizes of the zebrafish chromosomes with some exceptions. When the size of chromosomes was taken into consideration, chromosome 25 had the largest number of gene hits with 3.5 hits per Mb or one hit per 286 kb on average, followed by chromosome 5, 4, 20, 19, and 22 with 3.2, 3.1, 3.0, 2.9, and 2.9 hits per Mb, respectively (Table 2).

Table 2: Distribution of comparatively anchored BAC clones using protein encoding gene sequences only.

Zebrafish chromosome	Chromosome size (Mb)	No. of protein encoding genes*	No. of tBLASTx hits	Hits to unique genes	Unique gene hits per Mb	No. of contigs with single gene hits	No. of contigs with multiple gene hits	No. putative micro-syntenies
1	56.2	818	205	123	1.83	75	17	13
2	54.4	875	194	133	2.13	85	15	13
3	62.9	975	196	127	1.75	72	18	13
4	42.6	743	221	130	2.77	78	16	9
5	70.4	1,173	340	224	2.74	103	35	21
6	59.2	818	232	151	2.31	85	22	18
7	70.3	990	283	191	2.33	80	34	25
8	56.5	864	196	128	1.93	65	22	14
9	51.5	700	212	133	2.17	61	21	16
10	42.4	670	150	87	1.72	51	12	9

11	44.6	627	161	121	2.26	67	16	10
12	47.5	636	177	114	2.21	72	14	7
13	53.5	744	200	113	1.96	67	21	14
14	56.5	701	197	113	1.77	84	11	7
15	46.6	688	177	125	2.25	68	14	8
16	53.1	773	181	124	2.02	77	14	10
17	52.3	715	180	115	1.99	62	19	13
18	49.3	749	193	121	2.23	47	22	21
19	46.2	780	233	134	2.58	86	21	16
20	56.5	1,053	277	171	2.48	76	30	19
21	46.1	721	163	117	2.28	63	14	10
22	39.0	959	178	113	2.59	50	19	16
23	46.4	669	204	121	2.24	68	18	13
24	40.3	513	117	78	1.71	47	8	7
25	32.9	597	199	114	3.04	65	19	14
Total/ Average	**1,277.2**	**19,551**	**5,066**	**3,221**	**2.21**	**1,754**	**472**	**336**

*: Annotated genes only from ENSEMBL.

One particular finding of these BLAST searches is the observation of many highly repetitive genes. Out of 3,221 unique genes, 1,024 genes had hits from two or more BES. A single gene identity had hits from as many as 31 BES. A total of 14 genes had hits from at least 10 BES each (Table 3); an additional 139 genes had hits from 4-9 BES each; 230 genes had hits from 3 BES each, and 641 genes had hits from 2 BES each (Table 3). Some of the genes with hits from multiple BES may represent a whole array of related genes with similar functional domains. For instance, 18 BES hit NOD3-like gene of channel catfish, which was just recently characterized; NOD3 gene existed as a single copy gene in the catfish genome [45], and apparently the multiple BES contained many related genes harboring domains present within the NOD3 gene. Theoretically, a fraction of genes should have hits by more than one BES, simply because of the genome coverage of the BAC clones. We believe that overlapping (including identical) BAC clones does account for some of the observed hits of genes by more than one BES (data not shown), especially for those with 2-3 BES hits. However, the mathematical chances do not support multiple BES hits of a single gene unless the gene itself is repetitive in the catfish genome. Additional research is warranted to fully understand the nature of these genes/sequences in the catfish genome, but clearly many of these represent classes of repetitive gene families such as DNA polymerase gene that had hits from 31 BES.

Table 3: Distribution of genes with hits from multiple BAC end sequences, with details provided for genes with 10 or more hits from BAC end sequences

No. of Genes	**Putative Identities**	***No of BES hits***	**Presence in Zebrafish genome**	**Potential explanation**
1	Novel protein similar to DNA polymerases	31	28	Repetitive elements related to retroelements
1	Methionine amino-peptidase 1	22	2	Repetitive elements or multi-gene family
1	NOD3 protein-like	18	63	Common domains shared by many related proteins
1	Similar to tudor domain containing 7, hypothetical protein LOC393661	17	89	Repetitive elements or repetitive genes
1	Similar to porf2	16	81	Repetitive elements or multi-gene family
1	Similar to general transcription factor II-I repeat domain-containing protein 2A	16	82	Repetitive elements or multi-gene family

1	Similar to novel G protein-coupled receptor	13	92	Repetitive elements or multi-gene family
1	Similar to serine/threonine-protein kinase pim-3;	11	85	Repetitive elements or multi-gene family
1	Similar to novel protein from *Danio rerio*;	11	85	Repetitive elements or multi-gene family
1	Similar to Dynein heavy chain 6	11	20	Repetitive elements or multi-gene family
1	ORF2 [*Mus musculus domesticus*]	10	91	Repetitive elements or multi-gene family
1	PREDICTED: tubulin, alpha, ubiquitous isoform 8 [*Macaca mulatta*]	10	16	Repetitive elements or multi-gene family
1	PREDICTED: similar to vacuolar protein sorting 52 [*Danio rerio*]	10	69	Repetitive elements or multi-gene family
1	GF20795 [*Drosophila ananassae*]	10	4	Repetitive elements or multi-gene family
14		**Subtotal**		
68		5-9		Repetitive elements or multi-gene family
71		4		Repetitive elements or multi-gene family
139		**Subtotal**		
230		3		Potentially duplicated gene candidates
641		2		Potentially duplicated gene candidates
1024		**Total**		

Establishing Microsyntenies

Among the teleost genomes with high sequence coverage, zebrafish is the most closely related species to catfish [46]. Our initial BLAST searches of the catfish BES against the genome of the *T. nigroviridis* generated many fewer significant hits compared to those against the zebrafish genome. Therefore, we concentrated our comparative analysis efforts with the zebrafish genome in this study.

Conserved syntenies are most often established by comparing genome sequences of related species. However, the whole genome sequence is not yet available from catfish. In the absence of the whole genome sequence, we attempted to establish microsyntenies based on physical linkage of gene sequences. With the genome resources available in catfish, we have taken three approaches. First, if the genes were identified from both ends of a single BAC clone, they are physically linked with a distance of the BAC clone insert size. If the same two genes are found linked in the zebrafish genome in the same genome neighborhood, a microsynteny can then be established. These genes from mate-paired BES are physically linked with the average distances between them being the average insert size of the catfish BAC library, i.e., 161 kb. From the 63,387 BES, a total of 25,676 mate-paired BES were identified. Of these, 760 mate-paired BES had significant BLASTN hits against the zebrafish genome sequence. However, only 194 of the 760 significant hit pairs were on the same zebrafish chromosome, allowing syntenic comparison. Further tBLASTX searches against the ENSEMBL zebrafish cDNA database allowed identification of 95 mate-paired BES with genes on both sides. The genomic locations of these 95 mate-paired genes were determined from the zebrafish genome sequence. Fifty pairs were found to be present in neighboring genomic locations within one million base pairs, while the other 45 were present in more distant locations (> 1 Mb) on the same chromosomes. The vast majority of the 50 mate-paired genes were found to be within 500 kb on the zebrafish genome sequence; only 2 of the 50 pairs had a distance of 500-920 kb (Table 4), suggesting conserved syntenies of the involved genes.

Table 4: Summary of 50 conserved syntenies identified by comparison of 95 mate-paired genes of channel catfish with genomic locations of those on the zebrafish draft genome sequence.

Catfish BAC ID	SP6 hits	T7 hits	zebrafish Chr	Chr location (Mb)	Distance (bp)
035I16	[GenBank:NP_001038735]	[GenBank:XP_689146]	2	38.61	18,048
063L16	[GenBank:NP_001012377]	[GenBank:NP_001092217]	2	19.25	48,679
026B05	[GenBank:XP_001921249]	[GenBank:CAI20867]	3	31.82	4,418
098D05	[GenBank:CAI11873]	[GenBank:NP_001038462]	4	1.82	301,470
007M06	[GenBank:XP_001333162]	[GenBank:NP_571105]	5	5.46	379,579
028M04	[GenBank:XP_687685]	[GenBank:NP_958867]	5	22.23	133,063
074L07	[GenBank:XP_001334912]	[GenBank:XP_687570]	5	15.69	290,598
035N11	[GenBank:NP_001032187]	[GenBank:NP_958882]	6	18.25	231,524
040J24	[GenBank:XP_686613]	[GenBank:NP_991309]	6	7.37	443,524
077F04	[GenBank:NP_001034906]	[GenBank:NP_001038813]	6	11.42	484,877
032F20	[GenBank:XP_691291]	[GenBank:NP_001075159]	7	28.40	244,425
062B16	[GenBank:NP_001008651]	[GenBank:NP_001017550]	7	15.05	222,416
075K10	[GenBank:NP_956505]	[GenBank:NP_001070187]	7	13.34	268,455
103I21	[GenBank:NP_998033]	[GenBank:NP_001159825]	7	22.34	293,845
047N13	[GenBank:XP_699919]	[GenBank:XP_699627]	8	23.84	203,260
057N22	[GenBank:XP_001923800]	[GenBank:NP_997066]	9	36.69	163,677
076H22	[GenBank:NP_001004563]	[GenBank:XP_001921910]	9	13.84	313,352
105N24	[GenBank:NP_001122018]	[GenBank:NP_997992.2]	9	27.37	268,271
056A23	[GenBank:NP_998295]	[GenBank:NP_001103164]	10	7.50	486,745
068C21	[GenBank:NP_001019272]	[GenBank:XP_001920077]	10	17.72	104,067
093K02	[GenBank:XP_001344325]	[GenBank:XP_001919728]	10	41.67	152,479
096A14	[GenBank:NP_001120805]	[GenBank:NP_997775]	11	10.02	980,276
010B15	[GenBank:NP_956715]	[GenBank:NP_956756]	12	25.17	111,625
041B24	[GenBank:NP_001002607]	[GenBank:XP_693784]	12	13.59	398,886
109K22	[GenBank:XP_001920588]	[GenBank:XP_001920550]	12	25.55	449,096
018H11	[GenBank:NP_956915]	[GenBank:NP_997743]	13	23.23	293,181
026C08	[GenBank:NP_001119869]	[GenBank:NP_956611]	13	23.50	201,663
027E09	[GenBank:XP_001922173]	[GenBank:XP_001921791]	14	0.22	574,761
103F19	[GenBank:NP_571477]	[GenBank:XP_001919973]	14	22.90	256,143
022D09	[GenBank:NP_001096112]	[GenBank:XP_684421]	15	29.14	418,219
059B20	[GenBank:XP_682817]	[GenBank:XP_691794]	15	27.80	94,396

077H08	[GenBank:XP_001920240]	[GenBank:NP_001070609]	15	6.05	339,215
004O03	[GenBank:NP_001020707]	[GenBank:NP_001020642]	16	10.40	149,618
075M09	[GenBank:NP_001107266]	[GenBank:NP_001082899]	16	8.30	120,355
104I08	[GenBank:NP_571871]	[GenBank:NP_694503]	16	27.17	214,517
013P16	[GenBank:NP_001076304]	[GenBank:NP_001038173]	18	18.65	366,849
042J20	[GenBank:NP_001037796]	[GenBank:NP_001038370]	18	32.85	223,781
021L13	[GenBank:NP_001038343]	[GenBank:XP_688911]	19	17.40	286,019
052N18	[GenBank:XP_001921158]	[GenBank:NP_956134]	19	7.59	345,553
056A09	[GenBank:NP_001018463]	[GenBank:NP_001038416]	19	13.46	122,217
065M02	[GenBank:XP_001922809]	[GenBank:NP_001038686]	19	15.37	430,325
080O21	[GenBank:NP_956277]	[GenBank:NP_571262]	19	17.06	389,643
086C16	[GenBank:XP_001340912]	[GenBank:NP_001038562]	19	6.00	364,590
010H22	[GenBank:NP_998602]	[GenBank:XP_001920851]	21	2.24	240,019
034I14	[GenBank:NP_001038838]	[GenBank:NP_956334]	21	20.16	306,102
081J10	[GenBank:NP_001002411]	[GenBank:NP_001073439]	21	22.12	277,818
099H22	[GenBank:NP_001076277]	[GenBank:XP_699221]	22	7.03	115,754
053P11	[GenBank:NP_001002173]	[GenBank:NP_001035137]	23	30.73	197,703
068O10	[GenBank:XP_001339131]	[GenBank:XP_689922]	23	18.01	332,973
105H10	[GenBank:NP_001103636]	[GenBank:NP_001098596]	23	33.36	274,318
Average					**278,648**

We previously reported the relatively high levels of local region conservation. For instance, many genes within the bordering mate-paired genes were well conserved among catfish, zebrafish, and *Tetraodon*, as determined by direct sequencing of the catfish BAC DNA using primers predicted from known genes in zebrafish or Tetraodon [40]. We did not extend this part of the study, but all known genomic information suggested high levels of local genome conservation.

In addition to the 50 microsyntenies, we attempted to determine if significant gene hits in the same catfish BAC contigs also fall on the same chromosome locations comparable to the contig sizes. As shown in Table 2, of the contigs with gene hits, 1,754 contigs had only one gene hit, while 472 contigs had two or more gene hits within each contig. Because the genes in the same contig are physically linked, their linkage in a comparable distance in the zebrafish genome would indicate a conserved synteny. As shown in Figure 1, 2, 3, 4 &5, the vast majority of gene hits within the same contigs were found to be located on the same zebrafish chromosomes with comparable distances as estimated from the catfish BAC contigs. Using such an approach, a total of 336 conserved microsyntenies was identified (Table 2). Presence of multiple gene hits within large BAC contigs would allow identification of extended large conserved syntenic regions. Many of the microsyntenies were conserved with extended genomic distance to span over several million base pairs (Figure 1, 2, 3, 4 &5, for additional details, see Additional file 1). For instance, large conserved syntenies were identified from chromosomes 12, 13, 14, 22, 23, 24, and 25 (Figure 1, 2, 3, 4 &5). In spite of the identification of some relatively large conserved syntenic regions, the vast majority of the identified syntenies were microsyntenies. Such highly segmented microsyntenies are not very useful for genome-wide comparative analysis. However, if scaffolds can be established by determining the relationships among the microsyntenies, large-scale genome comparison should be possible. We, therefore, used two zebrafish chromosomes as the query to demonstrate if super scaffolds can be established. Chromosome 7, one of the chromosomes with the highest number of significant gene hits, and chromosome 13, one of the chromosomes with a large number of contigs having two or more hits (indicative of high level of syntenic conservation), were chosen for further analysis using genetic linkage mapping.

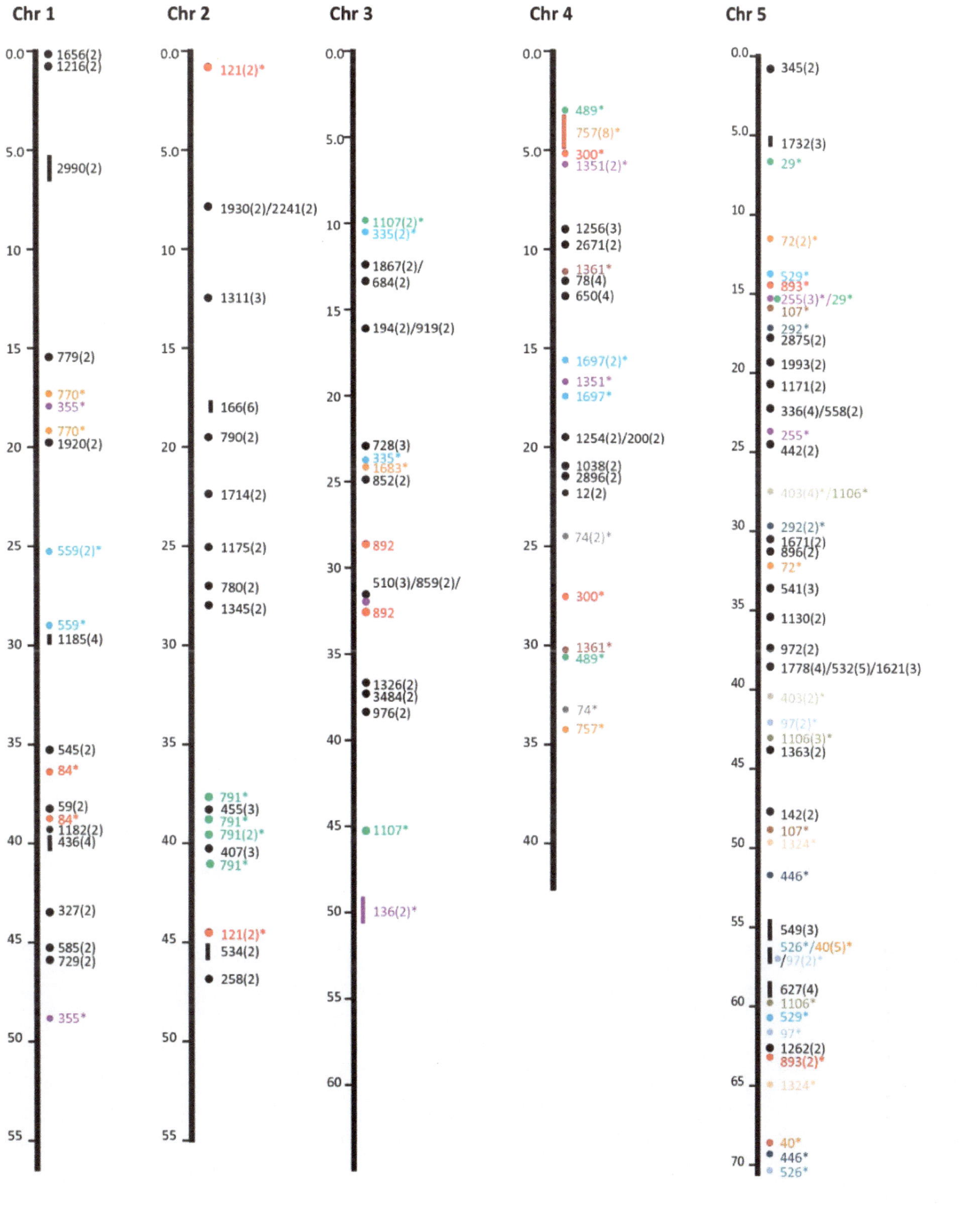

Figure 1: Identification of microsyntenies through comparative sequence analysis (chr 1 through chr 5). tBLASTX searches were conducted using BAC contig-associated BAC end sequences as queries against the zebrafish genome sequence. The putative conserved microsyntenies are presented along the 25 zebrafish chromosomes (chr 1 through chr 25). The position of the zebrafish sequence is shown on the left of each chromosome bars in million base pairs. The conserved microsyntenies are indicated on the right side of the chromosome bars, with the numbers representing the contig numbers of the BAC assembly of the catfish physical map [41]. Circles represent short syntenic regions and short vertical lines represent relatively longer conserved syntenic regions proportional to the length of the bar with a number in parenthesis representing the number of conserved sequences within the microsyntenies. The microsyntenies designated with asterisks (*) are those with duplicated conservation of the microsyntenies that are color-coded to facilitate the visualization of the duplicated syntenic regions along the chromosome. Duplicated syntenic regions refer to a conserved genomic segment between the catfish genome and the zebrafish genome that is duplicated in the zebrafish genome such that identical or nearly identical significant hits are generated from two chromosomal regions of the zebrafish genome using a single catfish genome segment (say it is a contig or a scaffold) as the query. In just few cases, this term is used in an extended fashion to include those that are tripled in the zebrafish genome.

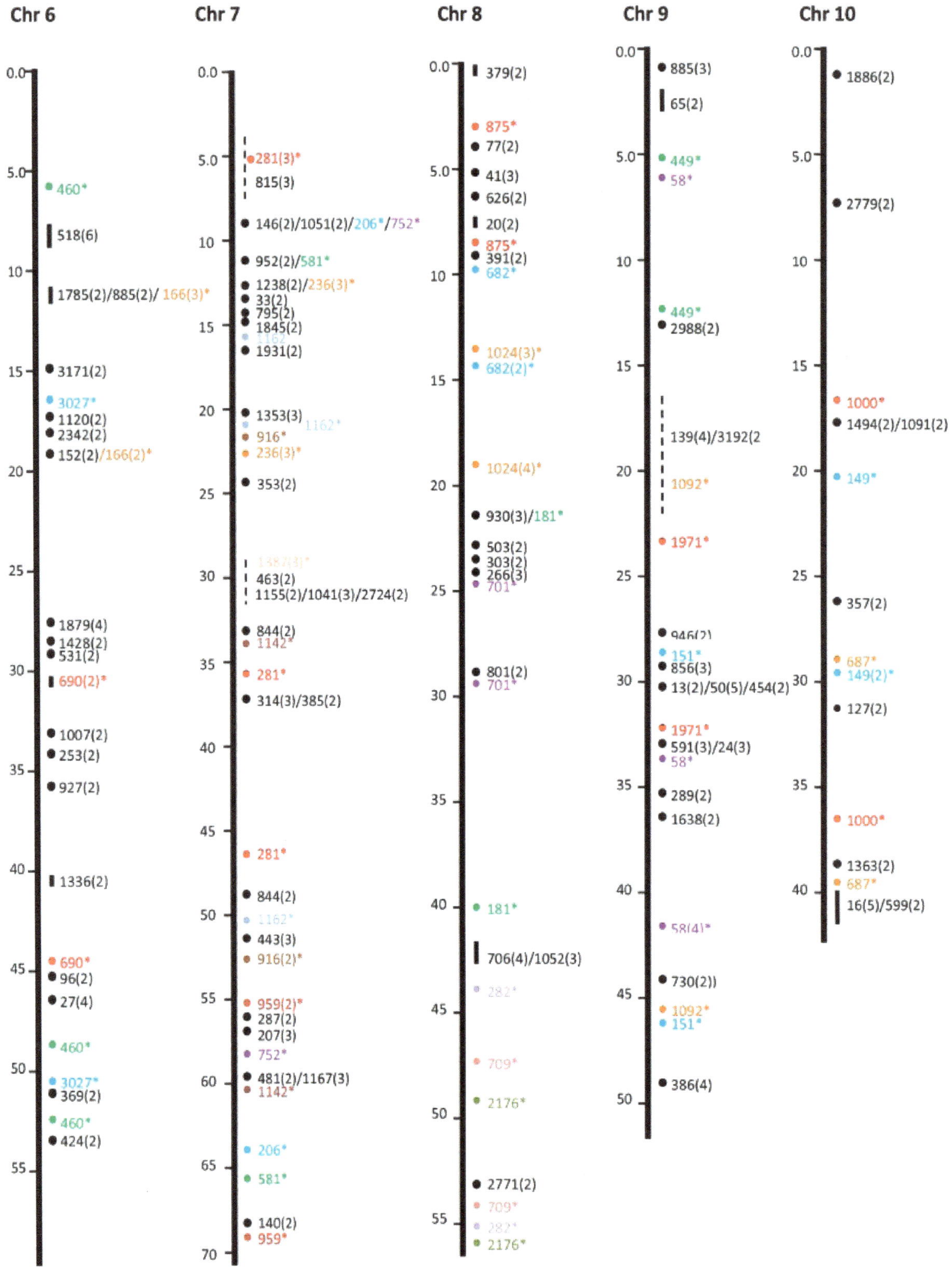

Figure 2: Identification of microsyntenies through comparative sequence analysis (chr 6 through chr 10). tBLASTX searches were conducted using BAC contig-associated BAC end sequences as queries against the zebrafish genome sequence. The putative conserved microsyntenies are presented along the 25 zebrafish chromosomes (chr 1 through chr 25). The position of the zebrafish sequence is shown on the left of each chromosome bars in million base pairs. The conserved microsyntenies are indicated on the right side of the chromosome bars, with the numbers representing the contig numbers of the BAC assembly of the catfish physical map [41]. Circles represent short syntenic regions and short vertical lines represent relatively longer conserved syntenic regions proportional to the length of the bar with a number in parenthesis representing the number of conserved sequences within the microsyntenies. The microsyntenies designated with asterisks (*) are those with duplicated conservation of the microsyntenies that are color-coded to facilitate the visualization of the duplicated syntenic regions along the chromosome. Duplicated syntenic regions refer to a conserved genomic segment between the catfish genome and the zebrafish genome that is duplicated in the zebrafish genome such that identical or nearly identical significant hits are generated from two chromosomal regions of the zebrafish genome using a single catfish genome segment (say it is a contig or a scaffold) as the query. In just few cases, this term is used in an extended fashion to include those that are tripled in the zebrafish genome.

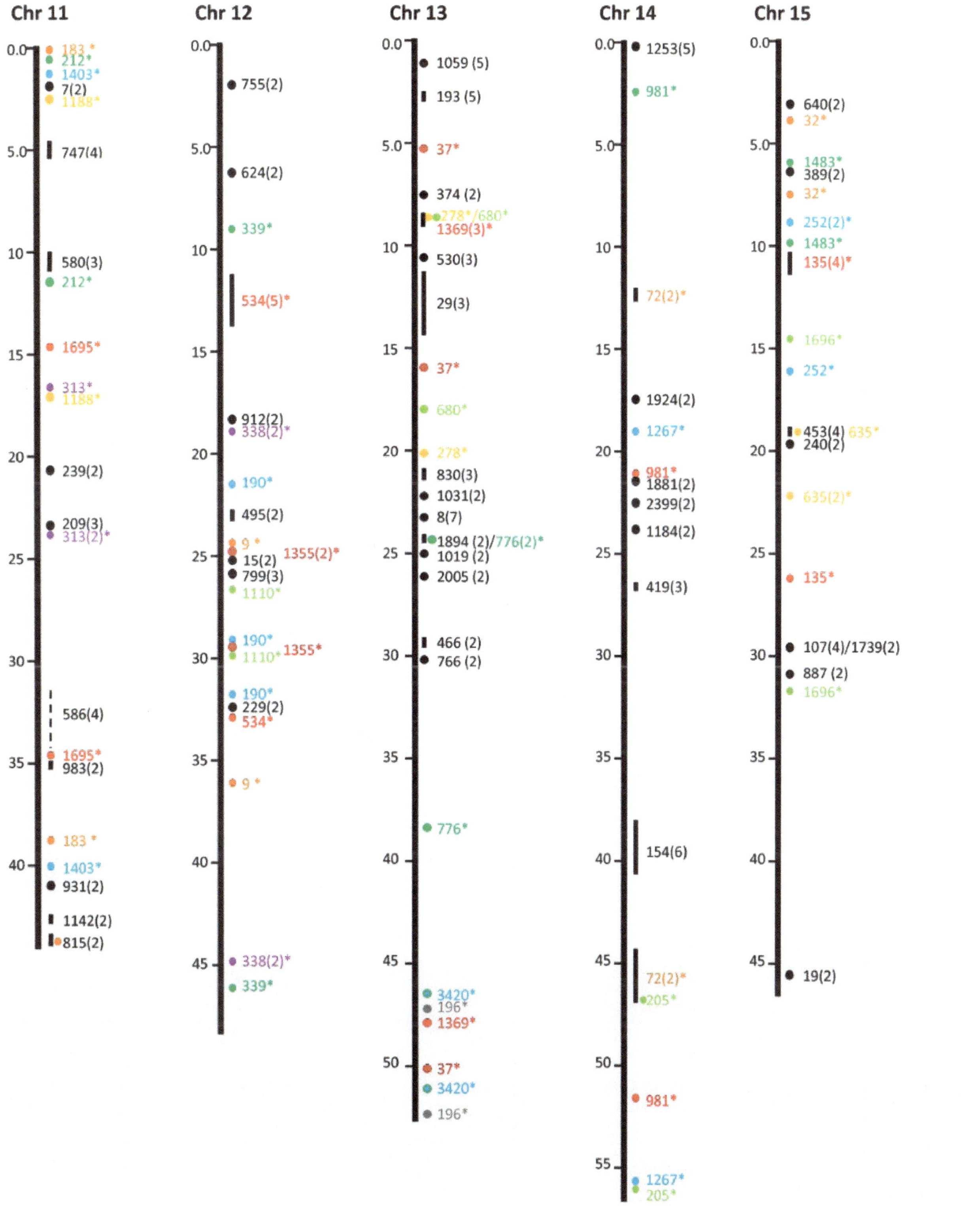

Figure 3: Identification of microsyntenies through comparative sequence analysis (chr 11 through chr 15). tBLASTX searches were conducted using BAC contig-associated BAC end sequences as queries against the zebrafish genome sequence. The putative conserved microsyntenies are presented along the 25 zebrafish chromosomes (chr 1 through chr 25). The position of the zebrafish sequence is shown on the left of each chromosome bars in million base pairs. The conserved microsyntenies are indicated on the right side of the chromosome bars, with the numbers representing the contig numbers of the BAC assembly of the catfish physical map [41]. Circles represent short syntenic regions and short vertical lines represent relatively longer conserved syntenic regions proportional to the length of the bar with a number in parenthesis representing the number of conserved sequences within the microsyntenies. The microsyntenies designated with asterisks (*) are those with duplicated conservation of the microsyntenies that are color-coded to facilitate the visualization of the duplicated syntenic regions along the chromosome. Duplicated syntenic regions refer to a conserved genomic segment between the catfish genome and the zebrafish genome that is duplicated in the zebrafish genome such that identical or nearly identical significant hits are generated from two chromosomal regions of the zebrafish genome using a single catfish genome segment (say it is a contig or a scaffold) as the query. In just few cases, this term is used in an extended fashion to include those that are tripled in the zebrafish genome.

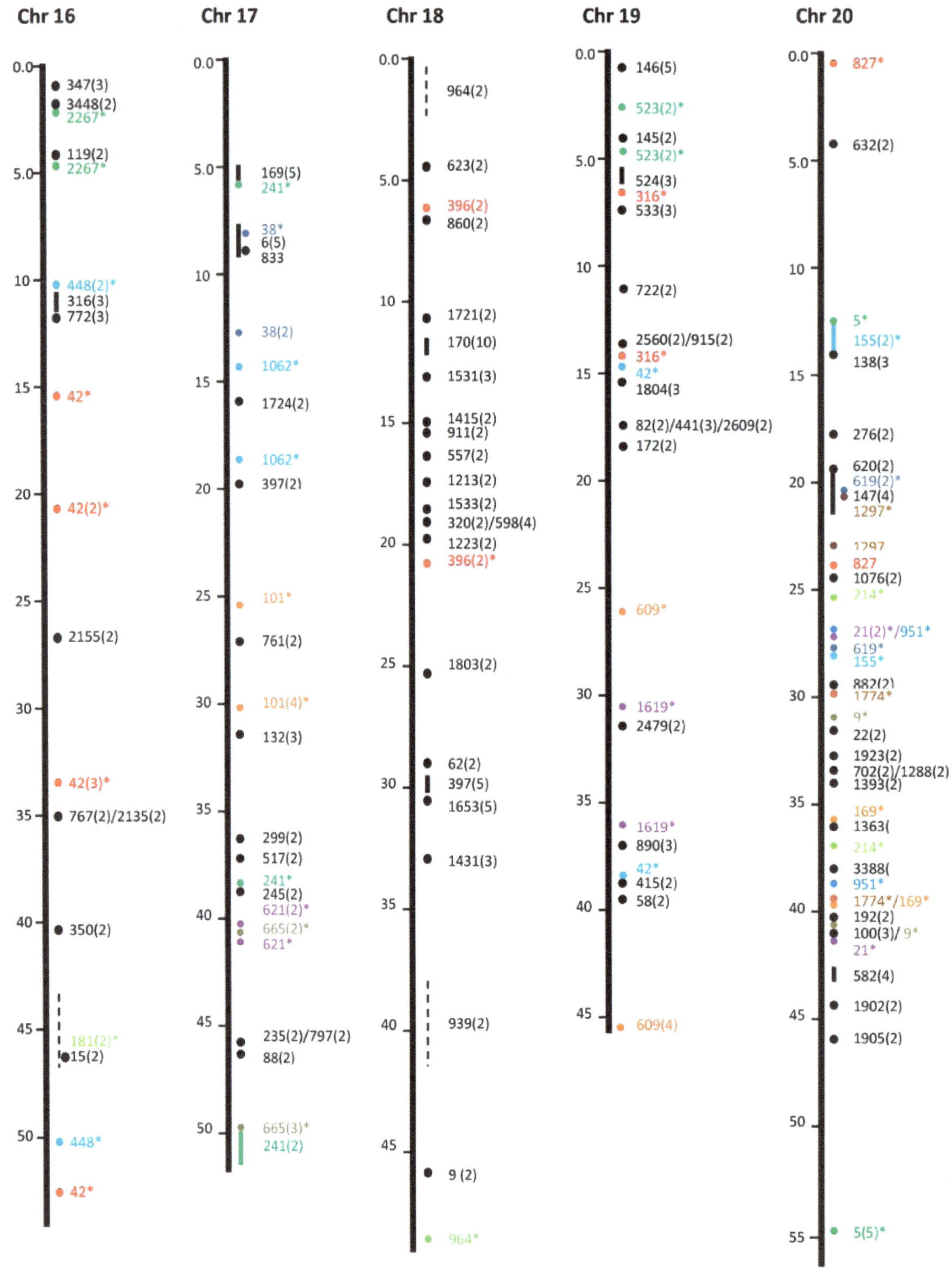

Figure 4: Identification of microsyntenies through comparative sequence analysis (chr 16 through chr 20). tBLASTX searches were conducted using BAC contig-associated BAC end sequences as queries against the zebrafish genome sequence. The putative conserved microsyntenies are presented along the 25 zebrafish chromosomes (chr 1 through chr 25). The position of the zebrafish sequence is shown on the left of each chromosome bars in million base pairs. The conserved microsyntenies are indicated on the right side of the chromosome bars, with the numbers representing the contig numbers of the BAC assembly of the catfish physical map [41]. Circles represent short syntenic regions and short vertical lines represent relatively longer conserved syntenic regions proportional to the length of the bar with a number in parenthesis representing the number of conserved sequences within the microsyntenies. The microsyntenies designated with asterisks (*) are those with duplicated conservation of the microsyntenies that are color-coded to facilitate the visualization of the duplicated syntenic regions along the chromosome. Duplicated syntenic regions refer to a conserved genomic segment between the catfish genome and the zebrafish genome that is duplicated in the zebrafish genome such that identical or nearly identical significant hits are generated from two chromosomal regions of the zebrafish genome using a single catfish genome segment (say it is a contig or a scaffold) as the query. In just few cases, this term is used in an extended fashion to include those that are tripled in the zebrafish genome.

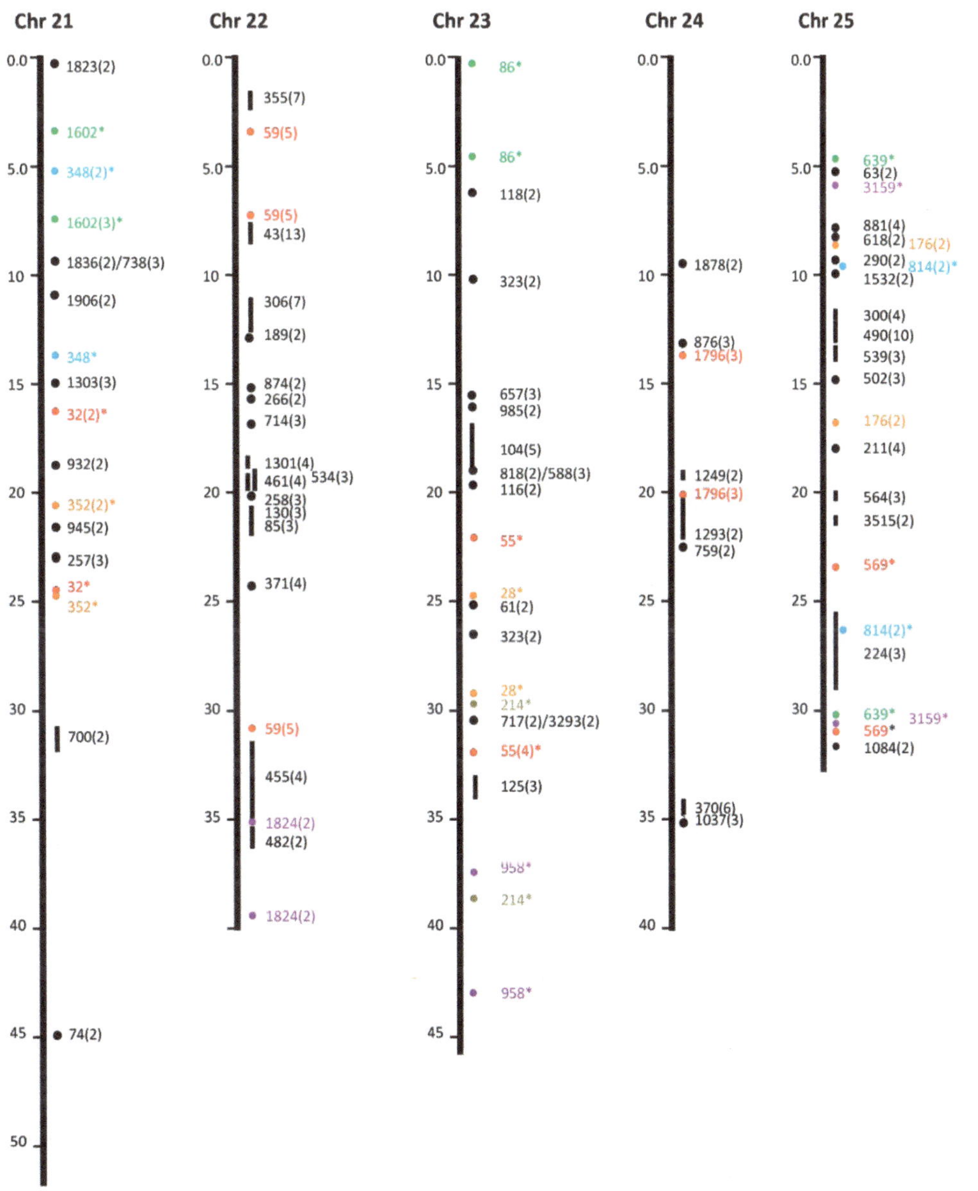

Figure 5: Identification of microsyntenies through comparative sequence analysis (chr 21 through chr 25). tBLASTX searches were conducted using BAC contig-associated BAC end sequences as queries against the zebrafish genome sequence. The putative conserved microsyntenies are presented along the 25 zebrafish chromosomes (chr 1 through chr 25). The position of the zebrafish sequence is shown on the left of each chromosome bars in million base pairs. The conserved microsyntenies are indicated on the right side of the chromosome bars, with the numbers representing the contig numbers of the BAC assembly of the catfish physical map [41]. Circles represent short syntenic regions and short vertical lines represent relatively longer conserved syntenic regions proportional to the length of the bar with a number in parenthesis representing the number of conserved sequences within the microsyntenies. The microsyntenies designated with asterisks (*) are those with duplicated conservation of the microsyntenies that are color-coded to facilitate the visualization of the duplicated syntenic regions along the chromosome. Duplicated syntenic regions refer to a conserved genomic segment between the catfish genome and the zebrafish genome that is duplicated in the zebrafish genome such that identical or nearly identical significant hits are generated from two chromosomal regions of the zebrafish genome using a single catfish genome segment (say it is a contig or a scaffold) as the query. In just few cases, this term is used in an extended fashion to include those that are tripled in the zebrafish genome.

Genetic mapping of BAC end-anchored microsatellites

In order to extend the scope of conserved microsyntenies, microsyntenies identified on zebrafish chromosomes 7 and 13 were genetically mapped to determine their chromosomal locations in the catfish genome. There were 373 significant BLASTN hits to zebrafish chromosome 13 involving 178 unique catfish BAC contigs; and 505 significant hits to zebrafish chromosome 7 involving 314 unique catfish BAC contigs. We, therefore, first identified microsatellites from these involved catfish BAC contigs, and then mapped them to the linkage groups when the microsatellites were polymorphic in the resource family. A total of 548 pairs of microsatellite primers were tested, of which 296 from 188 contigs (the details of the polymorphic markers are shown in the Additional file 2) were polymorphic in the resource family. Further analysis using JoinMap 4.0 allowed mapping of 290 microsatellite markers, of which 161 microsatellites were from BES with significant similarity to zebrafish chromosome 7, and 129 microsatellites were from BES with significant similarity to zebrafish chromosome 13.

Mapping of microsatellites from contigs with hits to zebrafish chromosome 13 allowed identification of a highly conserved chromosome between catfish and zebrafish. As shown in Figure 6, of the 129 microsatellites from BES with high similarities to the zebrafish chromosome 13, 57 microsatellites from 43 contigs were mapped into a single linkage group, spanning approximately 90 centi-Morgans, suggesting the conservation of a large segment of this chromosome. However, the entire chromosome is not conserved. The 129 microsatellites were mapped to a total of 24 linkage groups, with seven of the 24 linkage groups containing 4-12 markers (see Additional file 2).

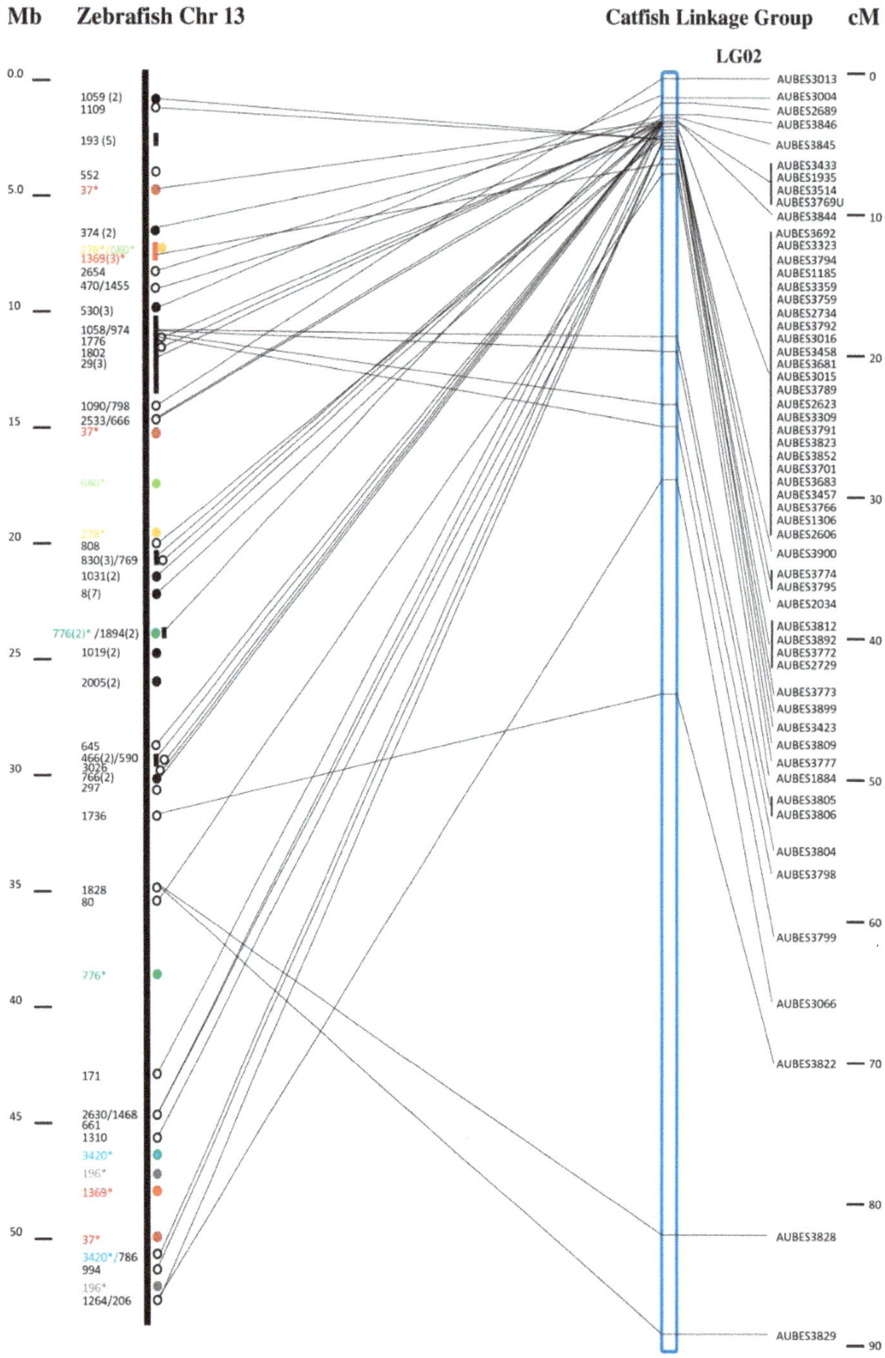

Figure 6: Scaffolds of conserved syntenic regions between the catfish and zebrafish genomes. Scaffolds of conserved syntenic regions were established by genetic linkage mapping of BAC contig-associated microsatellites. The zebrafish chromosome 13 (chr 13) is presented with its base positions on the far left in million base pairs; The second column of the numbers are catfish BAC contig numbers [41], with the identified syntenic regions shown immediately right of the chromosome bar. The numbers in the parenthesis are the number of conserved sequences; the circles and bars represent relatively short and long conserved syntenic regions; the asterisks represent duplicated syntenic regions with color coding to facilitate the visualization of duplicated regions, the same way as described under Figure 1 legend, except that the open circles represent conserved sequences coming from non-gene sequences while the solid circles represent conserved gene sequences. Microsatellites from the BAC contigs were genetically mapped to linkage groups as shown on the right, with the names of microsatellites being labeled on the second most right, e.g., AUBES1884. The positional relationship of the conserved syntenies on the zebrafish genome sequence and within the catfish linkage group is indicated by thin lines linking the zebrafish chromosome and the catfish linkage group positions. The positions of markers within the linkage group are shown on the furthest right in centi-Morgans.

Similarly but to a much lesser extent, microsatellites from BES with similarities to the zebrafish chromosome 7 were mapped to three major linkage groups (Figure 7). Once again, many smaller syntenic regions were mapped to various linkage groups, suggesting high levels of local conservation and low levels of chromosomal conservation. Nonetheless, the significant aspect of this is that scaffolds can be established by linking various contigs together through linkage mapping. This will allow integration of genetic linkage and physical maps once microsatellites are identified from most contigs of the physical map. Such scaffolds should guide genome sequence assembly in the future, and should also provide molecular length measurements of various polymorphic markers along the genome of catfish, providing guidance for the development of the SNP chip technology in catfish. Apparently, SNP chips constructed from evenly distributed SNPs provide the best coverage of the catfish genome when conducting the whole genome association studies.

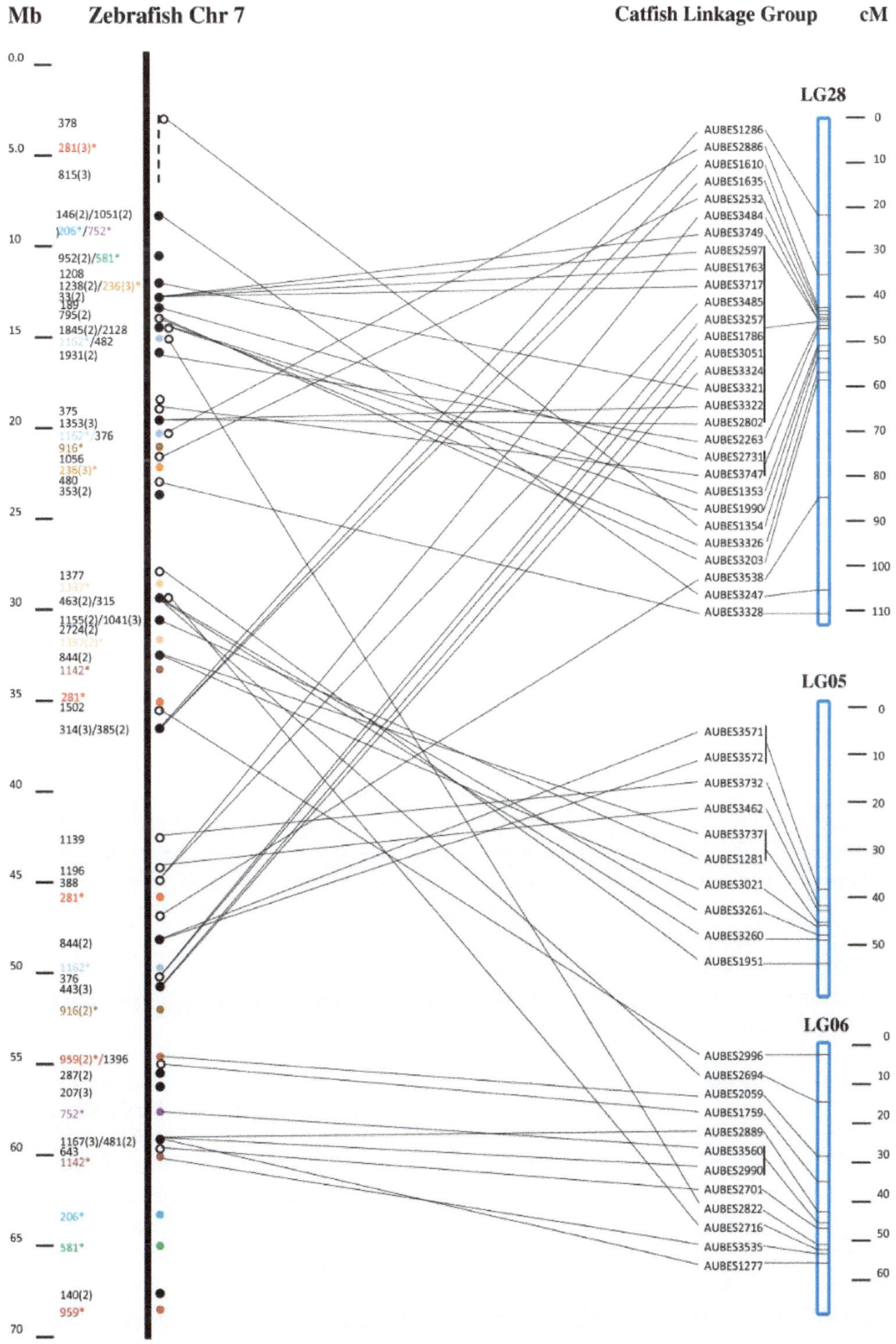

Figure 7: Scaffolds of conserved syntenic regions between the catfish linkage groups and zebrafish chromosome 7. Details of the methods used for the identification and presentation of the syntenic regions are the same as described under Figure 6.

Genetic linkage mapping of BAC end-anchored microsatellites provided a level of validation of the physical map. Discrepancies were found between the BAC assemblage and the linkage map. Of the 75 contigs with at least 2 markers, 54 contigs were mapped properly into the same linkage groups. However, 18 contigs were mapped into different linkage groups (see Additional file 2). Of these 18 contigs, 12 are large contigs with at least 40 BACs. Apparently, such discrepancy is indicative of mistakes in the BAC assemblage. Mapping additional BAC end-anchored microsatellites is under way to integrate the genetic linkage and physical maps, and to correct any additional mistakes in the assembly of the physical map [42].

CONCLUSION

Some highly conserved chromosomes or chromosomal regions exist between catfish and zebrafish. High levels of local conservation were found, but a high level of chromosomal shuffling and rearrangements exists between catfish and zebrafish genomes. Comparative genome analysis using zebrafish genome sequence is highly useful for regional comparisons, but not so useful at the chromosomal levels. The significance of comparative genome analysis in catfish is that it will allow more cost-effective structural genomic analysis, but more importantly, orthologues established through comparative genome analysis should facilitate functional assignment of genes. Given that functional genomics is more difficult with non-model fish species, inference from orthologues should be one of the most efficient and reliable approaches for functional analysis of the catfish genome.

Overall, the evolutionary syntenic conservation appeared to be relatively low between the catfish genome and the genomes of the zebrafish. This indicates many chromosome breakage and rearrangements among the fish genomes occurred during evolution. These findings are consistent with our previous findings that high levels of conservation were found within small genomic regions, whereas high levels of large-scale genome reshuffling were evident when comparing the genomes of catfish and zebrafish [26, 40]. These conclusions, however, are based on the assumption that the zebrafish genome assembly is correct. Apparently, due to the assembly mistakes in the zebrafish genome, some of the syntenic breaks may be due to the still poor assembly of the zebrafish genome. We also acknowledge that comparative genome analysis using a partial bank of sequences in catfish and a more complete databank in zebrafish could potentially lead to a bias. Caution should be exercised when establishing concrete syntenic relations. Such limitations themselves justify the need for whole genome sequencing in catfish.

METHODS

BAC culture and BAC-end sequencing

The CHORI-212 Channel Catfish BAC library [47] was used for BAC-end sequencing. BAC culture and sequencing reactions were conducted using standard protocols, and as previously described [25, 40]. Briefly, BAC clones were transferred from 384-well plates to 96-well culture blocks containing 1.5 ml of 2× YT medium with 12.5 µg/ml chloramphenicol and grown at 37°C overnight with shaking at 300 rpm. The blocks were centrifuged at 2000 × g for 10 min in an Eppendorf 5804R bench top centrifuge to collect bacteria. The culture supernatant was decanted and the blocks were inverted and tapped gently on paper towels to remove remaining liquid. BAC DNA was isolated using the Perfectprep™ BAC 96 kit (Eppendorf, Westbury, NY) according to the manufacturer›s specifications. BAC DNA was collected in 96-well plates and stored at -20°C until usage.

Sequencing of channel catfish BAC ends was conducted using the BigDye® Terminator v3.1 Cycle Sequencing Kit (Applied Biosystems, Foster City, CA), with modifications. Each sequencing reaction mix contained 2 µl of 5× sequencing buffer, 2 µl of primer (3 pmol/µl), 1.5 µl BigDye v3.1 dye terminator, and 4.5 µl of BAC DNA. BAC clones were sequenced from both ends using the primers T7 (5›-TAATACGACTCACTATAGGG-3›) and SP6 (5›-ATTTAGGTGACACTATAG-3›). Cycle sequencing was carried out in 96-well plate format using PTC-200 thermal cyclers (MJ Research/Bio-Rad, Hercules, CA) under the following thermal profile: an initial denaturing at 95°C for 5 min, followed by 100 cycles of 95°C for 30 s, 53°C for 10 s, and 60°C for 4 min. Products were purified using ethanol/EDTA precipitation according to the BigDye protocol (Applied Biosystems), with the following modifications. After thermal cycling, 1 µl of 125 mM EDTA and 30 µl chilled (-80°C) 100% ethanol were added to each reaction. Plates were gently mixed and incubated at room temperature for 15 min. Plates were then centrifuged at 2,250 × g at 4°C for 30 min, followed by washing in 30 µl of 70% ethanol at 2,000 × g for 15 min. Ethanol was decanted and 8 µl Hi-Di™ formamide (Applied Biosystems) was added to each well to re-suspend DNA. Products were denatured at 95°C for 5 min and sequenced on a 3130*xl* genetic analyzer (Applied Biosystems).

Sequence Processing and Analysis

The raw BES base calling were conducted by using Phred [48, 49] with Q20 as a cut-off. Lucy program [50] was used to remove the vector sequences and short sequence less than 200 bp. Repeats were masked using REPEATMASKER [51] before BLAST analysis. In order to anchor the catfish BES to the zebrafish genome, TBLASTX searches of the repeat-masked BES were conducted against the ENSEMBLE zebrafish cDNA database (Assembly 7).

Identification of Conserved Syntenies between Catfish and Zebrafish

In the absence of the whole genome sequence, we attempted to establish microsyntenies based on physical linkage of gene sequences. First, if the genes were identified from both sides of a single BAC clone (mate-paired BES), then they are physically linked with a distance of the BAC clone insert size. If the same two genes were found to be linked on the zebrafish genome in the same genome neighborhood, a microsynteny was established.

Initially, BES were analyzed by BLASTN (*E*-value ≤ -5) for the identification of mate-pairs with significant hits on both sides of the BAC insert. Mate-paired BES were analyzed by tBLASTX (*E*-value ≤ -5) for the identification of genes on both sides of the BAC insert. After identification, the two mate-paired genes in each BAC were used as queries to search for their chromosomal locations on the zebrafish genome. Conserved microsyntenies were declared when the mate-paired genes existed within a distance of 1.0 Mb within the zebrafish genome.

Syntenies were also established using genes within contiguous sequences (contigs) based on the catfish physical map [42]. Genes identified from BES were located along the catfish physical map. Genes identified within the same contig and located on the same zebrafish chromosome with comparable distances as estimated from the catfish BAC contig, an extended synteny was established.

Construction of the Catfish Syntenic Groups Using Linkagemaps

In order to assess the scope of microsyntenies, two zebrafish chromosomes, chromosome 7 and 13, were chosen for analysis. Chromosome 7 had the largest number of significant hits and chromosome 13 had a large number of contigs having two or more hits (suggestive of high level of syntenic conservation). Syntenies were established using microsatellite-based linkage mapping. A total of 548 microsatellite loci in the contigs which had significant BLASTN hits to the zebrafish chromosome 7 and 13 were tested using a hybrid catfish resource family, F_1-2 (female blue-channel catfish hybrid) × Ch-6 (male channel catfish) with 64 progeny.

Microsatellites were identified and analyzed using Msatfinder [52] and Vector NTI 10.0 (Invitrogen, Carlsbad, CA) as we previously described [24]. Polymerase chain reaction (PCR) primers were designed using Msatfinder [52]. Mononucleotide repeats were manually excluded. PCR amplification was conducted as previously described [24]. Briefly, each microsatellite PCR reaction contained 1× PCR buffer, 2 mM $MgCl_2$, 0.2 mM of each dNTP, 4 ng upper primer, 6 ng lower primer, 1 pmol labeled primer, and 0.25 U of JumpStart *Taq* polymerase (Sigma, St. Louis, MO), and 20 ng genomic DNA. PCR amplification was carried out using a touchdown program with the following thermal profile: an initial denaturation at 94°C for 3.5 min, followed by 94°C for 30 s, 57°C for 30 s, and 72°C for 30 s for 20 cycles as the first step, and at 94°C for 30 s, 53°C for 30 s, and 72°C for 30 s for 15 cycles as the second step. A final extension was performed at 72°C for 10 minutes. The PCR products were analyzed on 7% sequencing gels using the 4300 DNA Analyzer (LI-COR® Biosciences, Lincoln, NE). After gel electrophoresis, loci were manually genotyped to determine allele segregation patterns and polymorphisms in the resource family.

The catfish linkage map was constructed using JoinMap version 4.0 software as we previously described [26] using the cross-pollinating (CP) coding scheme, which handles the data containing various genotype configurations with unknown linkage phases [53]. Linkage between markers was examined by estimating LOD scores for recombination rate, and map distances were calculated using the Kosambi mapping function. Significance of marker linkage was determined at a final LOD threshold of 3.0.

Accession Numbers

The BES generated from this study have been deposited in GenBank and were assigned accession numbers from [GenBank:FI857756] to [GenBank:FI900776] and the existing BES from [GenBank:DX083364] to [GenBank:DX103729] were also used for comparative genome analysis in this study.

NOTES

Hong Liu, Yanliang Jiang contributed equally to this work.

DECLARATIONS

Acknowledgements

This project was supported by grants from USDA NRI Animal Genome Tools and Resources Program (USDA/NRICGP award# 2009-35205-05101, and partially award #2006-35616-16685).

Electronic Supplementary Material

12864_2009_2476_MOESM1_ESM.XLS Additional file 1: **tBLASTX search results**. A portion of the tBLASTX search results used for the identification of microsyntenies. Query ID are sequence identification of catfish BAC end sequences used as queries; Subject ID is the identification of the sequence with significant tBLASTX hits at the E-values (E-values are coded in this Table. For instance, 0.29 means e^{-29}, 0.06 means e^{-6}). Chromosome number is provided in the column Chr, with starting position (Chr SS) and ending position (Chr SE) provided. The potential gene identities are detailed under Description. (XLS 342 KB)

12864_2009_2476_MOESM2_ESM.XLS Additional file 2: Primers used to map the BAC end associated microsatellites. Ctg_ID is contig number on the catfish physical map; Contig size refers to the number of BAC clones within the contig; BAC_ID is BAC identification number; Upper and lower primer sequences, and the linkage groups the microsatellites were mapped are given in column E, F, and G, respectively. (XLS 78 KB)

REFERENCES

1. Fujiyama A, Watanabe H, Toyoda A, Taylor TD, Itoh T, Tsai S-F, Park H-S, Yaspo M-L, Lehrach H, Chen Z, Fu G, Saitou N, Osoegawa K, de Jong PJ, Suto Y, Hattori M, Sakaki Y: Construction and Analysis of a Human-Chimpanzee Comparative Clone Map. Science. 2002, 295: 131-134. 10.1126/science.1065199.
2. Gregory SG, Sekhon M, Schein J, Zhao S, Osoegawa K, Scott CE, Evans RS, Burridge PW, Cox TV, Fox CA, Hutton RD, Mullenger IR, Phillips KJ, Smith J, Stalker J, Threadgold GJ, Birney E, Wylie K, Chinwalla A, Wallis J, Hillier L, Carter J, Gaige T, Jaeger S, Kremitzki C, Layman D, Maas J, McGrane R, Mead K, Walker R, Jones S, Smith M, Asano J, Bosdet I, Chan S, Chittaranjan S, Chiu R, Fjell C, Fuhrmann D, Girn N, Gray C, Guin R, Hsiao L, Krzywinski M, Kutsche R, Lee SS, Mathewson C, McLeavy C, Messervier S, Ness S, Pandoh P, Prabhu A-L, Saeedi P, Smailus D, Spence L, Stott J, Taylor S, Terpstra W, Tsai M, Vardy J, Wye N, Yang G, Shatsman S, Ayodeji B, Geer K, Tsegaye G, Shvartsbeyn A, Gebregeorgis E, Krol M, Russell D, Overton L, Malek JA, Holmes M, Heaney M, Shetty J,

Feldblyum T, Nierman WC, Catanese JJ, Hubbard T, Waterston RH, Rogers J, de Jong PJ, Fraser CM, Marra M, McPherson JD, Bentley DR: A physical map of the mouse genome. Nature. 2002, 418: 743-750. 10.1038/nature00957.

3. Larkin DM, Wind Everts-van der A, Rebeiz M, Schweitzer PA, Bachman S, Green C, Wright CL, Campos EJ, Benson LD, Edwards J, Liu L, Osoegawa K, Womack JE, de Jong PJ, Lewin HA: A cattle-human comparative map built with cattle BAC-ends and human genome sequence. Genome Res. 2003, 13: 1966-1972.

4. Leeb T, Vogl C, Zhu B, de Jong PJ, Binns MM, Chowdhary BP, Scharfe M, Jarek M, Nordsiek G, Schrader F, Bloker H: A human-horse comparative map based on equine BAC end sequences. Genomics. 2006, 87: 772-776. 10.1016/j.ygeno.2006.03.002.

5. Meyers SN, Rogatcheva MB, Larkin DM, Yerle M, Milan D, Hawken RJ, Schook LB, Beever JE: Piggy-BACing the human genome: II. A high-resolution, physically anchored, comparative map of the porcine autosomes. Genomics. 2005, 86: 739-752. 10.1016/j.ygeno.2005.04.010.

6. Dalrymple BP, Kirkness EF, Nefedov M, McWilliam S, Ratnakumar A, Barris W, Zhao S, Shetty J, Maddox JF, O'Grady M, Nicholas F, Crawford AM, Smith T, de Jong PJ, McEwan J, Oddy VH, Cockett NE: Using comparative genomics to reorder the human genome sequence into a virtual sheep genome. Genome Biol. 2007, 8: R152-10.1186/gb-2007-8-7-r152.

7. Moen T, Hayes B, Baranski M, Berg PR, Kjoglum S, Koop BF, Davidson WS, Omholt SW, Lien S: A linkage map of the Atlantic salmon (*Salmo salar*) based on EST-derived SNP markers. BMC Genomics. 2008, 9: 223-10.1186/1471-2164-9-223.

8. Ng SHS, Artieri CG, Bosdet IE, Chiu R, Danzmann RG, Davidson WS, Ferguson MM, Fjell CD, Hoyheim B, Jones SJM, de Jong PJ, Koop BF, Krzywinski MI, Lubieniecki K, Marra MA, Mitchell LA, Mathewson C, Osoegawa K, Parisotto SE, Phillips RB, Rise ML, von Schalburg KR, Schein JE, Shin H, Siddiqui A, Thorsen J, Wye N, Yang G, Zhu B: A physical map of the genome of Atlantic salmon, *Salmo salar*. Genomics. 2005, 86: 396-404. 10.1016/j.ygeno.2005.06.001.

9. Rise ML, von Schalburg KR, Brown GD, Mawer MA, Devlin RH, Kuipers N, Busby M, Beetz-Sargent M, Alberto R, Gibbs AR, Hunt P, Shukin R, Zeznik JA, Nelson C, Jones SR, Smailus DE, Jones SJ, Schein JE, Marra MA, Butterfield YS, Stott JM, Ng SH, Davidson WS, Koop BF: Development and application of a salmonid EST database and cDNA microarray: data mining and interspecific hybridization characteristics. Genome Res. 2004, 14: 478-490. 10.1101/gr.1687304.

10. Guyomard R, Mauger S, Tabet-Canale K, Martineau S, Genet C, Krieg F, Quillet E: A type I and type II microsatellite linkage map of rainbow trout (*Oncorhynchus mykiss*) with presumptive coverage of all chromosome arms. BMC Genomics. 2006, 7: 302-10.1186/1471-2164-7-302.

11. Rexroad CE, Lee Y, Keele JW, Karamycheva S, Brown G, Koop B, Gahr SA, Palti Y, Quackenbush J: Sequence analysis of a rainbow trout cDNA library and creation of a gene index. Cytogenet Genome Res. 2003, 102: 347-354. 10.1159/000075773.

12. Katagiri T, Kidd C, Tomasino E, Davis JT, Wishon C, Stern JE, Carleton KL, Howe AE, Kocher TD: A BAC-based physical map of the Nile tilapia genome. BMC Genomics. 2005, 6: 89-10.1186/1471-2164-6-89.

13. Lee BY, Lee WJ, Streelman JT, Carleton KL, Howe AE, Hulata G, Slettan A, Stern JE, Terai Y, Kocher TD: A second-generation genetic linkage map of tilapia (*Oreochromis* spp.). Genetics. 2005, 170: 237-244. 10.1534/genetics.104.035022.

14. Franch R, Louro B, Tsalavouta M, Chatziplis D, Tsigenopoulos CS, Sarropoulou E, Antonello J, Magoulas A, Mylonas CC, Babbucci M, Patarnello T, Power DM, Kotoulas G, Bargelloni L: A genetic linkage map of the hermaphrodite teleost fish *Sparus aurata* L. Genetics. 2006, 174: 851-861. 10.1534/genetics.106.059014.

15. Sarropoulou E, Power DM, Magoulas A, Geisler R, Kotoulas G: Comparative analysis and characterization of expressed sequence tags in gilthead sea bream (*Sparus aurata*) liver and embryos. Aquaculture. 2005, 243: 69-81. 10.1016/j.aquaculture.2004.10.012.

16. Sarropoulou E, Franch R, Louro B, Power DM, Bargelloni L, Magoulas A, Senger F, Tsalavouta M, Patarnello T, Galibert F, Kotoulas G, Geisler R: A gene-based radiation hybrid map of the gilthead sea bream *Sparus aurata* refines and exploits conserved synteny with *Tetraodon nigroviridis*. BMC Genomics. 2007, 8: 44-10.1186/1471-2164-8-44.

17. Senger F, Priat C, Hitte C, Sarropoulou E, Franch R, Geisler R, Bargelloni L, Power D, Galibert F: The first radiation hybrid map of a perch-like fish: The gilthead seabream (*Sparus aurata* L). Genomics. 2006, 87: 793-800. 10.1016/j.ygeno.2005.11.019.

18. Chistiakov DA, Hellemans B, Haley CS, Law AS, Tsigenopoulos CS, Kotoulas G, Bertotto D, Libertini A, Volckaert FA: A microsatellite linkage map of the European sea bass *Dicentrarchus labrax* L. Genetics. 2005, 170: 1821-1826. 10.1534/genetics.104.039719.

19. Whitaker HA, McAndrew BJ, Taggart JB: Construction and characterization of a BAC library for the European sea bass *Dicentrarchus labrax*. Anim Genet. 2006, 37: 526-10.1111/j.1365-2052.2006.01514.x.

20. Liu Z: A review of catfish genomics: progress and perspectives. Comp Funct Genomics. 2003, 4: 259-265. 10.1002/cfg.265.

21. Liu Z: Catfish. Genome Mapping and Genomics in Fishes and Aquatic Animals. Edited by: Kocher T, Kole C. 2008, Berlin Heidelberg: Springer, 2: 85-100. [Kole C (Series Editor): Genome Mapping and Genomics in Animals, vol. 2]

22. He C, Chen L, Simmons M, Li P, Kim S, Liu ZJ: Putative SNP discovery in interspecific hybrids of catfish by comparative EST analysis. Anim Genet. 2003, 34: 445-448. 10.1046/j.0268-9146.2003.01054.x.

23. Serapion J, Kucuktas H, Feng J, Liu Z: Bioinformatic mining of type I microsatellites from expressed sequence tags of channel catfish (*Ictalurus punctatus*). Mar Biotechnol (NY). 2004, 6: 364-377. 10.1007/s10126-003-0039-z.
24. Somridhivej B, Wang S, Sha Z, Liu H, Quilang J, Xu P, Li P, Hu Z, Liu Z: Characterization, polymorphism assessment, and database construction for microsatellites from BAC end sequences of channel catfish (*Ictalurus punctatus*): A resource for integration of linkage and physical maps. Aquaculture. 2008, 275: 76-80. 10.1016/j.aquaculture.2008.01.013.
25. Xu P, Wang S, Liu L, Peatman E, Somridhivej B, Thimmapuram J, Gong G, Liu Z: Channel catfish BAC-end sequences for marker development and assessment of syntenic conservation with other fish species. Anim Genet. 2006, 37: 321-326. 10.1111/j.1365-2052.2006.01453.x.
26. Kucuktas H, Wang S, Li P, He C, Xu P, Sha Z, Liu H, Jiang Y, Baoprasertkul P, Somridhivej B, Wang Y, Abernathy J, Guo X, Liu L, Muir W, Liu Z: Construction of Genetic Linkage Maps and Comparative Genome Analysis of Catfish Using Gene-associated Markers. Genetics. 2009, 181: 1649-1660. 10.1534/genetics.108.098855.
27. Liu Z, Karsi A, Li P, Cao D, Dunham R: An AFLP-based genetic linkage map of channel catfish (*Ictalurus punctatus*) constructed by using an interspecific hybrid resource family. Genetics. 2003, 165: 687-694.
28. Waldbieser GC, Bosworth BG, Nonneman DJ, Wolters WR: A microsatellite-based genetic linkage map for channel catfish, *Ictalurus punctatus*. Genetics. 2001, 158: 727-734.
29. Cao D, Kocabas A, Ju Z, Karsi A, Li P, Patterson A, Liu Z: Transcriptome of channel catfish (*Ictalurus punctatus*): initial analysis of genes and expression profiles of the head kidney. Anim Genet. 2001, 32: 169-188. 10.1046/j.1365-2052.2001.00753.x.
30. Ju Z, Karsi A, Kocabas A, Patterson A, Li P, Cao D, Dunham R, Liu Z: Transcriptome analysis of channel catfish (*Ictalurus punctatus*): genes and expression profile from the brain. Gene. 2000, 261: 373-382. 10.1016/S0378-1119(00)00491-1.
31. Karsi A, Cao D, Li P, Patterson A, Kocabas A, Feng J, Ju Z, Mickett KD, Liu Z: Transcriptome analysis of channel catfish (*Ictalurus punctatus*): initial analysis of gene expression and microsatellite-containing cDNAs in the skin. Gene. 2002, 285: 157-168. 10.1016/S0378-1119(02)00414-6.
32. Kocabas AM, Li P, Cao D, Karsi A, He C, Patterson A, Ju Z, Dunham RA, Liu Z: Expression profile of the channel catfish spleen: analysis of genes involved in immune functions. Mar Biotechnol (NY). 2002, 4: 526-536. 10.1007/s10126-002-0067-0.
33. Li P, Peatman E, Wang S, Feng J, He C, Baoprasertkul P, Xu P, Kucuktas H, Nandi S, Somridhivej B, Serapion J, Simmons M, Turan C, Liu L, Muir W, Dunham R, Brady Y, Grizzle J, Liu Z: Towards the ictalurid catfish transcriptome: generation and analysis of 31,215 catfish ESTs. BMC Genomics. 2007, 8: 177-10.1186/1471-2164-8-177.
34. Ju Z, Dunham RA, Liu Z: Differential gene expression in the brain of channel catfish (*Ictalurus punctatus*) in response to cold acclimation. Mol Genet Genomics. 2002, 268: 87-95. 10.1007/s00438-002-0727-9.
35. Li RW, Waldbieser GC: Production and utilization of a high-density oligonucleotide microarray in channel catfish, *Ictalurus punctatus*. BMC Genomics. 2006, 7: 134-10.1186/1471-2164-7-134.
36. Liu Z, Li RW, Waldbieser GC: Utilization of microarray technology for functional genomics in ictalurid catfish. J Fish Biol. 2008, 72: 2377-2390. 10.1111/j.1095-8649.2008.01898.x.
37. Peatman E, Baoprasertkul P, Terhune J, Xu P, Nandi S, Kucuktas H, Li P, Wang S, Somridhivej B, Dunham R, Liu Z: Expression analysis of the acute phase response in channel catfish (*Ictalurus punctatus*) after infection with a Gram-negative bacterium. Dev Comp Immunol. 2007, 31: 1183-1196. 10.1016/j.dci.2007.03.003.
38. Peatman E, Terhune J, Baoprasertkul P, Xu P, Nandi S, Wang S, Somridhivej B, Kucuktas H, Li P, Dunham R, Liu Z: Microarray analysis of gene expression in the blue catfish liver reveals early activation of the MHC class I pathway after infection with *Edwardsiella ictaluri*. Mol Immunol. 2008, 45: 553-566. 10.1016/j.molimm.2007.05.012.
39. Quiniou SM, Katagiri T, Miller NW, Wilson M, Wolters WR, Waldbieser GC: Construction and characterization of a BAC library from a gynogenetic channel catfish *Ictalurus punctatus*. Genet Sel Evol. 2003, 35: 673-683. 10.1186/1297-9686-35-7-673.
40. Wang S, Xu P, Thorsen J, Zhu B, de Jong PJ, Waldbieser G, Kucuktas H, Liu Z: Characterization of a BAC library from channel catfish *Ictalurus punctatus*: indications of high levels of chromosomal reshuffling among teleost genomes. Mar Biotechnol (NY). 2007, 9: 701-711. 10.1007/s10126-007-9021-5.
41. Quiniou SM, Waldbieser GC, Duke MV: A first generation BAC-based physical map of the channel catfish genome. BMC Genomics. 2007, 8: 40-10.1186/1471-2164-8-40.
42. Xu P, Wang S, Liu L, Thorsen J, Kucuktas H, Liu Z: A BAC-based physical map of the channel catfish genome. Genomics. 2007, 90: 380-388. 10.1016/j.ygeno.2007.05.008.
43. Nandi S, Peatman E, Xu P, Wang S, Li P, Liu Z: Repeat structure of the catfish genome: a genomic and transcriptomic assessment of Tc1-like transposon elements in channel catfish (*Ictalurus punctatus*). Genetica. 2007, 131: 81-90. 10.1007/s10709-006-9115-4.
44. Sarropoulou E, Nousdili D, Magoulas A, Kotoulas G: Linking the genomes of nonmodel teleosts through comparative genomics. Mar Biotechnol (NY). 2008, 10: 227-233. 10.1007/s10126-007-9066-5.
45. Sha Z, Abernathy WJ, Wang S, Li P, Kucuktas H, Liu H, Peatman E, Liu Z: NOD-like subfamily of the nucleotide-binding domain and leucine-rich repeat containing

family receptors and their expression in channel catfish. Dev Comp Immunol. 2009, 31: 991-999. 10.1016/j.dci.2009.04.004.

46. Steinke D, Salzburger W, Meyer A: Novel relationships among ten fish model species revealed based on a phylogenomic analysis using ESTs. J Mol Evol. 2006, 62: 772-784. 10.1007/s00239-005-0170-8.

47. CHORI-212: Channel Catfish (*Ictalurus punctatus*) BAC Library. [http://bacpac.chori.org/library.php?id=103]

48. Ewing B, Green P: Base-calling of automated sequencer traces using phred. II. Error probabilities. Genome Res. 1998, 8: 186-194.

49. Ewing B, Hillier L, Wendl MC, Green P: Base-calling of automated sequencer traces using phred. I. Accuracy assessment. Genome Res. 1998, 8: 175-185.

50. Chou HH, Holmes MH: DNA sequence quality trimming and vector removal. Bioinformatics. 2001, 17: 1093-1104. 10.1093/bioinformatics/17.12.1093.

51. RepeatMasker Open-3.0. [http://www.repeatmasker.org]

52. Msatfinder: detection and characterisation of microsatellites. [http://www.genomics.ceh.ac.uk/msatfinder/]

53. Sekino M, Kobayashi T, Hara M: Segregation and linkage analysis of 75 novel microsatellite DNA markers in pair crosses of Japanese abalone (*Haliotis discus hannai*) using the 5'-tailed primer method. Mar Biotechnol (NY). 2006, 8: 453-466. 10.1007/s10126-005-6179-6.

CHAPTER 6

SCREEN FOR FOOTPRINTS OF SELECTION DURING DOMESTICATION/CAPTIVE BREEDING OF ATLANTIC SALMON

Anti Vasemägi, Jan Nilsson, Philip McGinnity, Tom Cross, Patrick O'Reilly, Brian Glebe, Bo Peng, Paul Ragnar Berg, and Craig Robert Primmer

INTRODUCTION

Over the last decade, advances in genomic technologies together with developments in methods of statistical analyses have resulted in considerable progress in the systematic detection of genomic regions influenced by artificial selection [1–5]. These studies rely on the concept of genetic hitchhiking which predicts that selection affects the genome in a locus-specific manner by leaving specific signatures surrounding the gene or genes under selection [6]. Specifically, the theory of genetic hitchhiking focuses on the spread of novel advantageous mutations in a population, generating characteristic footprints of selection [7, 8]. This type of selection is often referred to as a "hard sweep" in contrast to the "soft sweep", where selection acts on a standing variation [9]. The importance of hard sweeps versus soft sweeps in natural populations and domesticated strains remains an open question, although recent empirical work [10, 11] and theoretical models suggest that soft sweeps are likely to be widespread [9, 12, 13]. Nevertheless, hitchhiking-mapping efforts in domesticated animals and plants to date have mainly focused on identifying relatively old footprints of selection dating back hundreds or thousands of generations, during which time new advantageous mutations can spread to fixation (e.g., [1, 3, 14]) and very few studies have specifically focused on genetic mechanisms of the early stages of domestication (e.g., [15]).

Atlantic salmon (Salmo salar L.) represents a useful model for studying the genetic basis of recent, ongoing domestication since reared salmon strains have experienced intense artificial selection for only a short period of time (5–12 generations), and there are still many wild populations available for comparative purposes, including progenitor populations. The traits that have been targets of selective breeding in reared Atlantic salmon strains include higher growth rate, delayed sexual maturity, bacterial resistance, flesh colour, and fat content. The selection response of many of these traits has been as high as 10% per generation [16]. As a result, despite the short history of domestication, reared strains of Atlantic salmon often differ markedly from their wild counterparts [17]. In addition to farming, Atlantic salmon has also been a popular target for various restoration and supplementation/stocking programmes (here termed ranching), in which the species is captively bred and reared for subsequent release into the natural environment. Importantly, artificial propagation and rearing processes in the hatchery environment can involve intentional and/or inadvertent selection and relaxation of natural selection operating in the wild [18]. Inadvertent (nondeliberate) selection during captive breeding is particularly problematic for supportive breeding of endangered populations because it is essentially unavoidable and often associated with a reduction in fitness in the natural environment [19–21]. Importantly, the negative effects of domestication can be severe and even a few generations of captive breeding may have marked effects on natural reproduction in the wild [21–23].

Earlier studies on Atlantic salmon have demonstrated that 5–7 generations of domestication were sufficient to induce heritable changes in gene transcription level between domesticated and wild populations [24, 25]. Large proportion of these changes has been population specific while small number of genes has also shown to exhibit parallel changes in gene expression level in multiple wild-farmed population

Citation: Anti Vasemägi, Jan Nilsson, Philip McGinnity, et al., "Screen for Footprints of Selection during Domestication/Captive Breeding of Atlantic Salmon," Comparative and Functional Genomics, vol. 2012, Article ID 628204, 14 pages, 2012. doi:10.1155/2012/628204

comparisons [24, 25]. The presence of parallel changes in transcription level may indicate (i) that the same DNA sequence polymorphisms are responsible for the rapid changes in gene expression, or alternatively, (ii) that the same downstream pathways, involving the same or different DNA sequence polymorphisms, are being affected during the domestication process. However, currently very little is known about what genomic regions have been influenced by domestication and captive breeding in different Atlantic salmon populations. Here, we screened 261 SNPs and 70 microsatellite loci in three independent-reared strains (two of which were used for ranching (IRL, SWE) and one for farming (CAN)) and their wild progenitor populations of Atlantic salmon to identify recent footprints of selection related to domestication and captive breeding.

MATERIALS AND METHODS

Study Populations

Three independently reared strains and their wild progenitor populations from Europe and North America were used in the analyses: Burrishoole River (Mayo, Ireland, IRL), Ume/Vindelälven Rivers (Västerbotten, Sweden, SWE), and the Saint John River (New Brunswick, Canada, CAN), in an attempt to identify genomic regions affected by domestication/captive breeding. A short description of the history of each reared strain and sampling details are provided below.

Artificial rearing of juvenile salmon for ranching from the Burrishoole river began in 1960 and recruitment from the wild population continued until 1964. Apart from the addition of a small number of wild fish in the reared broodstock between 1970 and 1975, the breeding strain has been effectively closed since that time; the hatchery born fish were distinguished from their wild counterparts by the absence of a previously-clipped adipose fin. Assuming a generation time of 5 years for the wild fish, wild and reared (ranched) Burrishoole salmon have experienced different selection pressures for approximately 8 generations during the early freshwater phase (from fertilization to smoltification). Today, the Burrishoole-ranched strain is composed of three almost distinct coexisting cohorts with little overlap between year classes [23]. Burrishoole wild samples consisted of 16 individuals collected as descending smolts in 2006, while the samples from the reared (ranched) Burrishoole strain were obtained from the smolts collected in 2005 and 2006 (eight individuals per cohort).

The wild River Umeälven salmon were extirpated by power plant dam construction in the early 1950s but River Vindelälven (a tributary of Umeälven) continued to support wild salmon. The present hatchery stock of River Umeälven originates from the wild Vindelälven population and since 1971, all hatchery (ranched) smolts have had their adipose fin removed. Since 1974 all salmon without adipose fins (hatchery fish) or with intact adipose fins (wild salmon) have been recorded at the fish trap below the Norrfors hydroelectric dam and only salmon of wild origin are released above the dam to continue their upstream spawning migration to the Vindelälven. As a result, wild and captive (ranched) Ume/Vindelälven salmon have experienced different selection pressures for approximately 5-6 generations during the early freshwater phase (from fertilization until smoltification) assuming a generation time of 5-6 years. River Ume/Vindelälven wild samples consisted of 16 returning individuals in 2006, while the samples from the reared (ranched) strain were obtained in 2009 (16 1+ old juveniles).

The Saint John reared strain used for farming was founded in 1990 from wild Saint John River salmon, and in contrast to IRL and SWE populations, by 2006 it has experienced four generations of selection for higher growth and reduced early maturation [30]. As a result, wild and reared salmon from Saint John River have experienced different selection pressures due to selective breeding relative to wild salmon from the same system, for approximately 4-5 generations. The Saint John River sample collection analyzed here consisted of 16 returning adults of wild origin collected in 2000 (Nashwaak tributary, situated below Mactaquac Dam), while the samples from the Saint John river aquaculture strain came from Atlantic Salmon Broodstock Development Program (ASBDP) of New Brunswick (randomly selected breeders from the selection line SGRP 90 JC, sampled in 2007).

SNPs Genotyping

In total, 320 SNP developed by Moen et al. [28] were genotyped using Sequenom Mass Array TM system from (San Diego, USA). After quality control and assessment of levels of polymorphism, 59 markers were excluded from further analyses leaving a final set of 261 SNPs upon which all analyses were based (see Supplementary Material avaliable online at http://dx.doi.org/10.1155/2012/628204). Most of these SNPs were initially mapped by Moen et al. [28] and later many of them have been incorporated into a larger Atlantic salmon SNP linkage map [31]. The linkage group information used throughout the study corresponds to the SALMAP consortium nomenclature [32] while information about the corresponding chromosome numbers was derived from Phillips et al. [33]. Genotyping was performed using Sequenom iPLEX Gold chemistry according to the manufacturer's instructions. Detailed information about PCR conditions, PCR and extension primers, and post-PCR pooling procedures are similar to Moen et al. [28] and available on request. Genotypes were assigned in real time [34] using the MassARRAY SpectroTYPER RT v3.4 software (Sequenom) based on the mass peaks present. All results were manually inspected using the MassARRAY TyperAnalyzer v3.3 software (Sequenom).

Microsatellite Genotyping

A total of 70 variable number tandem repeat markers were also included in this study. They included two MHC linked mini- and microsatellite loci, 21 EST-derived microsatellite loci [35, 36], and 47 anonymous microsatellite loci (Supplementary Material). The microsatellite markers were PCR amplified in 19 multiplex reactions that were subsequently combined into six capillary electrophoresis runs. All reactions were carried out in 6 μL volumes including 10–100 ng of genomic DNA and 0.033–0.3 μM of locus specific primers, the forward primer being labelled with one of four fluorescent dyes (four colour set: PET, FAM, NED, and VIC, three colour set: FAM, HEX and NED), and 1 × QIAGEN multiplex PCR master mix. The thermal cycling conditions used were as follows: 15 min initial activation step at 95°C followed by 35 cycles of denaturation at 94°C for 30 s, annealing at 55–60°C for 90 s and extension at 72°C for 60 s and ending with a final extension at 60°C for 10–30 min. Amplifications were performed using an Applied Biosystems 2720, a PTC-100, or a PTC-200 (Bio-Rad) thermal cycler. The PCR products from separate multiplex reactions were pooled (1–2.5 μL) and made up to the final volume of 120 μL of distilled water, mixed with GS600LIZ size standard (Applied Biosystems), and formamide for electrophoresis on an ABI 3130 × 1 automated sequencer. Detailed multiplex amplification information, including specific primer concentrations, annealing temperatures, and post-PCR pooling procedures, are available on request.

Summary Statistics

Diversity indices were estimated using the program FSTAT 2.9.3 [37]. Deviations from Hardy-Weinberg equilibrium were tested with the exact test [38] as implemented in GENEPOP 3.4 [26]. Genetic differentiation between populations was measured with pairwise F_{ST} estimates [27]. The significance of F_{ST} estimates between pairwise comparisons was tested by permutation of individuals between samples using FSTAT 2.9.3 [37].

Analyses of Signatures of Selection

Standard Outlier Tests

Two different tests aimed at detection of loci subject to directional selection during domestication/captive breeding were used. Both approaches rely on the rationale that directional selection increases genetic differentiation between populations and reduces variability at linked loci. However, because both of the tests are based on different assumptions, identification of outlier loci simultaneously with both approaches provides additional support for the candidate status of a particular locus [29]. The first method, developed by Beaumont and Nichols [39] calculates F_{ST} for each locus in the sample, and coalescent simulations based on a symmetrical island model are used to generate data sets with the mean F_{ST} similar to the empirical distribution. To calculate approximate P values for each locus, 100,000 independent loci were generated and the simulated distribution of F_{ST} between two populations was then compared to the observed F_{ST} values, conditional on heterozygosity, to identify potential outliers as implemented in the software FDIST2. Sample sizes used to calculate F_{ST} from coalescent simulations were set to 16 individuals, corresponding to the empirical data.

The second method to detect footprints of selection also identifies loci that exhibit extreme differentiation compared to the rest of the genome using the Bayesian likelihood method implemented via reversible jump Markov Chain Monte Carlo [40]. Using this approach, implemented in the Bayescan software, F_{ST} is modelled using a logistic regression model implementing a locus effect and a population effect, relaxing the assumption of symmetrical island model by allowing for population structure asymmetries. The estimation of the model parameters was automatically adjusted on the basis of short pilot runs (10 pilot runs with 5000 iterations each). We used a burn-in of 50,000 iterations and the total chain length of 500,000 iterations (thinning interval of 50 and sample size of 10,000) to identify loci under selection. Altogether, three independent runs were performed for each wild and domesticated/captive population pair. Bayescan estimates the posterior probability that a locus is under selection by calculating a Bayes factor, which is given by the ratio of posterior model probabilities of two models (selected/neutral), given the data. For example, a Bayes factor of two indicates that the data favor one model over the other at odds of two to one. According to Jeffreys' interpretation, Bayes factor between 3 and 10 (log10 = 0.5–1) is considered "substantial evidence" to favour one model over the other and corresponds to a posterior probability between 0.76–0.91. A Bayes factor between 10 and 32 (log10 = 1–1.5) is considered "strong evidence" of different statistical support for the two models and corresponds to a posterior probability between 0.91–0.97. However, this probability cannot be compared directly to the P value from FDIST2, since Bayescan explicitly takes all loci into account in the analyses.

Clustering of Loci with Extreme F_{ST} and lnRH Values

As a complementary approach to the outlier tests described above, we evaluated whether there is a tendency for adjacent loci to exhibit high differentiation (F_{ST}) or large differences in genetic diversity (lnRH) [41] between wild and domesticated/captive population pairs. Such clustering of extreme loci would be indicative of genetic hitchhiking effects along a chromosome, while false positives from standard outlier tests are expected to occur randomly in the genome. Importantly, testing for clustering does not rely on specific assumptions about population history and thus provides complementary evidence to single-locus outlier tests. Specifically, we calculated the mean F_{ST} and mean absolute lnRH for pairs of markers along specific chromosomes and by using permutations evaluated whether the empirical pattern differs from random expectations (5,000 permutations). We subsequently identified locus pairs that exhibited elevated differentiation (F_{ST}) or large differences in genetic diversity (lnRH), against the permutation-derived 95% significance threshold. Absolute lnRH was used as a parameter for quantifying the difference in genetic diversity (lnRH) between populations, as very short-term directional selection can either reduce or increase the genetic variation of a particular locus depending on the frequency of the allele linked to an advantageous mutation. For example, short-term directional selection initially increases the diversity when the positively selected (or linked) allele is at low frequency, while it reduces the variation when the positively selected (or linked) allele is at high frequency in the population. Loss of diversity measured as lnRH has been commonly used to identify outliers in microsatellite datasets but this statistic can also be applied to SNPs markers [42] as lnRH is not based on a particular mutation model [41].

Estimation of False Discovery Rate (FDR)

To control for the multiple testing problem that arises when large numbers of markers are tested against the neutral null hypothesis, we employed the false discovery rate (FDR) methodology [43] implemented in the software q value to evaluate the proportion of putative outliers that are likely to be false positives (i.e., type I error); FDRs (qvalues) [44] were calculated for four neutrality tests based on the list of p values derived from (i) coalescent simulations (FDIST2 test), (ii) from expected normal distributions (lnRH test), and (iii) from permutation tests aimed at detecting clustering of loci with extreme F_{ST} and lnRH values. A bootstrap method was used for estimating the overall proportion of true null hypotheses as a more conservative approach compared to the smoother method as suggested by Storey and Tibshirani [43].

Simulation of Selection from Standing Variation

To evaluate whether the observed patterns of genetic variation were compatible with the effect of strong directional selection, we used an individual-based forward-time simulation environment, simu POP 1.0.2 [45], to explicitly simulate selection and hitchhiking processes based on the empirical data in hand. Specifically, we simulated selection on standing variation along the chromosome rather than from de novo mutations, as it is more likely that initial adaptation to an artificial environment occurs from standing variation. As a result, selection on preexisting genetic variation is expected to lead to immediate evolutionary response, fixation of more alleles of small effect and importantly, generally weaker footprints of selection compared to strong selective sweeps [46].

Given that the time since foundation of captive/domesticated strains are known and assuming equal population sizes and complete isolation (drift only) after the split, we calculated the effective population size using the estimated F_{ST} from $N_e = t/(2F_{ST})$. We then constructed a simple single population split model applying the derived population genetic parameters for each population pair (IRL, gen. time 8, $N_e = 330$; SWE, gen. time 5, $N_e = 65$; CAN, gen. time 5, $N_e = 238$). We subsequently applied forward simulations using the empirical genotype data from wild population and genetic linkage map information to simulate the effect of selection along specific chromosomes on identified outlier loci. We first created an ancestral wild population by using empirical genotype data from the wild progenitor population (wild Burrishoole, Vindelalven, or Nashwaak) by ¨ multiplication of existing genotypes and subsequent random mating for three generations. After the split of the ancestral population, we simulated positive selection on the minor allele in a single candidate region per population pair. We assumed that the fitness of three genotypes AA, Aa, and aa in captive environment is 1, 0.7 and 0.4 ($s = 0.6$; $h = 0.5$), while there is no selection acting in the wild populations. Note that we are aware of only one-published single locus selection coefficient for Atlantic salmon ($s = 0.49$ on MHCII allele conferring resistance for A. salmonicida) [47] but selection coefficients of this magnitude and even higher have been observed in domesticated crops [48–50]. We subsequently let the populations evolve for five (SWE and CAN) or eight (IRL) generations according to the history of wild population-reared strain pairing and repeated the whole domestication/captive breeding simulation 25 times for each population pair. We then calculated the genetic differentiation between wild populations and reared strains using Weir and Cockerham's F_{ST} estimator [27], by sampling 16 individuals per population to evaluate whether artificial selection from standing variation can create comparable genetic differentiation patterns compared to the empirical data. We also validated our simulation procedure without implying selection (pure drift model) which produced genetic differentiation estimates similar to the empirical data (data not shown). The effect of selection on F_{ST} was visualized by creating box plots using SPSS 14.0.

RESULTS

Summary Statistics

As expected, SNPs and microsatellites showed different levels of genetic variation (average gene diversity across populations 0.224 and 0.619, resp.; average number of alleles across populations 1.71 and 5.80, resp.; Mann-Whitney U test, all P values < 0.001). Contrary to earlier population genetic studies on natural and farmed Atlantic salmon populations [51–53], genetic diversity measures of wild and reared strain pairings did not differ significantly from each other (IRL and CAN; Wilcoxon signed ranks test, all P values > 0.05; Table 1) and for Swedish samples, both gene diversity and number of alleles were in fact higher in the reared strain compared with the wild population (Wilcoxon signed ranks test, both P values < 0.05; Table 1). Genetic differentiation, measured as θ, between reared strain/wild population pairs was rather low, ranging from ~0.01 in SWE and IRL to 0.039 in CAN but nevertheless, was highly significant (($P < 0.001$ for all three pairwise comparisons). When comparing wild population and reared strain pairs across different type of markers, the mean genetic differentiation estimates were rather similar (IRL: 0.017 versus 0.010; CAN: 0.031 versus 0.048; SWE: 0.012 versus 0.011 for microsatellites and SNPs, resp.), indicating that both types of markers are generally affected by a similar kind of evolutionary forces across the genome. Hardy-Weinberg disequilibrium was detected in 35 of 1370 loci by population tests at the 1% significance level before correction for multiple tests (data not shown). Approximately 14 of them are expected to emerge simply because of the large number of tests performed.

Table 1: Diversity indexes of the Atlantic salmon populations included in the study.

Pop.	Origin	Mean gene diversity			Mean number of alleles		
		Microsatellites (range)	SNPs (range)	All	Microsatellites (range)	SNPs (range)	All
IRE	Wild	0.661 (0–0.958)	0.319 (0–0.525)	0.385	6.08 (1–18)	1.90 (1-2)	2.71
	Reared	0.648 (0.063–0.944)	0.321 (0–0.523)	0.384	5.52 (2–17)	1.92 (1-2)	2.62
SWE	Wild	0.562 (0–0.948)	0.224 (0–0.542)	0.294	4.91 (1–15)	1.68 (1-2)	2.36
	Reared	0.587 (0–0.984)	0.233 (0–0.524)	0.307	5.23 (1–20)	1.72 (1-2)	2.45
CAN	Wild	0.638 (0.063–0.951)	0.125 (0–0.524)	0.219	6.73 (2–15)	1.54 (1-2)	2.50
	Reared	0.621 (0–0.963)	0.125 (0–0.533)	0.216	6.32 (1–20)	1.50 (1-2)	2.39

Pop.: population.

Standard Outlier Tests

Altogether, eleven loci were identified as outliers in three wild-reared comparisons (three each in SWE and CAN and five in IRL) at the 99% P-level using FDIST2 (Figure 1(a)). Those eleven loci consisted of nine SNPs and two microsatellite loci. Two outliers identified in SWE and CAN (loci 14053_0820 and 15556_147, resp.) were located on the same Atlantic salmon linkage group (LG23-Chrom. 16). The genetic distance between these markers in the female linkage map is 8 cM [28]. The Bayesian F_{ST} test (BAYESFST) identified nine outlier loci (three in IRL, four in SWE and two in CAN) with substantial evidence (Bayes factor > 3 corresponding to log10(BF) > 0.5) of directional selection according to Jeffreys' interpretation [27] (Figure 1(b)). Out of the fourteen outliers identified altogether, six loci were detected using both methods (Figure 1, Table 2).

Table 2: Ten loci/genomic regions supported by multiple lines of evidence of being affected by directional selection during domestication/captive breeding in Atlantic salmon. Microsatellite markers are given in italics.

Pop.	Linkage group (d—dam, s—sire)	Chromosome	Markers	Map position cM	FDIST2 $P < 0.01$	BayeScan log10(BF) > 0.5	Clustering of loci F_{ST}	Clustering of loci lnRH $P < 0.05$	Annotation[a] Top hit
IRL	d11	3	13699_0319	0	X	X		X	
			14711_157	0.5				X	Q5R5Y3|B3GT2_PONPY Beta-13-galactosyltransferase 2 (Beta-13-GalTase 2) (Beta-3-Gal-T2)
	d19	8	15784_1161	7.9	X	X	X		Q61425|HCDH_MOUSE Short chain 3-Hydroxyacyl-CoA dehydrogenase mitochondrial
			BHMS365	24.9			X		
	d14	21	*BHMS311*	1.5	X	X			
	s13	19	16938_271	25.9	X			X	gnl|CDD|29599 cd01417 Ribosomal_L19e_E Ribosomal protein L19e eukaryotic. L19e
			16938_888	25.9				X	gnl|CDD|29599 cd01417 Ribosomal_L19e_E Ribosomal protein L19e eukaryotic. L19e
SWE	d07	24	12320_396	46.3	X	X	X	X	sp|P20135|GSTT1_CHICK Glutathione S-Transferase theta-1 (GST-CL1) (GST class-theta)
			SSsp2216	46.3			X	X	
	s22	17	12302_129	7			X	X	sp|Q9JJL3|SO1B2_MOUSE Solute carrier organic anion transporter family member 1B2
			12302_196	7			X	X	sp|Q9JJL3|SO1B2_MOUSE Solute carrier organic anion transporter family member 1B2
	d33	28	14501_498	0	X				
			16532_248	52.5			X	X	sp|P63030|BR44L_MOUSE Brain protein 44-like protein
			12035_0440	54			X	X	sp|Q98967|CYT_ONCKE Cystatin precursor
CAN	d18	23	14186_163	25.7	X	X	X		sp|Q96KP1|EXOC2_HUMAN Exocyst complex component 2 (Exocyst complex component Sec5)
			Ssa85DU	36.4			X		
	d06	12	16937_940	57.4			X	X	sp|P24844|MLRN_HUMAN Myosin regulatory light chain 2 smooth muscle isoform (Myosin RLC)
			17580_1116	57.8			X	X	sp|Q6NWF6|K2C8_BRARE Keratin type II cytoskeletal 8 (Cytokeratin-8) (CK-8) (Keratin-8)
	d23	16	14053_0820	33.4	X	X			

[a]BLASTX search against CDD and Swissprot databases [28].

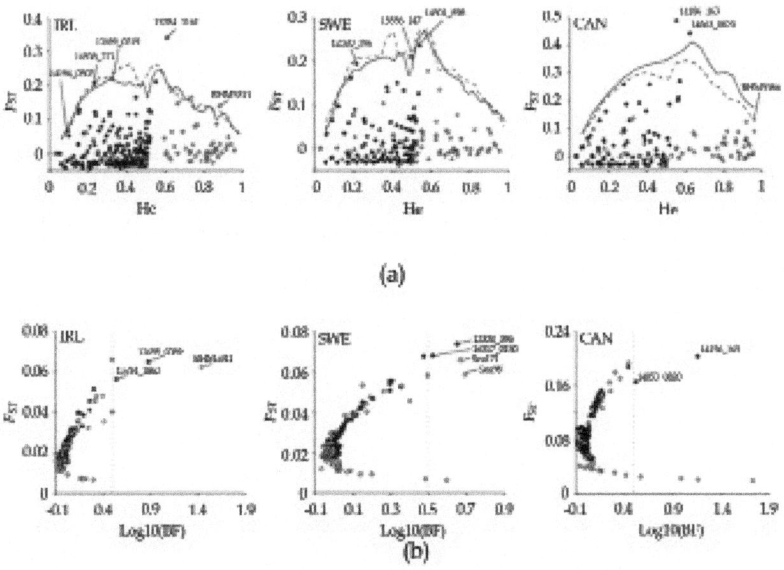

Figure 1: Results of two outlier tests in three independent pairs of wild-reared Atlantic salmon populations. Locus names of putative outliers potentially affected by selection (see Section 3) are indicated. (a) FDIST2: empirical distribution of F_{ST} against heterozygosity. The solid and dotted lines represent the upper 99% confidence interval for SNPs and tandem repeat markers, respectively; (b) Bayescan: The F_{ST} estimates plotted against Bayes factor. The dashed lines correspond to the Bayes factor 3 (log10(BF) = 0.5).

Clustering of Loci

By utilizing the Atlantic salmon genetic linkage map information, we evaluated the evidence for clustering of loci that exhibit elevated genetic differentiation (F_{ST}) or alternatively, show large differences in genetic diversity (lnRH) between wild-reared population pairs. Altogether, seven adjacent locus pairs (two in IRL, three in SWE, and two in CAN) exhibited evidence for clustering of markers (permutation-based significance threshold, P<0.05) with high genetic differentiation (F_{ST}) along the chromosome (Figure 2). However, in some cases high differentiation of adjacent markers was mostly driven by a single marker (IRL: LG19-Chrom. 8; SWE: LG7-Chrom. 24; CAN: LG18-Chrom. 23), while in other cases both loci exhibited elevated differentiation (IRL: LG21-Chrom. 26; SWE: LG22-Chrom. 17; CAN: LG6-Chrom. 12). When abs(lnRH) was used as a parameter to test the evidence for clustering of loci, nine adjacent locus pairs (three in IRL, four in SWE, and two in CAN) showed large changes in genetic diversity at 5% significance level (Figure 3).

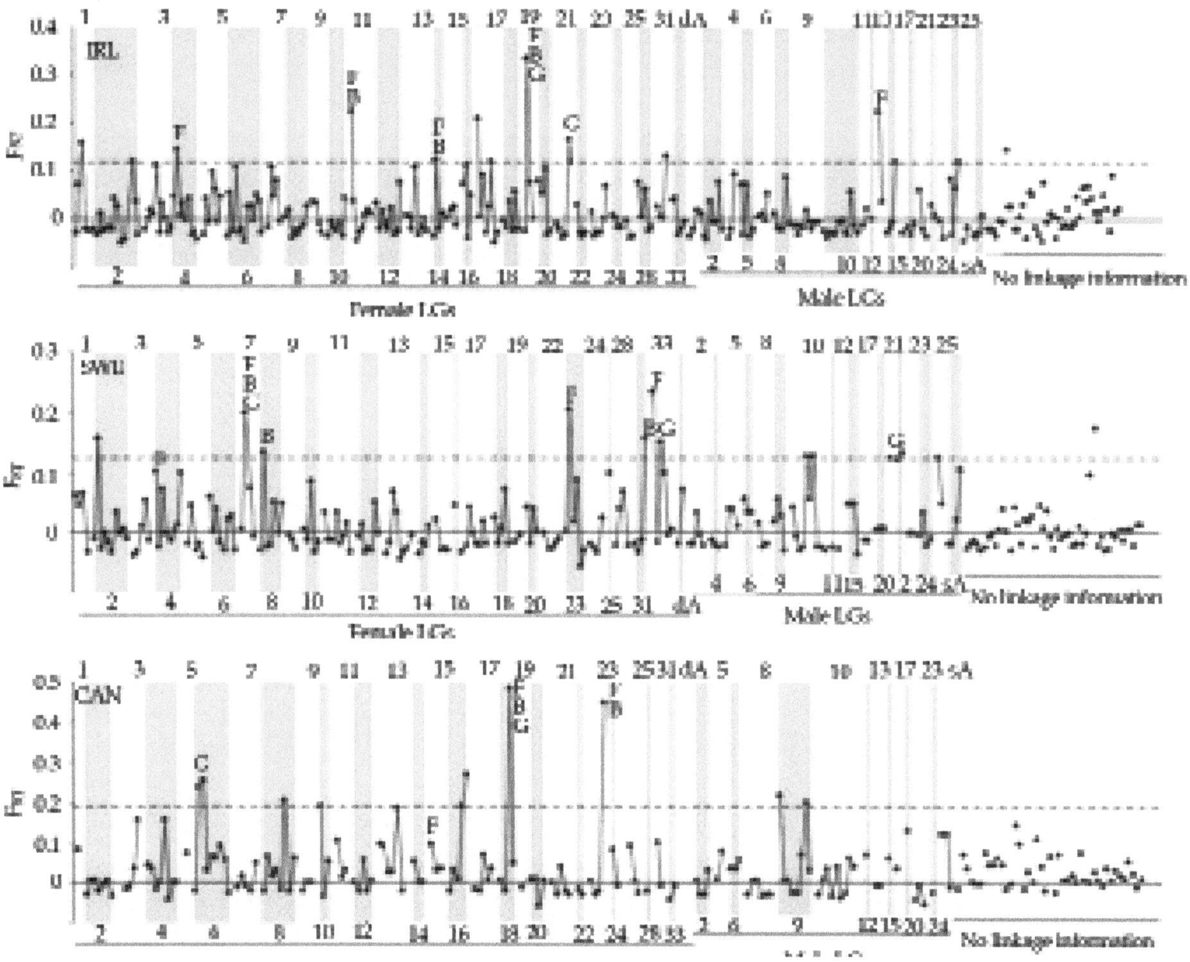

Figure 2: Distribution of F_{ST} along the female and male linkage groups (LGs) in three wild-reared Atlantic salmon pairwise comparisons (IRL, SWE, and CAN). F indicates markers that were identified as outliers at the 99% P level using FDIST2 [26]. B indicates markers that were identified as outliers (log10(Bayes factor) > 0.5) using Bayescan [27]. G indicates clustering of adjacent markers with elevated genetic differentiation (F_{ST}) (permutation-based significance threshold,P<0.05). Dashed lines correspond to the upper 2.5% of the empirical F_{ST} distribution and numbers along the x-axis indicate linkage groups as defined in [28].

Figure 3: Distribution of lnRH along the female and male linkage groups (LGs) in three wild-reared Atlantic salmon pairwise comparisons (IRL, SWE, CAN). G indicates clustering of adjacent markers (permutation-based significance threshold,P<0.05) that show large differences in genetic diversity measured as lnRH [29]. Dashed lines correspond to the 95% confidence limits (±1.96) of standardized lnRH estimates and numbers along the x-axis indicate linkage groups as defined in [28].

False Discovery Rate (FDR) and Putative Candidate Regions

To address the multiple testing problem that arises when large numbers of markers are tested against a null hypothesis, we estimated the false discovery rate (FDR) for the four previously-used neutrality tests: FDIST2, lnRH, clustering of adjacent loci with high F_{ST}, or abs(lnRH). As expected, a considerable proportion of the loci that fall to the tails of the distribution are probably false positives, that is, they emerge because of the large number of tests performed (Figure 4). As a result, calculation of FDR revealed that if we select for subsequent validation of five loci based on extreme values from a single neutrality test, it is likely that from 2 to 4 loci out of five, depending on the particular population comparison and neutrality test, are likely to be false positives (type I error). On the other hand, the estimation of FDR also suggests that a small but nonneglible proportion of loci from our genome scan might be truly deviating from the neutral expectations and hence, represent potentially real footprints of directional selection related to domestication/captive breeding. When considering four different neutrality tests simultaneously, directional selection in a total of ten genomic regions was supported by multiple sources of evidence (Table 2). Among these ten putative candidate loci, four loci/genomic regions (IRL: LG11-Chrom. 3, 0–0.5 cM; IRL: LG19-Chrom. 8, 7.9–24.9 cM; SWE: LG7-Chrom. 24, 46.3 cM; CAN: LG18-Chrom. 23, 25.7–36.4 cM) were detected as outliers in three or more tests. None of the putative candidate regions overlapped in the three wild-reared population pairs.

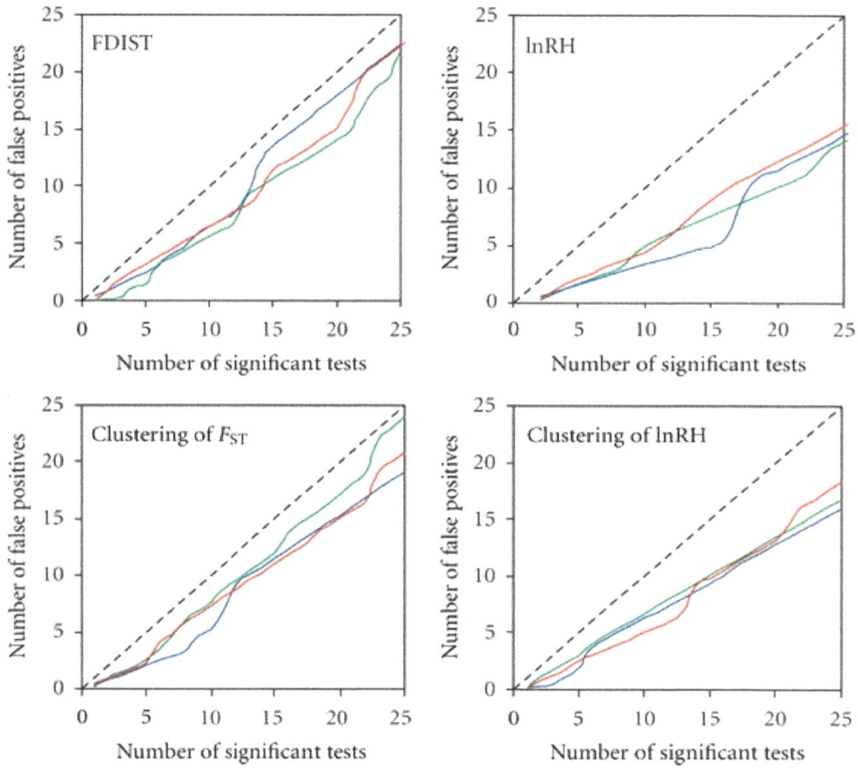

Figure 4: Estimated false discovery rate (FDR) for four different neutrality tests: FDIST2, lnRH, clustering of F_{ST} and lnRH. Green, blue and red lines correspond to q qvalue estimates for three wild-reared Atlantic salmon pairwise comparisons (IRL, SWE and CAN, resp.).

Simulation of Selection and Genetic Hitchhiking

We also evaluated whether strong directional selection over a very short period of time from standing variation also affects adjacent genetic markers along the chromosome, using the empirical genotype data from wild population as a source for subsequent forward simulations. We selected three linkage groups/chromosomes that contained previously-identified outlier loci (IRL: LG21-Chrom. 26, 15784_1161; SWE: LG33-Chrom. 28, 14501_498; CAN: LG6-Chrom. 12, 14186_163) with the closest adjacent marker 0 cM (CAN), 7.8 cM (IRL), and 33.6 cM (SWE) apart. Overall, strong selection on the minor allele resulted in high differentiation between wild population and reared strain pairs at selected markers, comparable with the empirical data (Figure 5). The hitchhiking effect of the adjacent markers, however, depended on the distance from the selected marker. For example, when the distance from the target of selection to the closest marker was relatively large (IRL: 7.8 cM; SWE: 33.6 cM), linked loci did not exhibit elevated differentiation between wild populations and reared strains. However, when markers were very closely linked to the target of selection, nearby loci also showed elevated differentiation (CAN: LG6-Chrom. 12).

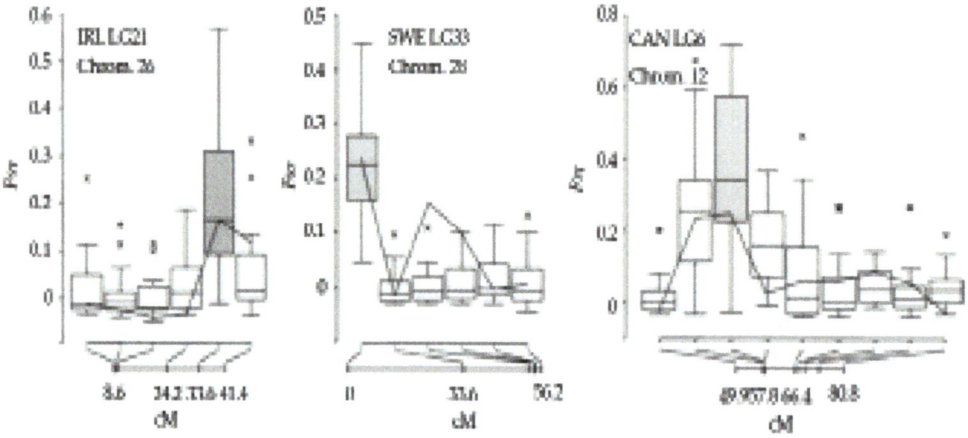

Figure 5: The effect of positive selection on standing variation on F_{ST} in three linkage groups (IRL: LG21-Chrom. 26; SWE: LG33-Chrom. 28, CAN: LG6-Chrom. 12). Bold line indicates the empirical distribution of F_{ST}. Box plots showing the median and interquartile range, nonoutlier range, and outliers and extreme cases from 25 simulation replicates are shown with a line within the rectangle, rectangle, whiskers, cross and an asterisk, respectively. Filled grey rectangles correspond to the markers under selection (positive selection on minor allele); the position of each marker is provided on the x-axis.

DISCUSSION

In recent years, there has been a considerably increased interest in searching for signatures of natural or artificial selection, both in model and nonmodel organisms [54–56]. In the present study, we screened 331 codominant markers in three population pairs and identified ten genomic regions potentially affected by artificial selection associated with domestication/captive breeding in Atlantic salmon. We also evaluated how adaptation from standing variation affects adjacent SNP and microsatellite variation along the chromosomes and, by using forward simulations with strong selection, we were able to generate genetic differentiation patterns comparable to the empirical data. However, similar to other first-pass genome scans, the actual causative polymorphisms that affect the performance of individuals in the captive environment remain to be identified. As such, this study represents useful step towards identifying functional variants involved in domestication of Atlantic salmon. Below, we discuss our findings, their biological significance, and also methodological issues relevant for future hitchhiking-mapping studies.

Overview of the Candidate Regions

By combining information from four different neutrality tests, in total, ten genomic regions showed signs of directional selection based on multiple sources of evidence (Table 2). When several linked markers exhibited evidence of directional selection, most of the identified candidate regions were relatively small ranging from zero to a few centimorgans (cM) in the female Atlantic salmon linkage map. In two cases, the adjacent outliers were located in the same contig [28] less than 1 kb from each other (IRL: LG13-Chrom. 19; SWE: LG22-Chrom. 17). In three cases, the putative candidate regions were larger, ranging from 10 to over 50 cM (CAN: LG18-Chrom. 23; IRL: LG19-Chrom. 8; SWE: LG33-Chrom. 28). These regions may represent false positives or alternatively contain multiple independent targets for selection, as our simulations indicate that the signatures of selection from standing variation do not necessarily extend over such large chromosomal regions (Figure 5). Examination of the genes within candidate regions revealed that two outlier loci in SWE LG33-Chrom. 28, corresponding to cystatin precursor and brain protein 44-like protein, were also reported in [24] as being among 1.7% of genes showing differential expression between the progeny of farmed and wild Atlantic salmon. Assuming that 3557 genes analysed by Roberge et al. [24] had an equal chance of ending up in our candidate list presented in Table 2, such overlap would be rarely emerge by chance alone (10,000 permutations,$P<0.012$). Interestingly, the same two genes also showed significant gene expression differences between domesticated- and wild-rainbow trout (Oncorhynchus mykiss) [57]. These findings, combined with the gene expression information, indicate that cystatin precursor and brain protein 44-like-protein might be interesting targets for more detailed genomic analysis.

We found that several of our candidate regions overlap with Atlantic salmon QTLs identified for various ecologically and economically important traits in earlier studies. For example, linkage groups 6 (Chrom. 12), 7 (Chrom. 24), and 11 (Chrom. 3) have been shown to harbour QTLs for body weight [58, 59], condition factor [58, 60], precoccious male parr maturation [Pederesen et al. unpublished], morphometry [61], and time of emergence [62]. Similarly, other linkage groups that included regions influenced by domestication have been shown to harbour QTLs for body weight (LG14-Chrom. 21; LG18-Chrom 23, LG19-Chrom.8, LG23-Chrom. 16), length (LG13-Chrom.19), LG14-Chrom. 21, LG18, LG19), condition factor (LG13-Chrom.19, LG18, LG23-Chrom. 16), and fillet color (LG 11-Chrom. 3, LG13-Chrom. 19, LG19-Chrom. 8) [58–62]. Therefore, it is possible that some of the identified regions under selection are associated with growth, morphology, and life history variation. However, it is important to remember that Atlantic salmon males show an extremely low recombination rate across the vast majority of the genome [31] and many of the QTLs have been identified using a male as a mapping parent. Hence, considerable uncertainty exists about the exact location of many quantitative trait loci in Atlantic salmon (but see also [63]). It is noteworthy that one of our outlier SNPs (LG11-Chrom. 3, 314711_157) has been identified as a strong outlier under diversifying selection among the North American Atlantic salmon populations [64].

Domestication in Salmonid Fishes: Same Pathways, Different Genes?

Our study provides little support for the hypothesis of independent parallel genetic changes at the DNA level in Atlantic salmon from different population pairs as none of the identified candidate regions overlapped in the three wild-population/reared strain pairs. This is in contrast with earlier studies at the gene expression level that have demonstrated the presence of parallel differences in transcription of genes between independent domesticated- and wild-Atlantic salmon [24]. However, a closer examination of the transcription patterns reveals that only a relatively small proportion of differentially expressed genes (16%) exhibited parallel changes in independent farmed reared strains, while the majority of changes in gene transcription are strain specific. Similarly, in rainbow trout (Oncorhynchus mykiss) and brook charr (Salvelinus fontinalis), differentially expressed genes between reared strains and wild populations have been shown to belong primarily to similar biological functions such as energy metabolism of lipids and protein synthesis, but the specific genes involved in these biological functions were found to be different [57, 65]. These results are also in accordance with the recent genome scan by Flori et al. [1] who found that the same physiological pathways but different genes are affected by artificial selection in dairy cattle. Such parallel effects at higher functional levels in domesticated animals strongly suggest that the same downstream pathways are being affected during the domestication process. As a result, studies focusing on analyses of gene expression alone provide only limited information about the number (and position) of causative allelic variants responsible for the rapid changes in gene expression of domesticated fish. This has been elegantly demonstrated by Devlin et al. [66] who showed how the insertion of growth hormone gene can generate whole genome transcriptome signatures very similar to the domesticated strain in coho salmon (Oncorhynchus kisutch).

It is also possible that several other factors contribute to the population-specific signatures of selection, such as (i) different selective regimes in farmed versus ranched strains, (ii) different selective regimes between separate ranched strains, or, (iii) allele frequency differences between populations/strains. In fact, it is possible that all of these factors contribute to the lack of overlap of outliers because (i) farmed strains experience strong intentional selection for growth, disease resistance, delayed maturity, and other commercially important traits while ranched strains experience inadvertent selection in captivity as well as natural selection at sea [22, 67]; (ii) the selection regimes at the feeding grounds of the Atlantic Ocean (IRL) and the Baltic Sea (SWE) may be different; and (iii) loci that are polymorphic among European salmon are often not variable in the North American populations and vice versa; even if polymorphism exists in all strains, when the allele that is linked to the positively selected site is already at high frequency in a particular population, selection does not necessarily induce a strong increase in the level of genetic differentiation and reduction in diversity.

Limitations and Future Directions

There is growing consensus that such genome scans typically suffer from low power and high-false positive rates [68–71]. As a consequence, it is important to note that our analyses, and hence interpretations, are subject to several limitations. First, our study is based on a relatively small number of individuals, which affects the accuracy of allele frequency estimates. Therefore, confirming the allele frequency estimates using larger sample sizes and multiple cohorts (including multiple tributary populations from the Saint John River) could be clearly beneficial. However, the forward simulations were able to generate strong footprints of selection when using similar numbers of individuals, thus suggesting that our study indeed have reasonable power. The variable outcome of individual simulations, however, might be at least partially influenced by the limited sample size. The limited sample size may have also some effect on the genetic diversity estimates. For example, in contrast to many earlier studies that have documented lower variability in farmed Atlantic salmon compared to wild populations (e.g., [51–53]), we did not observe lower levels of genetic diversity in reared strains compared with their wild progenitor populations. However, earlier studies have compared the diversity of wild populations with farmed Norwegian salmon which have the longest history of selective breeding [52, 53]. Therefore, it is not unexpected that reared strains of Burrishoole (IRL) and Umeälven (SWE) exhibit relatively similar levels of genetic variation compared to their wild counterparts, especially considering the relatively large number of fish used during breeding in recent years (pers. comm. P. McGinnity; J. Nilsson). This is also in accordance with the study of Vasemägi et al. [72] who reported relatively similar levels of genetic diversity in captive Umeälven and wild Vindeläven salmon.

Second, even though the number of codominant markers used in this study is higher than typically applied in nonmodel organism genome scans to date, it still covers only a very small proportion of the Atlantic salmon genome. For example, Pollinger et al. [73] estimated that for detecting strong recent artificial selection on novel mutation in dog, one would require a spacing of one highly polymorphic marker per 0.8 centimorgans (cM) or less. Assuming that the length of the female Atlantic salmon linkage map is close to 2000 cM [28], then approximately 2500 evenly spaced markers would be needed to cover the Atlantic salmon genome at this density. Therefore, a recently developed high density SNP chip in Atlantic salmon consisting of more than 5500 SNPs should provide a valuable tool for mapping regions under selection in the future [31]. On the other hand, it has been suggested that as many as 150,000 SNPs will be required for identifying genome-wide footprints of selection in the bovine genome which has the genome size comparable to Atlantic salmon [74]. To this end the use of next generation sequencing and genome complexity reduction methods hold great promise to gain whole-genome coverage for population genomic studies even for nonmodel species [11, 75]. The other alternative strategies for hitchhiking mapping include focusing on a priori defined different functional groups or pathways rather than single genes [76–78]. Finally, several major chromosomal differences between the European and North American Atlantic salmon have been recently demonstrated while the fine-scale order of genetic markers has been conserved [79]. Hence, our analysis using genetic maping information should be taken with caution until the sequence of the Atlantic salmon genome becomes available.

In conclusion, our results together with other recent studies [1, 11, 80, 81] illustrate the significance of standing genetic variation during the early stages of adaptation in leading to rapid evolutionary responses to novel environments. Consequently, one of the future challenges for hitchhiking mapping studies is to develop methods and approaches that consider alternatives to the classic theory of "hard" selective sweeps (adaptation from de novo mutation) in order to obtain more comprehensive understanding about molecular mechanisms of adaptation. Species with domesticated/captive strains are likely to remain popular models in this context, providing unique opportunities to identify molecular targets of artificial selection.

ACKNOWLEDGMENTS

This work was financed by the Academy of Finland, the Estonian Science Foundation (Grant nos. 6802, 8215, 7348), and ESF Research Networking Programme (ConGen). The authors would like to thank Arne Roseth (CIGENE) for his help during SNP genotyping. P. McGinnity and T. Cross were supported by the Beaufort Marine Research Award in Fish Population Genetics, funded by the Irish Government under the Sea Change Programme.

REFERENCES

1. L. Flori, S. Fritz, F. Jaffrézic et al., "The genome response to artificial selection: a case study in dairy cattle," PLoS ONE, vol. 4, no. 8, Article ID e6595, 2009
2. M. D. Purugganan and D. Q. Fuller, "The nature of selection during plant domestication," Nature, vol. 457, no. 7231, pp. 843–848, 2009.
3. M. Gautier, L. Flori, A. Riebler et al., "A whole genome Bayesian scan for adaptive genetic divergence in West African cattle," BMC Genomics, vol. 10, article 550, 2009
4. C. A. Driscoll, D. W. Macdonald, and S. J. O'Brien, "From wild animals to domestic pets, an evolutionary view of domestication," Proceedings of the National

Academy of Sciences of the United States of America, vol. 106, supplement 1, pp. 9971–9978, 2009.

5. J. Gu, N. Orr, S. D. Park et al., "A genome scan for positive selection in thoroughbred horses," PLoS ONE, vol. 4, no. 6, Article ID e5767, 2009.

6. J. Maynard Smith and J. Haigh, "The hitch-hiking effect of a favourable gene," Genetics Research, vol. 23, no. 1, pp. 23–35, 1974.

7. M. Przeworski, "The signature of positive selection at randomly chosen loci," Genetics, vol. 160, no. 3, pp. 1179–1189, 2002

8. Y. Kim and R. Neilsen, "Linkage disequilibrium as a signature of selective sweeps," Genetics, vol. 167, no. 3, pp. 1513–1524, 2004

9. J. Hermisson and P. S. Pennings, "Soft sweeps: molecular population genetics of adaptation from standing genetic variation," Genetics, vol. 169, no. 4, pp. 2335–2352, 2005

10. P. F. Colosimo, K. E. Hosemann, S. Balabhadra et al., "Widespread parallel evolution in sticklebacks by repeated fixation of ectodysplasin alleles," Science, vol. 307, no. 5717, pp. 1928–1933, 2005.

11. P. A. Hohenlohe, S. Bassham, P. D. Etter, N. Stiffler, E. A. Johnson, and W. A. Cresko, "Population genomics of parallel adaptation in threespine stickleback using sequenced RAD tags," PLoS Genetics, vol. 6, no. 2, Article ID e1000862, 2010

12. P. S. Pennings and J. Hermisson, "Soft sweeps II—molecular population genetics of adaptation from recurrent mutation or migration," Molecular Biology and Evolution, vol. 23, no. 5, pp. 1076–1084, 2006.

13. P. S. Pennings and J. Hermisson, "Soft sweeps III: the signature of positive selection from recurrent mutation," PLoS Genetics, vol. 2, no. 12, article e186, 2006

14. S. I. Wright, I. V. Bi, S. G. Schroeder, et al., "The effects of artificial selection on the maize genome," Science, vol. 308, no. 5745, pp. 1310–1314, 2005.

15. L. Trut, I. Oskina, and A. Kharlamova, "Animal evolution during domestication: the domesticated fox as a model," BioEssays, vol. 31, no. 3, pp. 349–360, 2009.

16. H. M. Gjøen and H. B. Bentsen, "Past, present, and future of genetic improvement in salmon aquaculture," ICES Journal of Marine Science, vol. 54, no. 6, pp. 1009–1014, 1997.

17. M. R. Gross, "One species with two biologies: Atlantic salmon (Salmo salar) in the wild and in aquaculture," Canadian Journal of Fisheries and Aquatic Sciences, vol. 55, supplement 1, pp. 131–144, 1998

18. H. Araki, B. Cooper, and M. S. Blouin, "Genetic effects of captive breeding cause a rapid, cumulative fitness decline in the wild," Science, vol. 318, no. 5847, pp. 100–103, 2007

19. F. Allendorf, "Delay of adaptation to captive breeding by equalizing family size," Conservation Biology, vol. 7, pp. 416–419, 1993.

20. R. S. Waples, "Dispelling some myths about hatcheries," Fisheries, vol. 24, no. 2, pp. 12–21, 1999.

21. P. McGinnity, E. Jennings, E. deEyto et al., "Impact of naturally spawning captive-bred Atlantic salmon on wild populations: depressed recruitment and increased risk of climate-mediated extinction," Proceedings of the Royal Society B, vol. 276, no. 1673, pp. 3601–3610, 2009.

22. H. M. Araki, B. A. Berejikian, M. J. Ford, and M. S. Blouin, "Fitness of hatchery-reared salmonids in the wild," Evolutionary Applications, vol. 1, no. 21, pp. 342–355, 2008.

23. P. McGinnity, P. Prodöhl, N. Ó. Maoiléidigh et al., "Differential lifetime success and performance of native and non-native Atlantic salmon examined under communal natural conditions," Journal of Fish Biology, vol. 65, supplement 1, pp. 173–187, 2004.

24. C. Roberge, S. Einum, H. Guderley, and L. Bernatchez, "Rapid parallel evolutionary changes of gene transcription profiles in farmed Atlantic salmon," Molecular Ecology, vol. 15, no. 1, pp. 9–20, 2006.

25. C. Roberge, É. Normandeau, S. Einum, H. Guderley, and L. Bernatchez, "Genetic consequences of interbreeding between farmed and wild Atlantic salmon: insights from the transcriptome," Molecular Ecology, vol. 17, no. 1, pp. 314–324, 2008.

26. M. Raymond and F. Rousset, "Genepop version 1.2, population genetics software for exact tests and ecumenicism," Journal of Heredity, vol. 86, no. 3, pp. 248–249, 1995.

27. B. S. Weir and C. C. Cockerham, "Estimating F-statistics for the analysis of population structure," Evolution, vol. 38, no. 6, pp. 1358–1370, 1984.

28. T. Moen, B. Hayes, M. Baranski et al., "A linkage map of the Atlantic salmon (Salmo salar) based on EST-derived SNP markers," BMC Genomics, vol. 9, article 223, 2008.

29. A. Vasemägi, J. Nilsson, and C. R. Primmer, "Expressed sequence tag-linked microsatellites as a source of gene-associated polymorphisms for detecting signatures of divergent selection in Atlantic salmon (Salmo salar L.)," Molecular Biology and Evolution, vol. 22, no. 4, pp. 1067–1076, 2005.

30. C. D. Quinton, I. McMillan, and B. D. Glebe, "Development of an Atlantic salmon (Salmo salar) genetic improvement program: genetic parameters of harvest body weight and carcass quality traits estimated with animal models," Aquaculture, vol. 247, no. 1–4, pp. 211–217, 2005.

31. S. Lien, L. Gidskehaug, T. Moen, et al., "A dense SNP-based linkage map for Atlantic salmon (Salmo salar) reveals extended chromosome homeologies and striking differences in sex-specific recombination patterns," BMC Genomics, vol. 12, article 615, 2011.

32. R. G. Danzmann, E. A. Davidson, M. M. Ferguson et al., "Distribution of ancestral proto-Actinopterygian chromosome arms within the genomes of 4R-derivative salmonid fishes (Rainbow trout and Atlantic salmon)," BMC Genomics, vol. 9, article 557, 2008.

33. R. B. Phillips, K. A. Keatley, M. R. Morasch et al., "Assignment of Atlantic salmon (Salmo salar) linkage groups to specific chromosomes: conservation of large syntenic blocks corresponding to whole chromosome

arms in rainbow trout (Oncorhynchus mykiss)," BMC Genetics, vol. 10, article 46, 2009.

34. K. Tang, D. J. Fu, D. Julien, A. Braun, C. R. Cantor, and H. Köstek, "Chip-based genotyping by mass spectrometry," Proceedings of the National Academy of Sciences of the United States of America, vol. 96, no. 18, pp. 10016–10020, 1999.

35. Vasemägi, R. Gross, T. Paaver, M. L. Koljonen, M. Säisä, and J. Nilsson, "Analysis of gene associated tandem repeat markers in Atlantic salmon (Salmo salar L.) populations: implications for restoration and conservation in the Baltic Sea," Conservation Genetics, vol. 6, no. 3, pp. 385–397, 2005.

36. Vasemägi, J. Nilsson, and C. R. Primmer, "Seventy-five EST-linked Atlantic salmon (Salmo solar L.) microsatellite markers and their cross-amplification in five salmonid species," Molecular Ecology Notes, vol. 5, no. 2, pp. 282–288, 2005.

37. J. Goudet, "FSTAT, version 1.2: a computer program to calculate F statistics," Journal of Heredity, vol. 86, no. 6, pp. 485–486, 1995.

38. S. W. Guo and E. A. Thompson, "Performing the exact test of Hardy-Weinberg proportion for multiple alleles," Biometrics, vol. 48, no. 2, pp. 361–372, 1992

39. M. A. Beaumont and R. A. Nichols, "Evaluating loci for use in the genetic analysis of population structure," Proceedings of the Royal Society B, vol. 263, no. 1377, pp. 1619–1626, 1996.

40. M. Foll and O. Gaggiotti, "A genome-scan method to identify selected loci appropriate for both dominant and codominant markers: a Bayesian perspective," Genetics, vol. 180, no. 2, pp. 977–993, 2008.

41. M. O. Kauer, D. Dieringer, and C. Schlötterer, "A microsatellite variability screen for positive selection associated with the "Out of Africa" habitat expansion of Drosophila melanogaster," Genetics, vol. 165, no. 3, pp. 1137–1148, 2003.

42. McEvoy, S. Beleza, and M. D. Shriver, "The genetic architecture of normal variation in human pigmentation: an evolutionary perspective and model," Human Molecular Genetics, vol. 15, no. 2, pp. R176–R181, 2006.

43. J. D. Storey and R. Tibshirani, "Statistical significance for genomewide studies," Proceedings of the National Academy of Sciences of the United States of America, vol. 100, no. 16, pp. 9440–9445, 2003

44. J. D. Storey, "A direct approach to false discovery rates," Journal of the Royal Statistical Society B, vol. 64, no. 3, pp. 479–498, 2002.

45. B. Peng and M. Kimmel, "SimuPOP: a forward-time population genetics simulation environment," Bioinformatics, vol. 21, no. 18, pp. 3686–3687, 2005.

46. R. D. H. Barrett and D. Schluter, "Adaptation from standing genetic variation," Trends in Ecology and Evolution, vol. 23, no. 1, pp. 38–44, 2008.

47. J. Lohm, M. Grahn, Å. Langefors, Ø. Andersen, A. Storset, and T. Von Schantz, "Experimental evidence for major histocompatibility complex-allele-specific resistance to a bacterial infection," Proceedings of the Royal Society B, vol. 269, no. 1504, pp. 2029–2033, 2002

48. R. L. Wang, A. Stec, J. Hey, L. Lukens, and J. Doebley, "The limits of selection during maize domestication," Nature, vol. 398, no. 6724, pp. 236–239, 1999

49. R. M. Clark, E. Linton, J. Messing, and J. F. Doebley, "Pattern of diversity in the genomic region near the maize domestication gene tb1," Proceedings of the National Academy of Sciences of the United States of America, vol. 101, no. 3, pp. 700–707, 2004

50. K. M. Olsen, A. L. Caicedo, N. Polato, A. McClung, S. McCouch, and M. D. Purugganan, "Selection under domestication: evidence for a sweep in the rice waxy genomic region," Genetics, vol. 173, no. 2, pp. 975–983, 2006.

51. B. Mjølnerød, U. H. Refseth, E. Karlsen, T. Balstad, K. S. Jakobsen, and K. Hindar, "Genetic differences between two wild and one farmed population of Atlantic salmon (Salmo salar) revealed by three classes of genetic markers," Hereditas, vol. 127, no. 3, pp. 239–248, 1997.

52. A. T. Norris, D. G. Bradley, and E. P. Cunningham, "Microsatellite genetic variation between and within farmed and wild Atlantic salmon (Salmo salar) populations," Aquaculture, vol. 180, no. 3-4, pp. 247–264, 1999.

53. Ø. Skaala, B. Høyheim, K. A. Glover, and G. Dahle, "Microsatellite analysis in domesticated and wild Atlantic salmon (Salmo salar L.): allelic diversity and identification of individuals," Aquaculture, vol. 240, no. 1–4, pp. 131–143, 2004.

54. J. F. Storz, "Using genome scans of DNA polymorphism to infer adaptive population divergence," Molecular Ecology, vol. 14, no. 3, pp. 671–688, 2005.

55. A. Vasemägi and C. R. Primmer, "Challenges for identifying functionally important genetic variation: the promise of combining complementary research strategies," Molecular Ecology, vol. 14, no. 12, pp. 3623–3642, 2005.

56. P. C. Sabeti, S. F. Schaffner, B. Fry et al., "Positive natural selection in the human lineage," Science, vol. 312, no. 5780, pp. 1614–1620, 2006.

57. W. Tymchuk, D. Sakhrani, and R. H. Devlin, "Domestication causes large-scale effects on gene expression in rainbow trout: analysis of muscle, liver and brain transcriptomes," General and Comparative Endocrinology, vol. 164, no. 2-3, pp. 175–183, 2009.

58. D. P. Reid, A. Szanto, B. Glebe, R. G. Danzmann, and M. M. Ferguson, "QTL for body weight and condition factor in Atlantic salmon (Salmo salar): comparative analysis with rainbow trout (Oncorhynchus mykiss) and Arctic charr (Salvelinus alpinus)," Heredity, vol. 94, no. 2, pp. 166–172, 2005.

59. R. D. Houston, S. C. Bishop, A. Hamilton et al., "Detection of QTL affecting harvest traits in a commercial Atlantic salmon population," Animal Genetics, vol. 40, no. 5, pp. 753–755, 2009.

60. M. Baranski, T. Moen, and D.-I. Våge, "Mapping of

quantitative trait loci for flesh colour and growth traits in Atlantic salmon (Salmo salar)," Genetics, Selection, Evolution, vol. 42, article 17, 2010.

61. E. G. Boulding, M. Culling, B. Glebe, P. R. Berg, S. Lien, and T. Moen, "Conservation genomics of Atlantic salmon: SNPs associated with QTLs for adaptive traits in parr from four trans-Atlantic backcrosses," Heredity, vol. 101, no. 4, pp. 381–391, 2008. · ·

62. A. Vasemägi, R. Gross, D. Palm, T. Paaver, and C. R. Primmer, "Discovery and application of insertion-deletion (INDEL) polymorphisms for QTL mapping of early life-history traits in Atlantic salmon," BMC Genomics, vol. 11, no. 1, article 156, 2010.

63. T. Moen, M. Baranski, A. K. Sonesson, and S. Kjøglum, "Confirmation and fine-mapping of a major QTL for resistance to infectious pancreatic necrosis in Atlantic salmon (Salmo salar): population-level associations between markers and trait," BMC Genomics, vol. 10, article 368, 2009.

64. H. Freamo, P. O›reilly, P. R. Berg, S. Lien, and E. G. Boulding, "Outlier SNPs show more genetic structure between two Bay of Fundy metapopulations of Atlantic salmon than do neutral SNPs," Molecular Ecology Resources, vol. 11, no. 1, pp. 254–267, 2011.

65. C. Sauvage, N. Derôme, E. Normandeau, J. St-Cyr, C. Audet, and L. Bernatchez, "Fast transcriptional responses to domestication in the brook charr Salvelinus fontinalis," Genetics, vol. 185, no. 1, pp. 105–112, 2010.

66. R. H. Devlin, D. Sakhrani, W. E. Tymchuk, M. L. Rise, and B. Goh, "Domestication and growth hormone transgenesis cause similar changes in gene expression in coho salmon (Oncorhynchus kisutch)," Proceedings of the National Academy of Sciences of the United States of America, vol. 106, no. 9, pp. 3047–3052, 2009.

67. M. M. McClure, F. M. Utter, C. Baldwin, et al., "Evolutionary effects of alternative artificial propagation programs: implications for viability of endangered anadromous salmonids," Evolutionary Applications, vol. 1, no. 2, pp. 356–375, 2008.

68. J. L. Kelley, J. Madeoy, J. C. Calhoun, W. Swanson, and J. M. Akey, "Genomic signatures of positive selection in humans and the limits of outlier approaches," Genome Research, vol. 16, no. 8, pp. 980–989, 2006.

69. K. M. Teshima, G. Coop, and M. Przeworski, "How reliable are empirical genomic scans for selective sweeps?" Genome Research, vol. 16, no. 6, pp. 702–712, 2006.

70. L. Excoffier, T. Hofer, and M. Foll, "Detecting loci under selection in a hierarchically structured population," Heredity, vol. 103, no. 4, pp. 285–298, 2009.

71. J. M. Akey, "Constructing genomic maps of positive selection in humans: where do we go from here?" Genome Research, vol. 19, no. 5, pp. 711–722, 2009.

72. Vasemägi, R. Gross, T. Paaver, M. L. Koljonen, and J. Nilsson, "Extensive immigration from compensatory hatchery releases into wild Atlantic salmon population in the Baltic sea: spatio-temporal analysis over 18 years," Heredity, vol. 95, no. 1, pp. 76–83, 2005.

73. J. P. Pollinger, C. D. Bustamante, A. Fledel-Alon, S. Schmutz, M. M. Gray, and R. K. Wayne, "Selective sweep mapping of genes with large phenotypic effects," Genome Research, vol. 15, no. 12, pp. 1809–1819, 2005.

74. W. Barendse, B. E. Harrison, R. J. Bunch, M. B. Thomas, and L. B. Turner, "Genome wide signatures of positive selection: the comparison of independent samples and the identification of regions associated to traits," BMC Genomics, vol. 10, article 178, 2009.

75. M. Bruneaux, S. E. Johnston, G. Herczeg, J. Merilä, C. R. Primmer, and A. Vasemägi, "Molecular evolutionary and population genetic analysis of ninespine stickleback using a modified RAD tag approach," Molecular Ecology. In press.

76. J. Kullo and K. Ding, "Patterns of population differentiation of candidate genes for cardiovascular disease," BMC Genetics, vol. 8, article 48, 2007.

77. A. M. Hancock, D. B. Witonsky, A. S. Gordon et al., "Adaptations to climate in candidate genes for common metabolic disorders," PLoS Genetics, vol. 4, no. 2, article e32, 2008.

78. Tonteri, A. Vasemägi, J. Lumme, and C. R. Primmer, "Beyond MHC: signals of elevated selection pressure on Atlantic salmon (Salmo salar) immune-relevant loci," Molecular Ecology, vol. 19, no. 7, pp. 1273–1282, 2010.

79. S. Brenna-Hansen, J. Li, M. P. Kent, et al., "Chromosomal differences between European and North American Atlantic salmon discovered by linkage mapping and supported by fluorescence in situ hybridization analysis," BMC Genomics, vol. 13, article 432, 2012.

80. M. K. Burke, J. P. Dunham, P. Shahrestani, K. R. Thornton, M. R. Rose, and A. D. Long, "Genome-wide analysis of a long-term evolution experiment with Drosophila," Nature, vol. 467, no. 7315, pp. 587–590, 2010.

81. J. K. Pritchard, J. K. Pickrell, and G. Coop, "The genetics of human adaptation: hard sweeps, soft sweeps, and polygenic adaptation," Current Biology, vol. 20, no. 4, pp. R208–R215, 2010.

CHAPTER 7

INTEGRATIVE TRANSCRIPTOME, GENOME AND QUANTITATIVE TRAIT LOCI RESOURCES IDENTIFY SINGLE NUCLEOTIDE POLYMORPHISMS IN CANDIDATE GENES FOR GROWTH TRAITS IN TURBOT

Diego Robledo, Carlos Fernández, Miguel Hermida, Andrés Sciara, José Antonio Álvarez-Dios, Santiago Cabaleiro, Rubén Caamaño, Paulino Martínez and Carmen Bouza,

Keywords: Scophthalmus maximus; turbot; growth; RNA-seq; candidate genes; SNP; integrative genomics

INTRODUCTION

Genetic dissection of polygenic traits is key for animal breeding and evolutionary genetics. Genomic studies offer the possibility of unraveling a huge amount of genetic polymorphisms useful for searching selection signatures in populations and for association studies with productive traits, such as growth or disease resistance, which could eventually be applied in marker-assisted selection (MAS) programs. The application of MAS has enabled increasing productivity in livestock previously subjected to traditional phenotype-based selection [1]. A more effective selection can be performed the closer the marker is to the DNA element responsible for phenotypic differences, or even better, if the marker itself causes that difference. Markers within genes physiologically related to the trait of interest may potentially allow for a more straightforward selection [2]. For example, an intronic single nucleotide polymorphism (SNP) in the chicken growth hormone gene was associated with a 16.9% increase in body weight [3], while microsatellite variation in the growth hormone receptor promoter in cattle could account for 6.9% production improvement (Hale *et al.*, 2000) [4]. Moreover, the identification of gene-associated markers has been applied for detecting adaptive variation and genetic structure in wild and farmed populations of commercially exploited species, which is particularly of interest in order to define conservation and management units in marine fisheries [5, 6].

These strategies require information on the target genes and also the discovery of suitable markers, a difficult task in non-model species with limited genomic resources. Methodologies using genomic information from related species allowed us to identify SNP markers associated with growth and reproduction in fish [7,8]. However, new generation sequencing has greatly facilitated marker identification in non-model species. In this regard, transcriptome sequencing using RNA-seq technology allows SNP identification at an affordable cost [9,10]. SNPs are abundant markers along the genomes (e.g., ~1 SNP/100 bp in fish [11]), suitable for assisted selection on a wide range of genomic scales relying on the linkage disequilibrium patterns. In addition, SNP variation may have functional meaning, involving not only non-synonymous amino acid substitutions but also affecting regulatory elements (transcription factor and microRNA binding sites, or splicing recognition sites) [12].

Growth is one of the most important traits in animal breeding and a target in most genetic breeding programs, also in aquaculture species [13]. While most terrestrial livestock such as cattle, pigs or chickens have been strongly selected for commercial traits, only around 10% of aquaculture

Citation: Diego Robledo, Carlos Fernández, Miguel Hermida, Andrés Sciara, José Antonio Álvarez-Dios, Santiago Cabaleiro, Rubén Caamaño , Paulino Martínez and Carmen Bouza, Integrative Transcriptome, Genome and Quantitative Trait Loci Resources Identify Single Nucleotide Polymorphisms in Candidate Genes for Growth Traits in Turbot, doi:10.3390/ijms17020243 species have undergone breeding programs [14]. Genomic technologies have increased our knowledge on the molecular basis of growth traits in fish [15]. Different genomic studies have been carried out to detect quantitative trait loci (QTL) for growth traits in finfish and identified candidate genes to be used in marker-assisted selection programs [16, 17, 18, 19, 20, 21]. Nevertheless, the genetic effects and regulatory pathways underlying the growth rate are still poorly understood in teleosts as compared to mammals [22, 23].

Turbot is a marine flatfish of commercial importance in Europe and China [24,25], whose culture in aquaculture facilities is a mature industry. The genomic resources available include a genetic map [26,27] and a recently assembled genome sequence [28]. Previous studies in turbot found several QTLs and associated markers for growth traits, major goals for selective breeding programs in this species [29,30]. In addition, genome scanning has been used for detecting loci involved in adaptive variation of turbot populations, some of them presumptively related to growth performance [31,32]. The integrative connection among these resources may have implications for fisheries management and genetic improvement in this species.

The main aim of this study was the detection and validation of SNP markers in turbot located in candidate genes involved in growth and associated with growth-related QTLs in this species. We used Illumina RNA-seq to obtain muscle and liver transcriptomes from control and fasted specimens enriched in growth regulation pathways, and used this information to identify growth-related genes and associated SNP markers. Integrating (positioning) our new transcriptome data with previous genetic map markers in the draft turbot genome sequence enabled us to locate the candidate transcripts and compare their positions with growth QTL markers reported in the turbot genetic map. SNPs located in 45 genes were selected based on their co-localization with QTLs and functional relevance for growth. These SNPs were validated in a turbot population and will be useful to perform association analysis with phenotypes for growth traits both in farmed broodstock and natural populations for fisheries management.

RESULTS AND DISCUSSION

RNA-Seq and SNP Calling

A total of 19,326,140 muscle and 17,466,901 liver reads were obtained from Illumina sequencing, which enriched the transcriptomic background for turbot, useful for functional and comparative genomics in flatfish and teleosts. After trimming, 89.6% of the reads remained and were aligned against the draft turbot genome [28]. Muscle and liver transcriptomes were successfully assembled from the aligned reads, identifying 19,147 and 15,928 genes, respectively, and a total of 20,447 genes (Table 1). The figures found in our study are comparable to those obtained using genome-guided alignment for *Takifugu rubripes* swim bladder (17,249 genes) and gill (16,836 genes), which together accounted for 19,388 different genes [33]. Genomes are critical tools for obtaining realistic transcriptomes since, at present, *de novo* assemblers usually produce a higher number of genes with lower mean length and N50 (length for which the collection of all genes of that length or longer contains at least half of the sum of the length of all the genes). For example, *de novo* assembly of the ray-finned fish *Coilia nasus* liver transcriptome resulted in 65,129 genes with a mean size of 607 bp and N50 of 813 bp [34].

Table 1: RNA-seq and transcriptome assembly statistics.

Statistic	Muscle	Liver	Total
Raw reads	19,326,140	17,466,901	36,793,041
Trimmed reads	17,596,236 (91.0%)	16,398,082 (93.9%)	33,994,318 (92.4%)
Concordant aligned reads	15,570,568 (88.5%)	14,805,040 (90.3%)	30,375,608 (89.4%)
Number of genes	19,147	15,928	20,447
Number of transcripts	27,664	22,061	33,795
Minimum transcript size	62	78	62
Maximum transcript size	67,989	17,880	67,989
Transcripts over 500 bp	26,659	21,319	32,650
Mean transcript length (bp)	2594.66	2428.24	2819.60
N50	3411	3154	3691
N90	1381	1293	1531

Concordant aligned reads are pairs of reads which were assembled at the same genomic location and the percentage shown is calculated to the trimmed reads.

Muscle and liver RNA-seq reads were used for SNP screening. A total of 93,558 polymorphisms were detected and 85,344 of them were SNPs (91.2%), similar to human estimates where SNPs account for 90% of genome variation [35]. The number of SNPs identified by next generation sequencing (NGS) technologies varies greatly among studies, probably due to different experimental designs (number of samples and tissues, sequencing depth, read length, genome availability, and SNP calling parameters, among others), but comparisons between transcriptome and genome data have been proven to improve experimental validation of SNPs [9,33,36,37]. As expected from previous studies in fish, most of the SNPs found in our work corresponded to A/G and C/T, each transition (ts) representing 28%, while transversion (tv) frequencies were below 13% (Figure 1). The ts/tv ratio of 1.346 is very similar to that previously reported in turbot based on 866 predicted SNPs obtained from two 454 Roche runs [38], and also consistent with the values found in common carp (*Cyprinus carpio*) (1.31) [39] and gilthead seabream (*Sparus aurata*) (1.37) [40]. Values were slightly higher in channel (1.56; *Ictalurus punctatus*) and blue (1.68; *Ictalurus furcatus*) catfish [37], but lower in other species, such as chum (*Oncorhynchus keta*) and sockeye (*Oncorhynchus nerka*) salmon (around 1) [41], close to the ratio estimated from intron regions in *C. carpio* (1.05) [39]. Still, these values are far from those in the human exome (3.2) [42] or in the whole nuclear genome (2.1) [43]. Besides the particular sampling scenario of each study, discrepancies may also be related to different selective pressures in the species under study [39].

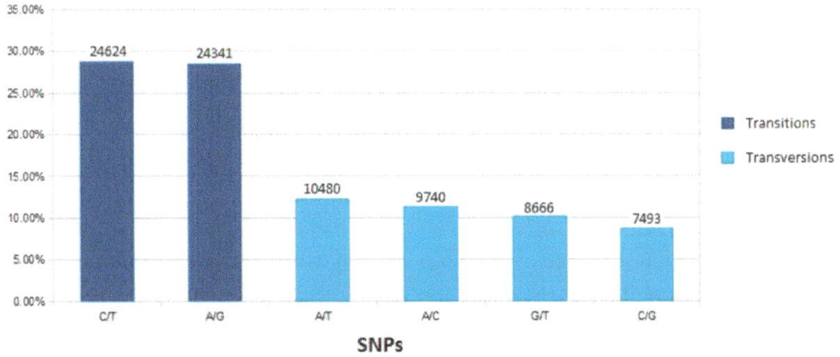

Figure 1: Single nucleotide polymorphisms (SNPs) detected in muscle and liver turbot RNA-seq. SNPs found by RNA-seq and aligned with turbot genomic sequences are separated according to their type.

Both muscle and liver transcriptomes were scanned for a large set of genes (~200) involved in growth regulation previously reported in fish [16,18,22,31,32]. We found 160 growth-related genes in muscle and 125 in liver, which account for a total of 174 different genes involved in growth processes. Furthermore, different isoforms were identified for some genes, totaling 325 different transcripts (1.86 isoforms per gene). Although in fish species the growth hormone (*gh*) has been described in liver [44], and the neuropeptide Y (*npy*) both in liver and muscle [45], we did not detect muscle or liver expression of these important growth-related hormones, nor did we find liver or muscle expression of prolactin (*prl*), consistent with the results in *Takifugu rubripes*, where *prl* was only found in the pituitary gland [46]. Hence, future growth-related studies should consider including brain as a target tissue.

All growth-related transcripts found in this study were located in the turbot genome and anchored to the genetic map using common markers [27,28]. Sixty-four genes were located in the vicinity of a previously reported growth QTL marker [29,30], within the same scaffold (Figure 2; Table S1, Supporting Information), and at an average distance of 1.9 Mb (~4 cM, assuming a relationship between genetic and physical distance of 0.5 Mb/cM [47]). The pooled population sample used for RNA sequencing in this study allowed the detection of SNPs for all these growth-related genes, which were selected for subsequent analyses. The number of SNPs per gene ranged from 1 to 25, with a mean of 7.81 ± 6.00.

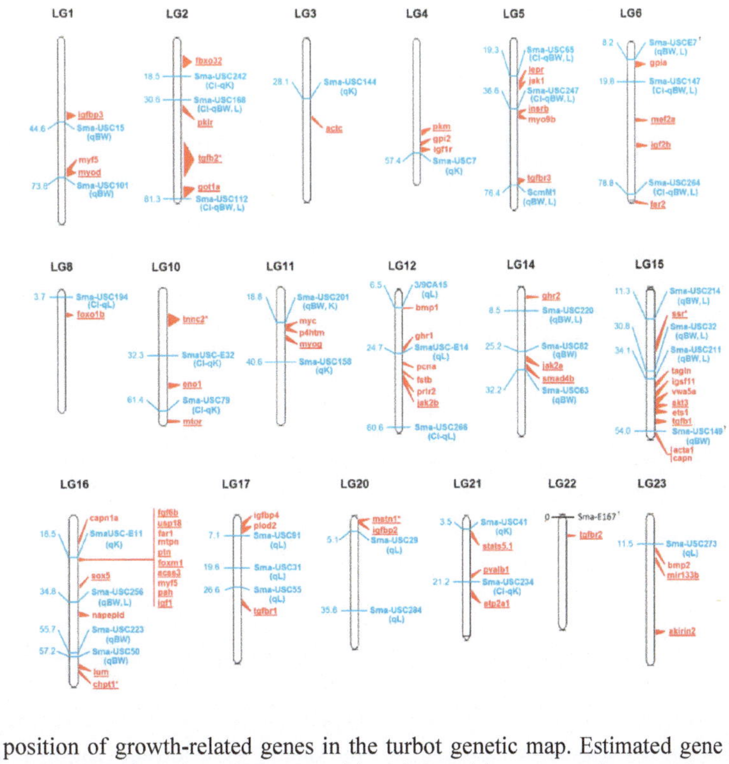

Figure 2: Predictive genome position of growth-related genes in the turbot genetic map. Estimated gene map positions are shown in red, underlining the 45 selected genes for SNP detection. (*) Map positions inferred by comparative mapping against model fish [26]. Reported growth-related markers [29,30,32] are shown in blue, either associated with growth traits (qBW, body weight; qL, length; qK, Fulton factor) or located within the confidence interval (CI) of growth QTLs. (†) Outlier markers proposed to be under selection in turbot [32].

We selected 45 out of 176 growth-related genes (those found in our transcriptome from an initial list of ~200) based on their relevance in growth-related processes in fish and vertebrates, their co-localization with growth QTL markers in turbot and the availability of technically suitable SNPs in the transcriptome (Table 2; Figure 2; Table S1, Supplementary Materials). According to Gene Ontology (GO) term classification, these genes were involved in different growth-related processes such as cell signaling, proliferation and growth (36%), energy metabolism (31%), muscle growth and development (26%) or cartilage and bone formation (7%) (Table S2, Supplementary Materials). We focused on the relationship between growth QTLs and candidate genes as a way to tackle the genetic basis underlying phenotypic effects, but we also focused on the gene relevance in growth-related processes [16,22] (Table 2). Previous QTL screening was performed with a limited amount of markers and families at early hatchery or at on-growing culture phases [29,30], so we would expect to find other genomic regions responsible for growth differences at different ages and under variable family or population genetic backgrounds. The selection included candidate genes associated with growth traits in different vertebrate and fish species, such as insulin growth factors 1 and 2 (*igf1* and *igf2*) and leptin receptor (*lpr*) [48,49,50], or myogenic regulatory factors such as myogenin (*myog*) associated with weight in pigs and chickens [51,52] and with indeterminate growth in fish [53]. Members of the transforming growth factor-beta (*tgfb*) pathway, which regulates cell growth and differentiation, were also included, since they were associated with growth and reproduction in vertebrates [54,55,56,57], such as, for example, myostatin 1 (*mstn1*). Null mutations in this negative regulator gene of muscle development in mammals lead to 30% increased growth in mice or to a "double muscle" phenotype in cattle [58,59]. SNP variation in the 3′ and 5′ untranslated regions (UTR) of myostatin was also associated with differences in growth traits in bighead carp (*Hypophthalmichthys nobilis*) [60] and Atlantic salmon (*Salmo salar*) [61], respectively. Other relevant genes selected were the parvalbumin 1 (*pvalb1*), where microsatellite variation at the 3′ UTR was associated with weight in Asian seabass (*Lates calcarifer*) [62] and growth hormone receptor 2 (*ghr2*), which has been associated with growth in cattle and chicken [4,63] and tilapia (*Oreochromis niloticus*, *O. aureus* and *O. mossambicus*) [23].

Table 2: Growth-related expressed genes selected for SNP screening in turbot.

Gene [1]	Annotation	Organ [2]	LG [3]	QTL Marker [4]	Distance [5] (kbp)	QTL Information [6] (VPE) †
acss3	Acyl-CoA synthetase 3	M&L	16	SmaUSC-E11	909	BW-QTL (13%)
actc	Actin, alpha cardiac muscle 1	M	3	Sma-USC144	1995	K-QTL (12%)
akirin2	Akirin2	M&L	23	nd	nd	L-QTL interval
akt3	Ser/Thr-protein kinase	M&L	15	Sma-USC211	1621	BW, L-QTL (13%)
atp2a1	Calcium transporting ATPase1	M	21	Sma-USC234	4364	K-QTL interval
chpt1	Choline phosphotransferase 1	M&L	16 *	nd	nd	BW-QTL (13%)
eno1	Alpha-enolase	M&L	10	SmaUSC-E32	5651	K-QTL interval
far2	Fatty acyl CoA reductase 2	M&L	6	Sma-USC264	823	BW, L-QTL interval
fbxo32	F-box protein 32	M&L	2	Sma-USC242	1924	K-QTL interval
fgf6b	Fibrobalst growth factor 6b	M	16	SmaUSC-E11	34	K-QTL (25%)
foxm1	Forkhead box M1	M&L	16	SmaUSC-E11	762	BW-QTL (13%)
foxo1b	Forkhead Box O1b	M&L	8	Sma-USC194	1843	L-QTL interval
ghr2	Growth hormone receptor 2	M&L	14	nd	nd	BW, L-QTL interval
got1a	Glu-oxaloacetic transaminase 1a	M&L	2	Sma-USC112	622	BW, L-QTL interval
igf1	Insulin-like growth factor 1	L	16	SmaUSC-E11	1453	K-QTL (25%)
igf2b	Insulin-like growth factor 2b	M&L	6	nd	nd	BW, L-QTL interval
igfbp2	Igf binding protein 2	L	20	Sma-USC29	7275	L-QTL (8%)
igfbp3	Igf binding protein 3	M	1	Sma-USC15	413	BW-QTL (14%)
insrb	Insulin Receptor	M&L	5	Sma-USC247	2321	BW, L-QTL interval
jak2a	Janus kinase 2a	M&L	14	Sma-USC82	152	BW-QTL (11%)
jak2b	Janus kinase 2b	M&L	12	SmaUSC-E14	2037	L-QTL (13%)
lepr	Leptin receptor	M&L	5	Sma-USC65	488	BW, L-QTL interval
lum	Lumican	M&L	16	Sma-USC223	504	BW-QTL (13%)
mef2a	Myocite enhancer factor a	M&L	6	nd	nd	BW, L-QTL interval
mir133b	Mir133b	M	23	Sma-USC273	5066	BW, L-QTL (11%)
mstn1	Myostatin 1	M	20 *	nd	nd	L-QTL interval
mtor	Ser/Thr-protein kinase mTOR	M&L	10	nd	nd	F-QTL interval
myod	Myoblast determination protein	M	1	Sma-USC101	351	BW-QTL (8%)
myog	Myogenin	M	11	Sma-USC158	3209	K-QTL (10%)
pah	Phenylalanine hydroxylase	M&L	16	SmaUSC-E11	1511	BW-QTL (13%)
pklr	Pyruvate kinase	M&L	2	Sma-USC168	443	K-QTL interval
pkm	Pyruvate kinase a, muscle	M&L	4	Sma-USC7	2610	K-QTL (10%)
ptn	Pleiotrpophin.	L	16	SmaUSC-E11	693	BW-QTL (13%)
pvalb1	Parvalbumin 1	M	21	Sma-USC234	1911	K-QTL interval
smad4b	Mothers- decapentaplegic 4	M&L	14	Sma-USC63	589	BW-QTL (11%)
sox5	SRY-box 5	M&L	16	Sma-USC256	560	BW, L-QTL (8%)
ssr3	Translocon protein gamma	M&L	15 *	nd	nd	BW, L-QTL interal
stat5.1	Activator of transcription 5	M&L	21	Sma-USC41	125	K-QTL (4%)
tgfb1	Transforming growth factor β1	M&L	15	nd	nd	BW-QTL interval
tgfb2	Transforming growth factor β2	M&L	2 *	nd	nd	BW, L-QTL inteval
tgfbr1	Tgf β1 receptor	M&L	17	Sma-USC55	1569	L-QTL (9%)
tgfbr2	Tgf β2 receptor	M&L	22	nd	nd	Gene function †
tgfbr3	Tgf β3 receptor	M&L	5	ScmM1	265	BW, L-QTL (11%)
tnnc2	Troponin C, skeletal muscle	M	10 *	nd	nd	K-QTL interval
usp18	Ubiquitin specific peptidase	M&L	16	SmaUSC-E11	84	BW-QTL (13%)

[1] Gene symbol (Table S1, Supplementary Materials); [2] Organ expression: muscle (M) and/or liver (L); [3] Predicted Linkage Group (LG) containing the gene-specific scaffold (Table S1, Supplementary Materials) obtained either from anchor markers of the genetic map into the turbot genome or by comparative mapping against model fish (*); [4] Closest growth QTL marker within scaffold, if applicable (nd: Gene and QTL maker/s are not in the same scaffold); [5] Physical distance between genes and QTL markers; [6] Selection criteria: Gene function and/or distance to growth QTL (BW: Body weight, L: Length, K: Fulton's factor [29,30]), either within QTL intervals or close to associated markers; VPE (%): Phenotypic variance explained by associated markers to growth traits, when applicable; † Outlier marker proposed to be under selection in turbot [32].

One SNP was selected for each gene among those available. Sixty percent of the 45 SNPs selected were located within the UTR (1:2 ratio in 5' and 3' UTR) and 31% within coding regions. Among the latter, five and nine SNPs were non-synonymous and synonymous mutations, respectively, whereas the others were located within introns of putative alternative transcript variants, as the alignment with orthologous genes from model fish suggested. These data could also be explained by the sequencing of immature transcripts, although intron retention seems to be a general phenomenon playing a role in the regulation of gene expression [64]. Non-synonymous mutations can directly alter the protein structure and function, since they cause substitutions in the amino acid sequence. Variation in the 5' UTR can affect transcription factor binding sites, thus altering gene expression, while 3' UTRs contain microRNA binding targets and can influence translation efficiency and mRNA stability [65]. Finally, some microRNA genes and consensus motifs required for correct splicing are found in introns [12,66]. Indeed, an intronic SNP in the growth hormone—Releasing hormone (*ghrh*) was associated with growth differences in Arctic charr (*Salvelinus alpinus*) [67].

SNP Variation

Sequenom assays were designed for the selected 45 SNPs into two multiplex reactions (Table S3, Supplementary Materials) for validation and genetic diversity estimation. Two SNPs were not technically feasible (*acss3*, *fbxo32*) and four were monomorphic in the tested population (*fgf6b*, *ghr2*, *got1a*, *myod*). Thus, a total of 43 SNPs were genotyped (95.6%) and 39 (86.7%) were finally polymorphic enough for diversity analysis in the Cantabrian population sample assayed (Table 3). Monomorphism at relevant growth-related loci in this wild population may be due to genetic differences between the Cantabric population and the samples used for RNA-seq SNP discovery, coming from a breeding strain of Atlantic origin. Accordingly, all feasible growth-related loci may be useful for further studies in other genetically divergent turbot populations of wild or farm origin [31,32]. Interestingly, some genic SNPs in this study were located close to outlier microsatellite and SNP markers proposed to be under divergent selection in turbot (e.g., SmaUSC-E7 at LG6; SmaUSC149 at LG15 or Sma-E167 at LG22; Figure 2) [32]. Gene-associated markers have become highly valuable tools for ascertaining origin assignment and detecting fine genetic structuration in fish [5,6]. Our panel of growth-related SNP markers will be a particularly useful tool to evaluate genetic differentiation patterns among wild Atlantic and Baltic populations with respect to farmed broodstock selected for growth in order to confirm changes in growth as an adaptation to differences in temperature and salinity conditions, which has implications for improving turbot aquaculture [32].

Table 3: SNP markers for growth-related genes in turbot.

SNP	Variant	Genomic Position [1]	Gene Region [2]	MAF [3]	*p* (HW) [4]	He [5]	Fis [6]
actc	T/C	Sm_46: 2,118,151	Exon syn	T = 0.368	1	0.471	−0.06
akirin2	C/G	Sm_26: 4,028,782	3' UTR	G = 0.324	0.7061	0.445	0.074
akt3	C/T	Sm_31: 2,270,529	3' UTR	T = 0.118	1	0.21	−0.119
atp2a1	A/T	Sm_11: 7,019,041	3' UTR	A = 0.485	1	0.507	0.014
chpt1	G/A	Sm_183: 505,467	Exon syn	T = 0.338	0.2538	0.453	−0.234
eno1	C/G	Sm_4: 2,026,813	5' UTR	C = 0.25	1	0.381	−0.005
far2	G/A	Sm_100: 790,917	Exon syn	A = 0.088	1	0.163	−0.082
fgf6b	G/A	Sm_49: 505,467	Exon R-Q	G = 1	-	-	-
foxm1	C/G	Sm_49: 1,237,830	5' UTR	C = 0.344	0.1129	0.461	0.322
foxo1b	C/G	Sm_14: 4,371,288	Exon A-G	C = 0.353	1	0.463	−0.015
ghr2	G/A	Sm_67: 9082	Exon syn	G = 1	-	-	-
got1a	C/A	Sm_35: 2,520,843	3' UTR	C = 1	-	-	-
igf1	G/T	Sm_49: 1,922,714	Intron	T = 0.029	1	0.058	−0.015
igf2b	G/A	Sm_15: 9,070,308	3' UTR	A = 0.265	1	0.395	−0.043
igfbp2	A/G	Sm_2: 14,079,242	3' UTR	G = 0.324	0.7061	0.445	0.074
igfbp3	C/A	Sm_1: 12,956,648	3' UTR	A = 0.103	1	0.187	−0.1
insrb	G/A	Sm_5: 4,670,818	Exon syn	A = 0.132	1	0.233	−0.138
jak2a	C/G	Sm_38: 2,890,626	Intron	G = 0.441	0.7412	0.501	0.06
jak2b	A/T	Sm_18: 4,376,249	3' UTR	A = 0.044	1	0.086	−0.031
lepr	A/G	Sm_5: 10,530,481	Exon V-A	A = G = 0.5	0.7387	0.508	0.074
lum	T/C	Sm_36: 3,958,348	5' UTR	C = 0.029	1	0.058	−0.015
mef2a	C/G	Sm_15: 4,752,470	3' UTR	G = 0.044	1	0.086	−0.031
mir133b	C/A	Sm_47: 179,652	5' UTR	A = 0.059	1	0.112	−0.048
mstn1	A/T	Sm_275: 53,874	3' UTR	A = 0.176	0.249	0.296	0.205
mtor	C/A	Sm_21: 2,840,205	Exon syn	A = 0.103	1	0.187	−0.1
myod	G/T	Sm_84: 1,867,326	Exon syn	G = 1	-	-	-
myog	A/T	Sm_9: 6,475,591	3' UTR	T = 0.059	1	0.112	−0.048
pah	G/C	Sm_49: 1,912,258	5' UTR	C = 0.162	1	0.275	0.039
pklr	C/T	Sm_6: 7,978,133	3' UTR	C = 0.029	1	0.058	−0.015
pkm	A/C	Sm_32: 2,688,948	Intron	A = 0.439	0.7262	0.501	0.093
ptn	T/G	Sm_49: 1,163,341	3' UTR	T = 0.25	0.0814	0.379	−0.32
pvalb1	C/T	Sm_11: 4,556,367	5' UTR	T = 0.044	1	0.086	−0.031
smad4b	C/T	Sm_38: 1,262,602	5' UTR	T = 0.074	1	0.138	−0.065
sox5	C/T	Sm_40: 1,284,941	Exon syn	C = 0.426	0.5019	0.496	−0.128
ssr3	C/T	Sm_55: 538,312	3' UTR	T = 0.176	1	0.295	0.003
stats5.1	G/C	Sm_85: 414,378	3' UTR	C = 0.309	0.686	0.434	0.119
tgfb1	A/C	Sm_28: 3,438,353	5' UTR	C = 0.426	1	0.496	−0.007
tgfb2	A/C	Sm_58: 12,144	3' UTR	C = 0.132	1	0.233	−0.138
tgfbr1	C/T	Sm_3: 2,957,935	3' UTR	T = 0.029	1	0.058	−0.015
tgfbr2	T/A	Sm_10: 2,181,357	3' UTR	T = 0.235	0.6469	0.365	−0.13
tgfbr3	T/C	Sm_34: 1,743,526	Intron	C = 0.338	0.7061	0.454	−0.102
tnnc2	T/C	Sm_174: 11,856	Exon syn	T = 0.368	1	0.471	−0.06
usp18	G/A	Sm_49: 555,569	Exon S-L	A = 0.191	0.3136	0.313	−0.222

[1] SNP genome position (scaffold code: Position in pb); [2] SNP genic position (within exons synonymous mutations (syn) or amino acid substitutions are indicated); [3] Minimum allele frequency; [4] *p*-value for Hardy-Weinberg equilibrium (HW) test; [5] Unbiased genetic diversity; [6] Deviation from HW expected heterozygosity (Fixation index, F_{IS}).

Among polymorphic SNPs, unbiased gene diversity (He) estimates ranged from 0.058 at *igf1*, *lum*, *pklr* and *tgfbr1* to 0.508 at *lepr* (mean: 0.304 ± 0.004). Minimum allele frequency (MAF) ranged from 0.029 (*igf1*, *lum*, *pklr* and *tgfbr1*) to 0.500 in *lepr* (mean: 0.221 ± 0.004). These values are in the range of those previously described in turbot [38,68] and other fish [39,69]. No departures from Hardy-Weinberg equilibrium (HW) were detected ($\alpha = 0.05$; Table 3). F_{IS} values per locus and globally were not significantly different from zero at $p < 0.05$ and showed a mean value of −0.032 ± 0.0003 (Table 3), congruent with conformance to HW in the Atlantic turbot population when tested simultaneously for all loci ($p = 1$). Significant linkage disequilibrium (LD) at $p < 0.05$ was detected between 37 pairs of SNPs out of the total number of pairs of loci tested in the population under study (741 G-tests), which is close to the 5% expected by chance. Seven out of these 37 significant tests involved pairs of closely linked loci (19%), whereas the other cases involved markers in different linkage groups (LG) (mostly LG5, LG6, LG10 and LG16; Table S4). These cases can be related to type I errors, although epistatic interactions between loci located in different QTL regions cannot be discarded, as reported in fish for growth traits [17]. Only two significant LD tests were retained after Bonferroni correction ($p = 0.00007$), corresponding to two pairs of closely linked loci, *tnnc2-actc1* and *mstn1-igfbp2*, as predicted by the mapping of genomic scaffolds (Table S1, Supplementary Materials; Figure 2). Some closely linked genes derived from mapping predictions (Figure 2) were prioritized in this study to increase the probability of detecting informative SNPs within relevant QTL regions in different genetic backgrounds, particularly important for genes with low polymorphism. In summary, the set of polymorphic gene SNPs developed and validated in this study represents useful tools to be used in further population and family association studies in this species. The validation rate based on integrated RNA-seq and structural genomic resources in turbot (~90%) was much higher than in other recent RNA-seq strategies in fish: 54% (26/48) in *Takifugu rubripes* from swimbladder RNA-seq [33]; 56.9% in rainbow trout from muscle RNA-seq [9].

Although the number of SNPs and the true SNP ratio in RNA-seq may vary due to technical parameters, this technique allows us to obtain a much higher amount of SNPs than previous technologies. SNP calling based on 454 previous transcriptome runs in turbot only detected 866 SNPs, although the efficiency was rather good (79.3%) [38]. In cod, a different method involving PCR in 71 DNA fragments achieved a 37.1% success [7]. Overall, RNA-seq performed very well for obtaining gene-targeted SNPs at affordable costs, and it is particularly efficient if integrated with available structural genomic resources, as in our study. More importantly, we could focus on candidate genes within target QTL regions for traits of productive or evolutionary interest by integrating RNA-seq with physical and genetic mapping data for SNP identification.

Towards a High Density SNP Array in Turbot

Although the aim of this work was to generate an affordable panel of gene-linked SNPs co-localizing with QTL for growth association in a large amount of fish with different genetic backgrounds, the resources presented here can be further exploited at a larger genomic scale. The total number of SNPs generated in this study can be used for designing a more extensive SNP panel as a highly informative tool for studying the architecture of quantitative traits and for increasing the efficiency of selective breeding. Our aim would be the design of a large turbot SNP array based on this information and previous reports [38,70], and on the huge SNP amount obtained in an ongoing restriction site–associated DNA (RAD) sequencing project (>10 k SNPs; unpublished data), which could integrate growth-, reproduction- and immune-related gene-linked SNPs. Although large SNP panels are routinely employed in cattle or pigs for genomic selection, high density genotyping in a large number of individuals is still expensive for aquaculture. However, the availability of new cost-effective genotyping technologies has rendered the first high density genome-wide association studies (GWAS) in aquaculture species [71,72,73]. Still, high density SNP panels are scarce, and currently have been reported only in two oyster species, the Pacific oyster and the European flat oyster [74], and in several fish, Atlantic salmon [75], catfish [76], common carp [77] and rainbow trout [78]. More aquaculture high density SNP arrays will be available in incoming years and the SNPs reported here will surely contribute to this goal in turbot.

EXPERIMENTAL SECTION

Sampling for RNA Sequencing

Turbot juveniles from six unrelated families coming from a breeding strain of Atlantic origin were reared in tanks at water temperature of 18 °C at the facilities of CETGA (Aquaculture cluster of Galicia; Ribeira, Spain). Animals were divided in two groups, one fed daily and the other one subjected to nutritional stress by fasting. This study was performed on muscle and liver because these organs play a key role in the regulation of somatic growth in fish, the first one also representing the main edible part of fish [22,79]. In order to obtain the widest gene expression range and SNP variation among individuals as possible, muscle and liver samples were obtained from muscle and liver tissues of treated and control animals at 15 and 30 days after the start of the treatment: three fed as usual and six fasted fish per sampling time (total sample of 18 individuals). Individual samples were pooled per organ and embedded in RNA later for preservation (Qiagen, Valencia, CA, USA). Animals were treated according to the Directive 2010/63/UE of the European Parliament and of the Council of 22 September 2010 on the protection of animals used for experimentation and other scientific purposes. Experimental protocols were approved by the Institutional Animal Care and Use Committee of the University of Santiago de Compostela (Spain).

RNA Sequencing

RNA extraction was performed using the RNeasy mini kit (Qiagen) with DNase treatment following manufacturer's instructions. RNA quality and quantity were evaluated in a Bioanalyzer 2100 (Agilent Technologies, Santa Clara, CA, USA) and in a NanoDrop® ND-1000 spectrophotometer (NanoDrop Technologies Inc., Wilmington, NC, USA), respectively. Muscle and liver pooled samples were barcoded and sequenced on an Illumina HiSeq 2000 (100 bp paired-end) using standard protocols. Sequencing output quality was assessed using FastQC v0.11.2 [80]. Quality filtering and removal of residual adaptor sequences was conducted on read pairs using Trimmomatic v0.32 [81]. Specifically, residual Illumina-specific adaptors were clipped from the reads, leading and trailing bases with a Phred score less than 15 were removed and the read trimmed if a sliding window average Phred score over four bases was less than 20. Only reads where both pairs had a length greater than 36 bp post-filtering were retained. The recently assembled turbot genome [28] was used as a reference for read mapping. The genome sequence spanned 544 Mb and consisted of 16,493 scaffolds with N50 of 4.3 Mb (European Nucleotide Archive (ENA) project: PRJEB11743). Eighty percent of the genome assembly (156 scaffolds) was anchored to linkage groups (LG) of the turbot genetic map [27], enabling integrative and comparative mapping. Filtered reads were mapped to the genome using Tophat2 v2.0.12 [82] that leverages the short read aligner Bowtie2 v2.2.3 [83] with a maximum intron length of 20 kb. Cufflinks v2.2.1 [84] was used to build gene transfer format (GTF) files and cufflinks gffread utility was employed to obtain final Fasta files for muscle and liver transcriptomes.

Growth-Related Turbot Sequences

Growth-related genes were selected based on gene ontology (GO) criteria using Blast2GO [85], and previous reports on growth traits in fish [16,18,22,32] and their sequences retrieved from phylogenetically close model teleosts (stickleback, medaka, tetraodon) in Ensembl release 75 [86]. These sequences were used to scan both muscle and liver transcriptomes for orthologous turbot sequences using local BLAST (e-value < E^{-10}) [77]. Once obtained, turbot growth-related sequences were blasted against the NCBI [87] non-redundant protein database to confirm their identity. Turbot gene sequences were also blasted against the turbot genome assembly to obtain their genomic position (unique significant hits with respect to specific genomic scaffolds).

Co-Localization of Candidate Genes with Growth QTLs

We checked for co-localization of the turbot growth-related genes in the genetic map with previously described growth QTL markers [29,30] also placed in the genome [28] using local BLAST (e-value < E^{-10}) [88]. Genes which were placed in the same scaffold as growth QTL markers, preferably with a gene-marker distance below 1 Mbp, were considered as candidates to explain growth associations. Since no associated markers were reported for some growth QTLs (LG 6, 10, 14, 15, 22, 23), the closest markers to the estimated QTL position in the genetic map were used as references to predict gene co-localization. For those genes placed on genomic scaffolds not anchored to any turbot LG, comparative mapping against the most informative model fish for turbot (*Gasterosteus aculeatus* and *Oryzias latipes*) [26] was used to infer their syntenic position in the turbot genetic map (LG2, LG10, LG15, LG16, LG22).

SNP Calling

SNP positions within the aligned reads were identified using the pileup function in SAMtools utilities v0.1.19 [89] with a Phredd base quality ≥20, sample read depth ≥10 and minor allele count ≥3. Reads from the two libraries (muscle and liver) were combined to increase coverage and confidence of SNP calling. SNPs located in turbot growth genes were inspected manually in the aligned reads using Tablet [90] to avoid genes with low minor allele frequencies, and their position within genes was investigated (UTR *vs.* coding regions) to select SNPs with higher chances of being biologically relevant.

SNP Genotyping and Validation

A total of 45 SNPs located in 45 different genes were selected based on previous specified criteria (functional relevance, co-localization with growth QTLs and SNP calling). DNA was extracted from a piece of caudal fin using standard phenol-chloroform procedures [91]. SNPs were validated and genotyped with the MassARRAY platform (Sequenom, San Diego, CA, USA) following the protocols and recommendations provided by the manufacturer. Briefly, the technique consists of an initial locus-specific polymerase chain reaction (PCR), followed by single-base extension using mass-modified dideoxynucleotide terminators of an oligonucleotide primer that anneals immediately upstream of the SNP [92,93]. The distinct mass of the extended primer identifies the SNP allele. MALDI-TOF mass spectrometry analysis in an Autoflex spectrometer was used for allele scoring. The 45 SNPs were combined in two multiplex reactions of 29 and 16 SNPs. SNPs were classified based on manual inspection as "failed assays" (in case that the majority of genotypes could not be scored and/or the samples did not cluster well according to genotype), and feasible SNPs (markers with proper and reliable genotypes), either monomorphic or polymorphic. All SNPs were genotyped in a population sample of 34 individuals from a wild Cantabrian population, different from the samples used for RNA-seq SNP detection (see above). The Cantabrian population had been previously used as reference for immune-related marker validation in turbot [38]. Genetic diversity (unbiased expected heterozygosity, He) and minimum allele frequency (MAF) were estimated using FSTAT 2.9.3 [94]. Departures from Hardy-Weinberg equilibrium (HW) were tested using GENEPOP 4.0 [95,96]. Linkage disequilibrium (association between genotypes at pairs of loci) was tested using the log-likelihood ratio G-statistic implemented in FSTAT 2.9.3. Conformance to HW was checked using the complete enumeration method implemented in GENEPOP 4.0, because all loci were biallelic. The fixation index FIS per locus was estimated using GENEPOP 4.0 and their significance using FSTAT 2.9.3. Bonferroni correction was applied when multiple tests were performed [97].

Data Accessibility

RNA sequencing project: ncbi BioSample accessions SAMN03740585 (muscle) and SAMN03740586 (liver). Sequenom assays for growth-related gene SNPs: Table S2, Supplementary Materials.

CONCLUSIONS

RNA-seq is an efficient technique to develop markers for candidate gene association studies related to targeted biological processes. A high number of SNPs were identified in the liver and muscle transcriptomes of turbot, which revealed enrichment in transcripts involved in growth regulation, and allowed for detection of genetic variation in several relevant growth-related genes of fish and vertebrate. Integrating genomic and linkage mapping resources allowed us to place the candidate genes in the turbot genetic map and to check for co-localization with growth QTLs in this species. SNP markers for 43 genes were validated in a turbot population, providing useful tools for fine mapping within QTL regions, population genomics studies, and functional and association analysis with growth phenotypes in turbot.

SUPPLEMENTARY MATERIALS

Supplementary materials can be found at http://www.mdpi.com/1422-0067/17/2/243/s1.

ACKNOWLEDGMENTS

This work was funded by Spanish Ministry of Economy and Competitiveness and European Regional Development Funds (AGL2012-35904), and Ministry of Science and Innovation (Consolider Ingenio, Aquagenomics, CSD200700002). DR was supported by a FPU fellowship funded by Spanish Ministry of Education, Culture and Sport. Thanks to Lucía Ínsua for technical assistance. We thank the High-Throughput Genomics Group at the Wellcome Trust Centre for Human Genetics for the generation of the sequencing data, and the Spanish National Genotyping Center (CEGEN-ISCIII)-USC node for SNP genotyping support. We acknowledge the support of the Centro de Supercomputación de Galicia (CESGA) in the completion of this work.

CONFLICTS OF INTEREST

The authors declare no conflict of interest.

REFERENCES

1. Havenstein, G.B.; Ferket, P.R.; Qureshi, M.A. Carcass composition and yield of 1957 versus 2001 broilers when fed representative 1957 and 2001 broiler diets. *Poult. Sci.* 2003, *82*, 1509–1518
2. Bonin, A. Population genomics: A new generation of genome scans to bridge the gap with functional genomics. *Mol. Ecol.* 2008, *17*, 3583–3584.
3. Nie, Q.; Sun, B.; Zhang, D.; Luo, C.; Ishag, N.A.; Lei, M.; Yang, G.; Zhang, X. High diversity of the chicken growth hormone gene and effects on growth and carcass traits. *J. Hered.* 2005, *96*, 698–703.
4. Hale, C.S.; Herring, W.O.; Shibuya, H.; Lucy, M.C.; Lubahn, D.B.; Keisler, D.H.; Johnson, G.S. Decreased growth in Angus steers with a short TG-microsatellite allele in the P1 promoter of the growth hormone receptor gene. *J. Anim. Sci.* 2000, *78*, 2099–2104.
5. Nielsen, E.E.; Cariani, A.; Mac Aoidh, E.; Maes, G.E.; Milano, I.; Ogden, R.; Taylor, M.; Hemmer-Hansen, J.; Babbucci, M.; Bargelloni, L.; *et al.* Gene-associated markers provide tools for tackling illegal fishing and false eco-certification. *Nat. Commun.* 2012, *3*.
6. Bekkevold, D.; Helyar, S.J.; Limborg, M.T.; Nielsen, E.E.; Hemmer-Hansen, J.; Clausen, L.A.; Carvalho, G.R.; FishPopTrace Consortium. Gene-associated markers can assign origin in a weakly structured fish, Atlantic herring. *ICES J. Mar. Sci.* 2015, *72*.
7. Hemmer-Hansen, J.; Nielsen, E.E.G.; Meldrup, D.; Mittelholzer, C. Identification of single nucleotide polymorphisms in candidate genes for growth and reproduction in a nonmodel organism; the Atlantic cod, *Gadus morhua*. *Mol. Ecol. Resour.* 2011, *11*, 71–80.
8. Diopere, E.; Hellemans, B.; Volckaert, F.A.; Maes, G.E. Identification and validation of single nucleotide polymorphisms in growth- and maturation-related candidate genes in sole (*Solea solea* L.). *Mar. Genom.* 2013, *9*, 33–38.
9. Salem, M.; Vallejo, R.L.; Leeds, T.D.; Palti, Y.; Liu, S.; Sabbagh, A.; Rexroad III, C.E.; Yao, J. RNA-Seq identifies SNP markers for growth traits in rainbow trout. *PLoS ONE* 2012, *7*, e36264.
10. Schunter, C.; Garza, J.C.; Macpherson, E.; Pascual, M. SNP development from RNA-seq data in a nonmodel fish: How many individuals are needed for accurate allele frequency prediction? *Mol. Ecol. Resour.* 2014, *14*, 157–165.
11. He, C.; Chen, L.; Simmons, M.; Li, P.; Kim, S.; Liu, Z.J. Putative SNP discovery in interspecific hybrids of catfish by comparative EST analysis. *Anim. Genet.* 2003, *34*, 445–448.
12. Soller, M. Pre-messenger RNA processing and its regulation: A genomic perspective. *Cell. Mol. Life Sci.* 2006, *63*, 796–819.
13. Gjedrem, T.; Baranski, M. *Selective Breeding in Aquaculture: An Introduction*; Springer Netherlands: Dordrecht, The Netherlands, 2010.
14. Gjedrem, T.; Robinson, N.; Rye, M. The importance of selective breeding in aquaculture to meet future demands for animal protein: A review. *Aquaculture* 2012, *350–353*, 117–129.
15. Garcia de la Serrana, D.; Estévez, A.; Andree, K.; Johnston, I.A. Fast skeletal muscle transcriptome of the gilthead sea bream (*Sparus aurata*) determined by next generation sequencing. *BMC Genom.* 2012, *13*.
16. De-Santis, C.; Jerry, D.R. Candidate growth genes in finfish—Where should we be looking? *Aquaculture* 2007, *272*, 22–38.
17. Wringe, B.F.; Devlin, R.H.; Ferguson, M.M.; Moghadam, H.K.; Sakhrani, D.; Danzmann, R.G. Growth-related quantitative trait loci in domestic and wild rainbow trout (*Oncorhynchus mykiss*). *BMC Genet.* 2010, *11*.
18. Valente, L.M.P.; Moutou, K.A.; Conceição, L.E.C.; Engrola, S.; Fernandes, J.M.O.; Johnston, I.A. What determines growth potential and juvenile quality of farmed fish species? *Rev. Aquac.* 2013, *5*, S168–S193.
19. Yue, G.H. Recent advances of genome mapping and marker-assisted selection in aquaculture. *Fish Fish.* 2014, *15*, 376–396.
20. Tsai, H.Y.; Hamilton, A.; Guy, D.R.; Houston, R.D. Single nucleotide polymorphisms in the insulin-like growth factor 1 (IGF1) gene are associated with growth-related traits in farmed Atlantic salmon. *Anim. Genet.* 2014, *45*, 709–715.
21. Kocmarek, A.L.; Ferguson, M.M.; Danzmann, R.G. Co-localization of growth QTL with differentially expressed

candidate genes in rainbow trout. *Genome* 2015, *58*, 393–403.

22. Johnston, I.A.; Bower, N.I.; Macqueen, D.J. Growth and the regulation of myotomal muscle mass in teleost fish. *J. Exp. Biol.* 2011, *214*, 1617–1628.
23. Liu, F.; Sun, F.; Xia, J.H.; Li, J.; Fu, G.H.; Lin, G.; Tu, R.J.; Wan, Z.Y.; Quek, D.; Yue, G.H. A genome scan revealed significant associations of growth traits with a major QTL and GHR2 in tilapia. *Sci. Rep.* 2014, *4*.
24. FAO: Food and Agriculture Organization of the United Nations. The State of World Fisheries and Aquaculture 2014. Available online: http://www.fao.org/3/a-i3720e.pdf (accessed on 21 May 2015).
25. APROMAR. La Acuicultura en España 2014. Available online: http://www.apromar.es/content/informes-anuales (accessed on 21 May 2015).
26. Bouza, C.; Hermida, M.; Pardo, B.G.; Vera, M.; Fernández, C.; de la Herrán, R.; Navajas-Pérez, R.; Álvarez-Dios, J.A.; Gómez-Tato, A.; Martínez, P. An expressed sequence tag (EST)-enriched genetic map of turbot (*Scophthalmus maximus*): A useful framework for comparative genomics across model and farmed teleosts. *BMC Genet.* 2012, *13*.
27. Hermida, M.; Bouza, C.; Fernández, C.; Sciara, A.A.; Rodríguez-Ramilo, S.T.; Fernández, J.; Martínez, P. Compilation of mapping resources in turbot (*Scophthalmus maximus*): A new integrated consensus genetic map. *Aquaculture* 2013, *414–415*, 19–25.
28. Figueras, A.; Corvelo, A.; Robledo, D.; Hermida, M.; Pereiro, P.; Gómez, J.; Carreté, L.; Bello, X.; Marcet-Houben, M.; Forn-Cuní, G.; *et al.* Genome sequencing of the turbot (*Scophthalmus maximus*; Pleuronectiformes) a flatfish of high aquaculture value. In Proceedings of the ISGA XII-The International Symposium on Genetics in Aquaculture XII, Santiago de Compostela, Spain, June 2015; p. 91.
29. Sánchez-Molano, E.; Cerna, A.; Toro, M.A.; Bouza, C.; Hermida, M.; Pardo, B.G.; Cabaleiro, S.; Fernández, J.; Martínez, P. Detection of growth-related QTL in turbot (*Scophthalmus maximus*). *BMC Genom.* 2011, *12*.
30. Rodríguez-Ramilo, S.T.; de la Herrán, R.; Ruiz-Rejón, C.; Hermida, M.; Fernández, C.; Pereiro, P.; Figueras, A.; Bouza, C.; Toro, M.A.; Martínez, P.; *et al.* Identification of quantitative trait loci associated with resistance to viral haemorrhagic septicaemia (VHS) in turbot (*Scopthalmus maximus*): A comparison between bacterium, parasite and virus diseases. *Mar. Biotechnol.* 2014, *16*, 265–276.
31. Vilas, R.; Bouza, C.; Vera, M.; Millán, A.; Martínez, P. Variation in anonymous and EST-microsatellites suggests adaptive population divergence in turbot. *Mar. Ecol. Prog. Ser.* 2010, *420*, 231–239.
32. Vilas, R.; Vandamme, S.G.; Vera, M.; Bouza, C.; Maes, G.E.; Volckaert, F.A.M.; Martínez, P. A genome scan for candidate genes involved in the adaptation of turbot (*Scophthalmus maximus*). *Mar. Genom.* 2015, *23*, 77–86.
33. Cui, J.; Liu, S.; Zhang, B.; Wang, H.; Sun, H.; Song, S.; Qiu, X.; Liu, Y.; Wang, X.; Jiang, Z.; *et al.* Transcriptome analysis of the gill and swimbladder of Takifugu rubripes by RNA-Seq. *PLoS ONE.* 2014, *9*, e85505.
34. Du, F.; Xu, G.; Nie, Z.; Xu, P.; Gu, R. Transcriptome analysis gene expression in the liver of *Coilia nasus* during the stress response. *BMC Genom.* 2014, *15*.
35. Brookes, A.J. *Single Nucleotide Polymorphism (SNP). In Encyclopedia of Life Sciences*; John Wiley & Sons, Ltd.: Chichester, UK, 2007.
36. Guryev, V.; Koudijs, M.J.; Berezikov, E.; Johnson, S.L.; Plasterk, R.H.A.; van Eeden, F.J.M.; Cuppen, E. Genetic variation in the zebrafish. *Genome Res.* 2006, *16*, 491–497.
37. Liu, S.; Zhou, Z.; Lu, J.; Sun, F.; Wang, S.; Liu, H.; Jiang, Y.; Kucuktas, H.; Kaltenboeck, L.; Peatman, E.; *et al.* Generation of genome-scale gene-associated SNPs in catfish for the construction of a high-density SNP array. *BMC Genom.* 2011, *12*.
38. Vera Rodríguez, M.; Álvarez-Dios, J.A.; Fernández, C.; Bouza da Costa, C.; Vilas, R.; Martínez, P. Development and validation of Single Nucleotide Polymorphisms (SNPs) markers from two transcriptome 454-runs of turbot (*Scophthalmus maximus*) using high-throughput genotyping. *Int. J. Mol. Sci.* 2013, *14*, 5694–5711.
39. Zhu, C.; Cheng, L.; Tong, J.; Yu, X. Development and characterization of new single nucleotide polymorphism markers from expressed sequence tags in common carp (*Cyprinus carpio*). *Int. J. Mol. Sci.* 2012, *13*, 7343–7353.
40. Cenadelli, S.; Maran, V.; Bongioni, G.; Fusetti, L.; Parma, P.; Aleandri, R. Identification of nuclear SNPs in gilthead seabream. *J. Fish Biol.* 2007, *70*, 399–405.
41. Smith, C.T.; Elfstrom, C.M.; Seeb, L.W.; Seeb, J.E. Use of sequence data from rainbow trout and Atlantic salmon for SNP detection in Pacific salmon. *Mol. Ecol.* 2005, *14*, 4193–4203.
42. DePristo, M.A.; Banks, E.; Poplin, R.; Garimella, K.V.; Maguire, J.R.; Hartl, C.; Philippakis, A.A.; del Angel, G.; Rivas, M.A.; Hanna, M.; *et al.* A framework for variation discovery and genotyping using next-generation DNA sequencing data. *Nat. Genet.* 2011, *43*, 491–498.
43. Lam, H.Y.; Pan, C.; Clark, M.J.; Lacroute, P.; Chen, R.; Haraksingh, R.; O'Huallachain, M.; Gerstein, M.B.; Kidd, J.M.; Bustamante, C.D.; *et al.* Detecting and annotating genetic variations using the HugeSeq pipeline. *Nat. Biotechnol.* 2012, *30*, 226–229.
44. Dong, H.; Zeng, L.; Duan, D.; Zhang, H.; Wang, Y.; Li, W.; Lin, H. Growth hormone and two forms of insulin-like growth factors I in the giant grouper (*Epinephelus lanceolatus*): Molecular cloning and characterization of tissue distribution. *Fish Physiol. Biochem.* 2010, *36*, 201–212.
45. MacDonald, E.; Volkoff, H. Neuropeptide Y (NPY), cocaine- and amphetamine-regulated transcript (CART) and cholecystokinin (CCK) in winter skate (*Raja ocellata*): cDNA cloning, tissue distribution and mRNA expression responses to fasting. *Gen. Comp. Endocrinol.* 2009, *161*, 252–261.
46. Lee, K.M.; Kaneko, T.; Aida, K. Prolactin and prolactin receptor expressions in a marine teleost, pufferfish *Takifugu rubripes*. *Gen. Comp. Endocrinol.* 2006, *146*, 318–328.

47. Bouza, C.; Hermida, M.; Pardo, B.G.; Fernández, C.; Fortes, G.G.; Castro, J.; Sánchez, L.; Presa, P.; Pérez, M.; Sanjuán, A.; *et al.* A microsatellite genetic map of the turbot (*Scophthalmus maximus*). *Genetics* 2007, *177*, 2457–2467.
48. Muñoz, G.; Ovilo, C.; Silió, L.; Tomás, A.; Noguera, J.L.; Rodríguez, M.C. Single- and joint-population analyses of two experimental pig crosses to confirm quantitative trait loci on *Sus scrofa* chromosome 6 and leptin receptor effects on fatness and growth traits. *J. Anim. Sci.*2009, *87*, 459–468.
49. Rimbault, M.; Beale, H.C.; Schoenebeck, J.J.; Hoopes, B.C.; Allen, J.J.; Kilroy-Glynn, P.; Wayne, R.K.; Sutter, N.B.; Ostranderm, E.A. Derived variants at six genes explain nearly half of size reduction in dog breeds. *Genome Res.* 2013, *23*, 1985–1995.
50. Feng, X.; Xu, X.; Tong, J. Novel single nucleotide polymorphisms of the insulin-like growth factor-I gene and their associations with growth traits in common carp (*Cyprinus carpio* L.). *Int. J. Mol. Sci.* 2014, *15*, 22471–22482.
51. Te Pas, M.F.; Soumillion, A.; Harders, F.L.; Verburg, F.J.; van den Bosch, T.J.; Galesloot, P.; Meuwissen, T.H. Influences of myogenin genotypes on birth weight, growth rate, carcass weight, backfat thickness, and lean weight of pigs. *J. Anim. Sci.* 1999, *77*, 2352–2356.
52. Genxi, Z.; Ying, T.; Tao, Z.; Jinyu, W.; Yongjuan, W. Expression profiles and association analysis with growth traits of the *MyoG* and *Myf5* genes in the Jinghai yellow chicken. *Mol. Biol. Rep.* 2014, *41*, 7331–7338.
53. Froehlich, J.M.; Galt, N.J.; Charging, M.J.; Meyer, B.M.; Biga, P.R. *In vitro* indeterminate teleost myogenesis appears to be dependent on Pax3. *In Vitro Cell. Dev. Biol. Anim.* 2013, *49*, 371–385.
54. Massagué, J. How cells read TGF-beta signals. *Nat. Rev. Mol. Cell Biol.* 2000, *1*, 169–178.
55. Ten Dijke, P.; Goumans, M.J.; Itoh, F.; Itoh, S. Regulation of cell proliferation by Smad proteins. *J. Cell. Physiol.* 2002, *191*, 1–16.
56. Li, H.; Deeb, N.; Zhou, H.; Mitchell, A.D.; Ashwell, C.M.; Lamont, S.J. Chicken quantitative trait loci for growth and body composition associated with transforming growth factor-β genes. *Poult. Sci.* 2003, *82*, 347–356.
57. Chen, K.; Hawken, R.; Flickinger, G.H.; Rodriguez-Zas, S.L.; Rund, L.A.; Wheeler, M.B.; Abrahamsen, M.; Rutherford, M.S.; Beever, J.E.; Schook, L.B. Association of the porcine transforming growth factor beta type I receptor (TGFBR1) gene with growth and carcass traits. *Anim. Biotechnol.* 2012, *23*, 43–63.
58. Kambadur, R.; Sharma, M.; Smith, T.P.L.; Bass, J.J. Mutations in myostatin (*GDF8*) in double-muscled Belgian Blue and Piedmontese cattle. *Genome Res.* 1997, *7*, 910–916.
59. McPherron, A.C.; Lee, S.J. Double muscling in cattle due to mutations in the myostatin gene. *Proc. Natl. Acad. Sci. USA* 1997, *94*, 12457–12461.
60. Liu, L.; Yu, X.; Tong, J. Molecular characterization of myostatin (MSTN) gene and association analysis with growth traits in the bighead carp (*Aristichthys nobilis*). *Mol. Biol. Rep.* 2012, *39*, 9211–9221.
61. Peñazola, C.; Hamilton, A.; Guy, D.R.; Bishop, S.C.; Houston, R.D. A SNP in the 5′ flanking region of the myostatin-1b gene is associated with harvest traits in Atlantic salmon (*Salmo salar*). *BMC Genet.* 2013, *14*.
62. Xu, Y.X.; Zhu, Z.Y.; Lo, L.C.; Wang, C.M.; Lin, G.; Feng, F.; Yue, G.H. Characterization of two parvalbumin genes and their association with growth traits in Asian seabass (*Lates calcarifer*). *Anim. Genet.* 2006, *37*, 266–268.
63. Huang, N.; Cogburn, L.A.; Agarwal, S.K.; Marks, H.L.; Burnside, J. Overexpression of a truncated growth hormone receptor in the sex-linked dwarf chicken: Evidence for a splice mutation. *Mol. Endocrinol.* 1993, *7*, 1391–1398.
64. Wong, J.J.; Ritchie, W.; Ebner, O.A.; Selbach, M.; Wong, J.W.H.; Huang, Y.; Gao, D.; Pinello, N.; Gonzalez, M.; Baidya, K.; *et al.* Orchestrated intron retention regulates normal granulocyte differentiation. *Cell* 2013, *154*, 583–595.
65. Barrett, L.W.; Fletcher, S.; Wilton, S.D. Regulation of eukaryotic gene expression by the untranslated gene regions and other non-coding elements. *Cell. Mol. Life Sci.* 2012, *69*, 3613–3634.
66. Ying, S.Y.; Lin, S.L. Current perspectives in intronic micro RNAs (miRNAs). *J. Biomed. Sci.*2006, *13*, 5–15.
67. Tao, W.J.; Boulding, E.G. Associations between single nucleotide polymorphisms in candidate genes and growth rate in Arctic charr (*Salvelinus alpinus* L.). *Heredity* 2003, *91*, 60–69.
68. Vera, M.; Álvarez-Dios, J.A.; Millán, A.; Pardo, B.G.; Bouza, C.; Hermida, M.; Fernández, C.; de la Herrán, R.; Molina-Luzón, M.J.; Martínez, P. Validation of single nucleotide polymorphism (SNP) markers from an immune Expressed Sequence Tag (EST) turbot, *Scophthalmus maximus*, database. *Aquaculture* 2011, *313*, 31–41.
69. Campbell, N.R.; Amish, S.J.; Pritchard, V.L.; McKelvey, K.S.; Young, M.K.; Schwartz, M.K.; Garza, J.C.; Luikart, G.; Narum, S.R. Development and evaluation of 200 novel SNP assays for population genetic studies of westslope cutthroat trout and genetic identification of related taxa. *Mol. Ecol. Resour.* 2012, *12*, 942–949.
70. Robledo, D.; Ronza, P.; Harrison, P.W.; Losada, A.P.; Bermúdez, R.; Pardo, B.G.; Redondo, M.J.; Sitjà-Bobadilla, A.; Quiroga, M.I.; Martínez, P. RNA-seq analysis reveals significant transcriptome changes in turbot (*Scophthalmus maximus*) suffering severe enteromyxosis. *BMC Genom.* 2014, *15*.
71. Geng, X.; Sha, J.; Liu, S.; Bao, L.; Zhang, J.; Wang, R.; Yao, J.; Li, C.; Feng, J.; Sun, F.; *et al.* A genome-wide association study in catfish reveals the presence of functional hubs of related genes within QTLs for columnaris disease resistance. *BMC Genom.* 2015, *16*.

72. Barson, N.J.; Aykanat, T.; Hindar, K.; Baranski, M.; Bolstad, G.H.; Fiske, P.; Jacq, C.; Jensen, A.J.; Johnston, S.E.; Karlsson, S.; *et al.* Sex-dependent dominance at a single locus maintains variation in age at maturity in salmon. *Nature* 2015, *528*, 405–408.
73. Tsai, H.Y.; Hamilton, A.; Guy, D.R.; Tinch, A.E.; Bishop, S.C.; Houston, R.D. Verification of SNPs associated with growth traits in 2 populations of farmed Atlantic salmon. *Int. J. Mol. Sci.* 2015, *17*.
74. Lapègue, S.; Harrang, E.; Heurtebise, S.; Flahauw, E.; Donnadieu, C.; Gayral, P.; Ballenghien, M.; Genestout, L.; Barbotte, L.; Mahla, R.; *et al.* Development of SNP-genotyping arrays in two shellfish species. *Mol. Ecol. Resour.* 2014, *14*, 820–830.
75. Houston, R.D.; Taggart, J.B.; Cézard, T.; Bekaert, M.; Lowe, N.R.; Downing, A.; Talbot, R.; Bishop, S.C.; Archibald, A.L.; Bron, J.E.; *et al.* Development and validation of a high density SNP genotyping array for Atlantic salmon (*Salmo salar*). *BMC Genom.* 2014, *15*.
76. Liu, S.; Sun, L.; Li, Y.; Sun, F.; Jiang, Y.; Zhang, Y.; Zhang, J.; Feng, J.; Kaltenboeck, L.; Kucuktas, H.; *et al.* Development of the catfish 250K SNP array for genome-wide association studies. *BMC Res. Notes* 2014, *7*.
77. Xu, J.; Zhao, Z.; Zhang, X.; Zheng, X.; Li, J.; Jiang, Y.; Kuang, Y.; Zhang, Y.; Feng, J.; Li, C.; *et al.* Development and evaluation of the first high-throughput SNP array for common carp (*Cyprinus carpio*). *BMC Genom.* 2014, *15*.
78. Palti, Y.; Gao, G.; Liu, S.; Kent, M.P.; Lien, S.; Miller, M.R.; Rexroad, C.E., III; Moen, T. The development and characterization of a 57K single nucleotide polymorphism array for rainbow trout. *Mol. Ecol. Resour.* 2015, *15*, 662–672.
79. Won, E.T.; Borski, R.J. Endocrine regulation of compensatory growth in fish. *Front. Endocrinol.* 2013, *4*.
80. Andrews, S. FastQC: A quality control tool for high throughput sequence data. *Ref. Source*2010. Available online: http://www.bioinformatics.babraham.ac.uk/projects/fastqc/ (accessed on 21 May 2015).
81. Bolger, A.M.; Lohse, M.; Usadel, B. Trimmomatic: A flexible trimmer for Illumina Sequence Data. *Bioinformatics* 2014, *30*, 2114–2120.
82. Kim, D.; Pertea, G.; Trapnell, C.; Pimentel, H.; Kelley, R.; Salzberg, S.L. TopHat2: Accurate alignment of transcriptomes in the presence of insertions, deletions and gene fusions. *Genome Biol.* 2013, *14*, R36.
83. Langmead, B.; Salzberg, S.L. Fast gapped-read alignment with Bowtie 2. *Nat. Methods* 2012, *9*, 357–359.
84. Trapnell, C.; Williams, B.A.; Pertea, G.; Mortazavi, A.; Kwan, G.; van Baren, M.J.; Salzberg, S.L.; Wold, B.J.; Pachter, L. Transcript assembly and quantification by RNA-Seq reveals unannotated transcripts and isoform switching during cell differentiation. *Nat. Biotechnol.*2010, *28*, 511–515.
85. Conesa, A.; Götz, S.; García-Gómez, J.M.; Terol, J.; Talón, M.; Robles, M. Blast2GO: A universal tool for annotation, visualization and analysis in functional genomics research. *Bioinformatics* 2005, *21*, 3674–3676.
86. Flicek, P.; Amode, M.R.; Barrell, D.; Beal, K.; Billis, K.; Brent, S.; Carvalho-Silva, D.; Clapham, P.; Coates, G.; Fitzgerald, S.; *et al.* Ensembl 2014. *Nucleic Acids Res.* 2014, *42*, D749–D755.
87. Geer, L.Y.; Marchler-Bauer, A.; Geer, R.C.; Han, L.; He, L.; He, S.; Liu, C.; Shi, W.; Bryant, S.H. The NCBI BioSystems database. *Nucleic Acids Res.* 2010, *38*, D492–D496.
88. Altschul, S.F.; Madden, T.L.; Schäffer, A.A.; Zhang, J.; Zhang, Z.; Miller, W.; Lipman, D.J. Gapped BLAST and PSI-BLAST: A new generation of protein database search programs. *Nucleic Acids Res.* 1997, *25*, 3389–3402.
89. Li, H.; Handsaker, B.; Wysoker, A.; Fennell, T.; Ruan, J.; Homer, N.; Marth, G.; Abecasis, G.; Durbin, R.; 1000 Genome Project Data Processing Subgroup. The sequence alignment/map format and SAMtools. *Bioinformatics* 2009, *25*, 2078–2079.
90. Milne, I.; Stephen, G.; Bayer, M.; Cock, P.J.; Pritchard, L.; Cardle, L.; Shaw, P.D.; Marshall, D. Using Tablet for visual exploration of second-generation sequencing data. *Brief. Bioinform.*2013, *14*, 193–202.
91. Sambrook, J.; Fritsch, E.F.; Maniatis, T. *Molecular Cloning: A Laboratory Manual*, 1st ed.; Cold Spring Harbor Laboratory Press: New York, NY, USA, 1989.
92. Buetow, K.H.; Edmonson, M.; MacDonald, R.; Clifford, R.; Yip, P.; Kelley, J.; Little, D.P.; Strausberg, R.; Koester, H.; Cantor, C.R.; *et al.* High-throughput development and characterization of a genomewide collection of gene-based single nucleotide polymorphism markers by chip-based matrix-assisted laser desorption/ionization time-of-flight mass spectrometry. *Proc. Natl. Acad. Sci. USA* 2001, *98*, 581–584.
93. Oeth, P.; del Mistro, G.; Marnellos, G.; Shi, T.; van den Boom, D. Qualitative and quantitative genotyping using single base primer extension coupled with matrix-assisted laser desorption/ionization time-of-flight mass spectrometry (MassARRAY). *Methods Mol. Biol.*2009, *578*, 307–343.
94. Goudet, J. FSTAT (version 1.2): A computer program to calculate F-statistics. *J. Hered.* 1995, *86*, 485–486.
95. Raymond, M.; Rousset, F. GENEPOP (Version 1.2): Population genetics software for exact tests and ecumenicism. *J. Hered.* 1995, *86*, 248–249.
96. Rousset, F. GENEPOP'007: A complete re-implementation of the GENEPOP software for Windows and Linux. *Mol. Ecol. Resour.* 2008, *8*, 103–106.
97. Rice, W.R. Analyzing tables of statistical tests. *Evolution* 1989, *43*, 223–225.

CHAPTER 8

DEVELOPMENT OF AN OLIGO DNA MICROARRAY FOR THE EUROPEAN SEA BASS AND ITS APPLICATION TO EXPRESSION PROFILING OF JAW DEFORMITY

Serena Ferraresso, Massimo Milan, Caterina Pellizzari, Nicola Vitulo, Richard Reinhardt, Adelino VM Canario, Tomaso Patarnello and Luca Bargelloni

BACKGROUND

Lower jaw protrusion or mandibular prognathism (MP) is a developmental malformation conferring a very distinctive facial phenotype. The most famous example of MP is the Habsburg family, one of the oldest European royal families, where prognathism has been observed in several successive generations [1], suggesting a strong genetic component for this disorder. In fact, although environmental factors appear to contribute to the development of MP, familiar aggregation of this character has been reported in several human populations, providing strong support to the hypothesis that heredity plays an important role in the etiology of MP. A recent study indicates the presence of a major gene that influences the expression of MP with clear signs of Mendelian inheritance (most likely autosomal dominant with incomplete penetrance), and a multifactorial component [2]. Mandibular prognathism is not limited to royal families nor to the human species, as it has been reported in several other vertebrates, *e.g.* iguanas, short-nosed dog breeds, and rabbits. In the latter species, pedigree analysis showed a simple autosomal recessive inheritance with incomplete penetrance for this condition [3]. Different types of lower jaw deformities have been reported also in several cultured as well as wild fish. In the European sea bass, *Dicentrarchus labrax*, a phenotype similar to prognatism is often observed [4]. In some cases lower jaw protrusion appears to be related to hypertrophy of the mandible (L. Bargelloni unpublished observations), in others it has been explained as an antero-posterior compression of the ethmoid region and upper jaws, with the resulting apparent protrusion of the lower jaw, and named "pugheadness". In *D. labrax*, this malformation has been attributed to a dietary excess of vitamin A as well as to absence or excess of Ω-3 poly-unsaturated fatty acids (PUFAΩ3) [5, 6]. Whilst external conditions (*e.g.* diet, water temperature) are considered the most likely causative factors of bone (cranial and spinal) deformities in cultured *D. labrax*, genetics does play a role as recently demonstrated for spine malformations [7]. Likewise, genetic analysis of a population of juvenile sea bass showed a highly significant bias in the frequency of lower-jaw protrusion across different full-sib families raised under communal rearing conditions (L. Bargelloni, unpublished data).

Gene expression analysis of jaw development/deformities in early ontogenetic stages has been reported for a few candidate genes in different fish species, including the European sea bass [8, 9]. Expression profiling of individuals showing alternative phenotypes has been suggested as a complementary approach to linkage analysis, in order to identify loci involved in the

Citation: Ferraresso, Massimo Milan, Caterina Pellizzari, Nicola Vitulo, Richard Reinhardt, Adelino VM Canario, Tomaso Patarnello and Luca Bargelloni, Development of an oligo DNA microarray for the European sea bass and its application to expression profiling of jaw deformity, doi: 10.1186/1471-2164-11-354 genetic determination of the trait (e.g. [9]). Furthermore, a transcriptomic approach might provide a broader picture of the molecular mechanisms underlying the development of cranial deformities, which could shed light also on environmental factors influencing this condition. Here, gene expression profiling of normal against jaw-deformed individuals from segregating families is reported. To this end, an oligo DNA microarray, specific for *D. labrax*, was developed, first constructing a database of unique sea bass transcripts, then annotating transcribed sequences through extensive data mining, finally designing two non-overlapping 60mer oligonucleotide probes for each transcript. Probes were synthesized *in situ* using the Agilent SurePrint™ technology to obtain a DNA microarray platform with over 40,000 probes. A similar approach has already provided robust and flexible microarray platform in other fish species [10, 11, 12, 13, 14, 15, 16, 17, 18]. While gene expression profiling of early developmental stages in the sea bass has been carried out using a salmonid microarray [19], to our knowledge, the present study represents the first report of a species-specific DNA microarray for *D. labrax*, either based on cDNA or oligo-DNA technology. This is quite surprising, taking into consideration the importance of the species in fisheries and aquaculture. The European sea bass is a euryhaline and eurythermic fish living in marine, estuarine, and lagoon habitats, from Scandinavia to Western Sahara in the North-eastern Atlantic Ocean and from the Western Mediterranean Sea to the Black Sea [20]. Although the sea bass supports local fisheries and recreational fishing, the main interest in this species is related to aquaculture. Sea bass hatcheries are present in several countries, from France to Greece, and sea bass farming is now widespread along the Atlantic and Mediterranean coasts from Portugal to Turkey, with a total annual production of over 134,000 metric tons (FEAP data relative to year 2008 [21]). Despite sea bass culture has greatly improved since the '80s, when the first sea bass hatcheries were established, significant bottlenecks still remain, including high frequency of cranial and skeletal malformations. These developmental abnormalities severely affect sea bass production as deformed juveniles show reduced fitness and/or marketability, requiring time-consuming and personnel-intensive manual selection to discard abnormal fish, with relevant consequences on farming costs, but also on animal welfare.

The aim of the present paper was thus to analyze mRNA expression in 38 and 58 days-old individuals, on whole heads and lower jaws respectively, using a newly developed transcriptomic platform. At the same time, expression profiling of jaw deformity represents a challenging test trial, to evaluate the performance of the platform itself.

RESULTS

DLPD database

A total of 19,048 unique sequences are present in the *Dicentrarchus labrax* Padova database (DLPD). Nearly half of DLPD entries (9,497) produced a significant blastx hit against either SwissProt or TrEMBL or NCBI protein databases (e-value $< e^{-3}$). A Venn diagram showing the number of matches with each database is presented in Figure 1. Further improvement in the annotation of DLPD sequences was obtained searching the NCBI nucleotide non redundant database using the blastn option (e-value $< e^{-5}$). This approach provided a significant match for additional 2,511 DLPD transcripts, which showed no correspondence with any known protein database, bringing the final number of DLPD entries associated with a known protein or transcript to 12,008 (63%). Fish-specific searches against five teleost transcriptomes (*Gasterosteus aculeatus, Oryzias latipes, Danio rerio, Tetraodon nigroviridis*, and *Takifugu rubripes)* identified 9,506 DLPD sequences matching against at least one species, while 4,504 ones found a significant match with all five databases (see Table 1). Additional functional information could be obtained using the Blast2GO software that allowed association of one or more GO terms to 4,692 DLPD entries. Of these, 3,267 were linked with Biological Process (BP) GO entries, 2,858 to Cellular Component (CC), and 3,879 to Molecular Function (MF). Unique GO terms represented in DLPD are 1,541 for BP, 381 for CC, and 998 for MF. A simplified view of these GO terms using a "Generic GO Slim" showed 47 BP classes, 28 for CC, and 36 for MF (see Additional File 1). All DLPD sequences are publicly available in a dedicated database (DLPD database: [22]), together with associated annotations, GO entries, and putative homologous genes in fish model species.

Table 1: Number of sea bass transcripts showing significant matches with transcriptome, publicly available on Ensembl, of five teleost species.

Ensemble transcriptomes	Species common name	Count (% of sea bass sequences)
Gasterosteus aculeatus	Stickleback	8448 (44.4%)
Oryzias latipes	Japanese medaka	7004 (36.8%)
Takifugu rubripes	Japanese pufferfish	7774 (40.8%)
Tetraodon nigroviridis	Green-spotted pufferfish	7739 (40.6%)
Danio rerio	Zebrafish	5579 (29.3%)

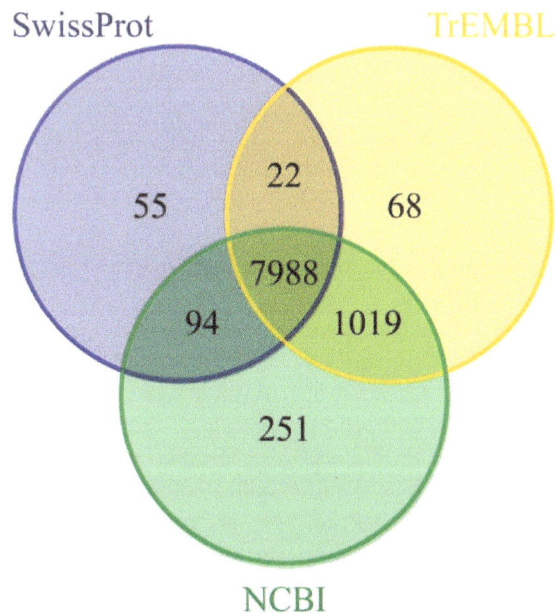

Figure 1: Individual contribution of three different protein databases to the number of annotated DLPD entries. Venn diagram showing the number of significant *matches* obtained with NCBI nr database, SwissProt and TrEMBL respectively.

Microarray quality assessment

Probe design was positively completed for 19,035 DLPD entries. Probe sequences and other details on the microarray platform can be found at the GEO database [23] under accession number GPL9663. Hybridization success for each probe was evaluated considering a total of 14 experiments (6 independent hybridizations for Stage 38 and 8 for Stage 58). Hybridization was considered successful when the value of flag "gls Found" was equal to 1 (see Methods). Across all experiments, only three probes (0.008%) never showed a signal higher than background, while 35,957, corresponding to 94.4% of the total number of target probes, successfully hybridized in at least half of the array experiments (see Figure S1 in Additional File 2). Six pools of Stage 38 were treated as biological replicates in order to evaluate the repeatability of array results. The degree of mutual agreement between replicates was assessed estimating Pearson correlation coefficients (r) on the entire set of expression values. Pair-wise comparisons of replicate experiments showed correlation coefficients with $r > 0.99$ and always significant (p-value < 0.01) (see Table 2).

Table 2: Correlation coefficients, on the entire set of expression values, across Stage 38 biological replicates.

	38d_N3	38d_N4	38d_N6	38d_P3	38d_P5	38d_P6
38d_N3	1.00					
38d_N4	0.996**	1.00				
38d_N6	0.996**	0.998**	1.00			
38d_P3	0.991**	0.994**	0.991**	1.00		
38d_P5	0.993**	0.996**	0.996**	0.994**	1.00	
38d_P6	0.994**	0.996**	0.997**	0.993**	0.996**	1.00

**p-value < 0.01

For each transcript two probes at non-overlapping positions as near as possible to the 3'-end of each transcript are present in the sea bass microarray platform. The variability between each two probes for the same transcript was evaluated using the ratio between their intensity levels (fold change, FC) as a measure of signal difference. This ratio is predicted to assume a value around 1 unless technical (*e.g.* poor probe quality) or biological factors (*e.g.* alternative splicing) are present, causing significant deviations from the expected. In Additional File 2 (Figure S2) each plot describes the distribution of observed fold changes between Probe_1 and Probe_2 for each array experiment in 38 days-old sea bass heads, which is symmetrical, centered around 1, and equal across all the experiments. Probe pairs show a small difference in terms of intensity values, with 68.1% of transcripts having a FC lower than 2. Pearson correlation coefficient between Probe_1 and Probe_2 for each gene across six experiments was always greater than 0.93 and highly significant ($p < 0.01$), further confirming the overall concordance of hybridization signal for probe pairs.

Analysis of Expression Data

Raw and normalized fluorescence data of both Stage 38 and Stage 58 microarray experiments have been deposited in the GEO database [23] under accession numbers GSE19041 and GSE19001 respectively. Statistical analysis on Stage 38 whole heads showed no significant differentially expressed genes. In contrast, even after complete exclusion of hybridization data for probes with an excess of missing values (6,404 probes), analysis of filtered data showed significant differential gene expression between normal and deformed mandibles of 58 days-old juveniles. After filtering, normalized fluorescence data for 17,694 (93%) DLPD transcripts, represented by either one probe (3,722) or two probes (13,972), were subjected to a two-class SAM test. Setting FDR to 5%, a list of 333 significant probes, corresponding to 242 unique transcripts, was obtained. For 91 DLPD sequences, both Probe_1 and Probe_2 were identified by SAM. The remaining 151 transcripts were represented by a single probe. For 14 DLPD entries identified by one probe the second probe was previously excluded in the filtering step. For all the other transcripts pointed out by SAM, all second probes (except for three) were differentially expressed in prognathous individuals compared to controls (46% with a Fold-change < -2), although not significant (p-value > 0.05). All 242 transcripts were down-regulated in pooled-samples of deformed mandibles compared to controls with a FC ranging from -1.12 to -15.7. Seventy transcripts were down-regulated four-fold, with two of them more than 10-fold (see Additional File 3). A putative annotation could be obtained for 122 differentially expressed transcripts. Of the 70 most differentially-regulated transcripts (fold change < -4), half had a significant match against a known protein/DNA sequence. A substantial fraction of putative protein homologs to this "short list" of DLPD entries appears to be specific to the nervous system, including genes involved in synapsis function and neuronal development (Table 3). This observation is confirmed when analyzing the complete set of differentially regulated sea bass RNAs (*e.g.*, Neuronal membrane glycoprotein M6-a, Carboxypeptidase E, Neural Wiskott-Aldrich syndrome protein, Schwannomin-interacting protein 1, Synapsin-2B, Neuroserpin, Ephrin type-B receptor 3, Metabotropic glutamate receptor 8, RUFY3, GMP-PDE delta). Other genes with a relevant role in the development of anatomical structures can be found in the full list of differentially expressed genes (*e.g.* GH-receptor, Pleiotrophin, SOX4, Retinoic acid receptor (RXR) gamma; Additional File 3).

Table 3: List of the most down-regulated genes (FC ≤ -4-fold) in prognathous individuals compared to controls specific of nervous system.

Probe ID	Fold Change	Gene Description
DLPD02513_1	0.23	Synaptophysin
DLPD02650_2	0.17	Synaptosomal-associated protein 25-A (SNAP-25a)
DLPD06508_1	0.08	Synaptosomal-associated protein 25-A (SNAP-25a)
DLPD06508_2	0.08	Synaptosomal-associated protein 25-A (SNAP-25a)
DLPD03740_1	0.16	Synaptosome-associated protein
DLPD03740_2	0.17	Synaptosome-associated protein
DLPD05555_1	0.24	Vesicle-fusing ATPase
DLPD05555_2	0.25	Vesicle-fusing ATPase
DLPD07764_1	0.24	Syntaxin-binding protein 1
DLPD07764_2	0.21	Syntaxin-binding protein 1
DLPD11576_2	0.22	Neuroplastin
DLPD02151_1	0.19	Microtubule-associated protein 2
DLPD02151_2	0.16	Microtubule-associated protein 2
DLPD02425_2	0.14	Contactin-1
DLPD02075_1	0.28	Neurogenic differentiation factor 2
DLPD03006_1	0.22	Chondroitin sulfate proteoglycan 5 precursor
DLPD06176_1	0.11	Fatty acid-binding protein brain isoform
DLPD06176_2	0.11	Fatty acid-binding protein brain isoform
DLPD04679_1	0.23	Visinin-like protein 1
DLPD04899_1	0.17	Beta-synuclein
DLPD04899_2	0.22	Beta-synuclein
DLPD13372_1	0.14	Plasticin
DLPD13559_1	0.18	Secretogranin II
DLPD06571_2	0.21	Zinc finger protein of the cerebellum 3 (Zic3)

Functional Annotation of Differentially Expressed Genes

In order to obtain a more systematic functional interpretation of the set of differentially expressed genes, GO enrichment analyses were performed following two alternative strategies. In the first one, GO enrichment analysis was performed using the GOStat tool [24]. Four GO terms with more than two gene counts were over-represented with significant uncorrected probability (see Table 4). In the second strategy, DLPD entries were linked either to human Ensembl Gene IDs or to zebrafish ZFIN IDs, following different approaches (see methods, Additional File 4), with a variable number of correspondences between sea bass and human/zebrafish sequences depending on the chosen method (see Table 5). Human or zebrafish IDs were then used in the bioinformatic tool DAVID (Database for Annotation, Visualization and Integrated Discovery) [25]. Results from direct comparison of DLPD entries against the human transcriptome yielded 60 putative human homologs of differentially expressed sea bass transcripts present in the DAVID 2008 knowledgebase. Enrichment analysis showed that 14 GO_BP terms are significantly over-represented (Table 6). Twenty six genes (43%) have a role in the regulation of biological processes (GO:0050789) and 14 are involved in the development of anatomical structures (GO:0048856). More specifically, five terms are directly related to nervous system development and function (neurological system process, synaptic transmission, regulation of neurotransmitter levels, nervous system development, transmission of nerve impulse). Two terms of GO Cellular Component (GO:0005856 "cytoskeleton", P = 0.01; GO:0019717 "synaptosome", P = 0.02; the latter one was significant, but with a low gene count), were found to be significantly over-represented. Both are related to the single entry for GO Molecular Function (GO:0008092 "cytoskeletal protein binding"), since cytoskeletal proteins are involved in synapsis genesis and function as well. Further evidence of the involvement of nervous system processes/ components comes from tissue specificity of differentially regulated genes, with 39 genes showing up significantly in the brain (UP_Tissue). Similar results were obtained using PANTHER definitions (see Additional File 5).

Table 4: GO terms significantly over-represented, among differentially expressed genes, based on GOStat analyses

GO TERM	Biological Process	Gene Count	p-value
GO:0006813	Potassium ion transport	3	0.116
GO:0030155	Regulation of cell adhesion	2	0.116
GO:0007399	Nervous system development	4	0.116
GO:0003008	System process	4	0.116
GO:0050877	Neurological system process	3	0.116

Table 5: Number of DAVID identifiers retrieved with different approaches.

	ENSEMBL HUMAN GENE ID		ZFIN ID	
	DLPD-GA-HUMAN	DLPD-HUMAN	DLPD-GA-DR	DLPD-DR
DLPD with identifier on DAVID (Unique entries)	9,156 (6,019)	9,277 (6,458)	7,217 (4,955)	7,010 (5,148)
Identifiers uploaded as "background"	5,385	5,654	4,058	4,150
Significant DLPD with identifier on DAVID (Unique entries)	78 (75)	70 (68)	65 (63)	56 (54)
Identifiers uploaded as "gene list"	71	60	55	46

Table 6: GO terms significantly over-represented, among differentially expressed genes, based on DAVID analyses

	Term	Count	%	*p-value*	Fold Enrichment	FDR
GOTERM BP_ALL	GO:0050877 neurological system process	8	13.33%	0.006574	3.468007	11.518
	GO:0048856 anatomical structure development	14	23.33%	0.026081	1.852929	38.759
	GO:0003008 system process	9	15.00%	0.022777	2.484961	34.786
	GO:0032501 multicellular organismal process	23	38.33%	2.19E-04	2.116818	0.4063
	GO:0050789 regulation of biological process	24	40.00%	0.01166	1.561735	19.557
	GO:0007154 cell communication	19	31.67%	0.034946	1.558464	48.316
	GO:0007399 nervous system development	7	11.67%	0.039474	2.695892	52.635
	GO:0007275 multicellular organismal development	15	25.00%	0.011784	1.970747	19.744

	GO:0007268 synaptic transmission	6	10.00%	0.002618	6.04953	4.7473
	GO:0006811 ion transport	7	11.67%	0.022343	3.079132	34.248
	GO:0048731 system development	13	21.67%	0.013046	2.117152	21.624
	GO:0001505 regulation of neurotransmitter levels	4	6.67%	0.002972	13.29403	5.372
	GO:0019226 transmission of nerve impulse	6	10.00%	0.005133	5.177002	9.1072
	GO:0065007 biological regulation	26	43.33%	0.011622	1.508146	19.499
GOTERM CC_ALL	GO:0005856 cytoskeleton	10	16.67%	0.013466	2.513273	18.459
GOTERM MF_ALL	GO:0008092 cytoskeletal protein binding	6	10.00%	0.046655	2.96898	56.079
UP_TISSUE	Brain	39	65.00%	1.78E-04	1.584613	0.2548

Natural Antisense Transcripts

All cDNA sequences represented in the sea bass microarray consist of public mRNA sequences or ESTs produced through 5' end sequencing of clones from directional cDNA libraries. For this reason as well as to exactly reproduce the available sea bass transcriptome without further assumptions, oligonucleotide probes were designed assuming that each unique transcript in DLPD were a "sense" strand. However, it is increasingly recognized that a substantial fraction of coding genes are transcribed on both strands, yielding sense as well as antisense transcripts. To assess whether this phenomenon occurs in the sea bass, the orientation of sea bass transcripts was compared with that of the best matching sequence in five fish species (see Methods). A conspicuous number of DLPD entries (640) showed opposite orientation against one or more putative homologs in model teleost species (see Methods). For 595 of these sea bass sequences, the matching gene(s) had always the same (opposite) orientation, while for 45 the orientation was discordant among putative homologous sequences in the matching species. A subset of DLPD entries (223) had a significant match against all five fish species, always with antisense orientation. At least, the latter likely represent natural antisense transcripts (NATs) in the sea bass transcriptome. These putative NATs appear to be generally expressed at significant levels, since 180 (80%) yielded a successful hybridization in more than half of the (14) experiments. Similar evidence was obtained for the complete set of candidate NATs (data not shown). To further explore this issue, 108 pairs of DLPD entries matching with opposite orientation a single transcript in the closest reference species, *G. aculeatus*, were identified. These sense/antisense sequences most likely represent either non-overlapping regions or alternative splicing isoforms of the same transcript, or two duplicated loci in *D. labrax* (an example is represented in Additional File 6). A comparison of hybridization signal for the corresponding probes to each pair of sense/antisense DLPD entries could therefore provide preliminary evidence on the ratio between sense and antisense transcription level at the same locus or for duplicated loci. After filtering out probe pairs with an excess of missing data (≥ 50%) across different experiments, in 38 days-old as well as 58 days-old sea bass samples sense transcripts showed higher levels of expression compared to the corresponding putative NATs, respectively in 86% (87/101) and 82% (75/91) of analyzed loci (Additional File 7). Functional annotation of the largest set of putative NATs indicated that over 50% are involved in cell metabolism, with over 20% being related to protein metabolism.

Real-time RT-PCR Validation

To cross-validate platform performance, a set of significant genes were tested using qRT-PCR assays. A total of 13 genes were selected for qRT-PCR analysis, within the annotated transcripts, encompassing the whole range of fold change values (1.06-13) observed in Stage 58 head samples. In order to increase the number of data points in correlation analysis between microarray and qRT-PCR results, Stage 38 samples were included in the validation procedure as well. Fold changes between Stage 38 and Stage 58 samples (an *ad hoc* quantile normalization was performed on all microarray data, 14 experiments) were calculated for 13 selected target genes. A Spearman rank correlation test was then performed considering a total of 14 experiments. A statistically significant correlation was obtained comparing expression levels for each target gene across all samples (see Table 7). Nine genes showed high correlation coefficients (Spearman rho >0.8) for both probes (p-value < 0.01) with qPCR data. Three genes exhibited a significant correlation (0.7 < rho >0.8 with p-value < 0.01) while no correlation was observed for Cyclin Dependent Kinase 2-associated protein 1 (CDK2) exhibiting a fold change < 1.2. Fold changes detected by gene-specific PCR assay and by both microarray probes (1 and 2) for the same target transcript were also compared. Fold change was calculated as the ratio of mean signal intensity between prognathous and controls on both Stage 38 and Stage 58 as well as between Stage 38 and Stage 58. For all tested targets, the direction of change in expression was concordant between qPCR and microarray results. For both probes, linear regression of microarray-estimated fold change against qPCR results demonstrated a strong positive correlation (Spearman rho >0.92, p-value < 0.01) between the two technologies (see Figure 2).

Table 7: Correlation between microarray and Real-time RT-PCR expression data.

DLPD ID	Gene Name	Spearman's rho		
		Probe_1/Probe_2	qPCR/Probe_1	qPCR/Probe_2
DLPD03006	Chondroitin sulfate proteoglycan 5 Precursor	0.867**	0.878**	0.912**
DLPD08826	Calcitonin-related peptide	0.887**	0.796**	0.846**
DLPD03840	Stathmin	0.978**	0.930**	0.974**
DLPD11717	Tetraspanin	0.991**	0.758**	0.734**
DLPD13372	Peripherin	0.979**	0.951**	0.895**
DLPD06176	Brain specific fatty acid protein	0.978**	0.960**	0.974**
DLPD02075	Neurogenic differentiation factor	0.863**	0.830**	0.907**
DLPD11576	Stromal cell derived receptor 1β	0.958**	0.762**	0.846**
DLPD03024	CDK2-associated protein 1	0.952**	0.162	0.086
DLPD06508	Synaptosomal associated protein	0.986**	0.902**	0.874**
DLPD12789	Pleiotropin	0.979**	0.958**	0.944**
DLPD07155	GRAM-domain containing protein 2 (splicing variant 1)	0.900**	0.934**	0.855**
DLPD09843	GRAM-domain containing protein 2 (splicing variant 2)	0.991**	0.855**	0.890**

** $p < 0.01$

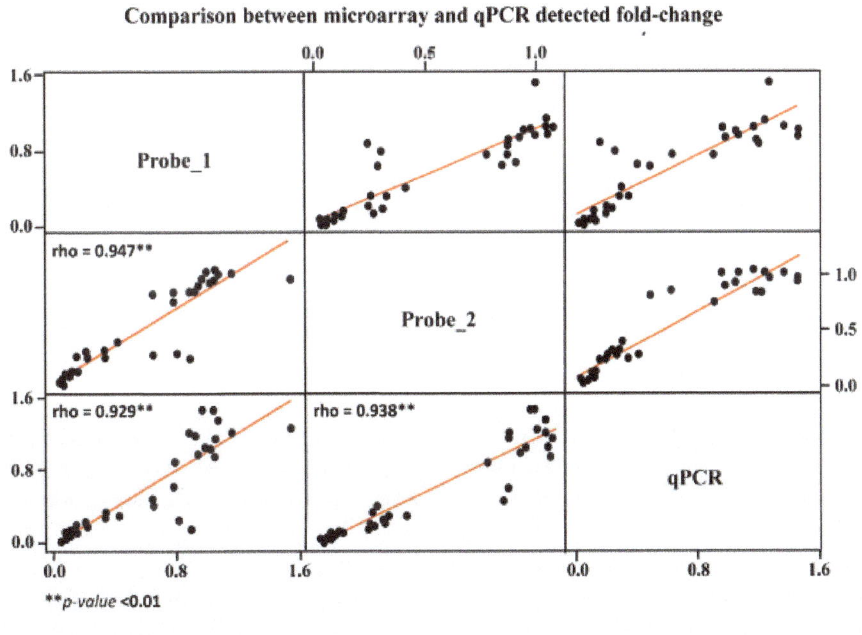

Figure 2: Comparison between microarray and qPCR results. Expression values for the 13 target genes were compared between microarray probes and Real-time RT-PCR data. In X and Y axis are reported microarray- and qPCR-estimated fold changes.

DISCUSSION

Four major points emerge from the results obtained in the present study. The first one regards the high level of flexibility, reliability, and reproducibility of the microarray platform developed for the European sea bass. Oligonucleotide probe design proved to be feasible on nearly all available transcribed sequences and almost every designed probe yielded a successful hybridization signal, providing a most complete representation of the current transcriptome for *D.labrax*. One of the most important requirements for a microarray platform is good reproducibility. Analysis of biological replicates showed extremely high correlation coefficients, demonstrating great reproducibility of microarray data. As already reported [16, 18], the use of a one color labeling scheme has apparently no negative effects on the quality of data, but it greatly simplifies the experimental design and allows easier analysis of novel samples. Finally, validation with qRT-PCR, a method that is based on a different technical approach, confirmed microarray hybridization data with the exception of a single gene, CDK2-associated protein 1, which showed a positive, but not significant correlation between results of the two different methodologies (Table 7). This is likely due to the small difference in expression for this gene (mean fold change estimated from array data is 1.3). Lack of correlation between microarray and qRT-PCR for genes exhibiting low levels of change (< 1.4 fold) has been commonly reported. Indeed a two-fold change is usually considered as the cut-off below which microarray and qRT-PCR data begin to lose correlation [26].

The second point concerns the annotation of the sea bass transcriptome. Whilst the design and validation of a custom oligo DNA microarray has become a straightforward procedure, a proper functional classification for the collection of unique transcripts that are represented onto the DNA microarray is certainly more difficult to achieve in non-model species, particularly in teleost fish.

High-throughput technologies have now prompted different comprehensive approaches toward functional analysis of lists of differentially expressed genes/proteins, Gene Ontology [27] being the most popular of them. These methods offer relevant advantages, although gene ontology annotations suffer from known limitations [28]. In non-model species these approaches are based on putative homology against better characterized organisms rather than on direct experimental evidence. There are here two major sources of error; the first related to the problem of identifying true orthology [29], the second to the fact that sequence homology does not always imply functional homology, due to inherent differences between the species compared. In the case of teleost fish, extracting gene functional annotations from mammalian models poses an additional problem. The ancestral teleost genome underwent a whole genome duplication (WGD) after the separation from the tetrapod lineage [30], which led to the presence of two copies of each gene present in higher vertebrates, a fraction of which genes was retained through different mechanisms (*e.g.*neofunctionalization, subfunctionalization, genomic inertia), whereas others have been lost by gene deletion or pseudogenization [31]. A ready example of such phenomenon is provided by the gene that showed the largest fold change difference in the present study, Arrestin 3 (Retinal X arrestin, ARR3) (see Additional file 3). ARR3 belongs to the cluster of beta-arrestins, which includes four paralogs (Arrestin beta 1, ARRB1; Arrestin beta 2, ARRB2, S-AG arrestin, ARR3) in the human genome [32] but at least seven paralogs (ARRB1a; ARRB1b, ARRB2a, ARRB2b, S-AG arrestin, ARR3a, ARR3b) in the stickleback genome. Despite all these limitations, functional annotation through sequence homology remains the best available method in non-model species. In the case of *D. labrax*, a good percentage (>60%) of DLPD entries has a significant match with a known gene/protein, generally higher than the value observed in similar studies on other species (gilthead sea bream 40% [16], turbot 50.7% [33] Senegal sole 40.6% [34], largemouth bass 46% [10], pre-smolt Atlantic salmon 50.3% [35], channel catfish 51% [36], and comparable to that reported for the Atlantic halibut 60% [37]). A smaller proportion (>4,500) of sea bass transcripts could be associated with a GO term, most likely as a result of the more stringent criteria enforced in Blast2GO and the lack of GO annotation for part of the protein sequences matching DLPD entries. Nevertheless, sufficient information was obtained to construct a custom background, which takes into account the fact that the sea bass transcriptome is only partly represented onto the microarray, and to identify biological processes that are significantly enriched among differentially expressed genes (see Table 4). A second approach of annotation by similarity, through direct or indirect comparison against a model species transcriptome appears to convey comparable information, with over 6,000 human or 5,000 zebrafish putative homologs. The use of the human transcriptome as a reference provides a larger number of significant functional annotations (see Table 5), which is somehow expected since the probability of identifying significantly enriched GO terms depends in part on the size of the gene list to be analyzed and the corresponding background. On the other hand, using human-centered annotations might be more prone to the risks described previously. Finally, the results obtained with different methods (GOStat compared to DAVID, DAVID human knowledgebase against zebrafish one) show a certain degree of overlap, yet there are terms that appear only in one or two analyses, therefore the obtained results should be interpreted with some caution.

While the majority of DLPD entries can be associated to a known gene, a substantial "silent" minority exists of transcripts that could not find any significant match against a variety of sequence databases. Even if not associated with a known coding gene, these transcripts might convey useful information. For instance, a large proportion (190/242 with a threshold of e^{-5}, and 169/242 with a threshold of e^{-10}) of all differentially expressed transcripts in jaw deformed sea bass find a significant match against a specific region in the stickleback genome. The strong conservation of large genome segments between *D. labrax* and *G. aculeatus* (R. Reinhardt, unpublished data) and the availability of a physical map of the sea bass genome (F. Galibert, personal communication) will allow the use of matching DLPD transcripts as positional candidates to identify loci involved in the genetic determination of mandibular prognathism.

The 40% of non-matching DLPDs likely represents different types of transcripts, *e.g.* 5' and 3' end untranslated regions (UTRs) or alternative splicing isoforms that have not been (yet) characterized in other species, extremely fast-evolving protein-coding regions, novel genes. In addition, a fraction of non-matching sea bass transcripts might belong to the expanding universe of non-coding RNAs, which appear to cover an increasing part of the animal genome [38, 39]. Although originally dismissed as transcriptional noise, evidence is accumulating for a functional role of these transcripts [40]. The third point of the present study, the presence of NATs in the sea bass transcriptome, is related to this issue. Natural antisense transcripts have been originally identified searching EST collections, and appear to be widespread across animal species, albeit at diverse frequency [41]. Various putative functions have been proposed for NATs [42], with an increasingly relevant role in the production of endogenous siRNAs [43]. Oligo DNA microarrays have been used to specifically detect sense-antisense gene expression [44, 45], although there are some caveats about the risk of false positive NAT detection due to genomic DNA contamination of RNA extracts or unintended labeling of both cDNA strands [46]. Microarray analysis of the sea bass transcriptome was not specifically aimed at investigating NATs yet it provided preliminary evidence for the existence of putative antisense RNAs in this species. Based on sequence homology with the stickleback genome, in most cases sea bass NATs represent non-overlapping regions of corresponding sense transcripts (see Additional File 6), and can derive from protein-coding regions as well as non-protein-coding sequences, including exonic, intronic and intergenic sequences. While the biological roles of NATs remain to be elucidated, putative sea bass NATs show lower levels of expression compared to the corresponding sense DLPD transcripts (see Additional File 7), as already reported for other species [46]. However, more than 10% of putative NATs showed higher expression levels compared to sense transcripts (*e.g.* 14 in 38 days-old larvae with fold change ranging between 1.4 and 71). Functional analysis of sea bass NATs also showed significant enrichment of certain molecular functions, although these results require further confirmation.

The fourth and last evidence from the present study was the discovery of a set of differentially regulated transcripts in the mandible of jaw-deformed juvenile sea bass compared to normally developed animals. Developmental defects are generally thought to trace back early in the ontogeny. For instance, mutations in master regulatory genes start to exert their effects during early patterning of craniofacial development (*e.g.*Tbx22 in mammals and zebrafish [47], Endothelin1 in several vertebrates [48]), while precocious treatment of Japanese flounder larvae (6-9 days post-hatching) with retinoid acid receptor agonists produces lower jaw deformities [49]. In the present study, transcriptional changes are observed at a much later stage, which might represent either the downstream effects of earlier changes in the expression of hierarchically higher genes and/or transcriptional perturbations that start at a less differentiated stage and are maintained until later stages. On the other hand, no statistically significant differences in the expression profile of 38-days old sea bass juveniles were observed comparing deformed against normal fish. For this stage, however, whole heads were analyzed, which might have caused a reduced sensitivity in detecting differential expression. Using image analysis on juvenile sea bass pictures it was estimated that the lower jaw represents less than 1/10 of the entire head, while the anterior region of the mandible, the dentary, is approximately 1/30 of the sampled tissues. This means that if for instance a certain gene has an average expression level of 10 in the whole head whereas it is up-regulated 10-fold (100) exclusively in the mandibular region of deformed animals, the observed overall fold change would be 1.9, and in case differential expression is limited to the dentary a 1.3 fold increase would be present, likely below the threshold of reliable detection using microarray analysis. Microdissection of the mandible or part of it will therefore be required to identify differentially expressed genes in deformed animals at early stages, as reported here for 58-days old juveniles. The relatively large set of down-regulated transcripts in the lower jaw of prognathous fish also suggests that mandibular prognathism observed in the present study might be distinct from pugheadness, which is reported to be the result of under-development of the upper jaw, with only apparent protrusion of the normally developed lower jaw. As in most teleost fish, mandible development in *D. labrax* is characterized by the presence of a cartilagenous component, the Meckel's cartilage, which is established early and subsequently regresses and ossifies, and a dermal bone component, the dentary, which appears later (25 days post hatching) and undergoes direct (membranous) ossification [50]. Both components receive an important contribution of cells originating from the neural crests [51]. The ossification process of the Meckel's cartilage is relatively well known, with a central role of the Hedgehog pathway, and it appears to be conserved in fish and mammals [52], whereas the molecular mechanisms controlling the dentary development are less characterized and seem to be at least partially distinct from those observed in the Meckel's cartilage [53]. How can this relate to the genes that were found to be differentially expressed in the present study? A significant enrichment in regulatory genes, especially in the development of anatomical structures was obtained in GO functional annotation, and gene-by-gene evaluation confirmed that some of them might be linked to bone formation. For instance, SOX4 has been reported to act downstream to Parathyroid Hormome (PTH) and PTH related protein (PTHrP) in osteoblast-like cells, being highly expressed in hypertrophic condrocytes during the mineralizing phase of endochondral ossification [54]. Pleiotrophin or HB-GAM is highly expressed in bone, where it seems to play a role in bone development and remodelling [55]. Retinoic acid receptor X gamma (RXRγ) forms heterodimers with vitamin D receptors, Peroxisome prolifireator-activated receptors (PPARs), and Retinoic acid receptors (RARs). All these nuclear receptors are somehow involved in bone formation and more in general in controlling skeletal growth. For instance, vitamin A and its metabolites, acting through RARs, have been reported to cause vertebral and craniofacial deformities in farmed fish [5]. Remarkably, high levels of dietary vitamin A significantly increased the frequency of cranial malformations (underdeveloped lower jaw) in *D. labrax* [5] as well as in other fish species [49, 56].

It seems, however, that bone development is not the most represented biological process when examining the functional roles of protein encoded by differentially expressed genes in deformed sea bass. A substantial fraction of them appears to affect nerve growth and function. This evidence might be explained by a temporal shift in the development of mandibular nerves, which in turn may be a consequence of the altered process of mandible bone formation. Down-regulation of several markers of neuronal development seems to suggest a delay in the differentiation of neuronal cells that constitute the nervous component of the lower jaw. However, there could be a closer relationship between mandibular bone formation and transcriptional changes for neuron-specific genes. Increasing evidence indicates that the nervous system participates in the regulation of bone physiology [57]. Peripherally-released neurotransmitters exert their actions on osteoblasts and osteoclasts, which have been demonstrated to express specific receptors for these mediators. Of particular interest for the present study is the role of calcitonin gene-related peptide (CGRP), which is a neuropeptide with a well-established function in bone metabolism, promoting bone formation and repressing bone resorption [58]. As other neuromediators, CGRP is also produced directly by osteoblasts as an autocrine factor [57]. The expression of neuronal-specific genes in bone cells is not limited to neurotransmitters and their receptors, but it extends to the molecular network for regulated glutamate exocytosis (*e.g.* SNARE, SNAP-25, syntaxin, synaptophysin, syntaghmin) generally described in pre-synaptic nerve terminals [59]. The role of glutamate signalling in osteoblasts and osteoclasts is complex and *in vivo* studies are still limited [60]. Evidence from conditional KO mice lacking components of the glutamate pathways showed reduced mineralization and delayed ossification. Several genes involved in exocytosis and glutamate signalling are represented among down-regulated transcripts in the present study (*e.g.* SNAP-25, Synaptophysin, Synapsin, Metabotropic glutamate receptor 8). A working hypothesis to hold together the above evidence might be that in jaw-protruding sea bass the bone formation process is delayed compared to normally developed animals, through down-regulation of different signalling pathways, which control bone formation/remodelling, either directly in osteoblasts and/or in neuronal terminals innervating the mandible. A delay in the ossification process might allow a prolonged growth of the mandibular bone components resulting in a protruding lower jaw.

Finally, it should be noted that part of the transcriptional differences observed in deformed animals might point, at least in part, to other regions of the dissected mandible. For instance, the formation of the tongue and/or the teeth might be indirectly affected by the deformity, and contribute to the differential gene profile. This in turn might provide a complementary/alternative hypothesis to explain the prevalence of neuron-related transcripts. In fact, tooth development has been shown to be tightly linked with nerve development.

Clearly, the above hypotheses await further confirmation from additional developmental stages and with the use of methods that allow better dissection of anatomical structures.

CONCLUSION

In conclusion, assembly and annotation of a first version of the European sea bass transcriptome led to the construction and validation of a species-specific oligonucleotide microarray. Microarray analysis of the sea bass transcriptome provided preliminary evidence for the existence of putative antisense RNAs in this species. This genomic platform was applied to detect differentially expressed genes in the mandible of jaw-deformed fish, revealing significant down-regulation of several transcripts involved in bone formation and neuronal function.

METHODS

Sample Collection and RNA Extraction

Two developmental stages, 38 days-old (Stage 38, average length 12 mm) and 58 days-old (Stage 58, average length 16 mm) sea bass juveniles, were included on the experimental design. The animals were collected at the fish farm "Impianto di Acquacoltura Ca' Zuliani" (Pila di Porto Tolle, Italy), not subjected to any experimental manipulation and sacrificed, using an excess of anesthetic, according to guidelines of the Italian law (DL 116/92) and the European legislation (Council Directive 86/609/EEC and subsequent amendments). For Stage 38, the cranial region was dissected under a stereomicroscope for 15 normal and 15 jaw-deformed individuals. For Stage 58, it was possible to dissect the lower jaw for a total of 40 individuals (20 normal fish and 20 deformed ones). Independent tissue pools (3 for Stage 38 and 4 for Stage 58) of five fish each were produced for each condition/stage combination.

Total RNA was extracted from pooled tissue samples using the RNAeasy Mini Kit (Qiagen, Hilden, Germany) following the manufacturer's instructions. RNA quality was previously checked by gel electrophoresis on a 1% agarose gel containing SYBR Safe™ DNA Gel stain 10,000× (Invitrogen™, Carlsbad, California). RNA concentration was determined using a UV-Vis spectrophotometer, NanoDrop® ND-1000 (NanoDrop Technologies, Wilmington, USA). RNA integrity and quality was finally estimated on an Agilent 2100 Bioanalyzer (Agilent Technologies, Palo Alto, CA). RNA Integrity Number (RIN) index was calculated for each sample using Agilent 2100 Expert software. RIN provides a numerical assessment of the integrity of RNA that facilitates the standardization of the quality interpretation. In order to reduce experimental biases in microarray analysis due to poor RNA quality, only RNA samples with RIN number >8 were further processed.

Database Construction, Annotation, and Probe Design

Unique sequences were assembled starting from an existing assembly (version 1.0 Adl1) that could be downloaded from the SIGENAE web site [61] at the starting time of the present study (February 2008), which originated from 14 normalized cDNA libraries representing different adult tissues (liver, ovary, testis, bone and cartilage, heart and vessel, brain and pituitary gland, adipose tissue, head kidney, trunk kidney, gill, spleen, intestine, muscle, skin), one larval cDNA library, and two subtracted libraries from sea bass challenges in freshwater and seawater. This initial set of 17,623 assembled unique sequences was clustered with 5,045 ESTs from a normalized cDNA library of the corpuscle of Stannius, 2,356 publicly available ESTs in GenBank, and 226 public full-length or partial mRNA sequences. Clustering was carried out using a strategy based on BLAST (Basic Local Alignment Search Tool) to identify candidate sequences (cut off e-value set to e-10) to be included in a cluster and Cap3 [62] to perform the assembly and produce the consensus sequences. ESTs were considered to belong to the same cluster if there was an overlap of at least 40 bp and an overlap identity of 90%. The clustering pipe-line produced a final set of 19,734 different clusters. Links between GenBank accession numbers and unique transcripts in the database presented here are reported in the GEO "Platform data table" (GPL9663).

Similarity searches for each unique transcript against different databases were performed using the BLAST. The procedure involved two different steps: i) BLAST search with blastx option (cut off e-value of < 1.0 e-3) against the NCBI (National Centre for Biotechnology Information) amino acid non-redundant database, SwissProt database, and TrEMBL database ii) BLAST search with blastn option (cut off e-value of < 1.0 e-5) against NCBI nucleic non-redundant database and Ensembl transcriptomes for the five fish species which have a high quality draft genome sequence zebrafish (*Danio rerio*), stickleback (*Gasterosteus aculeatus*), Japanese medaka (*Oryzias latipes*), Japanese pufferfish (*Takifugu rubripes*), and green-spotted pufferfish (*Tetraodon nigroviridis*)). Gene ontology (GO) associations for "Biological process", "Molecular function" and "Cellular component" were obtained using the blastx option against the NCBI amino acid non redundant database as implemented in the Blast2GO software [63]. A summary of the overall results of GO annotation was obtained with the CateGOrizer program [64], employing the "Generic GO slim" set [65]. All unique sea bass transcripts, and their corresponding annotations are stored in the *Dicentrarchus labrax* Padova Database (DLPD), which is based on the BIOMART environment and can be queried using different filters based on contig name, description, GO terms, or for a combination of these criteria. It is possible to visualize different attributes choosing among the sequence name, the sequence contig consensus, sequence description, GO annotation, and putative orthologous genes in the human genome as well as in five fish model species (*G. aculeatus, O. latipes, D. rerio, T. nigroviridis*, and *T. rubripes*).

Two non-overlapping probes for each unique transcript were designed to construct a high-density oligo-DNA microarray. Probe design was carried out using the Agilent eArray interface [66], which applies proprietary prediction algorithms to design 60mer oligo-probes. Microarrays were synthesized in situ using Agilent non-contact ink-jet technology with a 4 × 44K format. Each array included default positive and negative controls.

RNA Labeling and Hybridization

Sample labeling and hybridization were performed according to the Agilent One-Color Microarray-Based Gene Expression Analysis protocol. Briefly, for each sample 500 ng of total RNA were linearly amplified and labeled with Cy3-dCTP. A mixture of 10 different viral poly-adenilated RNAs (Agilent Spike-In Mix) was added to each RNA sample before amplification and labeling, to monitor microarray analysis work-flow. Labeled cRNA was purified with Qiagen RNAeasy Mini Kit, and sample concentration and specific activity (pmol Cy3/μg cRNA) were measured in a NanoDrop® ND-1000 spectrophotometer. A total of 1,650 ng of labeled cRNA was prepared for fragmentation adding 11 μl 10× Blocking Agent and 2.2 μl of 25× Fragmentation Buffer, heated at 60°C for 30 min, and finally diluted by addition with 55 μl 2× GE Hybridization buffer. A volume of 100 μl of hybridization solution was then dispensed in the gasket slide and assembled to the microarray slide (each slide containing four arrays). Slides were incubated for 17 h at 65°C in an Agilent Hybridization Oven, subsequently removed from the hybridization chamber, quickly submerged in GE Wash Buffer 1 to disassembly the slides and then washed in GE Wash Buffer 1 for approximately 1 minute followed by one additional wash in pre-warmed (37°C) GE Wash Buffer 2.

Data Acquisition and Analysis

Hybridized slides were scanned at 5 μm resolution using an Agilent G2565BA DNA microarray scanner. Default settings were modified to scan the same slide twice at two different sensitivity levels (XDR Hi 100% and XDR Lo 10%). The two linked images generated were analyzed together and data were extracted and background subtracted using the standard procedures contained in the Agilent Feature Extraction (FE) Software version 9.5.1. The software returns a series of spot quality measures in order to evaluate the goodness and the reliability of spot intensity estimates. All control features (positive, negative, etc.) except Spike-in (Spike-In Viral RNAs) were excluded from subsequent analyses. Spike-in control intensities were used to identify the best normalization procedure for each dataset. After normalization, spike intensities are expected to be uniform across the experiments of a given dataset. Normalization procedures were performed using R statistical software [67]. On our data quantile normalization always outperformed cyclic lowess and quantile-normalized data were used in all subsequent analyses.

Statistical tests implemented in the program Significance Analysis of Microarray (SAM) [68] were used to identify differentially expressed genes between normal and deformed animals for both stages. For Stage 58 analysis, probes with a high proportion of missing values (more than two missing values across the biological replicates of each condition) were removed from the dataset. The flag «glsFound» (set to 1 if the spot has an intensity value significantly different from the local background, 0 otherwise), obtained after data processing with Feature Extraction Software 9.5.1, was used to identify unreliable spots, probes with flag equal to 0 were considered as «missing».

Pearson correlation coefficients were estimated within and among arrays with Statgraphics Centurion XVI to evaluate repeatability and precision of the obtained microarray data. A non-parametric Spearman rank-correlation test was used to assess correlation between expression values measured respectively with real-time RT-PCR and microarray. The same test was performed separately for each microarray probe. Spearman correlation tests were implemented using SPSS 12.0.

Functional Annotation

Functional annotation of differentially expressed genes was performed using the DAVID web-server [69] and GOstat [70]. .For GOStat analysis, all significantly down-regulated transcripts with at least one associated GO description (39), previously obtained with Blast2GO, were compared against a background of 4,692 GO-annotated DLPD entries, using the GOStat program with default settings. Since DAVID databases contain functional annotation data for a limited number of species, it was necessary to link DLPD transcripts with sequence identifiers that could be recognized in DAVID (Ensembl Human Gene IDs and ZFIN-IDs). This was carried out through dedicated BLAST searches implemented as follows: i) blastx and blastn BLAST options were both used to search significant matches of DLPD sequences directly against human Ensembl proteins and transcripts respectively ii) the same search strategy was implemented first against Ensembl proteins and transcripts of *G. aculeatus*, and then using best-hits either from blastx (stickleback proteins) or blastn (predicted proteins from stickleback transcripts) search as queries in a second search, with blastp option, against all Ensembl human proteins, iii) a direct search using either blastn or blastx against all zebrafish Ensembl proteins, iv) an indirect search first through the stickleback protein and transcript databases in Ensembl and then recovering zebrafish putative orthologous proteins as described above for human proteins (ii). Finally, Ensembl Human Gene IDs and ZFIN-IDs were obtained from the corresponding Ensembl protein entries using the BIOMART data mining tool [71] A scheme summarizing the four different approaches is presented in Additional File 4. Human or zebrafish IDs genes corresponding to differentially expressed sea bass transcripts and to all genes represented on the array were then used to define respectively a "gene list" and a "background" in the bioinformatic tool DAVID. DAVID was then used with the following settings, gene count = 4 and ease = 0.05

Validation of Gene Expression data Using Quantitative (q) RT-PCR

Thirteen target genes and one reference gene (Malate dehydrogenase, MDH) were selected for qRT-PCR analysis. Gene-specific primers were defined for each transcritpt with the Primer 3 v.0.4.0 software. To design intron-spanning assays, intron-exon boundaries were identified through the alignment, whenever possible, with the corresponding genomic sequences available from the European sea bass genome sequencing project (R. Reinhardt, unpublished data). Otherwise, putative intron-exon junctions were inferred by comparison with homologous genes present in high-quality draft fish genomes.

One microgram of total RNA for each sample was reverse transcribed to cDNA using Superscript II (Invitrogen™, Carlsbad, California). An aliquot (2.5 μl) of diluted (2 ng/μl) cDNA template was amplified in a final volume of 10 μl, containing 5 μl Platinum SYBR Green qPCR SuperMix-UDG 2× (Invitrogen™) and 0.25 μl of each gene-specific primer (10 μM). The amplification protocol consisted of an initial step of 2 min at 50°C and 2 min at 95°C, followed by 45 cycles of 10 s at 95°C and 30 s at 60°C. All experiments were carried out in a LightCycler® 480 (Roche Diagnostics, Mannheim, Germany). To evaluate the efficiency of each assay, standard curves were constructed amplifying two-fold serial dilutions of the same cDNA (38d_N3), which was used as calibrator. For each sample, the Cp (Crossing point) was used to determine the relative amount of target gene; each measurement was made in duplicate, and normalized to the reference gene (MDH, contig name DLPD06340), which was also measured in duplicate. MDH was chosen as reference gene in qRT-PCR assays as it is considered a housekeeping gene, and it did not exhibit any significant change in microarray data between either the two developmental stages or the two conditions (normal and deformed) tested (percentage coefficient of variation for Probe_1 and Probe_2 was 13.4% and 13.5% respectively). Samples tested in qRT-PCR were the same of microarray experiments. One biological replicate of Stage 38 (38d_N3) was used as calibrator, the internal control for each amplification reaction.

NOTES

Serena Ferraresso, Massimo Milan contributed equally to this work.

DECLARATIONS

Acknowledgements

We would like to thank all the people at the fish farm "Valle Ca' Zuliani" for providing access to the samples and Clara Boglione for reviewing the ms. We thank also two anonymous reviewers for their useful comments on an earlier version of the manuscript. This work was partially supported by by a grant from European Union-funded Network of Excellence "Marine Genomics Europe" (6th Framework Programme, Contract n° 505403).

REFERENCES

1. Chudley AE: Genetic landmarks through philately - The Habsburg jaw. Clin Genet. 1998, 54: 283-284. 10.1034/j.1399-0004.1998.5440404.x.
2. Cruz RM, Krieger H, Ferreira R, Mah J, Hartsfield J, Oliveira S: Major gene and multifactorial inheritance of mandibular prognathism. Am J Med Genet A. 2008, 146: 71-77.
3. Huang CM, Mi MP, Vogt DW: Mandibular prognathism in the rabbit: discrimination between single-locus and multifactorial models of inheritance. J Hered. 1981, 72: 296-298.
4. Barahona-Fernandes MH: Body deformation in hatchery reared European sea bass *Dicentrarchus labrax* (L). Types, prevalence and effect on fish survival. J Fish Biol. 1982, 21: 239-249. 10.1111/j.1095-8649.1982.tb02830.x.
5. Cahu C, Zambonino Infante J, Takeuchi T: Nutritional components affecting skeletal development in fish larvae. Aquaculture. 2003, 227 (1-4): 254-258. 10.1016/S0044-8486(03)00507-6.
6. Mazurais D, Glynatsi N, Darias M, Christodoulopoulou S, Cahu C, Zambonino-Infante JL, Koumoundouros G: Optimal levels of dietary vitamin A for reduced deformity incidence during development of European sea bass larvae*(Dicentrarchus labrax*) depend on malformation type. Aquaculture. 2009, 294: 262-270. 10.1016/j.aquaculture.2009.06.008.
7. Bardon A, Vandeputte M, Dupont-Nivet M, Chavanne H, Haffray P, Vergnet A, Chatain B: What is the heritable component of spinal deformities in the European sea bass (*Dicentrarchus labrax*)?. Aquaculture. 2009, 294: 194-201. 10.1016/j.aquaculture.2009.06.018.
8. Villeneuve L, Gisbert E, Zambonino-Infante JL, Quazuquel P, Cahu CL: Effect of nature of dietary lipids on European sea bass morphogenesis: implication of retinoid receptors. Brit J Nut. 2005, 94: 877-884. 10.1079/BJN20051560.
9. Geay F, Darias MJ, Santigosa E, Desbruyères E, Quazuguel P, Zambonino-Infante JL, Cahu CL, Mazurais D: Cloning of endothelin-1 (ET-1) from European sea bass (*Dicentrarchus labrax*) and its gene expression analysis in larvae with retinoic acid-induced malformations. Aquaculture. 2009, 287: 169-173. 10.1016/j.aquaculture.2008.10.011.
10. Garcia-Reyero N, Griffitt R, Liu L, Kroll KJ, Farmerie WG, Barber DS, Denslow ND: Construction of a robust microarray from a non-model species largemouth bass, *Micropterus salmoides* (Lacepede), using pyrosequencing technology. J Fish Biol. 2008, 72: 2354-2376. 10.1111/j.1095-8649.2008.01904.x.
11. Villeneuve DL, Knoebl I, Larkin P, Miracle AL, Carter BJ, Denslow ND, Ankley GT: Altered gene expression in the brain and liver of female fathead minnows *Pimephales promelas* Rafinesque exposed to fadrozole. J Fish Biol. 2008, 72: 2281-2340. 10.1111/j.1095-8649.2008.01897.x.

12. Klaper R, Carter BJ, Richter CA, Drevnick PE, Sandheinrich MB, Tillitt DE: Use of a 15 k gene microarray to determine gene expression changes in response to acute and chronic methylmercury exposure in the fathead minnow *Pimephales promelas* Rafinesque. J Fish Biol. 2008, 72: 2207-2280. 10.1111/j.1095-8649.2008.01899.x.

13. Kane MD, Sringer JA, Iannotti NV, Gough E, Johns SM, Schlueter SD, Sepulveda MS: Identification of development and tissue-specific gene expression in the fathead minnow *Pimephales promelas*, Rafinesque using computational and DNA microarray methods. J Fish Biol. 2008, 72: 2341-2353. 10.1111/j.1095-8649.2008.01889.x.

14. Olohan LA, Li W, Wulff T, Jarmer H, Gracey AY, Cossins AR: Detection of anoxia-responsive genes in cultured cells of the rainbow trout *Oncorhynchus mykiss* (Walbaum), using an optimized, genome-wide oligoarray. J Fish Biol. 2008, 72: 2170-2186. 10.1111/j.1095-8649.2008.01877.x.

15. Salem M, Kenney PB, Rexroad CE, Yao J: Development of a 37 k high-density oligonucleotide microarray: a new tool for functional genome research in rainbow trout. J Fish Biol. 2008, 72: 2187-2206. 10.1111/j.1095-8649.2008.01860.x.

16. Ferraresso S, Vitulo N, Mininni AN, Romualdi C, Cardazzo B, Negrisolo E, Reinhardt R, Canario AVM, Patarnello T, Bargelloni L: Development and validation of a gene expression oligo microarray for the gilthead sea bream (*Sparus aurata*). BMC Genomics. 2008, 9: 580-10.1186/1471-2164-9-580.

17. Leder EH, Merilä J, Primmer CR: A flexible whole-genome microarray for transcriptomics in three-spine stickleback (*Gasterosteus aculeatus*). BMC Genomics. 2009, 10: 426-10.1186/1471-2164-10-426.

18. Millán A, Gómez-Tato A, Fernández C, Pardo GB, Álvarez-Dios JA, Calaza M, Bouza C, Vázquez M, Cabaleiro S, Martínez P: Design and Performance of a Turbot (*Scophthalmus maximus*) Oligo-microarray Based on ESTs from Immune Tissues. Mar Biotechnol. 2009

19. Darias MJ, Zambonino-Infante JL, Hugot K, Cahu CL, Mazurais D: Gene expression patterns during the larval development of European sea bass (*Dicentrarchus labrax*) by microarray analysis. Mar Biotechnol. 2008, 10: 416-428. 10.1007/s10126-007-9078-1.

20. Whitehead Pj, Bauchot Ml, Hureau JC: Fishes in the North-eastern Atlantic and the Mediterranean. 1987, UNESCO

21. FEAP. [http://www.aquamedia.org/Production/]

22. DLPD database. [http://enne.cribi.unipd.it:5556/biomart/martview/]

23. GEO database. [http://www.ncbi.nlm.nih.gov/geo/]

24. Beissbarth T, Speed TP: GOstat: find statistically overrepresented Gene Ontologies within a group of genes. Bioinformatics. 2004, 20: 1464-1465. 10.1093/bioinformatics/bth088.

25. Huang da W, Sherman BT, Lempicki RA: Systematic and integrative analysis of large gene lists using DAVID bioinformatics resources. Nat Protoc. 2009, 4: 44-57. 10.1038/nprot.2008.211.

26. Morey JS, Ryan JC, VanDolah FM: Microarray validation: factors influencing correlation between oligonucleotide microarrays and real-time PCR. Biol Proced Online. 2006, 8: 175-193. 10.1251/bpo126.

27. Ashburner M, Ball CA, Blake JA, Botstein D, Butler H, Cherry JM, Davis AP, Dolinski K, Dwight SS, Eppig JT, Harris MA, Hill DP, Issel-Tarver L, Kasarskis A, Lewis S, Matese JC, Richardson JE, Ringwald M, Rubin GM, Sherlock G: Gene ontology: tool for the unification of biology. The Gene Ontology Consortium. Nat Genet. 2000, 25: 25-29. 10.1038/75556.

28. Rhee SY, Wood V, Dolinski K, Draghici S: Use and misuse of the gene ontology annotations. Nat Rev Genet. 2008, 9: 509-515. 10.1038/nrg2363.

29. Kuzniar A, van Ham RC, Pongor S, Leunissen JA: The quest for orthologs: finding the corresponding gene across genomes. Trends Genet. 2008, 24: 539-551. 10.1016/j.tig.2008.08.009.

30. Kasahara M, Naruse K, Sasaki S, Nakatani Y, Qu W, Ahsan B, Yamada T, Nagayasu Y, Doi K, Kasai Y, Jindo T, Kobayashi D, Shimada A, Toyoda A, Kuroki Y, Fujiyama A, Sasaki T, Shimizu A, Asakawa S, Shimizu N, Hashimoto S, Yang J, Lee Y, Matsushima K, Sugano S, Sakaizumi M, Narita T, Ohishi K, Haga S, Ohta F, Nomoto H, Nogata K, Morishita T, Endo T, Shin-I T, Takeda H, Morishita S, Kohara Y: The medaka draft genome and insights into vertebrate genome evolution. Nature. 2007, 447: 714-719. 10.1038/nature05846.

31. Kassahn KS, Dang VT, Wilkins SJ, Perkins AC, Ragan MA: Evolution of gene function and regulatory control after whole-genome duplication: comparative analyses in vertebrates. Genome Res. 2009, 19: 1404-18. 10.1101/gr.086827.108.

32. Alvarez CE: On the origins of arrestin and rhodopsin. BMC Genomics. 2008, 8: 222-

33. Pardo BG, Fernández C, Millán A, Bouza C, Vázquez-López A, Vera M, Alvarez-Dios JA, Calaza M, Gómez-Tato A, Vázquez M, Cabaleiro S, Magariños B, Lemos ML, Leiro JM, Martínez P: Expressed sequence tags (ESTs) from immune tissues of turbot (*Scophthalmus maximus*) challenged with pathogens. BMC Vet Res. 2008, 4: 37-10.1186/1746-6148-4-37.

34. Cerdà J, Mercadé J, Lozano JJ, Manchado M, Tingaud-Sequeira A, Astola A, Infante C, Halm S, Viñas J, Castellana B, Asensio E, Cañavate P, Martínez-Rodríguez G, Piferrer F, Planas JV, Prat F, Yúfera M, Durany O, Subirada F, Rosell E, Maes T: Genomic resources for a commercial flatfish, the Senegalese sole (*Solea senegalensis*): EST sequencing, oligo microarray design, and development of the Soleamold bioinformatic platform. BMC Genomics. 2008, 9: 508-10.1186/1471-2164-9-508.

35. Adzhubei AA, Vlasova AV, Hagen-Larsen H, Ruden TA, Laerdahl JK, Høyheim B: Annotated expressed sequence tags (ESTs) from pre-smolt Atlantic salmon (*Salmo

salar) in a searchable data resource. BMC genomics. 2007, 8: 209-10.1186/1471-2164-8-209.

36. Liu R, Li RW, Waldbieser GC: Utilization of microarray technology for functional genomics in ictalurid catfish. J Fish Biol. 2008, 72: 2377-2390. 10.1111/j.1095-8649.2008.01898.x.

37. Douglas SE, Knickle LC, Kimball J, Reith ME: Comprehensive EST analysis of Atlantic halibut (*Hippoglossus hippoglossus*), a commercially relevant aquaculture species. BMC genomics. 2007, 8: 144-10.1186/1471-2164-8-144.

38. Carninci P: Constructing the landscape of the mammalian transcriptome. Exp Biol. 2007, 210: 1497-1506. 10.1242/jeb.000406.

39. Jacquier A: The complex eukaryotic transcriptome: unexpected pervasive transcription and novel small RNAs. Nat Rev Genet. 2009, 10: 833-844. 10.1038/nrg2683.

40. Amaral PP, Mattick JS: Noncoding RNA in development. Mamm Genome. 2008, 19: 454-492. 10.1007/s00335-008-9136-7.

41. Zhang Y, Liu XS, Liu QR, Wei L: Genome-wide in silico identification and analysis of cis natural antisense transcripts (cis-NATs) in ten species. Nucleic Acids Res. 2006, 34: 3465-3475. 10.1093/nar/gkl473.

42. Lapidot M, Pilpel Y: Genome-wide natural antisense transcription: coupling its regulation to its different regulatory mechanisms. EMBO Rep. 2006, 7: 1216-1222. 10.1038/sj.embor.7400857.

43. Werner A, Sayer JA: Naturally occurring antisense RNA: function and mechanisms of action. Curr Opin Nephrol Hypertens. 2009, 18: 343-349. 10.1097/MNH.0b013e32832cb982.

44. Werner A, Schmutzler G, Carlile M, Miles CG, Peters H: Expression profiling of antisense transcripts on DNA arrays. Physiol Genomics. 2007, 28: 294-300.

45. Okada Y, Tashiro C, Numata K, Watanabe K, Nakaoka H, Yamamoto N, Okubo K, Ikeda R, Saito R, Kanai A, Abe K, Tomita M, Kiyosawa H: Comparative expression analysis uncovers novel features of endogenous antisense transcription. Hum Mol Genet. 2008, 17: 1631-1640. 10.1093/hmg/ddn051.

46. Beiter T, Reich E, Williams RW, Simon P: Antisense transcription: a critical look in both directions. Cell Mol Life Sci. 2009, 66: 94-112. 10.1007/s00018-008-8381-y.

47. Jezewski PA, Fang PK, Payne-Ferreira TL, Yelick PC: Alternative splicing, phylogenetic analysis, and craniofacial expression of zebrafish tbx22. Dev Dyn. 2009, 238: 1605-1612. 10.1002/dvdy.21962.

48. Miller CT, Yelon D, Stainier DY, Kimmel CB: Two endothelin 1 effectors, hand2 and bapx1, pattern ventral pharyngeal cartilage and the jaw joint. Development. 2003, 130: 1353-1365. 10.1242/dev.00339.

49. Haga Y, Suzuki T, Takeuchi T: Retinoic acid isomers produce malformations in postembryonic development of the Japanese flounder, *Paralichthys olivaceus*. Zoolog Sci. 2002, 10: 1105-12. 10.2108/zsj.19.1105.

50. Kuzir S, Kozaric' Z, Nejedli S: Development of mandibular arch in European sea bass, *Dicentrarchus labrax* (Linnaeus, 1758) from the "Cenmar" hatchery, Croatia. Veter arc. 2004, 74: 321-330.

51. Calloni GW, Le Douarin NM, Dupin E: High frequency of cephalic neural crest cells shows coexistence of neurogenic, melanogenic, and osteogenic differentiation capacities. Proc Natl Acad Sci USA. 2009, 106: 8947-8952. 10.1073/pnas.0903780106.

52. Hammond CL, Schulte-Merker S: Two populations of endochondral osteoblasts with differential sensitivity to Hedgehog signalling. Development. 2009, 136: 3991-4000. 10.1242/dev.042150.

53. Abzhanov A, Rodda SJ, McMahon AP, Tabin CJ: Regulation of skeletogenic differentiation in cranial dermal bone. Development. 2007, 134: 3133-3144. 10.1242/dev.002709.

54. Reppe S, Rian E, Jemtland R, Olstad OK, Gautvik VT, Gautvik KM: Sox-4 messenger RNA is expressed in the embryonic growth plate and regulated via the parathyroid hormone/parathyroid hormone-related protein receptor in osteoblast-like cells. J Bone Miner Res. 2000, 15: 2402-2412. 10.1359/jbmr.2000.15.12.2402.

55. Imai S, Heino TJ, Hienola A, Kurata K, Büki K, Matsusue Y, Väänänen HK, Rauvala H: Osteocyte-derived HB-GAM (pleiotrophin) is associated with bone formation and mechanical loading. Bone. 2009, 44: 785-794. 10.1016/j.bone.2009.01.004.

56. Takeuchi T, Dedi J, Haga Y, Seikai T, Watanabe T: Effect of vitamin A compounds on bone deformity in larval Japanese flounder (*Paralichthys olivaceus*). Aquaculture. 1998, 169: 155-165. 10.1016/S0044-8486(98)00373-1.

57. Elefteriou F: Neuronal signaling and the regulation of bone remodeling. Cell Mol Life Sci. 2005, 62: 2339-2349. 10.1007/s00018-005-5175-3.

58. Huebner AK, Keller J, Catala-Lehnen P, Perkovic S, Streichert T, Emeson RB, Amling M, Schinke T: The role of calcitonin and alpha-calcitonin gene-related peptide in bone formation. Arch Biochem Biophys. 2008, 473: 210-217. 10.1016/j.abb.2008.02.013.

59. Bhangu PS, Genever PG, Spencer GJ, Grewal TS, Skerry TM: Evidence for targeted vesicular glutamate exocytosis in osteoblasts. Bone. 2001, 29: 16-23. 10.1016/S8756-3282(01)00482-3.

60. Skerry TM: The role of glutamate in the regulation of bone mass and architecture. J Musculoskelet Neuronal Interact. 2008, 8 (2): 166-73.

61. Sigenae. [http://www.sigenae.org/]

62. Huang X, Madam A: CAP3: a DNA sequence assembly program. Genome Res. 1999, 9: 868-877. 10.1101/gr.9.9.868.

63. Götz S, García-Gómez JM, Terol J, Williams TD, Nagaraj SH, Nueda MJ, Robles M, Talón M, Dopazo J, Conesa A: High-throughput functional annotation and data mining with the Blast2GO suite. Nucleic Acids Res. 2008, 36: 3420-35. 10.1093/nar/gkn176.

64. Hu ZL, Bao J, Reecy JM, CateGOrizer: A web-based program to batch analyze gene ontology classification categories. Online J Bioinform. 2008, 9: 108-112.
65. Geneontology. [http://www.geneontology.org/GO.slims.shtml]
66. Agilent eArray. [https://earray.chem.agilent.com/earray/]
67. R statistical software. [http://www.r-project.org]
68. Thusher V, Tibshirani R, Chu G: Significance analysis of microarrays applied to the ionizing radiation response. Proc Natl Acad Sci USA. 2001, 98: 5116-5121. 10.1073/pnas.091062498.
69. DAVID 2008. [http://david.abcc.ncifcrf.gov/]
70. GOstat. [http://gostat.wehi.edu.au/]
71. Ensembl BioMart. [http://www.ensembl.org/biomart/martview/]

CHAPTER 9

GENOME WIDE RESPONSE TO DIETARY TETRADECYLTHIOACETIC ACID SUPPLEMENTATION IN THE HEART OF ATLANTIC SALMON (SALMO SALAR L)

Fabian Grammes, Kjell-Arne Rørvik, Magny S Thomassen, Rolf K Berge and Harald Takle

BACKGROUND

High levels of dietary lipids are used in commercial Atlantic salmon diets to promote rapid growth and as a inexpensive source of energy. These high lipid levels may promote excess lipid deposition in the viscera and the muscle, thereby reducing the market quality of the fish. Thus, tetradecylthioacetic acid (TTA: CH_3-$(CH_2)_{13}$-S–CH_2-COOH) has been tested for aquaculture nutrition, initially to increase lipid catabolism and thereby reducing lipid deposition [1]. However, beneficial effects on cardiac growth and disease resistance have also been addressed.

TTA is a modified fatty acid (FA) that possesses a sulfur atom in the β position. Like a normal FA, TTA can be converted to co-enzyme A thioester, but further catabolism by β-oxidation does not occur. This lack of metabolism is likely to determine the biological effects of TTA. Biological effects of TTA have been the focus of extensive research in rodents and also in humans. The most important findings from these experiments are that TTA increases the mitochondrial and peroxisomal B-oxidation and possesses hypolipidemic effects. In addition, TTA acts as an antioxidant in vivo and can modulate the inflammatory response (reviewed in[2]). Cell culture experiments demonstrated that TTA can act as a ligand for all Peroxisome proliferator activated receptors (PPARs) [3, 4], which are ligand-activated transcription factors. Upon ligand activation PPARs heterodimerizes with retinoic-x-acid receptor (RXR) and have been shown to regulate the expression of genes involved in fatty acid metabolism, cell differentiation, development and inflammation (reviewed in[5]). Arguably most of the biological effects of TTA are mediated through activation of PPARs.

Studies addressing the biological effects of TTA in Atlantic salmon have demonstrated that TTA increases β-oxidation in the liver[1] and white muscle[6]. Furthermore, TTA reduces the secretion of triacylglycerides from Atlantic salmon hepatocytes in vitro[7] and has been shown to increase the expression of genes associated with fat metabolism in the liver and the heart ventricle[1, 8]. Previous results also suggested that TTA stimulates the transcription of PPARα and β in the heart[8, 9], thus indicating that TTA affects the metabolism in Atlantic salmon through activation of PPARs, similar to the mechanism known from rodents. Interestingly, this activation of PPARs may have been related to increased survival after a natural outbreak of a heart related viral disease in Atlantic salmon[8, 9]. In mammals, cardiac activation of PPARs has yielded substantial attention due to the fact that PPARs have been proven to be major regulators of cardiac metabolism[10,

Citation: Grammes, Kjell-Arne Rørvik, Magny S Thomassen, Rolf K Berge and Harald Takle, Genome wide response to dietary tetradecylthioacetic acid supplementation in the heart of Atlantic Salmon (Salmo salar L), doi: 10.1186/1471-2164-13-180

11, 12, 13]. In addition, PPAR agonists have been reported to exert beneficial effects by attenuating the pathogenesis of heart failure and atherosclerosis[14, 15].

Poor development of the outer muscle layer, atherosclerosis and metabolic dysfunction have been related to under-dimensioned hearts and reduced cardiac function in Atlantic salmon aquaculture, consequently resulting in increased mortality[16]. Therefore, methods to improve cardiac metabolism and performance in fish are needed; something which has been sparsely studied. It appears that, similar to the mammalian heart, the oxidative cardiac metabolism in fish depends on the metabolism of fatty acids and glucose[17].

This study aims to characterize the cardiac transcriptional response of Atlantic salmon to a TTA supplemented diet. A feeding trial was conducted in sea, feeding a control and TTA supplemented diet during the first 8 weeks and only control diet for the subsequent 9 weeks of the experiment. Fish were sampled both at the end of the TTA feeding period (8.weeks) and at the end of the experiment (17.weeks). Our results show that administration of TTA to Atlantic salmon resulted in a marked change of cardiac gene expression. The expression profile suggests that TTA induces cardiac fatty acid oxidation, glycolysis, TCA cycle and contractility as well as cardiac growth.

RESULTS

Production data

Atlantic salmon that were fed with 0.25% TTA had significantly lower fat content in the muscle at the 8.weekssampling point, and showed a tendency for increased mean relative heart weight (Table1). No significant effect of TTA on fish weight was detected. During the experiment none of the dietary groups showed higher mortality than the control group. In the heart ventricles, 120.2 μgTTA/gTissue was detected at the 8.weekssampling point in the TTA group, while 1.2 μgTTA/gTissue was detected in the control group. Based on the TTA measurements of a group fed a higher (0.5% w/w) TTA diet from the same trial, we can assume that the TTA levels in the heart ventricles at the 17.weeks sampling point were no different to the control group (see Additional file1: Table S1).

Table 1: Effect of TTA on Atlantic salmon production parameters

	Start sampling	***8.weeks***		***17.weeks***	
		Control	**0.25% TTA**	**Control**	**0.25% TTA**
Weight[g]	102 ± 5	166 ±1	165 ±1	438 ±4	440 ±2
CF[1]	1.2 ± 0.1	1.1 ±0.1	1.1 ± 0.1	1.2 ± 0.1	1.2 ± 0.1
LI[2]	0.82 ±0.05	1.00 ±0.02	1.07 ±0.07	1.43 ±0.04	1.41 ± 0.04
CSI[3]	0.074 ±0.002	0.092 ±0.001	0.094 ±0.001	0.101±0.002	0.103 ± 0.002
Mortality[%]		0.74 ± 0.18	0.41 ± 0.27	2 ± 0.23	2 ± 0.21
Muscle fat content[4][%]		4.2^{a}±0.1	3.9^{b}±0.1	6.5±0.4	6.8±0.1

Values within the same row with different subscripts are significantly different ($p \leq 0.05$, t-test). Mean ± SEM, the statistical unit is the mean of the net pen (n = 3). 19 fish were used to calculate the net pen mean.

[1]Condition factor (CF) =1000×weight[g]length[g]3.

[2]Liver index (LI) =liver−weight[g]weight[g]× 100.

[3]Cardio somatic index (CSI) =heart−weight[g]weight[g]× 100.

[4]Muscle samples (NQC: Norwegian quality cut) from 10 fish/cage were pooled to analyse the muscle fat content.

Microarray Analysis

RNA cardiac samples from six individual fish from each dietary group and sampling point were used in the microarray analysis, utilizing the Atlantic salmon SIQ2 microarray [18] in a one-color setup, resulting in a total of 24 arrays. After normalization and filtering, 13166 probes (63%) were classified as present. To obtain a global overview of the general structure of the dataset we applied correspondence analysis (CA) as an explorative technique [19]. The first 2 components of the CA are displayed, together explaining 67% of the total inertia of the different samples (Figure1). The analysis shows a clear distinction between the two sampling points at 8.weeks and 17.weeks. It further shows a relatively dense cluster at the 8.weeks point with minor separation between samples from the TTA and control group. Even though it was impossible to draw a straight line to separate between the TTA and control samples at the 17.weeks sampling point, we observed a clear tendency of separation.

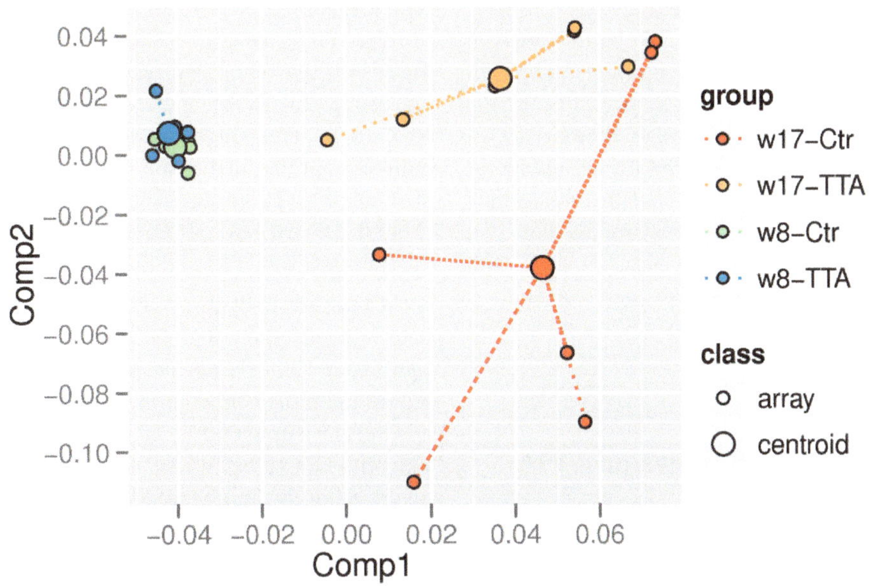

Figure 1: Correspondence Analysis (CA). CA of arrays from the two dietary groups for both sampling points (8.weeks and17.weeks), n = 6.

To identify differentially expressed (DE) probes in the data set we used moderated t-statistics[20], comparing samples from TTA to control fed Atlantic salmon for each sampling point.

Sampling Point: *8.Weeks*, end of TTA Feeding

At the 8.weeks sampling point, five genes were found to be DE between the TTA and the control fed group (Figure2). The genes were: Ephrin-b2, arf gtpase-activating protein (git2), f-box only protein 11 (fbx11),angiopoietin-related protein 4 (ANGPTL4) and sodium- and chloride-dependent creatine transporter 1 (sc6a8).Ephrin-b2 was the only gene found to be down-regulated in the TTA fed group.

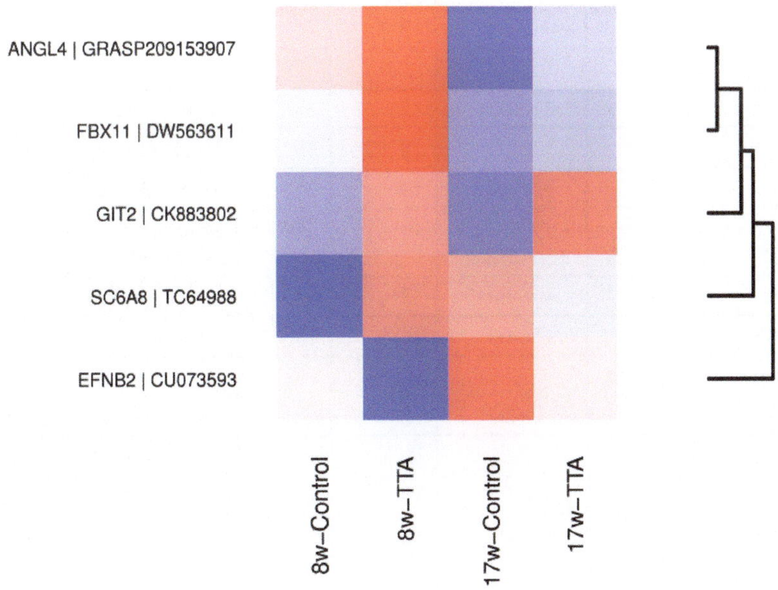

Figure 2: Heatmaps of genes DE at 8.weeks. Columns display the mean $\log_2$ signal of the biological replicates (n = 6), rows display the genes (probe sets) showing DE. Rows were scaled and ordered by hierarchical clustering using euclidean distances (indicated by the dendrogram).

Ephrin-b2 (Efnb2) in mammals has been reported to be highly expressed in the heart and serves also as a marker for angiogenesis[21]. The protein GIT-2 participates in pleiotropic cellular processes like cell migration and T-cell activation; however, a function affecting the structure of the cytoskeleton[22] may be relevant in our study. Cardiomyocytes rely solely on the creatine transporter sc6a8 for the uptake of creatine from the plasma. Over-expression of the creatinfabe transporter in mice has been reported to correlate with the myocardial creatine content, but also to be associated with cardiac hypertrophy[23]. Angiopoietin-related protein 4 possesses a role in regulating angiogenesis and is also known as a target gene for PPARs and acts as an important stimulator of lipid metabolism[24].

Sampling point: *17.weeks*, 9 weeks post TTA feeding

1198 probes (930 genes) were found to be DE between the TTA and the control fed group at the 17.weekssampling point. In order to facilitate a functional interpretation of the vast number of DE genes, we tested them for enrichment (over-representation) of GO terms from the category "biological process" [25], using conditional hypergeometric testing [26]. To ensure that one gene was represented by a maximum of one probe [26], probes matching the same gene were collapsed prior to hypergeometric testing. This step reduced the total number of probes in the data set from 11143 to 7659 probes and the number of DE probes from 1198 to 930 probes (930 genes). From these 930 genes, 90% showed higher gene expression in the samples from TTA fed fish.

Conditional hypergeometric testing revealed significant over-representation of 36 GO terms. To simplify interpretation, significant GO terms were grouped into five categories according to their function in the heart (Table2). The grouping was further supported by a strong gene overlap between the different GO terms within the categories (Additional file2: Figure S1 and Additional file3: Table S2). Overall, the results from the enrichment analysis suggests an increased capacity of heart ventricles from TTA fed Atlantic salmon to catabolize lipids and glycogen. Further, an increased capacity for cardiac contractility and cardiac tissue morphogenesis is indicated.

Table 2: Gene Ontology enrichment analysis for sampling point 17.weeks

GOID	*p*-value	Count[1]	Size[2]	Gene ontology term
Fat metabolism				
GO:0046395	5.33e-04	24	96	carboxylic acid catabolic process
GO:0034440	5.69e-03	13	49	lipid oxidation
GO:0044242	8.41e-03	19	86	cellular lipid catabolic process
Heart performance				
GO:0006936	1.56e-03	37	182	muscle contraction
GO:0008015	5.56e-03	38	202	blood circulation
GO:0003015	6.45e-03	22	102	heart process
GO:0008016	7.18e-03	17	73	regulation of heart contraction
GO:0055008	8.94e-03	12	46	cardiac muscle tissue morphogenesis
Citrate cycle (TCA)				
GO:0009109	1.87e-07	15	28	co-enzyme catabolic process
GO:0006084	2.12e-07	18	39	acetyl-CoA metabolic process
GO:0006099	2.32e-07	14	25	tricarboxylic acid cycle
GO:0006091	9.39e-05	46	214	generation of precursor metabolites and energy
GO:0045333	3.49e-04	23	88	cellular respiration
GO:0044248	1.34e-03	36	180	cellular catabolic process
carbohydrate metabolism				
GO:0006112	6.21e-04	16	54	energy reserve metabolic process
GO:0016052	1.17e-03	19	73	carbohydrate catabolic process
GO:0006073	1.27e-03	14	47	cellular glucan metabolic process
GO:0046164	1.77e-03	16	59	alcohol catabolic process
GO:0019320	2.97e-03	14	51	hexose catabolic process
GO:0006096	3.26e-03	12	41	glycolysis
Other				
GO:0007338	1.27e-03	14	47	single fertilization
GO:0001824	1.59e-03	12	38	blastocyst development

GO:0051246	2.00e-03	86	514	regulation of protein metabolic process
GO:0010171	2.83e-03	20	84	body morphogenesis
GO:0055114	3.60e-03	40	210	oxidation reduction
GO:0007050	3.67e-03	19	80	cell cycle arrest
GO:0043009	5.00e-03	48	267	chordate embryonic development
GO:0001822	5.64e-03	19	83	kidney development
GO:0040010	6.35e-03	33	171	positive regulation of growth rate
GO:0009790	7.54e-03	124	815	embryo development
GO:0001655	8.41e-03	19	86	urogenital system development
GO:0070585	8.44e-03	10	35	protein localization in mitochondrion
GO:0044265	8.45e-03	66	399	cellular macromolecule catabolic process
GO:0007018	9.35e-03	16	69	microtubule-based movement
GO:0006839	9.35e-03	16	69	mitochondrial transport
GO:0006402	9.66e-03	13	52	mRNA catabolic process

[1]Number of times the GO term is represented in the list of DE genes, only GO terms having $\geq$ 10 genes were considered.

[2]Number of times the GO term is represented in the filtered list of genes on the array.

Heart Performance

The group fed TTA showed an up-regulation in the expression of genes encoding contractile proteins like myosin heavy chain 6 (MYH6), myosin light chain (MYL9), cardiac myosin binding protein (MYBPC3), cardiac troponin (TNNT2), myomesin-1 (Myom1) and actin (ACTA1, ACTA2).

Moreover, the same group showed increased expression of the cardiac homeodomain factor Nkx2.5 and the iroquois-related homeobox factors 3 and 5 (irx3, irx5). Nkx2.5 and the iroquois transcription factors have been reported to control cardiac morphogenesis and growth[27, 28]. Furthermore, we observed increased expression of FK506 binding protein 1A and 1B (FKBP1A, FKBP1B) and N^{a+}/K^{+}-transporting ATPase subunit α 3(ATP2A2), encoding an ion-pump responsible for establishing and maintaining the electrochemical gradients at the plasma membrane of the cardiomyocyte. Decreased amounts of this transporter were found in biopsies from humans suffering heart failure [29]. FKBP1A and B are known to interact with intracellular calcium-release channels. In cardiomyocytes FKBP1B is a binding partner for the major Ca^{2+} release channel ryanodine receptor 2 (RyR2). RyR2 is required for the Ca^{2+} -induced Ca^{2+} release from the sacroplasmatic reticulum (SR) causing activation of the contractile proteins. Binding of FKBP1B to RyR2 results in channel closure. Mice deficient for FKBP1B showed no divergence in the normal cardiac phenotype under normal conditions but showed exercise-induced arrhythmias [30]. In relation to Ca^{2+} signaling, we also found the Na^{+}/Ca^{2+}exchanger SLC8A1 to be up-regulated. We also found an increased expression of the SR Ca^{2+} ATPase 2 (ATP2A2 also known as SERCA2), encoding an SR calcium pump that is a key component of the cardiac excitation-contraction mechanism [31].

The Kv channel interacting protein 1 (Kcnip1) was found to be down-regulated. The protein Kcnip1 is an integral part of the multimeric Kv4 channel complex, and important for modulating the K-flux across this channel by causing a shortening of the cardiac action potential [32]. Prolongation of the cardiac action potential on the other hand, potentially caused by decreased Kcnip1 expression, is associated to cardiac hypertrophy [33].

In summary, the results suggest an increased cardiac hypertrophy together with increased potential for cardiac contractility, as indicated by the higher transcription of the various ion channels/pumps and contractile proteins.

Fat Metabolism

Nearly all of the genes in this group were up-regulated (Figure3). We found up-regulation of the mitochondrial trifunctional protein HADHA, the mitochondrial fatty acid transporter carnitine palmitoyltransferase (Cpt1a),lipoprotein lipase (Lpl), the mitochondrial acyl-CoA dehydrogenases (ACADS, ACADV and ACADSB) and ofperoxisomal multi-functional enzyme type 2 (Hsd17b4). We also observed an up-regulation of malonyl-CoA decarboxylase (MLYCD). Malonyl-CoA is a potent inhibitor of CPT1 and thus crucial in regulating the transport of fatty acids into the mitochondria for catabolism. Malonyl-CoA decarboxylase has been reported to function as a positive regulator of cardiac fatty acid oxidation by decreasing the levels of the CPT1 inhibitor malonyl-CoA[34]. Thus, the results indicate increased fatty acid oxidation capacity in cardiac ventricles from TTA fed Atlantic salmon.

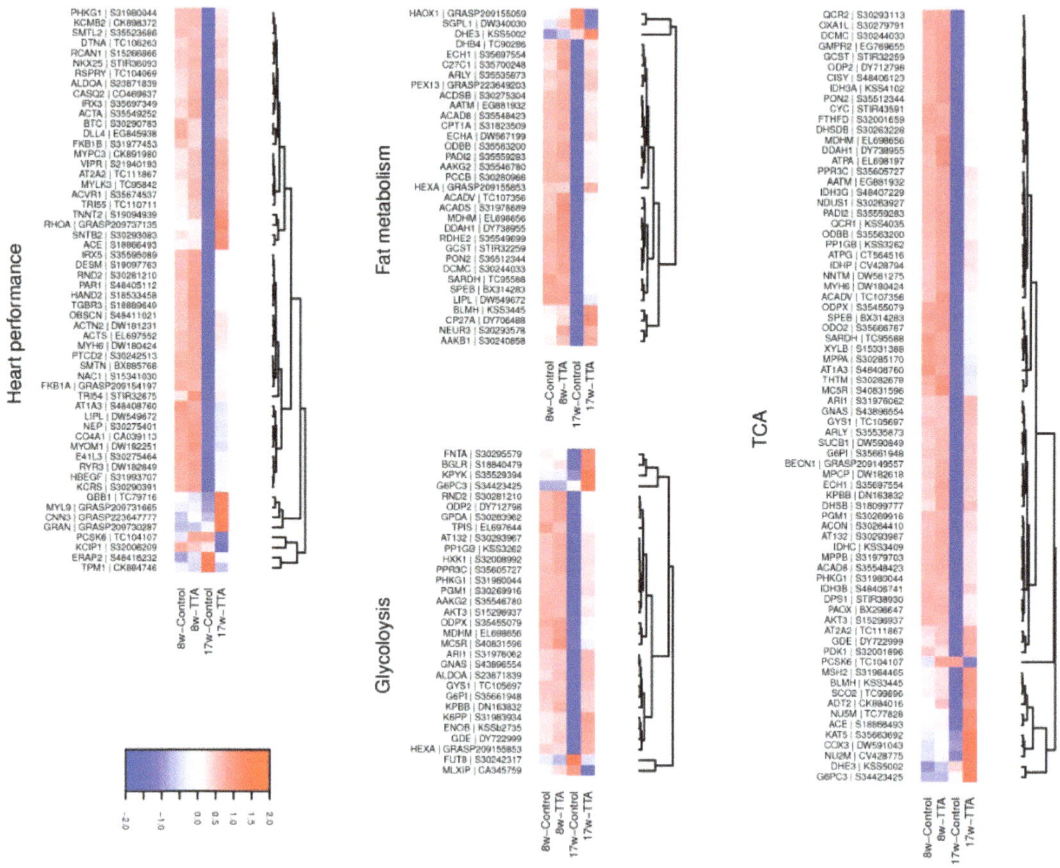

Figure 3: Heatmaps of genes DE at 17.weeks. Unique genes from the different categories (Table2). Columns display the mean log_2 signal of the biological replicates (n = 6), rows display the genes (probe sets) showing DE. Rows were scaled and ordered by hierarchical clustering using euclidean distances (indicated by the dendrogram).

Glycolysis

The genes of the six GO terms that were grouped together contained almost entirely genes encoding enzymes or subunits participating in glycolysis. We found increased expression of hexokinase 1 (HXK1), the phospho-fructokinases aldolase A (ALDOA) and 6-phosphofructokinase type C (K6PP), glycerol-3-phosphate dehydrogenase (GPDA/GAPDH) and the pyruvate dehydrogenases DLAT and OPDX. In accordance, we also observed an increased expression of MLX-interacting protein (MLXIP), which has been suggested to be an essential regulator of cellular glycolysis[35]. All of the genes showed increased transcription, therefore clearly indicating increased glycolysis in the hearts from TTA fed Atlantic salmon.

Tricarboxylic acid (TCA) cycle

As in the previously described categories, almost all of the genes in this category showed an increased expression. The proteins encoded by nearly all of the genes in this category are part of the TCA-cycle. For a graphical representation of the genes within the TCA cycle, see Additional file4: Figure S2.

qRT-PCR

To validate the microarray results, six genes were analyzed by qRT-PCR between the TTA and control fed group for the 8.weeks and 17.weeks samples, using the same RNA samples that were used in the microarray experiment. The results showed a significant correlation between the logFCs obtained by qRT-PCR and those obtained by microarray (Pearson correlation r = 0.8314; p = 0.0008; Figure4).

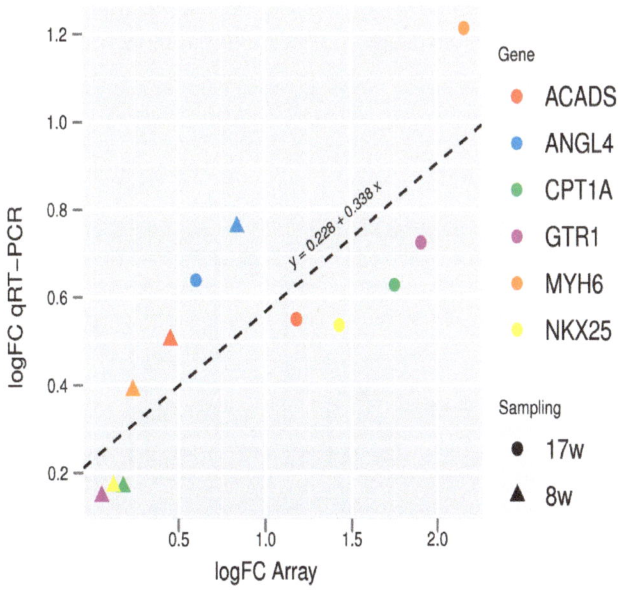

Figure 4: qRT-PCR verification. Comparison of the log_2FC data for the two sampling points between Microarray and qRT-PCR measured samples. qRT-PCR data was normalized to the expression of the housekeeping gene EF1α. Pearson correlation; r = 0.8314, p = 0.0008. The dotted line indicates the linear regression line.

In addition, we measured the gene expression of the three PPAR subtypes α,β and γ in control and TTA fed Atlantic salmon in the four different tissues: Heart, muscle, liver and gut (pyloric caeca) from both sampling points (Figure5A). Analyzing the expression levels using analysis of variance (ANOVA) showed that only thePPAR α expression in the heart was significantly increased in the TTA group. The gene expression levels of PPARγ in heart and muscle were too low to allow reliable quantification. Since the microarray data revealed quite clearly that TTA affected genes are involved in regulating the heart performance, we used qRT-PCR to measure the expression of the cardiac transcription factors GATA4, Mef2C and osteonectin (Osx). For all three transcription factors we observed a trend of higher mean transcription in the TTA group (Figure5B), however, only Mef2C showed statistical significance.

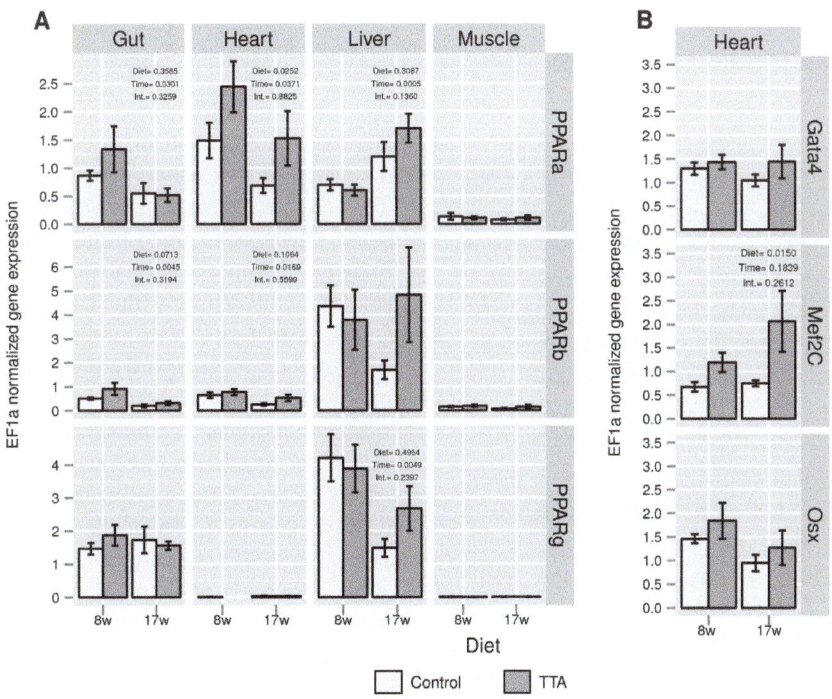

Figure 5: Gene expression. A: Expression of PPARα, β and γ in Atlantic salmon fed TTA or a control diet. The different tissues liver, heart, muscle and gut were sampled at two sampling points (8.weeks and 17.weeks). **B**: Cardiac expression of***Gata4***, Myocyte-specific enhancer factor 2C **(Mef2C)** and ***Nkx2.5*** in Atlantic salmon fed TTA or a control diet. Gene expression was normalized to the expression of the housekeeping gene EF1α. The p-values for the effects of Diet, Time and Interaction from the two way ANOVA are displayed in the upper right corner. Data are presented as means ± SEM, n = 6.

TTA has previously been reported to stimulate mitochondrial biogenesis in mammals [36]. In this study TTA had no effect on mitochondrial biogenesis as measured by the ratio of mt/nDNA (Figure6). In the liver we found a significant interaction between dietary treatment and time for mitochondrial biogenesis.

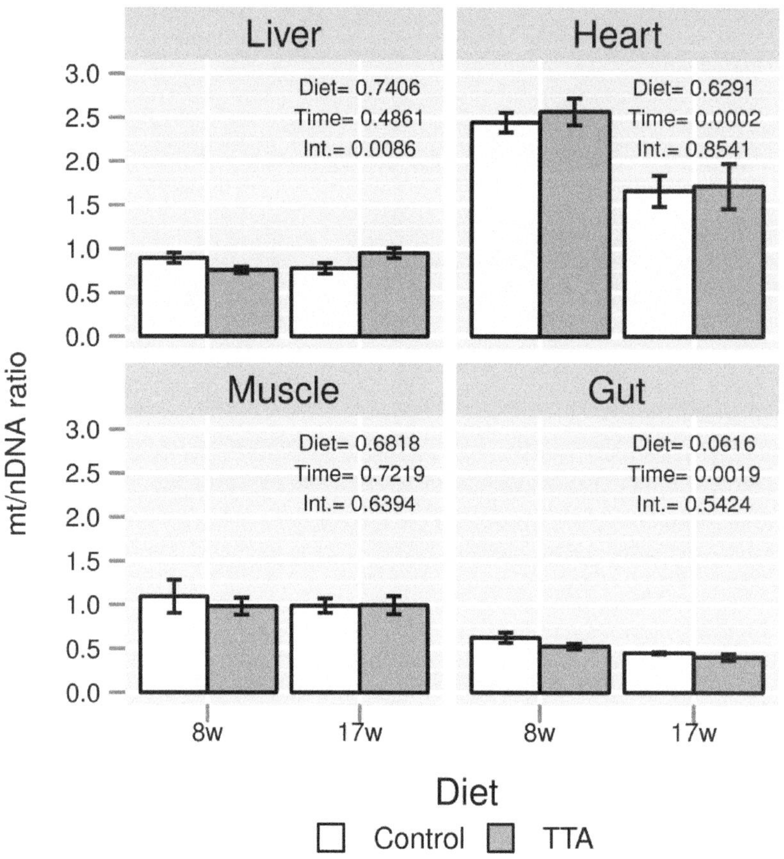

Figure 6: Mitochondrial biogenesis. Effect of feeding TTA on the ratio between mitochondrial and nuclear DNA in the different tissues. The p-values for the effects of Diet, Time and Interaction from the two way anova are displayed in the upper right corner. Data are presented as means ± SEM, n = 6.

DISCUSSION

In the present study we investigated the response of Atlantic salmon to TTA during the seawater phase. The results from our study show that feeding TTA had profound effects on the cardiac gene expression at sampling point 17.weeks, 9 weeks after TTA feeding ended. The level of TTA applied in the study was, with 0.25%, lower than previous studies that have been conducted in Atlantic salmon (compare [1, 6, 37]). The mortality rates previously observed in Atlantic salmon in response to high TTA levels was not observed in this study.

Gene set over-representation of the transcription profile at 17.weeks shows an increased capacity of fat catabolism, glycolysis and activity of the TCA cycle as well as cardiac contractility and cardiac hypertrophy. Overall, the results suggest a scenario where cardiac ventricles of TTA pre-fed fish are able to generate more energy via a TCA-cycle that is fueled by metabolites from fat catabolism and glycolysis. TTA functions as a ligand for all three PPAR subtypes [3, 4], which have crucial functions in the transcriptional regulation of cardiac metabolism. In mice the transcriptional effects of TTA in the heart have been shown to be mediated almost exclusively via PPARα[38]. Gain-of-function and loss-of-function mutations have shown that PPARα is a crucial transcription factor in the cardiac metabolism, regulating mainly cardiac fatty acid uptake and oxidation [10, 11]. Furthermore, activation of PPARα has been demonstrated to shift cardiac energy utilization away from glucose and towards fatty acid oxidation, actually mimicking the cardiac phenotype observed in diabetic hearts [11]. Interestingly the cardiac phenotype of PPARβ differs from that of PPARα, indicating that both transcription factors regulate, at least partly, different subsets of genes in the heart. PPARβ loss-of-function hearts suffer from myocardial lipid accumulation and cardiomyopathy [13]. Gain-of-function mutations on the other hand clearly show that PPARβ positively regulates cardiac glucose utilization [12], and also stimulates cardiac growth [39]. Thus, the significantly higher cardiac transcription of PPARα and the elevated mean transcription of PPARβ/ in concert with the activation of their down-stream pathways, fat catabolism and the glycolysis pathway suggest that cardiac effects of TTA in Atlantic salmon are mediated by both PPARα and PPARβ. Intriguingly, over-expression of a constitutively active form of PPARβ in murine skeletal muscle has been reported to mimic training-based muscle adaptation [40]. Hence, it has been speculated, in accordance with the results from PPARβ over-expression in mice[12], that PPARβ causes "physiological" cardiac hypertrophy[39].

Between the 8.weeks and 17.weeks sampling points, the hearts grew by a considerable portion in absolute and relative terms. The gene expression profile in hearts of TTA fed fish at 17.weeks suggests that the cardiac growth of TTA fed fish is shifted towards "physiological" hypertrophy, which may translate to an increased cardiac output. This notion is supported by the expression profile found for the category "cardiac performance" at 17.weeks, unanimously pointing to an increased cardiac contractility and also showing up-regulation of crucial cardiac transcription factors. In particular the higher transcription of the cardiac transcription factors Nkx2.5 and Mef2C can be regarded as markers for cardiac hypertrophy/growth. It has been demonstrated in mice that over-expression of Mef2C is sufficient to induce cardiac hypertrophy [41]. Furthermore, both Mef2C and Nkx2.5 have been shown, in vitro, to be regulated by PPARα in cardiomyocytes[42].

It should also be noted that although we did not find significant differences in relative heart weight in this study, in other studies we found that TTA significantly increases heart size in Atlantic salmon[8, 9], and that the effect seems to be correlated to the dose of TTA (Rørvik, unpublished data). Thus, it is tempting to speculate that the increase in relative heart weight may be related to the cardiac transcriptional changes induced by TTA. A "cardiac exercise" stimulating effect is of high relevance for salmonid aquaculture. Atlantic salmon, having a circulatory system that is naturally adapted to long migration routes and high activity, show alteration in cardiac morphology and a reduced relative heart weight in captivity[43]. In addition, circulatory failure has been identified as an important cause of mortality in salmon farming[44]. Thus, using TTA may be one way to support the cardiac performance of fish in captivity.

The highest tissue concentrations of TTA in Atlantic salmon, as well as in mice, can be found in the heart[1, 45]. In accordance, the heart was also the tissue where the strongest transcriptional response of PPARα was detected. The main transcriptional effects were found nine weeks after the TTA feeding stopped and where our data suggested that the cardiac tissue levels of TTA were neglectable. However, we have no information about the course of gene expression between both sampling points, thus it might very well be that the effects of sampling at 17.weeks are the remains of earlier, stronger transcriptional effects. It is remarkable that a similar, delayed response in expression of lipid metabolism related genes to TTA has been observed in our previous Atlantic salmon studies[8, 9], indicating a common underlying mechanism. It is possible that the delay in transcriptional response is caused by a common, yet unknown, mechanism.

CONCLUSIONS

In conclusion, based on results from microarray analysis, this study demonstrates that TTA increases cardiac fatty acid oxidation and glycolysis as well as contractility and cardiac hypertrophy in Atlantic salmon. The gene expression profiles further favor a scenario of "physiological" hypertrophy in response to TTA. This increased cardiac efficiency may offer significant benefits in situations with increased oxygen demand.

METHODS

Feeding Trial

The experiment was conducted at Nofima Marin sea-water research station, Averøy, western Norway. Atlantic salmon used in this experiment were hatched at Nofima Marin research station (Sunndaløra, Norway) one year earlier (S1/1+ Salmon). The experiment started with the seawater transfer of the fish on the 15th of May 2007 and lasted until the 27th of September 2007. A randomized block design with triplicate seawater net-pens and 400 fish per pen (pen = 125 m^3) was used for the experiment. Control and TTA diets (0.25% (w/w) TTA (Thia Medica, Norway)) were produced by Biomar (Biomar AS, Myre, Norway). Both TTA and control diets were fed to the fish until the 16th of July 2007, from this point until the end of the experiment only the control diet was fed to the fish. Low levels of TTA (0.25%) and a short feeding period were chosen in order to avoid negative TTA effects (mortality, altered kidney morphology[1, 46]). TTA was fed for the first eight weeks after sea transfer, a period where the physiology of the salmon alters due to changing from a fresh to a saltwater environment; and we speculated that an increased capacity for energy utilization may be beneficial. Fish were sampled from the cages to represent the average fish weight for the cage. Sampling was done on the following dates: 16-18th of July 2007 (sampling point: 8.weeks, end of the TTA feeding period) and 25-27th of September 2007 (sampling point: 17.weeks). For each sampling point, fish were sampled for: heart ventricle, liver, muscle and gut (pyloric caeca). The tissue samples were snap frozen in liquid nitrogen and stored at -80°C.

Fat Analysis

Fat content in the muscle (Norwegian quality cut–NQC, Norwegian standard procedure - NS 9401 1994) was measured in pooled samples (10 fish) from each net pen as described in[47]. TTA was measured within the total cardiac lipids. For the analysis 10 ventricles from Atlantic salmon out of the same net pen were pooled. Total heart lipids were extracted with chloroform-methanol[48] and fatty acid methyl esters (FAME) were obtained by heating of lipids with methanol at 90°C/1 hour, where $H_2S\ O_4$ was used as a catalyst[49]. After extraction into an organic solvent, the FAME were analyzed by gas-liquid chromatography. A gas chromatograph GC 8000 TOP (Finnigan, USA) was equipped with a programmed temperature vaporization (PTV) injector, flame-ionization detector (FID), AS 800 autosampler and a fused silica capillary column coated with dimethylpolysiloxane stationary phase, DB1-ms (J & W Scientific, USA). Hydrogen was used as a carrier gas. Column temperature was programmed from 110 to 310°C with a gradient 2.5°C/min. The GC signal was acquired using Chromeleon software (Dionex, USA). Peaks were identified by means of known FA standards (Larodan Fine Chemicals, Sweden and Sigma-Aldrich, USA) and by means of mass spectra, obtained by GC/MS analysis (GCQ, Finnigan, USA) on the same column. An internal standard (C21:0) was used for quantitation after calibration with known mixtures of FA standards.

RNA Extraction

Two individual samples from each one of the 3 net-pens were samples for heart ventricle, liver, muscle and gut. The samples were randomly chosen (n=6 per dietary group and sampling point) and homogenized using a rotor tissue lyser (Precellys 24, Bertin technologies, France). Total RNA was extracted and purified using column purification (96 universal Tissue Kit, Qiagen, Hilden, Germany) according to the manufacturer's instructions. Traces of genomic DNA in the samples were eliminated by on-column-DNase digestion (Qiagen). RNA concentrations were measured for all samples using a NanoDrop 1000 Spectrophotometer (Thermo Fisher Scientific, Wilmington, USA). RNA quality for samples later used in the microarray was determined using a Agilent 2100 Bioanalyzer (RNA 6000 NanoLabChip, Agilent, Waldborn, Germany).

Microarray Hybridization

A customized oligo (60-mer) Atlantic salmon microarray in the 4x44K format (Agilent,[18]) was used to detect differential gene expression between samples from the heart ventricles of control and 0.25% TTA fed fish for the 8.weeks and 17.weeks sampling points. The array contained 21012 different probes spotted in duplicates. RNA samples from individual fish were hybridized to the microarray. 24 individual microarrays were performed using 12 fish (6 control and 6 TTA fish) at the 8 week sampling point, and similarly at the 17 week sampling point. All RNA samples used in the hybridization had RIN values ranging from 9.5 to 10. 500ng RNA were amplified and labeled with Cy3 using the Quick Amp Labeling Kit (One Color-Agilent). After purification the cRNA was quantified using NanoDrop. Subsequently 1.65μ g Cy-dye-labeled cRNA was fragmented (mean size, approximately 50-100 nucleotides) with fragmentation buffer (Agilent Technologies) at 60°C for 30 min; cRNA was subsequently hybridized to the microarray at 65°C for 17 h. All steps were conducted according to the Agilent protocol (One-Color Quick Amp Labeling, Version 5.7). The microarray chips were scanned using a Agilent Microarray Scanner (G2565CA) and analysis of the microarray images was done in Agilent's Feature Extraction Software (Version 10.5.1.1) using the one-color (GE1_105_Dec08) protocol.

Microarray Analysis

Normalization and analysis of the data was performed in R/ Bioconductor[50, 51] using the "limma" package[20]. The background corrected fluorescence signals (gProcessedSignal) were obtained from Feature Extraction (Agilent). Spots were filtered according to the following criteria provided by Feature Extraction: gIsFound, gIsPosAndSignif and gIsWellAboveBG (a description of the parameters can be found in the Feature Extraction Software Reference Guide). The mean signal of the duplicated probes was calculated and all control spots together with probe sets showing more than three missing values were removed from the dataset. The data was subsequently normalized using quantile normalization in order to adjust the scale of intensities across arrays[52]. After normalization the signals were $\log_2$ transformed. The normalized/filtered dataset then contained 13166 probe sets (63% of the total). The raw and normalized data is publicly available at NCBI's GEO repository (http://www.ncbi.nlm.nih.gov/geo/, AccNr: **GSE25305**). Differential expression of probe sets was assessed by fitting a linear model, including the effects of feeding (2 levels: Control and TTA) and the effects of sampling point (2 levels: 8.weeks and 17.weeks) and their interaction. The specific comparisons: TTA vs.Control at sampling point 8.weeks and TTA vs. Control at sampling point 17.weeks were made by extracting the appropriate contrasts from the linear model. For each contrast moderated t-statistics were calculated using an empirical Bayes method[53]. Probes without annotation were removed from the dataset before controlling the false discovery rate[54] simultaneously across probe sets and contrasts (method: " global" in the limma function "decideTests"). Probe sets with a q-value ≤ 0.05 and a $\log_2 FC \geq 0.5$ were declared DE for the corresponding contrast. A comprehensive list of all DE probes for each contrast can be found in Additional file5: Table S3.

Probe annotation and GOs were retrieved using the top Blast function implemented in Blast2GO[55]. Full length probe sequences were blasted against protein sequences from the Swissprot database in a BlastX search. The E-value cut off was set to 10^{-6}. Hypergeometric testing for over-representation of GO terms from the category biological process[25] among the genes DE for the contrast TTA vs. Control at the 17.weekssampling point was conducted using the GOstats package[26]. Before testing, probes matching to the same gene were collapsed to the probe showing the largest variance.

Correspondence analysis was conducted using the R package "made4"[56]. Probe sets with missing values were removed from the dataset prior to correspondence analysis.

Quantitative RT-PCR

Single strand cDNA was synthesized from 500ng of total RNA using oligo dT primers and the Taq Man reverse transcription Kit (Applied Biosystems, CA, USA). qRT-PCR was performed on a Light-Cycler 480 (Roche, Switzerland). For the PCR reaction, 2x SYBR green I master Mix (Roche), 0.41nM of each primer and the cDNA template were mixed in a total reaction volume of 10μ l. Primer sequences are listed in Table3. A three step PCR protocol with 45 cycles (15s 95°C, 15s 60°C, 15s 72°C) was used. To verify specific amplification, a melting curve analysis step was done at the end of the program. In order to verify the results obtained through the microarray experiment, the same 24 samples used in the array were used in a qRT-PCR approach. Six genes were then randomly picked and samples were analyzed in duplicates. The expression level was calculated using the standard curve method (Applied Biosystems User Bulletin 2). The standard curve was produced from a serial dilution of a pool consisting of all cDNA samples. The expression levels were standardized to the expression of the housekeeping gene elongation factor 1 α(EF1 α,[57]).

Table 3: qRT-PCR primer sequences

Gene	Accession no.	Forward primer (3' - 5')	Reverse primer (5' - 3')	
mt D Loop B (gDNA)	NC001960	CCCCTGAAAGCCGAATGTAA	CGACCTTGTTAGACTTCTTTGCTTG	
MyoD2 (gDNA)	AJ557150	CAGAGCCAGGATTACACTCGTTACA	GCATGTCGCTGGTGTTGAAG	
PPARα	DQ294237	TCCTGGTGGCCTACGGATC	CGTTGAATTTCATGGCGAACT	
PPARβ	AJ416953	GAGACGGTCAGGGAGCTCAC	CCAGCAACCCGTCCTTGTT	
PPARγ	AJ416951	CATTGTCAGCCTGTCCAGAC	TTGCAGCCCTCACAGACATG	
EF1α	AF321836	CACCACCGGCCATCTGATCTACAA	TCAGCAGCCTCCTTCTCGAACTTC	
MYH6	DW559270	CAGGTCCTCTATGTGCTGGTGTG	TCCTCATTGTAGTTGCTGTCCTCAC	
ANGL4	GRASP209147493	CCGTATGGGGGATGATGCTAA	GGTAGTATGCTGACGACTGACACCT	
GTR1	S30269700	GCCATGGATGTCCTACGTGA	CTCCGCTACATACGGGAAGG	
CPT1A	S31823509	TCCCACATCATCCCCTTCAACT	TGTCCCTGAAGTGAGCCAGCT	
ACADS	S31978689	CTGGGGAAGAAGGAGGACAAG	TCTAGAGCAGCCTGAGCAATACC	
NKX2.5	DW550500	CCCAGTACGTCCACACCCTT	GGAGGTCGGTAAGGCACAGT	
Mef2C	GU252207	CACCGTAACTCGCCTGGTCT	GCTTGCGGTTGCTGTTCATA	
GATA4	HM475152	TCTCCATTCGACAGCTCCGT	CATCGCTCCACAGTTCACACA	
Osx	FJ195614	ATTACTGAGGAGGAGCCCATCATT	CCTCATCCACCTCACACACCTT	

mt/nDNA Ratio

Genomic DNA was isolated from tissue samples from the same individuals as the ones used for the total RNA extraction (totally 96 samples, n=6). DNA was isolated using DNAeasy kit (Qiagen) according to the manual. The DNA quality for all samples was checked on a 1% agarose gel and concentration was measured using a NanoDrop Spectrophotometer. The MyoD gene (intron-exon spanning primers) and the mitochondrial D-loop were amplified by qRT-PCR. For the PCR reaction 1x SYBR green I master Mix (Roche), 0.41nM of each primer and the 6.4ng DNA template were mixed in a reaction volume of 10μ l. PCR amplification was conducted as described above. All reactions were run in duplicates. Absolute concentrations for mt- and nDNA samples were obtained using the standard curve method. The ratio was calculated by dividing the absolute mtDNA by the absolute nDNA concentration.

Statistical Analysis

All data are presented as means ± SEM with an n value as stated. The effect of dietary treatment on the production parameters and qRT-PCR were analyzed by 2-way analysis of variance (ANOVA), using dietary treatment and sampling point as fixed factors and block as a random factor. TTA effects on gene expression and mt/nDNA ratio were calculated using EF1 α standardized expression values in a 2-way ANOVA with dietary treatment and time as fixed factors. Unless otherwise stated the statistical unit is the individual fish. All analyses were conducted using R[50], plots were produced using the R package ggplot[58] and the heatmaps were produced with the R package lattice/latticeExtra[59].

DECLARATIONS

Acknowledgements

We would like to thank Dr. Aleksei Krasnov for designing the Atlantic salmon microarray. We would also like to thank Thia Medica for supplying the TTA used in this experiment.

Competing Interests

The authors declare that they have no competing interests.

REFERENCES

1. Moya-Falcón C, Hvattum E, Dyrøy E, Skorve J, Stefansson SO, Thomassen MS, Jakobsen JV, Berge RK, Ruyter B: Effects of 3-thia fatty acids on feed intake, growth, tissue fatty acid composition, beta-oxidation and Na+,K+-ATPase activity in Atlantic salmon. Comp Biochem Physiol B , Biochem Mol Biol. 2004, 139 (4): 657-68. 10.1016/j.cbpc.2004.08.009.
2. Berge RK, Skorve J, Tronstad KJ, Berge K, Gudbrandsen OA, Grav H: Metabolic effects of thia fatty acids. Curr Opin Lipidol. 2002, 13 (3): 295-304. 10.1097/00041433-200206000-00010.
3. Westergaard M, Henningsen J, Svendsen ML, Johansen C, Jensen UB, Schrøder HD, Kratchmarova I, Berge RK, Iversen L, Bolund L, Kragballe K, Kristiansen K: Modulation of keratinocyte gene expression and differentiation by PPAR-selective ligands and tetradecylthioacetic acid. J Invest Dermatol. 2001, 116 (5): 702-12. 10.1046/j.1523-1747.2001.01329.x.

4. Forman BM, Chen J, Evans RM: Hypolipidemic drugs, polyunsaturated fatty acids, and eicosanoids are ligands for peroxisome proliferator-activated receptors alpha and delta. Proc Natl Acad Sci USA. 1997, 94 (9): 4312-7. 10.1073/pnas.94.9.4312.
5. Berger JP, Akiyama TE, Meinke PT: PPARs: therapeutic targets for metabolic disease. Trends Pharmacol Sci. 2005, 26 (5): 244-51. 10.1016/j.tips.2005.03.003.
6. Rørvik KA, Alne H, Gaarder M, Ruyter B, Måseide NP, Jakobsen JV, Berge RK, Sigholt T, Thomassen MS: Does the capacity for energy utilization affect the survival of post-smolt Atlantic salmon, Salmo salar L., during natural outbreaks of infectious pancreatic necrosis?. J Fish Dis. 2007, 30 (7): 399-409. 10.1111/j.1365-2761.2007.00823.x.
7. Vegusdal A, Gjøen T, Berge RK, Thomassen MS, Ruyter B: Effect of 18:1n-9, 20:5n-3, and 22:6n-3 on lipid accumulation and secretion by Atlantic salmon hepatocytes. Lipids. 2005, 40 (5): 477-86. 10.1007/s11745-005-1407-z.
8. Alne H, Thomassen MS, Takle H, Terjesen BF, Grammes F, Oehme M, Refstie S, Sigholt T, Berge RK, Rørvik KA: Increased survival by feeding tetradecylthioacetic acid during a natural outbreak of heart and skeletal muscle inflammation in S0 Atlantic salmon, Salmo salar L. J Fish Dis. 2009, 32 (11): 953-61. 10.1111/j.1365-2761.2009.01078.x.
9. Grammes F, Rørvik KA, Takle H: Tetradecylthioacetic acid modulates cardiac transcription in Atlantic salmon, Salmo salar L., suffering heart and skeletal muscle inflammation. J Fish Dis. 2012, 35 (2): 109-17. 10.1111/j.1365-2761.2011.01326.x.
10. Sambandam N, Morabito D, Wagg C, Finck BN, Kelly DP, Lopaschuk GD: Chronic activation of PPARalpha is detrimental to cardiac recovery after ischemia. Am J Physiol Heart Circ Physiol. 2006, 290: H87—95-
11. Finck BN, Lehman JJ, Leone TC, Welch MJ, Bennett MJ, Kovacs A, Han X, Gross RW, Kozak R, Lopaschuk GD, Kelly DP: The cardiac phenotype induced by PPARalpha overexpression mimics that caused by diabetes mellitus. J Clin Invest. 2002, 109: 121-30.
12. Burkart EM, Sambandam N, Han X, Gross RW, Courtois M, Gierasch CM, Shoghi K, Welch MJ, Kelly DP: Nuclear receptors PPARbeta/delta and PPARalpha direct distinct metabolic regulatory programs in the mouse heart. J Clin Invest. 2007, 117 (12): 3930-9.
13. Cheng L, Ding G, Qin Q, Huang Y, He N, Evans RM, Schneider MD, Brako FA, Xiao Y, Chen YE, Yang Q: Cardiomyocyte-restricted peroxisome proliferator-activated receptor-[delta] deletion perturbs myocardial fatty acid oxidation and leads to cardiomyopathy. Nat Med. 2004, 10 (11): 1245-50. 10.1038/nm1116.
14. Chinetti G, Fruchart JC, Staels B: Peroxisome proliferator-activated receptors (PPARs): nuclear receptors at the crossroads between lipid metabolism and inflammation. Inflamm Res. 2000, 49 (10): 497-505. 10.1007/s000110050622.
15. Finck BN: The PPAR regulatory system in cardiac physiology and disease. Cardiovasc Res. 2007, 73 (2): 269-77. 10.1016/j.cardiores.2006.08.023.
16. The health situation in Norwegian aquaculture 2009. Tech. rep., Norwegian Veterinary Institute 2009.Http://www.vetinst.no/eng/Research/Publications/Fish-Health-Report,
17. Driedzic WR: Cardiac energy metabolism. Fish Physiology, Volume 7. Edited by: Hoar WS, Randall DJ, Farrell AP. 1992, Academic Press, New York, 219-266.
18. Krasnov A, Timmerhaus G, Afanasyev S, Jørgensen SM: Development and assessment of oligonucleotide microarrays for Atlantic salmon (Salmo salar L.). Comp Biochem Physiol Part D Genomics Proteomics. 2010
19. Fellenberg K, Hauser NC, Brors B, Neutzner A, Hoheisel JD, Vingron M: Correspondence analysis applied to microarray data. Proc Natl Acad Sci USA. 2001, 98 (19): 10781-6. 10.1073/pnas.181597298.
20. Smyth GK: Limma: linear models for microarray data. Bioinformatics and Computational Biology Solutions using R and Bioconductor. Edited by: Gentleman R, Carey V, Dudoit S, R Irizarry WH. 2005, Springer, New York, 397-420.
21. Gerety SS, Anderson DJ: Cardiovascular ephrinB2 function is essential for embryonic angiogenesis. Development. 2002, 129 (6): 1397-410.
22. Hoefen RJ, Berk BC: The multifunctional GIT family of proteins. J Cell Sci. 2006, 119 (Pt 8): 1469-75.
23. Wallis J, Lygate CA, Fischer A, ten Hove, Schneider JE, Sebag-Montefiore L, Dawson D, Hulbert K, Zhang W, Zhang MH, Watkins H, Clarke K, Neubauer S: Supranormal myocardial creatine and phosphocreatine concentrations lead to cardiac hypertrophy and heart failure: insights from creatine transporter-overexpressing transgenic mice. Circulation. 2005, 112 (20): 3131-9. 10.1161/CIRCULATIONAHA.105.572990.
24. Mandard S, Zandbergen F, van Straten E, Wahli W, Kuipers F, Müller M, Kersten S: The fasting-induced adipose factor/angiopoietin-like protein 4 is physically associated with lipoproteins and governs plasma lipid levels and adiposity. J Biol Chem. 2006, 281 (2): 934-44.
25. Ashburner M, Ball CA, Blake JA, Botstein D, Butler H, Cherry JM, Davis AP, Dolinski K, Dwight SS, Eppig JT, Harris MA, Hill DP, Issel-Tarver L, Kasarskis A, Lewis S, Matese JC, Richardson JE, Ringwald M, Rubin GM, Sherlock G: Gene ontology: tool for the unification of biology. The Gene Ontology Consortium. Nat Genet. 2000, 25: 25-9. 10.1038/75556.
26. Falcon S, Gentleman R: Using GOstats to test gene lists for GO term association. Bioinformatics. 2007, 23 (2): 257-8. 10.1093/bioinformatics/btl567.
27. Christoffels VM, Habets PE, Franco D, Campione M, de Jong, Lamers WH, Bao ZZ, Palmer S, Biben C, Harvey RP, Moorman AF: Chamber formation and morphogenesis in the developing mammalian heart. Dev Biol. 2000, 223 (2): 266-78. 10.1006/dbio.2000.9753.

28. Chen JN, Fishman MC: Zebrafish tinman homolog demarcates the heart field and initiates myocardial differentiation. Development. 1996, 122 (12): 3809-16.

29. Schwinger RHG, Bundgaard H, Müller-Ehmsen J, Kjeldsen K: The Na, K-ATPase in the failing human heart. Cardiovasc Res. 2003, 57 (4): 913-20. 10.1016/S0008-6363(02)00767-8.

30. Wehrens XHT, Lehnart SE, Huang F, Vest JA, Reiken SR, Mohler PJ, Sun J, Guatimosim S, Song LS, Rosemblit N, D'Armiento JM, Napolitano C, Memmi M, Priori SG, Lederer WJ, Marks AR: FKBP12.6 deficiency and defective calcium release channel (ryanodine receptor) function linked to exercise-induced sudden cardiac death. Cell. 2003, 113 (7): 829-40. 10.1016/S0092-8674(03)00434-3.

31. Frank KF, Bölck B, Erdmann E, Schwinger RHG: Sarcoplasmic reticulum Ca2+-ATPase modulates cardiac contraction and relaxation. Cardiovasc Res. 2003, 57: 20-7. 10.1016/S0008-6363(02)00694-6.

32. Beck EJ, Bowlby M, An WF, Rhodes KJ, Covarrubias M: Remodelling inactivation gating of Kv4 channels by KChIP1, a small-molecular-weight calcium-binding protein. J Physiol. 2002, 538 (Pt 3): 691-706.

33. Wickenden AD, Kaprielian R, Kassiri Z, Tsoporis JN, Tsushima R, Fishman GI, Backx PH: The role of action potential prolongation and altered intracellular calcium handling in the pathogenesis of heart failure. Cardiovasc Res. 1998, 37 (2): 312-23. 10.1016/S0008-6363(97)00256-3.

34. Dyck JR, Barr AJ, Barr RL, Kolattukudy PE, Lopaschuk GD: Characterization of cardiac malonyl-CoA decarboxylase and its putative role in regulating fatty acid oxidation. Am J Physiol. 1998, 275 (6 Pt 2): H2122—9-

35. Sans CL, Satterwhite DJ, Stoltzman CA, Breen KT, Ayer DE: MondoA-Mlx heterodimers are candidate sensors of cellular energy status: mitochondrial localization and direct regulation of glycolysis. Mol Cell Biol. 2006, 26 (13): 4863-71. 10.1128/MCB.00657-05.

36. Totland GK, Madsen L, Klementsen B, Vaagenes H, Kryvi H, Frøyland L, Hexeberg S, Berge RK: Proliferation of mitochondria and gene expression of carnitine palmitoyltransferase and fatty acyl-CoA oxidase in rat skeletal muscle, heart and liver by hypolipidemic fatty acids. Biol Cell. 2000, 92 (5): 317-29. 10.1016/S0248-4900(00)01077-7.

37. Kleveland EJ, Ruyter B, Vegusdal A, Sundvold H, Berge RK, Gjøen T: Effects of 3-thia fatty acids on expression of some lipid related genes in Atlantic salmon (Salmo salar L.). Comp Biochem Physiol B , Biochem Mol Biol. 2006, 145 (2): 239-48. 10.1016/j.cbpb.2006.07.012.

38. Hafstad AD, Khalid AM, Hagve M, Lund T, Larsen TS, Severson DL, Clarke K, Berge RK, Aasum E: Cardiac peroxisome proliferator-activated receptor-alpha activation causes increased fatty acid oxidation, reducing efficiency and post-ischaemic functional loss. Cardiovasc Res. 2009, 83 (3): 519-26. 10.1093/cvr/cvp132.

39. Wagner N, Jehl-Piétri C, Lopez P, Murdaca J, Giordano C, Schwartz C, Gounon P, Hatem SN, Grimaldi P, Wagner KD: Peroxisome proliferator-activated receptor beta stimulation induces rapid cardiac growth and angiogenesis via direct activation of calcineurin. Cardiovasc Res. 2009, 83: 61-71. 10.1093/cvr/cvp106.

40. Narkar VA, Downes M, Yu RT, Embler E, Wang YX, Banayo E, Mihaylova MM, Nelson MC, Zou Y, Juguilon H, Kang H, Shaw RJ, Evans RM: AMPK and PPARdelta agonists are exercise mimetics. Cell. 2008, 134 (3): 405-15. 10.1016/j.cell.2008.06.051.

41. Xu J, Gong NL, Bodi I, Aronow BJ, Backx PH, Molkentin JD: Myocyte enhancer factors 2A and 2C induce dilated cardiomyopathy in transgenic mice. J Biol Chem. 2006, 281 (14): 9152-62.

42. Sharifpanah F, Wartenberg M, Hannig M, Piper HM, Sauer H: Peroxisome proliferator-activated receptor alpha agonists enhance cardiomyogenesis of mouse ES cells by utilization of a reactive oxygen species-dependent mechanism. Stem Cells. 2008, 26: 64-71. 10.1634/stemcells.2007-0532.

43. Poppe TT, Johansen R, Gunnes G, Tørud B: Heart morphology in wild and farmed Atlantic salmon Salmo salar and rainbow trout Oncorhynchus mykiss. Dis Aquat Org. 2003, 57 (1-2): 103-8.

44. Tørud B, Hillestad M: "Hjerte-rapporten" Rapport om hjertelidelser hos laks og regnbueørret. 2004, 1-69. [www.fiskerifond.no/files/projects/attach/hjerterapporten.pdf]

45. Asiedu DK, Frøyland L, Vaagenes H, Lie O, Demoz A, Berge RK: Long-term effect of tetradecylthioacetic acid: a study on plasma lipid profile and fatty acid composition and oxidation in different rat organs. Biochim Biophys Acta. 1996, 1300 (2): 86-96. 10.1016/0005-2760(95)00235-9.

46. Gjøen T, Kleveland EJ, Moya-Falcón C: Effects of dietary thia fatty acids on lipid composition, morphology and macrophage function of Atlantic salmon (Salmo salar L.) kidney. Comp Biochem Physiol B , Biochem Mol Biol. 2007, 148: 103-11. 10.1016/j.cbpb.2007.04.021.

47. Alne H, Oehme M, Thomassen M, Terjesen B, Rørvik KA: Reduced growth, condition factor and body energy levels in Atlantic salmon Salmo salar L. during their first spring in the sea. Aquacult Res. 2010, 42 (2): 248-259.

48. Bligh EG, Dyer WJ: A rapid method of total lipid extraction and purification. Can J Biochem Physiol. 1959, 37: 911-917. 10.1139/o59-099.

49. Dates M: General analytical procedures. Techniques in Lipidology. Edited by Kates M. Elsevier; 1986

50. R Development CoreTeam: R: A Language and Environment for Statistical Computing. 2009, R Foundation for Statistical Computing, Vienna, http://www.R-project.org. [ISBN 3-900051-07-0]

51. Gentleman RC, Carey VJ, Bates DM: Bioconductor: Open software development for computational biology and bioinformatics. Genome Biology. 2004, 5: R80-10.1186/gb-2004-5-10-r80.

52. Bolstad BM, Irizarry RA, Astrand M, Speed TP: A comparison of normalization methods for high density

oligonucleotide array data based on variance and bias. Bioinformatics. 2003, 19 (2): 185-93. 10.1093/bioinformatics/19.2.185.

53. Smyth GK: Linear models and empirical bayes methods for assessing differential expression in microarray experiments. Stat Appl Genet Mol Biol. 2004, 3:

54. Benjamini Y, Yosef H: Controlling the false discovery rate: a practical and powerful approach to multiple testing. J Roy Stat Soc B. 1995, 57: 289-300.

55. Conesa A, Götz S: Blast2GO: a universal tool for annotation, visualization and analysis in functional genomics research. Bioinformatics. 2005, 21 (18): 3674-6. 10.1093/bioinformatics/bti610.

56. Culhane AC, Thioulouse J, Perrière G: MADE4: an R package for multivariate analysis of gene expression data. Bioinformatics. 2005, 21 (11): 2789-90. 10.1093/bioinformatics/bti394.

57. Olsvik PA, Lie KK, Jordal AEO, Nilsen TO, Hordvik I: Evaluation of potential reference genes in real-time RT-PCR studies of Atlantic salmon. BMC Mol Biol. 2005, 6: 21-10.1186/1471-2199-6-21.

58. Wickham H: Ggplot: Elegant graphics for data analysis. 2009, Springer, New York

59. Sarkar D: Lattice: Multivariate Data Visualization with R. 2008, Springer, New York

60. Moriya Y, Itoh M, Okuda S, Yoshizawa AC, Kanehisa M: KAAS: an automatic genome annotation and pathway reconstruction server. Nucleic Acids Res. 2007, 35 (Web Server issue): W182-5.

CHAPTER 10

IDENTIFICATION AND ANALYSIS OF IMMUNE-RELATED TRANSCRIPTOME IN ASIAN SEABASS LATES CALCARIFER

Jun Hong Xia and Gen Hua Yue

BACKGROUND

Fish diseases caused by viruses, bacteria and parasites are recognized as a significant constraint on aquaculture production and trade hence affecting the economy seriously [1, 2]. A global estimate of disease losses in aquaculture surpassed US$ 9 billion per year, which is about 15% of the value of world farmed fish and shellfish production [3]. Successful defence against pathogenic infection is dependent on the ability to detect the presence of the invading pathogen [4, 5, 6]. Teleost fish possess the elements of both the innate defence system and the acquired specific immune system [7]. However, the adaptive immune response in fish is less developed than that in higher vertebrates [5]. Therefore, innate immune system is quite important in fish and believed to be the first line of host defence in opposing pathogenic organisms or any other foreign material [4, 7]. In aquaculture, the basic data on fish-pathogen interactions have been effectively applied for large scale vaccination to aid in the generation of robust and long-lasting immune responses [8, 9]. However, the development of an effective vaccine is a complex process. The prerequisites for developing vaccines are the understanding of the basic epidemiology of diseases and the immune system of the target species and identifying the genes and pathways involved in transcript response of a fish upon infection [10, 11].

Expressed sequence tags (ESTs) generated by cDNA cloning have proven to be a powerful and rapid tool for identifying genes [12, 13, 14]. ESTs also form the basis for subsequent microarray design, SNP detection and the placement of novel markers on genetic linkage maps [13, 15, 16, 17, 18]. Currently ESTs are available from a large number of fish species, such as European seabass [19], halibut [13], cod [20], trout [16], salmon [15], European and Japanese flounders [21] and catfish [17] which enabled the identification of immune-related genes in these species. However, screening for immune-related genes in EST databases using bioinformatic tools allows identification of only those genes that share sequence similarities with known immune-related proteins from other organisms [22], but misses novel genes related to immune responses. Suppression subtractive hybridization (SSH) [23] can be applied to identify differentially expressed genes in different tissues or conditions and thus is also a highly effective method for identifying novel genes related to important biological processes. This technique has been proven to be a suitable tool for identification of novel immune-related genes in a variety of teleost fish species, including flounder [21, 24, 25], rainbow trout [26], salmon [27,28, 29], grouper [30], croaker [31], cod [32], sea bream [14, 33], turbot [34], dogfish [35] and European seabass [36].

Citation: Jun Hong Xia and Gen Hua Yue, Research article Identification and analysis of immune-related transcriptome in Asian seabass Lates calcarifer, doi: 10.1186/1471-2164-11-356

The Asian seabass *Lates calcarifer* distributed in the tropical and sub-tropical areas of Asia is an important marine foodfish species in Southeast Asia and Australia. This species has been cultured for more than 20 years in brackish-water ponds and in recent years in floating cages. The global annual production of Asian seabass was currently 400,000 metric tons according to FAO statistics [37, 38]. In the past few years, we conducted a breeding program for Asian seabass [37] and developed a number of genomic tools such as microsatellites [39], SNPs in genes [40], linkage maps [37], BAC and cDNA libraries [40, 41] to facilitate the selective breeding program. Massive mortalities caused by bacterial or viral infections in intensive aquaculture production have caused serious economic losses in Asian seabass aquaculture. Some kinds of bacteria, such as *Cytophaga johnsoniae* and *Streptococcus iniae* have currently been isolated from sick seabass in Singapore [42], Thailand [43] and Australia [44, 45]. In order to shield the aquaculture loss by pathogenic diseases, Asian seabass aquaculture urgently requires effective disease prevention strategies. Although some studies had shown that Asian seabass exhibited strong immune responses against bacteria based on the antibody activities in sera [46], little information on host--pathogen interaction during infection with pathogenic microorganisms is available for this species.

Immune response can be experimentally stimulated by bacterial lipopolysaccharides (LPS) [47, 48, 49]. cDNA libraries of liver, kidney and spleen have been proven to be an excellent source of genetic information concerning immune function in fish [13]. The aim of this study was to identify and analyze immune-response genes in Asian seabass by challenging individuals with LPS, the SSH technology and bioinformatic tools.

RESULTS

SSH efficiency

The efficiency of subtraction can be estimated with PCR analysis by comparing the abundance of housekeeping genes and differentially expressed genes before and after subtraction (Clontech manual). In this experiment, an immune-related gene (the immunoglobulin heavy constant mu: IGHM) and a house-keeping gene (í-actin) were used to examine the efficiency of subtraction. Quantitative real-time PCR analysis of IGHM and í-actin revealed that SSH efficiency was 3.4 fold for the forward subtractive library and 4.7 fold for the reverse subtractive library.

Generation of subtractive cDNA libraries and assembly of ESTs

Two subtractive cDNA libraries, a forward subtractive library (genes expected to be up-regulated in response to immune challenge in this library) and a reverse subtractive library (genes expected to be down-regulated in response to immune challenge) were constructed using subtractive cDNA from spleen of Asian seabass sampled at 24 hour post challenge with LPS and control samples. A total of 1527 and 1824 randomly picked clones for the forward subtractive library and the reverse subtractive library, respectively, were sequenced. After trimming of end and vector sequences and eliminating sequences with low quality and/or shorter than 100 bases, a total of 2887 high quality sequences were obtained. Among the 2887 high quality sequences 1168 were derived from the forward subtractive library whereas 1719 from the reverse subtractive library (Table 1).

Table 1: EST sequences and assembly statistics for two suppression subtractive hybridization libraries constructed from spleen cDNA of Asian seabass

	Forward subtractive library	Reverse subtractive library	Sequences combined
Total number of sequences	1527	1824	3351
Clean sequences (>100 bp)	1168	1719	2887
Low quality sequences	359	105	464
Total number of unique sequences	761	1093	1692
Singlets	588	797	1207
Contigs	173	296	485
2 ESTs	118	196	292
3-4 ESTs	41	72	130
≥5 ESTs	14	28	63
SNP-contained contigs	--	--	303
Microsatellite -contained unique sequences (≥8 repeats)	--	--	20

After assembly and clustering, 1692 unique transcripts consisting of 1207 singletons and 485 contigs (Table 1) were obtained. Among the 485 contigs, 290 (60.2%) contained 2 ESTs, 130 (26.8%) had 3-4 ESTs and 63 (13.0%) consisted of 5 or more sequences. The largest contig was formed by 97 ESTs.

Annotation of ESTs

Blast searches against known sequences in public databases using the programs BLASTx and BLASTn revealed that 1074 unique transcripts were similar to 743 known genes and 105 unannotated mRNA sequences with high confidence (E value $< 10^{-4}$) in the database. The remaining 618 unique transcripts were potentially novel sequences or UTRs of known genes. Three hundred and thirty-one of these known genes and unannotated mRNA sequences were represented by multiple sequences (see Additional file 1). Among the known genes and mRNA sequences represented by EST clones, 542 were found in the reverse subtractive library and 467 were found in the forward subtractive library, respectively. Of which, 161 were present in both libraries. The percentage (~9%) of unique transcripts presented both in reverse and forward subtractive libraries was slightly higher than that in some previous studies (3-5.95%), such as on grouper *Epinephelus coioides* [50]. This might be related to the nature of SSH technique used in this study. Although differentially expressed genes were enriched significantly with the approach, some of the unwanted genes might not have been eliminated completely in the libraries. When more clones of the SSH libraries were sequenced, as in our study, more rare transcripts would be found in both of the forward and reverse libraries. This should increase the percentage of unique transcripts that are present both in reverse and forward subtractive libraries.

The globin gene family was among the predominant EST sequence clusters in both libraries. Of the 2887 EST sequences, 13.8% (397) matched the globin family (hbaa1, ba1, hbb, Hba-x), 5.7% (165) were ribosomal proteins (12 S, 16 S, 18 S, 40 S and 60 s), 1.6% (47) were initiation factors and 1.2% (34) belonged to elongation factors (Additional file 1). These results probably reflect that the fish were in an active growth stage. Interestingly, the expression of the globin gene family ($p < 0.0001$) and elongation factors ($p = 0.0023$) were significantly differentiated between two libraries, indicating some globin proteins (up-regulated upon LPS challenge) and elongation factors (down-regulated upon LPS challenge) might play important roles in host defence.

The complete results of BLAST searches of the Asian seabass ESTs from the forward subtractive and reverse subtractive libraries are summarized in Additional file 1.

Functional Classification

Available bioinformatic tools, such as the Gene Ontology (GO) GO Slimmer and Kyoto Encyclopedia of Genes and Genomes (KEGG) provide useful information to analyze the functional profile of annotated genes [51]. A total of 743 annotated genes were used in functional classification (Additional file 1). According to GO Slimmer results, genes encoding for proteins associated with cellular process, metabolic process and biology regulation in category of biological processes (Additional file 2) were the three largest annotated subcategories. The second largest group was genes encoding products related to regulation of biological process, response to stimulus and developmental process in both of the forward subtractive library and the reverse subtractive library. The remaining genes encoded products were involved in many other diverse biological processes. With respect to cellular components (Additional file 3), we observed that a large proportion of sequences were classified into cell, cell part and organelle followed by organelle part, macromolecular complex, extracellular region and membrane-enclosed lumen in the forward and the reverse subtractive libraries. As expected, a remarkably high proportion of annotated sequences were categorized as binding and catalytic activity followed by transcription regulator activity and transporter activity in both of the libraries (Additional file 4). Analysis of GO categories showed that the functional distribution of the genes of the two libraries in the three categories was similar ($P > 0.05$). The detailed information of functional classification is shown in Additional file5 and 6.

KO (KEGG Orthology) is a pathway-based classification of orthologous gene groups. Four hundred and sixty seven of the 1692 unique transcripts with KO assignments were composed of 402 unique genes. Of which, 257 unique genes belonged to 160 KEGG pathways. Interestingly, 52 annotated genes were found in 10 immune-related pathways, including complement and coagulation cascades (13 hits), chemokine signaling pathway (12 hits), antigen processing and presentation (10 hits), Toll-like receptor signaling pathway (9 hits), leukocyte transendothelial migration (8 hits), T cell receptor signaling pathway (7 hits), hematopoietic cell lineage (6 hits), B cell receptor signaling pathway (5 hits), natural killer cell mediated cytotoxicity (4 hits) and Fc epsilon RI signaling pathway (3 hits). Most of the annotated genes were associated with cellular process (213 hits) and diseases (171 hits) and metabolism (167 hits), followed by environmental (79 hits) and genetic (70 hits) information processing. Our data suggest that several important immune processes were likely to be affected by the LPS challenge.

Immune-Related Genes and Differentially Expressed Genes in Subtractive Libraries

Genes in the GO subcategories 'response to stimulus 'and' immune system processes are likely to be involved in the interaction of bacteria with its host. We identified a total of 161 genes (78 in the forward subtractive library and 111 in the reverse subtractive library) with response to the LPS challenge, whereas only 28 genes were shared between two libraries. These genes were mainly associated with response to stress (81), response to chemical stimulus (80), response to external stimulus (48), cellular response to stimulus (40), regulation of response to stimulus (35) and immune response (36). Most of the responsive genes were under represented in the library with only one copy in each library. Although the percentage of unique genes responding to stimulus contained in both libraries was not significant, the actual number of the EST clones representing these genes (396 in the forward subtractive library and 384 in the reverse subtractive library) was quite different between two libraries ($P < 0.0001$). In addition, 13.3% of the EST clones (383) matched the genes in immune system process (115 unique genes) based on GO term and KAAS data. The EST clones of immune genes were found statistically different between two libraries ($P < 0.0001$). Majority of the immune and stress-related EST sequences identified in this study (Additional file 5 and 6) were reported for the first time in the Asian seabass.

Based on the number of homologous ESTs in the two libraries, Fisher's exact test found a significant increase in abundance for a total of 25 genes (including 2 unknown genes) and a significant decrease in abundance for 4 genes upon challenge with LPS, suggesting a strong transcriptional regulation upon LPS challenge (Table 2). Homologues of some immune-related genes such as LY6 D and EEF2L were found exclusively in one library, and some others such as C1QL4L and HMOX1 were found highly expressed not only in one library but also in another library at a much lower frequency. Only 6 (IGH-1B, C3, FGG, LGMN, B2 M and HMOX1) of the 29 genes that differentially expressed in spleen of the Asian seabass upon challenge with LPS were classified as immune response-related genes in the gene ontology database [52]. However, 15 of the remaining 23 differentially expressed genes could be identified as immune response-related genes in InnateDB non-redundant list [53] except BA1, AGC1, HDR, HBAA1, RPS7, EEF and two unknown genes.

Table 2: Immune response-related genes differentially expressed in Asian seabass at 24 h post challenge by bacterial lipopolysaccharides ($P < 0.05$)

Cluster ID	Gene ID and description	Count-Up	Count-Down	P-value
Up-regulated genes				
T1	BA1; ba1 globin (*Danio rerio*)	162	141	0.0000
T2	HBAA1; hemoglobin alpha adult-1 (*Danio rerio*)	48	36	0.0021
T4	C1QL4L; complement component 1, q subcomponent-like 4 like (*Danio rerio*)	25	1	0.0000
T23	LY6D; lymphocyte antigen 6 complex, locus D (*Mus musculus*)	9	0	0.0003
T32	IGH-1B; immunoglobulin heavy chain 1b (*Homo sapiens*)	7	1	0.0091
T33	C3; complement component 3 (*Mus musculus*)	8	0	0.0007
T34	SAA3; serum amyloid A 3 (*Mus musculus*)	8	0	0.0007
T44	APOE; apolipoprotein E (*Canis lupus*)	7	0	0.0016
T45	NCAN; neurocan (*Mus musculus*)	7	0	0.0016
T59	RPS7; 40 S ribosomal protein S7 (*Homo sapiens*)	5	1	0.0429
T60	RRT15; uncharacterized protein YLR162W-A (*Saccharomyces cerevisiae*)	5	1	0.0429
T61	AGC1; aggrecan 1(*Danio rerio*)	6	0	0.0044
T82	FGG; fibrinogen, gamma polypeptide (*Rattus norvegicus*)	5	0	0.0108
T83	HDR; hematopoietic death receptor (*Danio rerio*)	5	0	0.0108
T84	LCP1; lymphocyte cytosolic protein 1 (*Mus musculus*)	5	0	0.0108
T85	LGMN; legumain (*Danio rerio*)	5	0	0.0108
T127	APO; apolipoprotein (*Lates calcarifer*)	4	0	0.0267
T128	B2M; beta-2 microglobulin (*Mus musculus*)	4	0	0.0267
T129	CFL2; cofilin-2 (*Gallus gallus*)	4	0	0.0267
T130	GUK1;guanylate kinase (*Homo sapiens*)	4	0	0.0267
T131	mt-ND5; NADH-ubiquinone oxidoreductase chain 5 (*Homo sapiens*)	4	0	0.0267
T132	TKTL2; transketolase-like protein 2 (*Rattus norvegicus*)	4	0	0.0267
T133	WBP2; WW domain-binding protein 2 (*Homo sapiens*)	4	0	0.0267
Cg0105	Unknown mRNA sequence	4	0	0.0267
Cg0526	Unknown mRNA sequence	4	0	0.0267
Down-regulated genes				
T6	HMOX1; heme oxygenase (decycling) 1 (*Danio rerio*)	1	19	0.0008
T17	EEF2L; eukaryotic translation elongation factor 2, like (*Danio rerio*)	0	9	0.0134
T35	DBPHT; DNA binding protein with his-thr domain (*Mus musculus*)	0	7	0.0463
T36	EEF; eukaryotic translation elongation factor (*Siniperca chuatsi*)	0	7	0.0463

Count-Up, EST count of a given gene in the forward subtractive library; EST count in the reverse subtractive library. $P < 0.05$ indicates statistically significant.

By the GO classification scheme, the significantly differentially expressed genes were divided into subsets based on their functions. Among the genes with increased abundance, 6 major groups can be identified, including binding (IGH-1B, C3, APOE, WBP2, B2 M, LY6 D, CFL2, LCP1, FGG, RPS7), catalytic activity (LGMN, mt-ND5, GUK1), transporter activity (APOE and AGC1), enzyme regulator (APOE, C1QL4L), antioxidant activity (APOE) and structural molecule activity (LCP1 and RPS7) and with no hit for the remaining 10 genes (BA1, SAA3, NCAN, HDR, APO, TKTL2, HBAA1, RRT15 and 2 unknown genes). For the abundance-decreased genes, GO annotation reveals their involvement in several subcategories of molecular function including binding (HMOX1, DBPHT1), catalytic activity (HMOX1) and molecular transducer activity (HMOX1) and no assignment for two remaining genes (EEF, EEF2L).

ESTs Containing Microsatellites and SNPs

Screening of all EST sequences for microsatellites (2-6 bp) identified 18 microsatellites with ≥8 repeats (Additional file 7). Out of these 18 microsatellites, 6 had significant hits with known genes (CCL25, KRT13, C4BP, CAHZ, LSM4 and EIF3H) by BLAST (e-value ≤ $1e^{-4}$). Twelve (67%) of the microsatellites were dinucleotide and 4 were trinucleotide (22%), while only 2 tetranucleotide (11%) were found. Because 8 individuals were used in library construction in this study and around 3000 EST sequences were available, it was possible to identify high quality SNPs. Three hundred and three of the 485 contigs (62.5%) contained SNPs, indicating that a very high polymorphism exists in genome of the Asian seabass among individuals.

Validation of subtractive library data by quantitative RT-PCR (qRT-PCR) and expression profiles of genes in response to LPS challenges

To validate the subtractive library data, qRT-PCR was performed on 13 randomly selected annotated genes and 2 unknown gene sequences. Of which, 9 genes were considered to be up-regulated and 6 genes were presumably down-regulated in spleen upon LPS challenge, based on the differences of homologous EST counts in both libraries (Fig. 1). The expression levels for these genes were evaluated on the spleen RNA samples at 24 h post challenge with LPS and in the control group. The expression levels were normalized with a house-keeping gene, elongation factor 1-alpha (EF1A) and are presented as in Fig. 2. In well agreement with the EST data, all genes with higher counts in the forward subtractive library tested in the qRT-PCR assay showed clear induction in the spleen upon LPS challenge and three (HMOX1, G0s2 and KRT8) of the remaining genes with higher counts in the reverse subtractive library showed clear suppression after injection with LPS. However, the expression levels for the remaining genes MYD88 and CMKLR1 in two groups were significantly different from the EST counts in two libraries. This inconsistency might be caused by the random sampling errors in sequencing since their expressions were very low in spleen with only one copy in reverse subtractive library (Fig. 1).

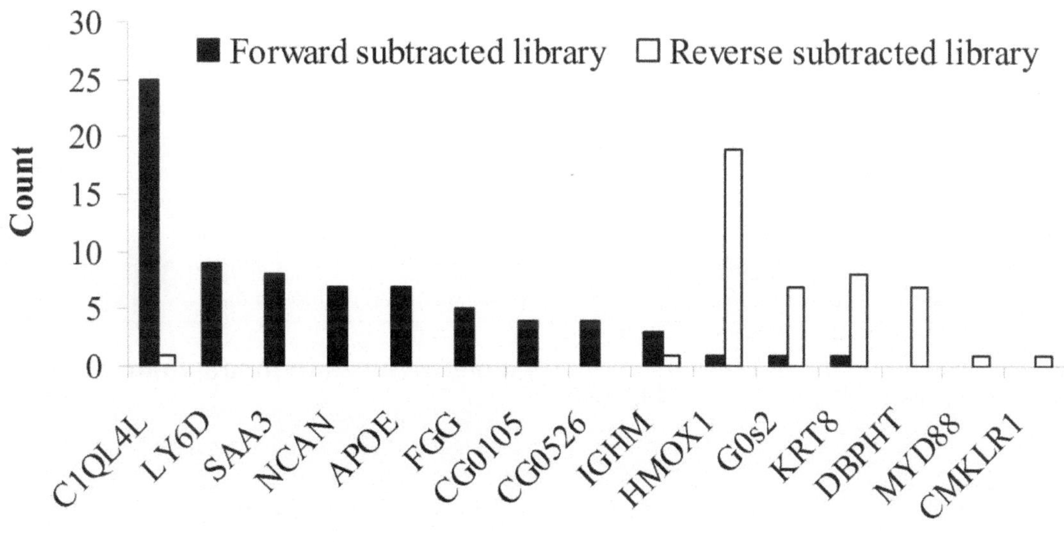

Figure 1: The EST counts in two libraries for selected genes used in quantitative RT-PCR analysis.

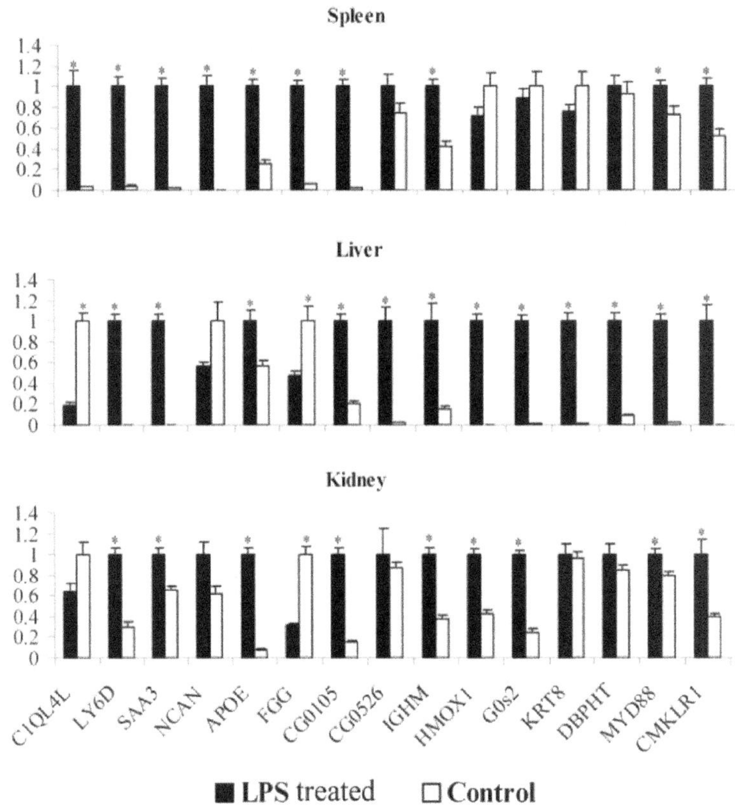

Figure 2: Analysis of gene expression in spleen, liver and kidney of Asian seabass by quantitative RT-PCR. Expression levels in three tissues were normalized using EF1A gene as the reference gene. Normalized fold expression data (mean ± s.e.) represented the average of three independent experiments. A star indicated that the expression of the corresponding gene was significantly different ($P < 0.01$) between LPS treated seabass and control.

In addition to the quantitative analysis of the expressions of these genes in spleen by qRT-PCR, expressions of these genes in liver and kidney in experimental (at 24 h post challenge with LPS) and control groups were analyzed using qRT-PCR to add information about the spatial expression of these genes. As shown in Fig. 2, it was surprising that all of the 6 genes down-regulated in spleen were highly up-regulated in liver and 4 of which were significantly up-regulated in kidney, except gene KRT8 and DBPHT. But for the 9 genes up-regulated in spleen, only 6 were significantly up-regulated in liver and 5 were significantly up-regulated in kidney as detected by qRT-PCR assay.

DISCUSSION

Discovery of novel ESTs from the subtractive cDNA libraries of Asian seabass spleen

SSH is an effective method to study functional expression levels of specific transcripts in various cells and tissues [54]. This study identified a total of 1692 unique transcripts, of which 1074 were similar to 743 known genes and 105 unannotated mRNA available in public databases and the remaining 618 unique transcripts were potentially novel sequences or UTRs of known genes. Some of these genes might play important roles in host-pathogen interaction during infection. Future functional studies of these genes could improve our understanding of innate defence system of fish against pathogenic infection.

Currently, total EST collection for the Asian seabass in NCBI EST database [55] is approaching to 5637 (dated at April 24, 2010) and all of these data originated from a brain cDNA library [56]. The addition of 1692 unique transcripts from Asian seabass to the existing database would not only make a contribution to functional genomic studies, but also help the annotation of genome and the comparative analysis of gene expression profiles in the near future.

Genes involved in innate immune responses

Fish represents the earliest class of vertebrates possessing the elements of both innate and acquired immunity [57, 58]. The soluble mediators released by fish immune cells can regulate inflammatory responses and have a fully functional complement system and unique receptors that recognize pathogens [59, 60]. We identified a total of 161 annotated genes (19%) in response to stimulus and 115 genes (13.6%) matching genes in immune system process. It was much higher than those reported in European seabass (*Dicentrarchus labrax*) with only 79 genes (6%) categorized to the GO category "immune system process" after infected with *V. anguillarum* [19] and with 8.7% of the ESTs showing significant similarities with immune genes after Nodavirus infection [36]. The majority of immune and stimulus-related EST sequences detected in this study were reported for the first time in the Asian seabass. Of which, some homologous genes found in this study were previously reported to play an important role in the innate immune response of vertebrate animals upon bacterial infection, e.g., Hepcidin

[61]. Based on our EST data, we identified hepcidin-1 (hep-1) gene in the Asian seabass. The cDNA sequence with a length of 710 bp consisted of the whole ORF and partials of the UTRs for Hepcidin and encoded a peptide of 89 amino acids with a molecular weight of 9914 Daltons. This represents the first antimicrobial peptide gene discovered in the Asian seabass.

Many pro-inflammatory cytokines such as tumor necrosis factor (TNF)-like, interleukin-1 (IL-1) and IL-1-like receptors have been identified in a number of fish species [59]. In our study some pro-inflammatory cytokines and related receptors have also been identified, such as TNF receptor super family member 14 (TNFRSF14), IL-31 receptor A (IL31RA), chemokine receptor-like 1 (CMKLR1), chemokine (C-X-C motif) receptor 3 (CXCR3), chemokine (C-C motif) receptor 7 (CCR7) and chemokine (C-C motif) ligand 25 (CCL25). Complement components are also known to play a role in linking innate and adaptive immunity. A number of complement and complement receptor homologs such as C1R, C1 inhibitor, C2, C4, C3, C5, C6, C7, C8 and C9 have also been identified from a variety of fish species [59, 62, 63, 64]. From our EST data, we identified complement component C1 (c1ql4l, C1QC, C1qb), C2 and CR2, C3 and C3B and complement factor (CFD and CFP). To our knowledge, the complement receptor CR2 (CD21) has not been previously reported in fish.

Lectins are also known to play an important role in the innate immunity of fish [59]. Two C-type lectins (TCL-1 and TCL-2) have been identified in rainbow trout [65, 66] and C-type lectin receptors have also been identified in two species of cichlid fish, *Paralabidochromis chilotes* and *Oreochromis niloticus* [67]. We identified C-type lectin domain family 4 member C (CLEC4C), C-type lectin domain family 4 member E (CLEC4E) and a kind of F-type lectins (fucolectins) in our libraries. In addition, the Toll-like receptors (TLR) were reported to play a critical role in innate immunity against fungal and bacterial infections by initiating intracellular signal transduction that results in the expression of genes involved in inflammation, antiviral responses and maturation of dendritic cells [7, 68]. We found at least 9 genes were involved in Toll-like receptor signaling pathway such as phosphoinositide-3-kinase, regulatory subunit (PIK3R), tumor necrosis factor receptor superfamily, member 5 (TNFRSF5, CD40), proto-oncogene protein c-fos (FOS) and Ras-related C3 botulinum toxin substrate 1 (RAC1).

Differentially expressed genes associated with immune processes

A number of similarities in immune response exist between mammals and fish [59, 69]. Analysis of the immune system of the zebrafish revealed a fully developed adaptive and innate immune system showing notable similarities to the mammalian equivalent [70]. By comparative analysis of ESTs, many conserved genes responded upon the bacterial infection were revealed between Asian seabass and mammals. For example, apolipoprotein E (ApoE) was found to work as an immune modulator in humans [71], European seabass, carp and medaka [19]. In this study we also detected a substantial increase in the expression of this gene upon LPS challenge. Based on the approximate expression patterns inferred from spleen EST sources of mammals [55], we concluded that more than half of the differentially expressed genes detected in this study were also found to be highly expressed in the spleen of mammals. Some of the genes detected in this study have highly homologous counterparts of well-known mammalian spleen genes such as BA1, EEF2L and C1QL4L, being a homologue of Hbb-b1, EEF2 and C1QA, respectively. All together, these data indicate that there are many conserved features in expression and function of genes in spleen between fish and mammals.

Our data also suggest that 8 (HDR, BA1, AGC1, HBAA1, RPS7, EEF and two unknown genes) out of the 29 significantly differentially expressed transcripts could not be identified as immune response-related genes in current public databases. The hematopoietic death receptor (ZH-DR) and the BA1 globin gene were highly differentially expressed upon LPS challenge. ZH-DR was the first TNFR family member to be identified in zebrafish and specific to haematopoietic tissue [72]. However, little information of the function of this gene in innate immune response is available currently. It was reported that death receptors belong to the TNF (tumor necrosis factor)/NGF (nerve growth factor) receptor superfamily and contribute to regulation of the adaptive immune response [73]. Some reports also suggested that hemoglobin can be induced in macrophages stimulated with LPS and interferon-γ that synergistically increases the release of pro-inflammatory cytokines from the innate immune system in response to LPS [74, 75]. Our result suggested that similar to other DR and hemoglobin genes, ZH-DR and BA1 globin gene might have potential immune regulatory functions in response to bacterial infection in fish. Alternatively, mammalian heme oxygenase 1 (HMOX1) was up-regulated strongly during stress and following pathogen entry [76, 77]. To our surprise, HMOX1 was found to be significant down-regulated in spleen of seabass during the acute inflammatory response in this study. Another study of Atlantic salmon (*Salmo salar* L.) also showed that HMOX1 was down-regulated in spleen to infection with the salmon louse (*Lepeophtheirus salmonis*) at 22 day post infection [78]. Taken together, these data suggest that some differences might exist in immune-mediated inflammatory responses during host--pathogen interaction between fish and mammal.

Putative markers in innate immunity

The development of fish comparative immunology has been hampered by the lack of specific markers for immunoregulatory peptides [79]. In this study at least 29 genes that significantly responded on the challenge of LPS were detected and almost all of the genes could be classified as immune response-related genes. In addition, a reliable predictive of changes in gene expression with the EST data in two libraries was confirmed by qPCR experiments; our study also showed that the expression of these genes was highly conserved between fish and mammals. Therefore, these significantly differentially expressed genes might be considered as candidate markers of bacterial infections in spleen of the Asian seabass. These genes provide the basis for further research into the identification of specific markers for immuno-regulatory peptides and for understanding the immune response of the Asian seabass.

Microsatellites and SNPs in immune-related ESTs

In this study, we identified 18 microsatellites and over 300 SNPs in immune-related ESTs. These markers could be used for many genetic studies [80, 81], such as linkage and QTL mapping [37], examining population dynamics [82, 83], analyzing pedigree [84], managing genetic diversity of broodstock [85] and conducting comparative genome analysis [86, 87].

CONCLUSIONS

By challenging Asian seabass with bacterial lipopolysaccharide (LPS) and using suppression subtractive hybridization (SSH) technique and bioinformatic tools, we identified a large number of potential immune-related genes. Some of which are vertebrate homologues and others are hitherto unreported putative defence proteins. These genes will supply us a solid basis for a better understanding of immunity in Asian seabass, for conducting detailed functional analysis of these genes and for developing effective strategies for immune protection against infections in the Asian seabass.

METHODS

Fish

Around 100 small Asian seabass at the age of 15 dph were transported from a commercial fish farm to TLL animal house. The fish were maintained in a large tank containing 500 L seawater at 25°C for acclimatization of 3 weeks. Fish were fed twice daily with palliated feed.

Challenging with LPS and sampling

One day prior to challenge, 16 healthy fish individuals of average weight of 5 g were transferred to two smaller tanks holding 10 L of sea water. For 8 fishes in tank 1, each fish was injected intra-peritoneally with 0.1 ml of 2 mg/ml of *Escherichia coli* LPS (Sigma-Aldrich, Saint Louis, USA) by dilution with phosphate buffered saline (PBS) at RT. In tank 2 (control), each of the 8 fishes received an intra-peritoneal injection of 0.1 ml of PBS. Just before injection and sampling, the fishes were anaesthetized using AQUI-S® with a concentration of 15 mg/L (AQUI-S New Zealand Ltd, Lower Hutt, New Zealand). Eight fishes from each tank were sacrificed at 24 h post challenge. Tissues including spleen, kidney and liver were taken from every fish of each tank and kept in Trizol reagent (Invitrogen, Carlsbad, USA) at -80°C until use.

RNA isolation

Total RNA from spleen was isolated using the TRIZOL kit (Invitrogen, Carlsbad, USA) according to the manufacturer's instructions. Purification of mRNA from total RNA was carried out using Oligotex mRNA Midi Kit (Qiagen, Valencia, USA). The concentration and purity of mRNA were measured using a NanoDrop Spectrophotometer, ND-1000 (NanoDrop Technologies, Wilmington, USA).

Subtractive library construction and sequencing of clones

cDNA suppression subtractive libraries enriched for differentially expressed genes were constructed using PCR-Select cDNA subtraction kit (Clontech, Mountain view, USA) according to the manufacturer's protocol. Isolation of pure poly A+ mRNA from total RNA was performed using Oligotex® mRNA Mini Kit (Qiagen, Valencia, USA) according to the manufacturer›s protocol. The resulting mRNA from 8 LPS-challenged fishes and the mRNA from 8 PBS-treated fishes (control group) was mixed in equal quantity separately. The forward and reverse subtraction experiments of 1 µg of the mixed spleen mRNA were performed for both samples. The efficiency of subtraction was estimated by comparing the abundance of known cDNA (í-actin and IGHM) before and after subtraction with quantitative RT-PCR. The PCR products of subtractive cDNA were directly inserted into a pGEM-T vector (Promega, Madison, USA) and transformed into *E. coli* strain XL-1 (Stratagene, La Jolla, CA) to make two subtractive cDNA libraries, a forward subtractive library (genes expected to be up-regulated in response to immune challenge in this library) and a reverse subtractive library (genes expected to be down-regulated in response to immune challenge). A total of 3551 randomly picked clones from two libraries were sequenced in both directions with M13 forward and M13 reverse primers using BigDye chemicals and ABI 3730 × 1 Genetic Analyzer (Applied Biosystems, Foster city, CA).

Sequence analysis and functional annotation

Base calling from chromatogram traces and trimming of vector and adaptor sequences and low-quality regions from EST sequences were performed by using commercial software Sequencher 4.9 (Gene Codes, Ann arbor, MI, USA). Then, high quality ESTs (≥100 bp) of both forward and reverse subtractive libraries were used to form contigs. These contigs were manually revised to detect possible errors. Singletons and consensus sequences of each contig were referred to as unique sequences and were compared against the gene ontology database[52] using BLASTx. The remaining sequences without significant assignments were compared against the NCBI database [55] using BLASTn. The significant UniGene information (e value cut off was $\leq 1e^{-4}$) of the query set of unique sequences subsequently were mapped into several level 1 subcategories of the three broad categories of 'cellular component', 'molecular function' and 'biological process' respectively, with software GO Slimmer [52]. The KO (KEGG Orthology) assignments and KEGG pathway reconstruction were performed in KAAS (Automatic Annotation Server Ver. 1.6a) [88]. Antimicrobial peptides were identified based on the Antimicrobial Peptide Database (APD) [89] and the assignments for the 29 differentially expressed genes were also carried out in InnateDB non-redundant list [53]. All of the EST sequences were submitted in GenBank with accession nos GT219120-GT222006.

Mining of microsatellites and SNPs

All of the unique sequences were searched for microsatellites using the program Tandem Repeats Finder (ver. 4.00) [90].

The repeat units were set to 2-10 and other parameters were set to default. The microsatellite-containing ESTs (with ≥8 repeat number) were identified as candidates for future marker development. Single nucleotide polymorphisms (SNPs) in contigs were detected manually.

Statistical analysis

A web tool IDEG6 was used for detection of differentially expressed genes between two libraries using Fisher exact test [91]. The *P* value of less than 0.05 was considered statistically significant for these analyses. For testing the null hypothesis that the two means of gene expressions between LPS treated seabass and control were equal, a two-tailed T-test was performed by using the web calculator 'Independent groups T-TEST for means calculator' [92]. A confidence level of 99% was used in the test.

Analysis of gene expression by quantitative real-time PCR

Spleen, liver and kidney samples were collected from three seabass individuals challenged with LPS and three individuals from the control group (only injected with PBS, see details in the section "Challenging with LPS and sampling") at 24 h post challenges. Total RNA was isolated using the TRIZOL kit (Invitrogen, Carlsbad, USA) according to the manufacturer's instructions. After DNase treatment with DNase I recombinant (Roche, Branchburg, USA) and purified by phenol-chroform, around 1 µg aliquot of the DNase-treated total RNA were reverse transcribed to cDNA by M-MLV reverse transcriptase (Promega, Madison, USA) with 0.67 µM poly dT as RT primer in 15 µl volume following the manufacturer›s protocol. The reaction mixture of the RNA template and RT primer was heated at 70°C for 5 min to denature the RNA and then incubated on ice for 5 min. The remaining reagents were added as specified in the thermoscript protocol and the reaction proceeded for 1 hour at 42°C. Finally, the reverse transcriptase was inactivated by incubation at 70°C for 15 min.

For the analysis of expression patterns, the resulting single strand cDNA were 10 times diluted and assayed as DNA template by real-time PCR using primers (see Additional file 8) for 15 different genes. The EF1A gene was used as control (Fig. 2) as suggested by Olsvik et al. [93] and Jorgensen et al. [94]. PCR was performed with the iQ SYBR Green Supermix (Bio-Rad, Hercules, CA, USA) in an iQ™ 5 Real Time PCR Detection Systems (Bio-Rad, Hercules, CA, USA). Briefly, the 25 µl of reaction including 12.5 µl SYBR Green Supermix, 200 nmol each primer and 1 µl diluted cDNA was initially denaturated at 95°C for 3 min, then amplified for 40 cycles (95°C, 5 s, 55 or 60°C, 10 s and 72°C, 20 s). PCR was performed in triplicates. Values shown in Fig. 2 were the average of triplicate real-time PCR reactions, normalized to EF1A gene expression. A qRT-PCR MIQE checklist is shown in Additional file 9.

All procedures conducted with Asian seabass fish were adhered to animal care guidelines (Guidelines on the Care and Use of Animals for Scientific Purposes) as outlined by the National Advisory Committee For Laboratory Animal Research in Singapore. An approval is not required as both authors have the certificate of responsible care and use of laboratory animals. The remaining fish were used for other experiments, e.g., construction of cDNA libraries, isolation and identification of microRNAs.

DECLARATIONS

Acknowledgements

This study is funded by Ministry of National Development for the project "To ensure self-sufficiency of safe seafood for Singapore through the development of aquaculture genomic tools for marker-assisted selective breeding of tropical marine foodfish". We thank our colleagues Dr Mamta Chauhan and Ms Rahila Qureshi for editing English of this paper.

REFERENCES

1. Bondad-Reantaso MG: Assessment of freshwater fish seed resources for sustainable aquaculture. 2007, Rome: Food and Agriculture Organization of the United Nations
2. Verschuere L, Rombaut G, Sorgeloos P, Verstraete W: Probiotic bacteria as biological control agents in aquaculture. Microbiol Mol Biol Rev. 2000, 64 (4): 655-671. 10.1128/MMBR.64.4.655-671.2000.
3. Tan Z, Komar C: Enright WJ: Health management practices for cage aquaculture in Asia - a key component for sustainability. 2007, Intervet Norbio Singapore Pte Ltd
4. Ellis AE: Innate host defense mechanisms of fish against viruses and bacteria. Dev Comp Immunol. 2001, 25: 827-839. 10.1016/S0145-305X(01)00038-6.
5. Gómez GD, Balcázar JL: A review on the interactions between gut microbiota and innate immunity of fish. FEMS Immunol Med Microbiol. 2008, 52: 145-154. 10.1111/j.1574-695X.2007.00343.x.
6. Evans TJ: Bacterial triggering of inflammation by intracellular sensors. Future Microbio. 2009, 4 (1): 65-75. 10.2217/17460913.4.1.65.
7. Whyte SK: The innate immune response of finfish: A review of current knowledge. Fish Shellfish Immun. 2007, 23: 1127-1151.
8. Markestad A, Grave K: Reduction of antibacterial drug use in Norwegian fish farming due to vaccination. Dev Biol. 1997, 90: 365-369.
9. Evelyn TPT: Finfish immunology and its use in preventing infectious diseases in cultured finfish. Diseases in Asian Aquaculture IV (Fish Health Section). Edited by: Lavilla-Pitogo CR, Cruz-Lacierda ER. 2002, Manila: Asian Fisheries Society, 303-324.
10. Grisez L, Tan Z: Vaccine development for Asian Aquaculture. Diseases in Asian Aquaculture V Fish health section, Proceedings of the Fifth Symposium in Asian Aquaculture. Edited by: Walker P, Lester R, Bondad-Reantaso MG. 2005, Goldcoast, Australia: Asian Fisheries Society, 483-494.
11. Dumetz F, Duchaud E, LaPatra SE, Marrec CL, Claverol S, Urdaci M, He'naff ML: A protective immune response is generated in rainbow trout by an OmpH-Like surface antigen (P18) of *Flavobacterium psychrophilum*. Appl Environ Microbiol. 2006, 4845-4852. 10.1128/AEM.00279-06.

12. Liew CC, Hwang DM, Fung YW, Laurenssen C, Cukerman E, Tsui S, Lee CY: A catalogue of genes in the cardiovascular system as identified by expressed sequence tags (ESTs). Proc Natl Acad Sci USA. 1994, 91: 10645-10649. 10.1073/pnas.91.22.10645.
13. Douglas SE, Knickle LC, Kimball J, Reith ME: Comprehensive EST analysis of Atlantic halibut (*Hippoglossus hippoglossus*), a commercially relevant aquaculture species. BMC Genom. 2007, 8: 144-154. 10.1186/1471-2164-8-144.
14. Pinto PIS, Teodósio HR, Galay-burgos M, Power DM, Sweeney GE, AVM C: Identification of Estrogen-Responsive Genes in the Testis of Sea Bream (*Sparus auratus*) Using Suppression Subtractive Hybridization. Mol Reprod Dev. 2006, 73: 318-329. 10.1002/mrd.20402.
15. Hayes B, Laerdahl JK, Lien S, Moen T, Berg P, Hindar K, WS D, Koop BF, Adzhubei A, Hoyhem B: An extensive resource of single nucleotide polymorphism markers associated with Atlantci Salmo (*Salmo salar*) expressed sequences. Aquaculture. 2007, 265: 82-90. 10.1016/j.aquaculture.2007.01.037.
16. Sánchez CC, Smith TPL, Wiedmann RT, Vallejo RL, Salem M, Yao J, Rexroad III CE: Single nucleotide polymorphism discovery in rainbow trout by deep sequencing of a reduced representation library. BMC Genom. 2009, 10: 559-566. 10.1186/1471-2164-10-559.
17. Wang S, Peatman E, Abernathy J, Waldbieser G, Lindquist E, Richardson P, Lucas S, Wang M, Li P, Thimmapuram J: Assembly of 500,000 inter-specific catfish expressed sequence tags and large scale gene-associated marker development for whole genome association studies. Genome Biol. 2010, 11: R8-10.1186/gb-2010-11-1-r8.
18. Cerdà J, Mercadé J, Lozano JJ, Manchado M, Tingaud-Sequeira A, Astola A, Infante C, Halm S, Viñas J, Castellana B: Genomic resources for a commercial flatfish, the Senegalese sole (*Solea senegalensis*): EST sequencing, oligo microarray design, and development of the Soleamold bioinformatic platform. BMC Genom. 2008, 9: 508-521. 10.1186/1471-2164-9-508.
19. Sarropoulou E, Sepulcre P, Poisa-Beiro L, Mulero V, Meseguer J, Figueras A, Novoa B, Terzoglou V, Reinhardt R, Magoulas A: Profiling of infection specific mRNA transcripts of the European seabass *Dicentrarchus labrax*. BMC Genom. 2009, 10: 157-174. 10.1186/1471-2164-10-157.
20. Olsvik PA, Holen E: Characterization of an Atlantic cod (*Gadus morhua*) embryonic stem cell cDNA library. BMC Res Notes. 2009, 2: 74-81. 10.1186/1756-0500-2-74.
21. Aoki T, Hirono I, Kim MG, Katagiri T, Tokuda Y, Toyohara H, Yamamota E: Identification of viral induced genes in Ig+ leucocytes of Japanese flounder *Paralichthys olivaceus*, by differential hybridisation with subtracted and un-subtracted cDNA probes. Fish Shellfish Immun. 2000, 10: 623-630. 10.1006/fsim.2000.0279.
22. Altincicek B, Vilcinskas A: Analysis of the immune-related transcriptome of a lophotrochozoan model, the marine annelid *Platynereis dumerilii*. Front Zool. 2007, 4: 18-10.1186/1742-9994-4-18.
23. Diatchenko L, Lau YF, Campbell AP, Chenchik A, Moqadam F, Huang B: Suppression subtractive hybridization: a method for generating differentially regulated or tissue-specific cDNA probes and libraries. Proc Natl Acad Sci USA. 1996, 93: 6025-6030. 10.1073/pnas.93.12.6025.
24. Dong CW, Zhang YB, Zhang QY, Gui JF: Differential expression of three *Paralichthys olivaceus* Hsp40 genes in responses to virus infection and heat shock. Fish Shellfish Immun. 2006, 21: 146-158. 10.1016/j.fsi.2005.11.002.
25. Marchand J, Tanguy A, Charrier G, Quiniou L, Plee-Gauthier E, Laroche J: Molecular identification and expression of differentially regulated genes of the european flounder, *Platichthys flesus*, submitted to pesticide exposure. Mar Biotechnol. 2006, 8: 275-294. 10.1007/s10126-005-0099-3.
26. Bayne CJ, Gerwick L, Fujiki K, Nakao M, Yano T: Immune-relevant (including acute phase) genes identified in the livers of rainbow trout, *Oncorhynchus mykiss*, by means of suppression subtractive hybridization. Dev Comp Immunol. 2001, 25: 205-217. 10.1016/S0145-305X(00)00057-4.
27. Tsoi SC, Ewart KV, Penny S, Melville K, Liebscher RS, Brown LL, Douglas SE: Identification of immune-relevant genes from Atlantic salmon using suppression subtractive hybridization. Mar Biotechnol. 2004, 6: 199-214. 10.1007/s10126-002-0101-2.
28. Matejusova I, Felix B, Sorsa-Leslie T, Gilbey J, Noble LR, Jones CS, Cunningham CO: Gene expression profiles of some immune relevant genes from skin of susceptible and responding Atlantic salmon (*Salmo salar* L.) infected with*Gyrodactylus salaris* (Monogenea) revealed by suppressive subtractive hybridization. Int J Parasitol. 2006, 36: 1175-1183. 10.1016/j.ijpara.2006.04.009.
29. Rhodes LD, Wallis S, Demlow SE: Genes associated with an effective host response by Chinook salmon to *Renibacterium salmoninarum*. Dev Comp Immunol. 2009, 33: 176-186. 10.1016/j.dci.2008.08.006.
30. Wang L, Wu X: Identification of differentially expressed genes in lipopolysaccharide-stimulated yellow grouper Epinephelus awoara spleen. Fish Shellfish Immun. 2007, 23: 354-363. 10.1016/j.fsi.2006.11.009.
31. Huang y, Lou h, Wu x, Chen Y: Characterization of the BPI-like gene from a subtracted cDNA library of large yellow croaker (*Pseudosciaena crocea*) and induced expression by formalin-inactivated *Vibrio alginolyticus* and *Nocardia seriolae* vaccine challenges. Fish Shellfish Immun. 2008, 25: 740-750. 10.1016/j.fsi.2008.02.012.
32. Rise ML, Hall J, Rise M, Hori T, Gamperl AK, Kimball J, Hubert S, Bowman S, Johnson SC: Functional genomic analysis of the response of Atlantic cod (*Gadus morhua*) spleen to the viral mimic polyriboinosinic polyribocytidylic acid (pIC). Dev Comp Immunol. 2008, 32: 916-931. 10.1016/j.dci.2008.01.002.

33. Dios S, Poisa-Beiro L, Figueras A, Novoa B: Suppression subtraction hybridization (SSH) and macroarray techniques reveal differential gene expression profiles in brain of sea bream infected with nodavirus. Mol Immunol. 2007, 44: 2195-2204. 10.1016/j.molimm.2006.11.017.

34. Wang C, Zhang XH, Jia A, Chen J, Austin B: Identification of immune-related genes from kidney and spleen of turbot,*Psetta maxima* (L.), by suppression subtractive hybridization following challenge with *Vibrio harveyi*. J Fish Dis. 2008, 31: 505-514. 10.1111/j.1365-2761.2008.00914.x.

35. Inoue Y, Saito T, Endo M, Haruta C, Nakai T, Moritomo T, Nakanishi T: Molecular cloning and preliminary expression analysis of banded dogfish (*Triakis scyllia*) CC chemokine cDNAs by use of suppression subtractive hybridization. Immunogenetics. 2005, 56: 722-734. 10.1007/s00251-004-0730-x.

36. Poisa-Beiro L, Dios S, Ahmed H, Vasta GR, Martínez-López A, Estepa A, Alonso-Gutiérrez J, Figueras A, Novoa B: Nodavirus infection of sea bass (*Dicentrarchus labrax*) induces up-regulation of galectin-1 expression with potential anti-inflammatory activity. J Immunol. 2009, 183: 6600-6611. 10.4049/jimmunol.0801726.

37. Wang CM, Zhu ZY, Lo LC, Feng F, Lin G, Yang WT, Li J, Yue GH: A Microsatellite Linkage Map of Barramundi, *Lates calcarifer*. Genetics. 2006, 175: 907-915. 10.1534/genetics.106.059972.

38. Pillay TVR, Kutty MN: Aquaculture: Principles and Practices. Technology & Engineering. Edited by: Pillay TVR, Kutty MN. 2005, 442-444.

39. Zhu ZY, Lin G, Lo LC, Xu YX, Feng F, Chou R, Yue GH: Genetic analyses of Asian seabass stocks using novel polymorphic microsatellites. Aquaculture. 2006, 256: 167-173. 10.1016/j.aquaculture.2006.02.033.

40. Xu YX, Zhu ZY, Lo LC, Wang CM, Lin G, Feng F, Yue GH: Characterization of two parvalbumin genes and their association with growth traits in Asian seabass (*Lates calcarifer*). Anim Genet. 2006, 37: 266-268. 10.1111/j.1365-2052.2006.01423.x.

41. Wang CM, Lo LC, Feng F, Gong P, Li J, Zhu ZY, Lin L, Yue GH: Construction of a BAC library and mapping BAC clones to the linkage map of Barramundi, *Lates calcarifer*. BMC Genom. 2008, 9: 139-152. 10.1186/1471-2164-9-139.

42. Cheong I, Chou R, Mec CT: Quarantine and Fish Diseases in Southeast Asia. Report of Workshop. 1983, Jakarta, Indonesia: UNDP, FAO, IDRC

43. Chinabut S, Danayadol Y: Quarantine and Fish Diseases in Southeast Asia. Report of Workshop. 1983, Jakarta, Indonesia: UNDP, FAO, IDRC

44. Carson J, Schmidtke LM, Munday BL: *Cytophaga johnsoniae*:a putative skin pathogen of juvenile farmed barramundi,*Lates calcarifer* Bloch. J Fish Dis. 1993, 16: 209-218. 10.1111/j.1365-2761.1993.tb01250.x.

45. Bromage EG, Thomas A, Owens L: *Streptococcus iniae*, a bacterial infection in barramundi *Lates calcarifer*. Dis Aquat Org. 1999, 36: 177-181. 10.3354/dao036177.

46. Crosbie PBB, Nowak BF: Immune responses of barramundi, *Lates calcarifer* (Bloch), after administration of an experimental *Vibrio harveyi* bacterin by intraperitoneal injection, anal intubation and immersion. J Fish Dis. 2004, 27 (11): 623-632. 10.1111/j.1365-2761.2004.00575.x.

47. Fearon DT, Locksley RM: Elements of immunity -The instructive role of innate immunity in the acquired immune response. Science. 1996, 272: 50-54. 10.1126/science.272.5258.50.

48. Medzhitov R, Janeway CA: Innate immunity: Impact on the adaptive immune response. Curr Opin Immuno. 1997, 19: 4-9. 10.1016/S0952-7915(97)80152-5.

49. Watzke J, Schirmer K, Scholz S: Bacterial lipopolysaccharides induce genes involved in the innate immune response in embryos of the zebrafish (*Danio rerio*). Fish Shellfish Immun. 2007, 23: 901-905. 10.1016/j.fsi.2007.03.004.

50. Dan X, Wei QQ: Singapore grouper iridorirus (SGIV) infected grouper revealed by suppression subtractive hybridization-reverse Library. dbEST of GenBank database. 2009

51. Pardo BG, Fernández C, Millán A, Bouza C, Vázquez-López A, Vera M, Alvarez-Dios JA, Calaza M, Gómez-Tato A, Vázquez M: Expressed sequence tags (ESTs) from immune tissues of turbot (*Scophthalmus maximus*) challenged with pathogens. BMC Vet Res. 2008, 4: 37-48. 10.1186/1746-6148-4-37.

52. Carbon S, Ireland A, Mungall CJ, Shu S, Marshall B, Lewis S: AmiGO Hub, Web Presence Working Group. AmiGO: online access to ontology and annotation data. Bioinformatics. 2009, 25 (2): 288-289. 10.1093/bioinformatics/btn615.

53. InnateDB. [http://www.innatedb.com/resources.jsp]

54. Adams MD, Kelley JM, Gocayne JD, Dubnick M, Polymeropoulos MH, Xiao H: Complementary DNA sequencing: expressed sequence tags and human genome project. Science. 1991, 252: 1651-1656. 10.1126/science.2047873.

55. NCBI. [http://www.ncbi.nlm.nih.gov/]

56. Tan SL, Mohd-Adnan A, Mohd-Yusof NY, Forstner MR, Wan KL: Identification and analysis of a prepro-chicken gonadotropin releasing hormone II (preprocGnRH-II) precursor in the Asian seabass, *Lates calcarifer*, based on an EST-based assessment of its brain transcriptome. Gene. 2008, 411 (1-2): 77-86. 10.1016/j.gene.2008.01.008.

57. Hildemann WH, Reddy AL: Phytogeny of immune responsiveness: Marine invertebrates. Fed Proc. 1973, 32: 2188-2194.

58. Warr GW: The immunoglobulin genes of fish. Dev Comp Immunol. 1995, 19: 1-12. 10.1016/0145-305X(94)00052-H.

59. Plouffe DA, Hanington PC, Walsh JG, Wilson EC, Belosevic M: Comparison of select innate immune mechanisms of fish and mammals. Xenotransplantation. 2005, 12: 266-277. 10.1111/j.1399-3089.2005.00227.x.

60. Randelli E, Buonocore F, Scapigliati G: Cell markers and determinants in fish immunology. Fish Shellfish Immun. 2008, 25: 326-340. 10.1016/j.fsi.2008.03.019.

61. Chen SL, Xu M, Ji X, Yu G, Liu Y: Cloning, Characterization, and Expression Analysis of Hepcidin Gene from Red Sea Bream (*Chrysophrys major*). Antimicrob Agents Chemother. 2005, 49 (4): 1608-1612. 10.1128/AAC.49.4.1608-1612.2005.

62. Kuroda N, Wada H, Naruse K, Simada A, Shima A, Sasaki M, Nonaka M: Molecular cloning and linkage analysis of the Japanese medaka fish complement Bf/C2 gene. Immunogenetics. 1996, 44 (6): 459-467. 10.1007/BF02602808.

63. Wang T, Secombes CJ: Complete sequencing and expression of three complement components, C1r, C4 and C1 inhibitor, of the classical activation pathway of the complement system in rainbow trout *Oncorhynchus mykiss*. Immunogenetics. 2003, 5 (9): 615-628. 10.1007/s00251-003-0622-5.

64. Chondrou M, Papanastasiou AD, Spyroulias GA, Zarkadis IK: Three isoforms of complement properdin factor P in trout: Cloning, expression, gene organization and constrained modeling. Dev Comp Immunol. 2008, 32 (12): 1454-1466. 10.1016/j.dci.2008.06.010.

65. Zhang H, Nichols K, Thorgaard GH, Ristow SS: Identification, mapping, and genomic structural analysis of an immunoreceptor tyrosine-based inhibition motif-bearing C-type lectin from homozygous clones of rainbow trout (*Oncorhynchus mykiss*). Immunogenetics. 2001, 53: 751-759. 10.1007/s00251-001-0386-8.

66. Zhang H, Robison B, Thorgaard GH, Ristow SS: Cloning, mapping and genomic organization of a fish C-type lectin gene from homozygous clones of rainbow trout (*Oncorhynchus mykiss*). Biochim Biophys Acta. 2000, 1494 (1514): 14-22.

67. Sato A, Mayer WE, Overath P, Klein J: Genes encoding putative natural killer cell C-type lectin receptors in teleostean fishes. Proc Natl Acad Sci USA. 2003, 13: 7779-7784. 10.1073/pnas.1235938100.

68. Lemaitre B, Nicolas E, Michaut L, Reichhart JM, Hoffmann JA: The dorsoventral regulatory gene cassette spatzle/toll/cactus controls the potent antifungal response in Drosophila adults. Cell. 1996, 86: 973-983. 10.1016/S0092-8674(00)80172-5.

69. Van Muiswinkel WB: A history of fish immunology and vaccination I. The early days. Fish Shellfish Immun. 2008, 25: 397-408. 10.1016/j.fsi.2008.02.019.

70. Stockhammer OW, Zakrzewska A, Hegedu^s Z, Spaink HP, Meijer AH: Transcriptome profiling and functional analyses of the zebrafish embryonic innate immune response to Salmonella infection. J Immunol. 2009, 182: 5641-5653. 10.4049/jimmunol.0900082.

71. Tenger V, Zhou X: Apolipoprotein E modulates immune activation by acting on the antigen-presenting cell. Immunology. 2003, 109 (3): 392-397. 10.1046/j.1365-2567.2003.01665.x.

72. Long Q, Huang H, Shafizadeh E, Liu N, Lin S: Stimulation of erythropoiesis by inhibiting a new hematopoietic death receptor in transgenic zebrafish. Nat Cell Biol. 2000, 2: 549-552. 10.1038/35019592.

73. Fas SC, Fritzsching B, Suri-Payer E, Krammer PH: Death receptor signaling and its function in the immune system. Curr Dir Autoimmun. 2006, 9: 1-17.

74. Liu L, Zeng M, Stamler JS: Hemoglobin induction in mouse macrophages. Proc Natl Acad Sci USA. 1999, 96 (12): 6643-6647. 10.1073/pnas.96.12.6643.

75. Yang H, Wang H, Bernik TR, Ivanova S, Wang H, Ulloa L, Roth J, Eaton JW, Tracey KJ: Globin attenuates the innate immune response to endotoxin. Shock. 2002, 17 (6): 485-490. 10.1097/00024382-200206000-00008.

76. Poss KD, Tonegawa S: Reduced stress defense in heme oxygenase 1-deficientcells. Proc Natl Acad Sci USA. 1997, 94 (20): 10925-10930. 10.1073/pnas.94.20.10925.

77. Nairz M, Theurl I, Ludwiczek S, Theurl M, Mair SM, Fritsche G, Weiss G: The co-ordinated regulation of iron homeostasis in murine macrophages limits the availability of iron for intracellular *Salmonella typhimurium*. Cell Microbiol. 2007, 9 (9): 2126-2140. 10.1111/j.1462-5822.2007.00942.x.

78. Skugor S, Glover KA, Nilsen F, Krasnov A: Local and systemic gene expression responses of Atlantic salmon (*Salmo salar*L.) to infection with the salmon louse (*Lepeophtheirus salmonis*). BMC Genom. 2008, 9: 498-515. 10.1186/1471-2164-9-498.

79. Mulero I, Sepulcre MP, Roca FJ, Meseguer J, García-Ayala A, Mulero V: Characterization of macrophages from the bony fish gilthead seabream using an antibody against the macrophage colony-stimulating factor receptor. Dev Comp Immunol. 2008, 32 (10): 1151-1159. 10.1016/j.dci.2008.03.005.

80. Chistiakov DA, Hellemans B, Volckaert FAM: Microsatellites and their genomic distribution, evolution, function and applications: A review with special reference to fish genetics. Aquaculture. 2006, 255: 1-29. 10.1016/j.aquaculture.2005.11.031.

81. Liu ZJ, Cordes JF: DNA marker technologies and their applications in aquaculture genetics. Aquaculture. 2004, 238 (1-4): 1-37. 10.1016/j.aquaculture.2004.05.027.

82. Yue GH, Zhu ZY, Lo LC, Wang CM, Lin G, Feng F, Pang HY, Li J, Gong P, Liu HM: Genetic variation and population structure of Asian seabass (*Lates calcarifer*) in the Asia-Pacific region. Aquaculture. 2009, 293: 22-28. 10.1016/j.aquaculture.2009.03.053.

83. Slate J, Hale MC, Birkhead TR: Simple sequence repeats in zebra finch (*Taeniopygia guttata*) expressed sequence tags: a new resource for evolutionary genetic studies of passerines. BMC Genom. 2007, 8: 52-63. 10.1186/1471-2164-8-52.

84. Hara M, Sekino M: Efficient detection of parentage in a cultured Japanese flounder *Paralichthys olivaceus* using microsatellite DNA marker. Aquaculture. 2003, 217 (1-4): 107-114. 10.1016/S0044-8486(02)00069-8.

85. Yue GH, Li Y, Chao TM, Chou R, Orban L: Novel microsatellites from Asian sea bass (*Lates calcarifer*) and their application to broodstock analysis. Mar Biotechnol. 2002, 4 (5): 503-511.

86. Yue GH, Ho MY, Orban L, Komen J: Microsatellites within genes and ESTs of common carp and their applicability in silver crucian carp. Aquaculture. 2004, 234 (1-4): 85-98. 10.1016/j.aquaculture.2003.12.021.

87. Luro FL, Costantino G, Terol J, Argout X, Allario T, Wincker P, Talon M, Ollitrault P, Morillon R: Transferability of the EST-SSRs developed on Nules clementine (*Citrus clementina* Hort ex Tan) to other Citrus species and their effectiveness for genetic mapping. BMC Genom. 2008, 9: 287-299. 10.1186/1471-2164-9-287.

88. KAAS - KEGG Automatic Annotation Server. [http://www.genome.jp/tools/kaas/]

89. Antimicrobial Peptide Database. [http://aps.unmc.edu/AP/about.php]

90. Benson G: Tandem repeats finder: a program to analyze DNA sequences. Dev Comp Immunol. 1999, 27 (2): 573-580.

91. Romualdi C, Bortoluzzi S, d'Alessi F, Danieli GA: IDEG6: a web tool for detection of differentially expressed genes in multiple tag sampling experiments. Physiol Genom. 2003, 12: 159-162.

92. Independent groups T-TEST for means calculator. [http://www.dimensionresearch.com/resources/calculators/ttest.html]

93. Olsvik PA, Lie KK, Jordal AO, Nilsen TO, Hordvik I: Evaluation of potential reference genes in real-time RT-PCR studies of Atlantic salmon. BMC Mol Biol. 2005, 6: 21-29. 10.1186/1471-2199-6-21.

94. Jorgensen SM, Kleveland EJ, Grimholt U, Gjoen T: Validation of reference genes for real-time polymerase chain reaction studies in Atlantic salmon. Mar Biotechnol. 2006, 8 (4): 398-408. 10.1007/s10126-005-5164-4.

CHAPTER 11 PROFILING OF INFECTION SPECIFIC MRNA TRANSCRIPTS OF THE EUROPEAN SEABASS DICENTRARCHUS LABRAX

Elena Sarropoulou, Pilar Sepulcre, Laura Poisa-Beiro, Victoriano Mulero, José Meseguer, Antonio Figueras, Beatriz Novoa, Vasso Terzoglou, Richard Reinhardt, Antonios Magoulas and Georgios Kotoulas

BACKGROUND

The European seabass *Dicentrarchus labrax* is one of the most extensively aquacultured fish species in the Mediterranean, resulting in steadily increasing pressure on producers. Consequently, it is important to acquire new techniques and knowledge in order to improve aquaculture practices. Detailed information concerning growth, health, disease resistance and flesh quality benefit from the molecular as well as from the physiological point of view can provide illuminating new findings leading to improved aquaculture techniques. Several efforts have been made up till now to enrich the genomic resources in aquaculture production in the Mediterranean (chiefly for the gilthead sea bream *Sparus aurata* and for the European seabass *Dicentrarchus labrax*), e.g. Marine Genomics Europe (Network of Excellence) (CT-2003-505403), [1, 2, 3, 4, 5] as well as in the Atlantic (e.g. Atlantic halibut *Hippoglossus hippoglossus*, Salmon *Salmo salar*) e.g. [6, 7, 8, 9, 10, 11, 12, 13]. These studies focused mainly on non-challenged tissues in order to obtain a first unigene catalogue. Aquaculture production however is affected by viral and pathogenic bacteria, particularly in respect of *D. labrax* which has been shown to be the species most sensitive to pathogenic bacteria such as *Vibrio anguillarum* [14] and to viral infections such as Nodavirus [15, 16]. There are several commercial vaccines which provide protection against infection from *V. anguillarum* although the mechanism of immune response still remains unknown. Nodavirus can cause massive mortalities [17] and cannot be controlled so far because the production of commercial vaccines here is still in its infancy. In the present study we have generated a collection of EST sequences from tissues of European seabass infected with *V. anguillarum* and Nodavirus. Within this collection we were able to isolate immune relevant genes, and have gone on to compare gene expression in different tissues after viral and pathogenic bacteria infection. Additionally we determined *in silico* differential expression between the two infections. In this context the construction and analysis of a total of ten cDNA libraries are described; six cDNA libraries were from tissues of the European seabass infected with *V. anguillarum* (liver, spleen, head kidney, peritoneal exudate, gill and intestine) with peritoneal exudate, gill and intestine as target organs for *V. anguillarum* infections, and four cDNA libraries were from tissues of the European seabass infected with Nodavirus (liver, spleen, head kidney and brain) with the brain as target organ of the virus. Comparisons between the predicted European seabass peptide data set and the zebrafish, medaka, stickleback, tetraodon and human proteomes were performed. Genes showing *in silico* differential expression between Nodavirus infection and *V. anguillarum* infection were further analysed by real-time PCR.

RESULTS

Summary of ESTs from the cDNA libraries infected with Nodavirus and V. anguillarum

The amplified libraries contained insert size from approximately 0.5 to 2.0 kb. Single pass sequencing was performed resulting

Citation: Elena Sarropoulou, Pilar Sepulcre, Laura Poisa-Beiro, Victoriano Mulero, José Meseguer, Antonio Figueras, Beatriz Novoa, Vasso Terzoglou, Richard Reinhardt, Antonios Magoulas and Georgios Kotoulas, Profiling of infection specific mRNA transcripts of the European seabass Dicentrarchus labrax, doi: 10.1186/1471-2164-10-157.

in 9605 high quality sequences. All sequences were submitted to the EST database (dbESThttp://www.ncbi.nlm.nih.gov/projects/dbEST/ with the accession numbers FK939975 – FK944381, FL484477 – FL488763 and FL501471 – FL502381. A set of 3075 unique sequences was generated. Among the unique sequences (3075) [see Additional file 1] 371 [12%, see Additional file 2] sequences contained Simple Sequence Repeats (SSR). Cluster analyses performed for each library separately (Table 1a and Table 1b) revealed redundancy rates which varied from 72% (28% unique sequences) in intestine cDNA library infected with *V. anguillarum* to 37% (63% unique sequences) in spleen cDNA library infected with *V. anguillarum* (Table 1a). The set of unique EST sequences was annotated with Blast2GO which carries out BLASTX searches and attempts to assign function and GO classification. Out of the 3075 unique EST sequences submitted to GO2Blast for annotation and GO classification, 1521 sequences fell into 14 categories of biological process function at GO annotation level 2 (Fig. 1), where two categories, cellular process and metabolic process, were predominant. The category "immune system process" was represented by 79 transcripts.

Table 1: Summary of sequences derived of cDNA libraries of *D. labrax* tissue infected with V. anguillarum (a) and Nodavirus (b)

A					
Tissue	**singletons**	**contigs**	**unique**	**total sequences**	**% of unique sequences**
Liver	503	190	693	2140	32.38
Spleen	326	38	364	651	63.14
Kidney	266	66	332	911	36.24
Peritoneal exudate	386	103	489	827	59.13
Gill	92	15	107	343	31.19
Intestine	88	4	92	326	28.22
B					
Tissue	**singletons**	**contigs**	**unique**	**total sequences**	**% of unique sequences**
Liver	253	127	380	1126	33.75
Spleen	698	118	816	1284	63.55
Brain	617	41	658	1099	59.87
Kidney	321	55	376	934	40.26

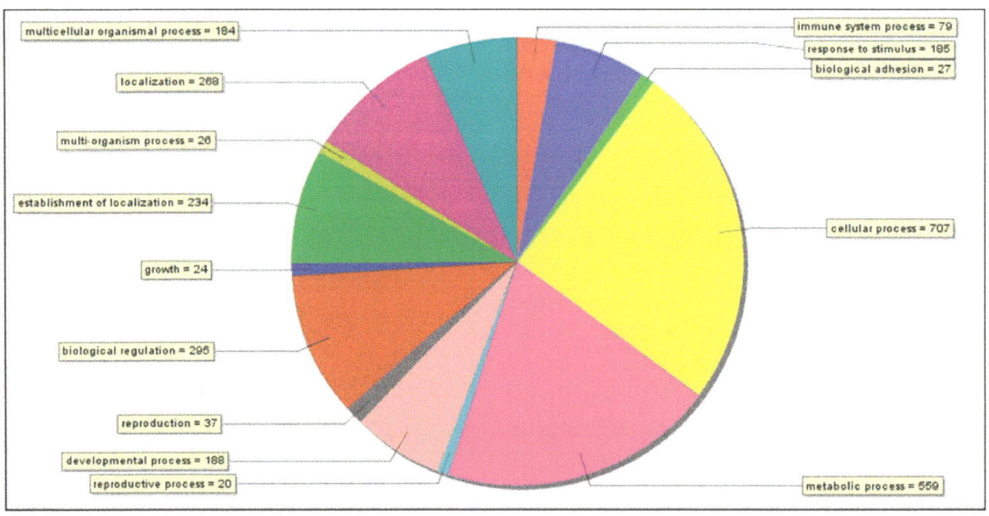

Figure 1: Summary of GO category Biological process of unique ESTs obtained from seabass cDNA libraries infected with Nodavirus and *Vibrio anguillarum.*

EST matches with known function

Out of the 3075 EST sequences, 1246 (~ 41%) had a positive hit after submission to BLASTX database search. Among those EST sequences with a known function, 128 homologues were found to be involved in the immune response and 79 of these were grouped into the GO category "immune system process". The remaining 49 transcripts were manually determined to be involved in the immune response (see Additional file 3). Immune related transcripts isolated for the first time for seabass amounted to 115 (Table 2). Among transcripts of interest, the transcript encoding for an important antimicrobial protein, hepicidin, was isolated. Aligning EST sequences grouped into one contig can provide additional data. In the case of hepcidin it is probable that different isoforms are grouped together. Alignments of other cDNA sequences either showed alternative polyadenylation or they showed *in silico* polymorphism of microsatellite DNA as for instance the transcript coding for cysteine-rich protein 1-I (see Additional file 4).

Table 2: Transcripts isolated for the first time in *D. labrax* and grouped to the GO category" immune system process"

Contig ID	BLASTX Hit	Accession number
Contig_265, Contig_3	alpha globin	Q9PVM4 BAA86218
Contig_681	alpha-1-microglobulin bikunin precursor	CAA45294
Contig_3061	b-cell leukemia lymphoma 6	XP_001340785
Contig_1838	bcl2 adenovirus e1b 19 kda interacting protein 3	NP_001012245, AAR83676
Contig_276	beta-2 microglobulin	ABB60035, ABB60037
Contig_773	beta-2 microglobulin precursor	AAC64994
Contig_2731	blood thirsty	AAX12162
Contig_2210	blood thirsty	XP_699830
Contig_1617	c1 inhibitor	NP_001117851, CAD58653
Contig_936	cathepsin s	AAQ01147
Contig_2779	cc chemokine	AAY79324
Contig_616	ccaat enhancer binding protein (c ebp) beta	BAB40971
Contig_1810	cd59-like protein 2	NP_001117969, AAT94063
Contig_1441	cell division cycle 42	NP_956159, AAH48035, AAH75761, AAX20139, CAM56524, AAI64988
Contig_2676	chemokine (c-c motif) ligand 13	BAC20610
Contig_2058	chemokine (c-c motif) ligand 21b	AAT52146, ABA54959
Contig_241	chemokine (c-c motif) ligand 25	ABC69050
Contig_2858	chemokine (c-x-c motif) ligand 12b (stromal cell-derived factor 1)	NP_840092, AAN64414, AAS92649, AAI09418
Contig_524	chemokine (c-x-c motif) ligand 9	ABC69049
Contig_627	chemokine (c-x-c motif) receptor 4	ABP48751
Contig_525, Contig_1672	chemokine cxc-like protein	ABC69049
Contig_983	complement c4-2	CAD45003
Contig_2306	complement component 1 q subcomponent	ABV57766
Contig_513	complement component 1 qb chain	XP_001110783
Contig_986	complement component 5	BAC23058
Contig_1044	complement component 7	BAA88899
Contig_2558	complement component alpha polypeptide	NP_001118096, CAH6548
Contig_1413	complement component c3	BAA88901
Contig_1114	complement component c4	CAD45003
Contig_2843	complement component c5-1	BAC23057
Contig_2534, Contig_2600	complement component c9	P79755, AAC60288
Contig_1499	complement component factor h	NP_001117882, CAF25505
Contig_1496	complement component gamma polypeptide	NP_001117880, CAF22027
Contig_927	complement component1, q gamma polypeptide	XP_544508
Contig_2432	complement component beta subunit	Q9PVW7, BAA86877
Contig_2605	complement component q subcomponent binding protein	EDM05067

Contig_1536	complement component r subcomponent	AAR20889
Contig_868	complement factor b	CAD21938
Contig_1607	complement factor d preproprotein	XP_001117186
Contig_2013	complement factor h	NP_001117876, CAF05664, CAF05665
Contig_1481	complement factor h-related 1	AAA92556
Contig_1375	cornichon homolog	O35372, AAC15828
Contig_1198	c-reactive protein	NP_999009, O19062 BAA21473
Contig_1657	c-type lectin	BAE45333
Contig_1596	cu zn superoxide dismutase	AAW29025
Contig_395	deah (asp-glu-ala-his) box polypeptide 16	NP_956318, AAH45393, AAI65206
Contig_2814	ets-1 transcript variant ets-1 delta(iii-vi)	AAY19514
Contig_626	ferritin heavy chain	NP_001117129, P49946, AAB34575
Contig_280	fth1 protein	CAL92185
Contig_2392	g-protein coupledGedg6	NP_001112363
Contig_2662	heat shock 10 kda protein 1 (chaperonin 10)	AAV37068
Contig_2975	heat shock 70 kda protein 4	AAH65970
Contig_2939	heme oxygenase1	ABL74501
Contig_275	hemoglobin alpha chain	CAP69820
Contig_731	Hephaestin	NP_579838
Contig_1713	hypoxanthine phosphoribosyltransferase 1	NP_001002056, AAH71336
Contig_1745, Contig_1876	integrin beta 2	BAB39130, NP_990582, CAA50671
Contig_2452	interleukin 18 receptor accessory protein	XP_001371334
Contig_2261	interleukin 1 type i	XP_416914
Contig_1718	interleukin 1 type ii	NP_001015713, AAH89644
Contig_1506	interleukin 2 receptor gamma chain	CAJ38407
Contig_1835	interleukin enhancer binding factor 3	AAH47175
Contig_966	interleukin-1 receptor type ii	ABP99035
Contig_852	interleukin-1 receptor type ii	CAL30143
Contig_2196	loc559360 protein	AAI51869
Contig_2044	macrophage migration inhibitory factor	ABG54276
Contig_1348	major histocompatibility complex class i a chain	BAD13369
Contig_17	mflj00348 protein	BAD90390
Contig_889	mhc class i alpha antigen	ABB04088
Contig_1349	mhc class i antigen	BAD13366
Contig_2797	mitochondrial ribosomal protein s18b	NP_001106610, AAI52129, AAI55448
Contig_3008	natural resistance-associated macrophage protein	AAG31225
Contig_576	neurofibromatosis 1	AAD15839
Contig_1716	novel protein vertebrate complement component 3	NP_001116778, CAQ13357
Contig_1327	otuubiquitin aldehyde binding 1	NP_001002500, AAH76301
Contig_2661	Proteasome activator subunit 1 (pa28 alpha)	ABE60902, ABK41199
Contig_1948	protein kinase alpha	AAI51472
Contig_2319	protein tyrosinereceptorc	XP_547374
Contig_2009	purinergic receptorg-protein13	XP_001516794

Contig_1715	rhamnose binding lectin	NP_001117668, BAA92256
Contig_934	ribosomal protein s19	P61155, AAP20214
Contig_2680	sam domain- and hd domain-containing protein 1	XP_001097562
Contig_684	serum amyloid p-component	P12246 AAA40093, CAA34774, AAH61125, AAY88178, BAE25796, BAE38344, EDL39002
Contig_2165	sffv proviral integration 1	NP_035485
Contig_1473	sh2 containing inositol-5-phosphatase	XP_687502
Contig_2675	skin mucus lectin	BAD90686
Contig_840	strawberry notch homolog 2	EDL31603
Contig_1540	tnf superfamily member 14	NP_001118039, ABC84585
Contig_469	transcription factor 3 isoform cra_b	NP_571169, CAA54305
Contig_2299	transforming growth beta receptor ii (70 80 kda)	XP_534237
Contig_2692	trypsin 10	BAF76146
Contig_1814	vascular endothelial growth factor	NP_001038320, AAY89335
Contig_1660	x-box binding protein 1	AAQ08005

Similarity Relationships

Figs. 2 and 3 show SimiTri representation of predicted seabass transcripts compared to *Danio rerio, Homo sapiens, Oncorhynchus mykiss, Gasterosteus aculateus* and *Tetraodon nigroviridis* proteomes. Of 3075 isolated unique transcripts 1040, 1051, 1122 1159, 1103 had Blast hits with a score > 50 against the *H. sapiens, D. rerio, O. mykiss, G. aculateus and T. nigroviridis* protein databases respectively.

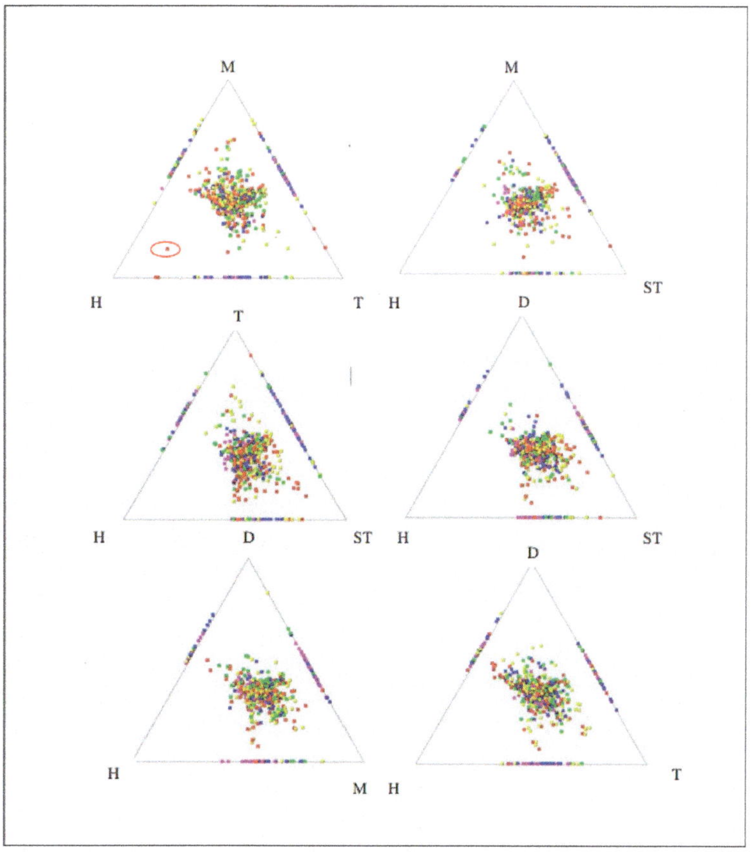

Figure 2: Similarity of *D. labrax* ESTs to the proteomes of *Homo sapiens* (H), *Danio rerio* (D), *Oryzias latipes* (M),*Gasterosteus aculateus* (ST) and *Tetraodon nigroviridis* (T). SimiTri plots show the graphical similarity i) between putative *D. labrax* peptides and *H. sapiens, O. latipes* and *T. nigroviridis*, cytochrome b is circled in red ii) between putative *D. labrax* peptides and *H. sapiens, O. latipes* and *G. aculateus*, iii) between putative *D. labra x* peptides and *H. sapiens, D. rerio* and *T. nigroviridis*, iv) between putative *D. labrax* peptides and *H. sapiens, D. rerio, G. aculateus*, v) between putative *D. labrax* peptides and *H. sapiens, D. rerio, O. latipes*, as well as vi) between putative *D. labrax*-peptides and *H. sapiens, D. rerio, T. nigroviridis*.

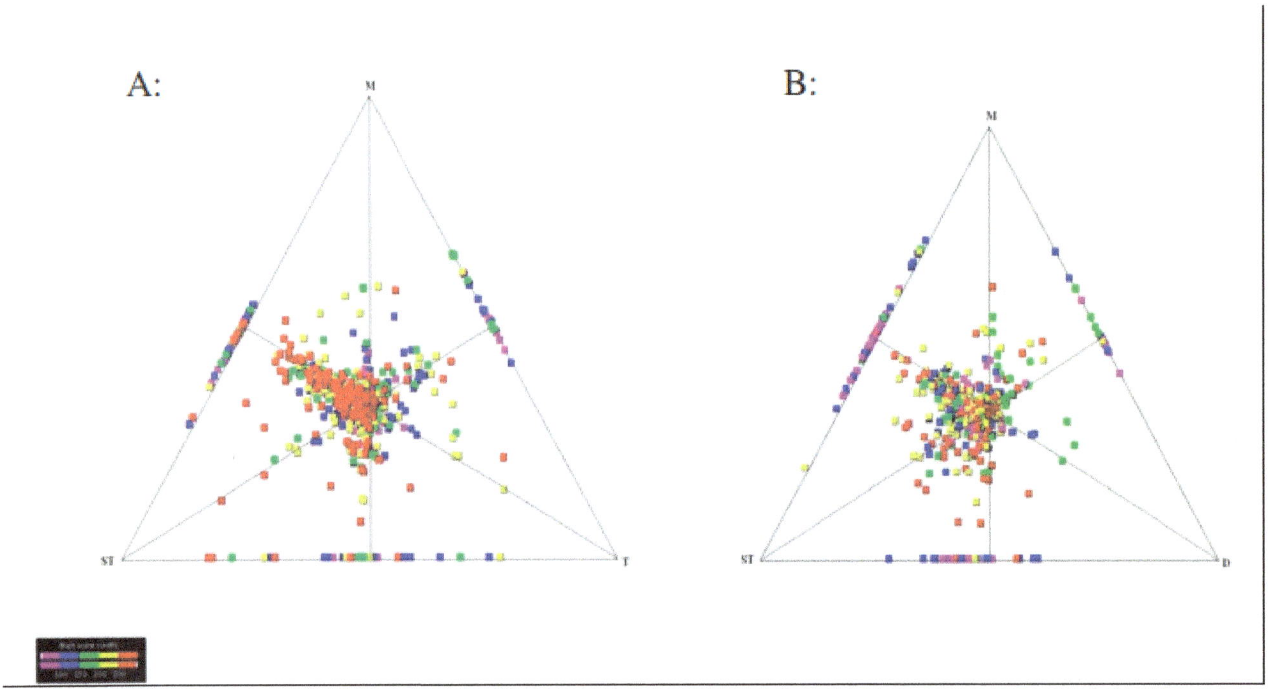

Figure 3: Similarity of *D. labrax* ESTs to the proteomes of *Danio rerio* (D), *Oryzias latipes* (M), *Gasterosteus aculateus* (ST) and *Tetraodon nigroviridis* (T). SimiTri plots show the graphical similarity A: between putative *D. labrax* peptides and *G. aculateus, O. latipes* and *T. nigroviridis*, B: between putative *D. labrax* peptides and *G. aculateus*, *O. latipes*and *D. rerio*.

Expression Analysis

The test to compare multiple cDNA libraries with each other [22] revealed that the genes with the value > 6 of the test statistic R can be confidently considered as genes with true variation with the slope of 1.081 and are therefore not significantly different from -1 at the 5% level (see Additional file 6). The hits with R > 6 are in total 109 out of 2234 contigs resulting from EST sequences of liver, spleen and head kidney infected with Nodavirus and *V. anguillarum*. The list of the 109 transcripts with R > 6 and their putative homologues are shown in Additional file 5. It is interesting to note that although most transcripts were abundantly expressed in both bacterial and viral infected tissues, not all of them could be considered as specific markers of a specific infection. For example, fructose-1,6-biphosphate aldolase A, hepcidin, apolipoprotein A1 precursor, ferritin heavy chain and chemokine receptor 4 transcripts were found in *V. anguillarum*-infected tissues, though rarely in Nodavirus-infected tissues (see Additional file 5). Conversely, fructose-1, 6-biphosphate aldolase B and 14 kDa apolipoprotein transcripts were frequently observed in Nodavirus infected tissues compared with *V. anguillarum*-infected tissues. The above results were further validated by determining the expression of putative markers for each infection in key tissues using real-time PCR. Here also control tissues were included in order to determine the expression of untreated fish. The real-time PCR confirmed the results obtained with the *in silico* analysis for selected genes. Taking into account the fold inductions of the real-time PCR experiments the correlations between *in silico* and qPCR are uniform. For instance the transcript for hepicidin precursor revealed *in silico* (R = 298.16) high expression only in liver tissues infected with *V. anguillarum*. The real-time PCR results show higher expression in all three tissues infected with *V. anguillarum*. However fold induction in liver is 20,000 times more than in spleen tissue, therefore theoretically 20,000 more cDNA clones had to be sequenced to obtain the sequence for hepicidin precursor in spleen infected with *V. anguillarum*. This correlation of high fold induction with *in silico* results can be observed for each transcript examined in this study. Thus, while the mRNA levels of hepcidin were found to increase considerably 24 h post-infection in the liver, spleen and head kidney of *V. anguillarum*-infected fish, they increased only slightly in Nodavirus-infected fish (Fig. 4). Notably, although the mRNA levels of transferrin and ferritin, both involved in iron metabolism with spleen and liver as the two main organs, increased in the liver after infection with both pathogens, they increased only in the spleen of *V. anguillarum*-infected animals (Figs. 5a and 5b).

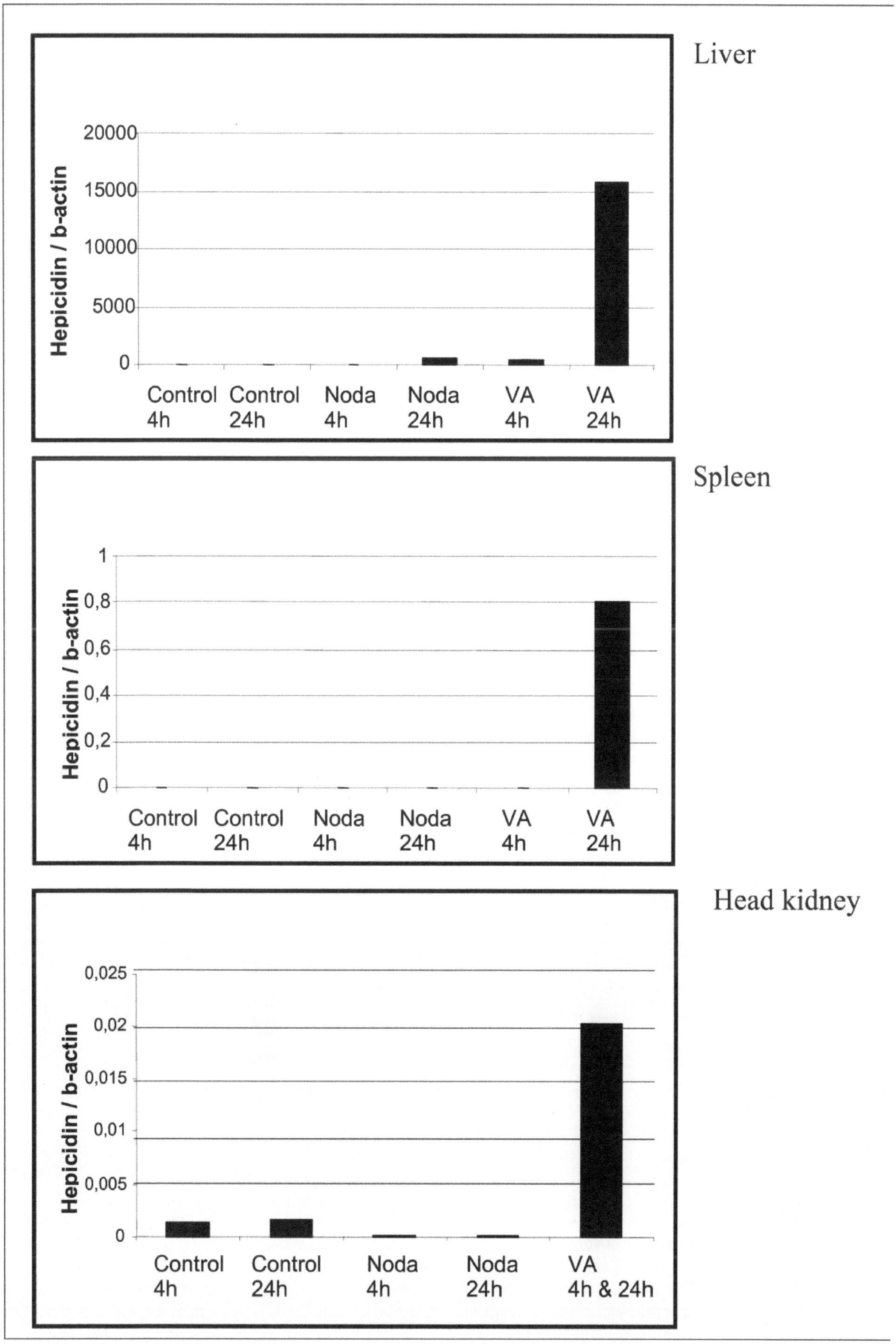

Figure 4: Hepicidin precursor expression index of liver, spleen, and head kidney infected with Nodavirus or with *V. anguillarum* for 4 h and 24 h. Infection for 4 h and 24 h of head kidney with *V. anguillarum* is pooled as not enough material was available. Each bar represents the mean of two technical duplicates of cDNA originating out of three individuals pooled prior to RNA extraction.

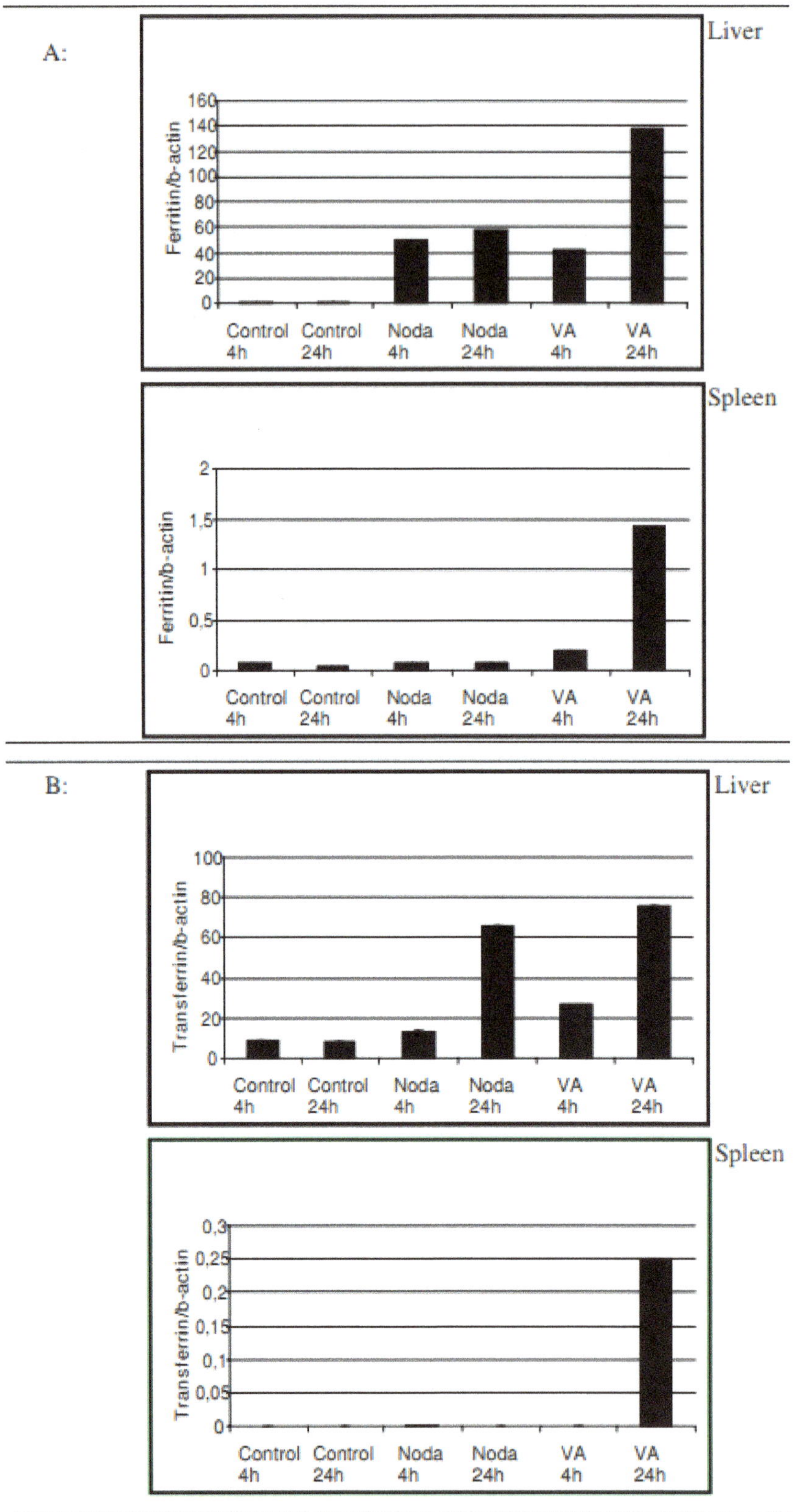

Figure 5: (A) Ferritin expression index of liver and spleen infected with Nodavirus or with *V. anguillarum* for 4 h and 24 h.(B) Transferrin expression index of liver and spleen infected with Nodavirus or with *V. anguillarum* for 4 h and 24 h. Each bar represents the mean of two technical duplicates of cDNA originating out of three individuals pooled prior to RNA extraction.

The mRNA levels of the chemokine receptor 4 were not affected or were slightly reduced in the head kidney and spleen of Nodavirus-infected fish but were considerably increased in these two tissues after *V. anguillarum* infection (Fig. 6). On the other hand, the mRNA levels of the 14 kDa apolipoprotein increased in the fish livers infected with both pathogens, but at 4 h and 24 h post-infection in the case of the Nodavirus and at 4 h post-infection in the case of *V. anguillarum* (Fig. 7). Here the expression in the liver is studied, as the liver is the major organ in the production of apoliprotein. Finally, although the mRNA levels of fructose-1,6-bisphosphate aldolase B decreased in the liver and head kidney following infection with both pathogens, they increased at 24 h post-infection in the spleen of *V. anguillarum*-infected fish (Fig. 8).

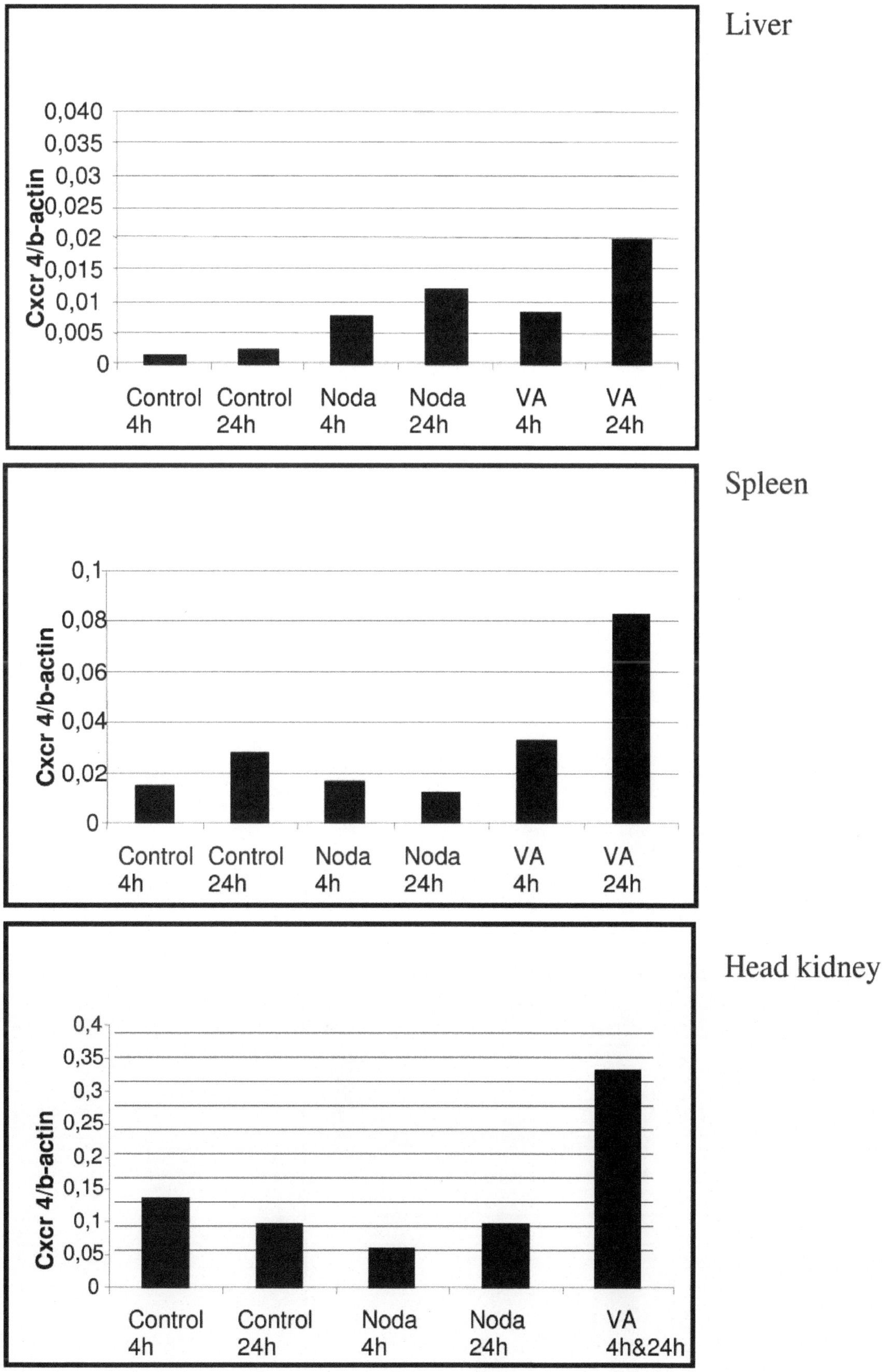

Figure 6: Chemokine (c-x-c motif) receptor 4 (Cxcr 4) expression index of liver, spleen, and head kidney infected with Nodavirus or with *V. anguillarum* for 4 h and 24 h. Infection for 4 h and 24 h of head kidney with *V. anguillarum* is pooled as not enough material was available. Each bar represents the mean of two technical duplicates of cDNA originating out of three individuals pooled prior to RNA extraction. Between the two technical replicates of Noda 4 h and of VA 24 h a greater variation was detected. The values of the two replicates of Noda 4 h are 0.014 and 0.00095 and the values of the two replicates of VA 24 h are 0.03 and 0.009.

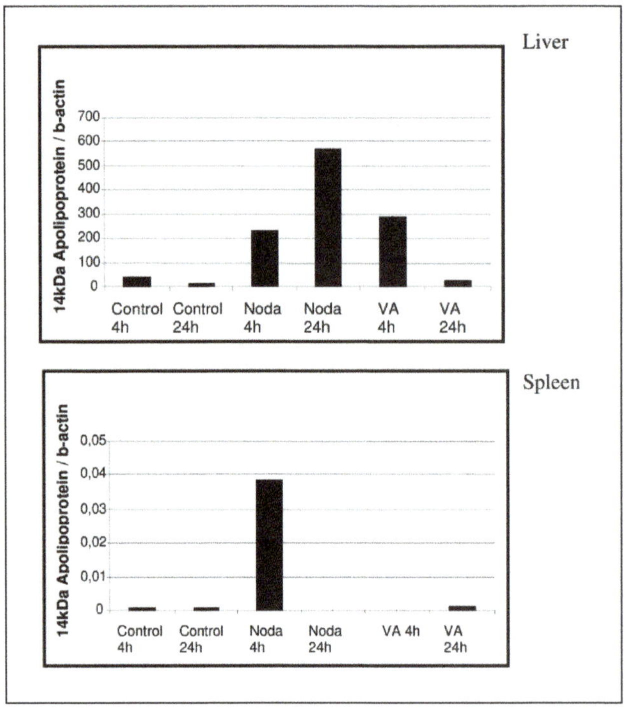

Figure 7: 14 KDa apolipoprotein expression index of liver infected with Nodavirus or with *V. anguillarum* for 4 h and 24 h. Each bar represents the mean of two technical duplicates of cDNA originating out of three individuals pooled prior to RNA extraction.

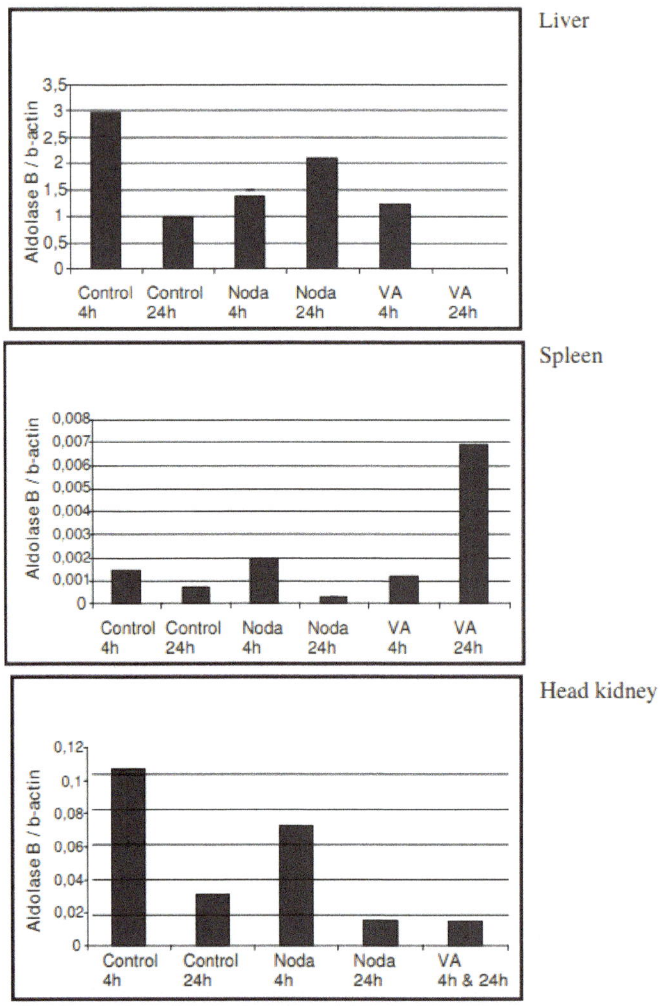

Figure 8: Fructose-1,6-bisphosphatase aldolase B expression index of liver, spleen, and head kidney infected with Nodavirus or with *V. anguillarum* for 4 h and 24 h. Infection for 4 h and 24 h of head kidney with *V. anguillarum* is pooled as not enough material was available. Each bar represents the mean of two technical of cDNA originating out of three individuals pooled prior to RNA extraction.

DISCUSSION

Although viral and bacterial infections are among the key challenges in fish aquaculture, nevertheless today the immune response of fish against *V. anguillarum* and Nodavirus remains largely unknown. Identification of genes involved in the immune response as well as the detection of differentially expressed genes between the two infections can make a significant contribution to future research leading to a better understanding of the biological system of immune response after fish infection. In the present study ten cDNA libraries, six from tissues infected with *V. anguillarum* and four from tissues infected with Nodavirus were analysed. Analysis of EST sequences coming from infected tissues will enhance the construction of an immune specific microarray chip containing already known transcripts involved in immune-related biological processes, such as the immune response as well as transcripts for which no annotation is available so far. Furthermore, transcripts indicating a higher expression level in one of the infections can be taken for future functional studies at RNA or DNA level as well as at protein level.

Over the past 30 years cDNA cloning for gene discovery and transcriptome analysis has become a very important molecular technique. Various techniques have been developed to address several scientific issues such as the cloning of rare transcripts, the construction of libraries with a wider cloning range, etc. (for review [26]). Construction of non-normalized libraries in the present study gave a first insight into the tissue-specific manner of transcript abundance according to their origin. Besides the possibility of identifying higher expressed transcripts, the percentages of unique sequences can also be assessed. In this study the redundancy of the cDNA libraries of liver, spleen and kidney infected with Nodavirus and *V. anguillarum* was in agreement with all three tissues (~33%, ~63% and ~38% respectively). This result is in line with other cDNA libraries of various fish species where the redundancy ranges between 40% and 60% depending on the tissues of origin [e.g. [12, 13]]. Besides the identification and characterization of ESTs for components of the immune system, detection of microsatellite sequences will help in the completion of quantitative trait locus (QTL) scans currently being performed. Microsatellite sequences, also called Simple Sequence Repeats (SSR), are frequent in non-coding regions and are used as molecular markers. Detection of SSR within ESTs (exonic microsatellites or EST-SSRs) presents a shortcut to obtaining microsatellite markers. Since EST-SSRs are exonic they have two advantages over intergenic microsatellites. First, it is expected that their flanking regions are more conserved, so that the primers can be used even in related species, and second, it is assumed that they are in strong linkage disequilibrium with functionally important sites. Therefore they are frequently used in population genomics or in mapping of genes of economic significance identified as candidate markers for QTL and/or quantitative trait nucleotide (QTN). For EST similarity search in the present study a homologue of a known gene is defined as a cDNA whose similarity to a gene of any other organism in the database exceeds a certain fixed threshold. The identification of orthologues is outside the scope of this study. In total 1246 (41%) were assigned to a known transcript, with 79 (6%) categorized to the GO category "immune system process". Separate examination of these 79 transcripts by GO annotation reveals their involvement in 11 other categories of biological function (Fig. 9), with three dominant categories of response to stimulus, cellular process, and biological regulation. This collection should provide the base material for further research into understanding the immune response of European seabass as well as for the isolation of putative biomarkers.

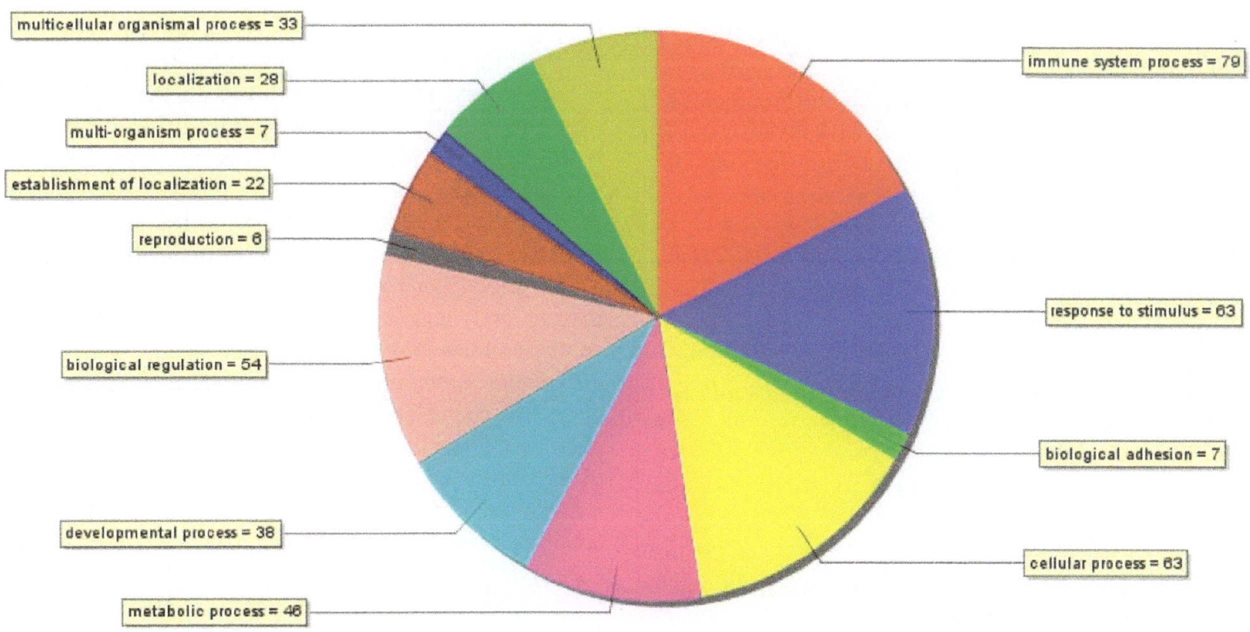

Figure 9: GO categorization of only the EST sequences grouped into the category of immune system process.

Similarity Relationships

Comparison of predicted seabass genes compared to the genomes of zebrafish, medaka, tetraodon, stickleback and human (Fig. 2) showed that the majority of putative proteins were located in the centre. From separate examination of the different triads a bias towards the top and right sections is revealed. This bias is not unexpected as seabass is more closely related to medaka, zebrafish, stickleback and tetraodon. However it is worth noting, that the seabass cytochrome b seems to be more similar to human cytochrome b than to the tetraodon and medaka cytochrome b as shown in Fig. 2. This was not the case with stickleback and zebrafish cytochrome b. Interestingly, results from comparisons of putative proteins of the Atlantic halibut (*Hippoglossus hippoglossus*) with the human, zebrafish and tetraodon protein database showed that the halibut cytochrome c oxidase subunit 3 (Cox3) is more similar to human COX3 than the zebrafish and tetraodon Cox3 [13]. Comparison of predicted proteins with only the protein database of fish genomes shows a slight bias towards medaka and stickleback looking at the triad medaka, stickleback and tetraodon (Fig. 3A) and again a slight bias towards medaka and stickleback looking at the triad stickleback, medaka and zebrafish (Fig. 3B). These results give a first insight towards the evolution of immune related genes as the relatively equal distribution indicate that sequence variation between the clade Percomorpha is comparable to that between the clade Percomorpha and Ostariophysi.

Expression Analysis

For *in silico* expression analysis transcript appearing more than once in the cDNA libraries were selected and their relative abundance were submitted to expression analysis after Stekel et al. [22]. Validation of *in silico*analysis was performed by qPCR. Individual variation may be masked in this approach as pooling strategy was chosen for qPCR experiments. The differential expressed transcripts detected in the present study can be further put forward for analysis of individual expression pattern. Nonetheless in order to study individual expression pattern the sampling frame has to be extended. In the present study the pooling strategy for qPCR was chosen in order to show cross-method consistency. However since results are consistent between the two approaches, influence of between individuals variability in response to infection has been addressed to some extent. In addition total RNA for qPCR analysis was extracted out of different individuals than the once used for cDNA library construction and patterns appear to be consistent between the different samples for all the selected candidate genes, which reflect the robustness of the approach and the small, if any bias, contributed by individual outliers. *In silico* expression analysis revealed a number of genes for $R > 6$ that are considerably above the exponential curve (see Additional file 6). Genes with $R > 6$ can be considered as significant and thus are candidate genes for further studies. Several of those transcripts including transcripts involved in iron metabolism such as ferritin and transferrin are also reported as differential expressed genes in the catfish *Ictalurus punctatus* and *Ictalurus furcatus* infected with the gram negative bacterium *Edwardsiella ictaluri* [27,28]. One of the main mechanisms whereby gram-negative bacteria pathogens like *V. anguillarum* obtain iron is the use of free heme or heme proteins from the host tissues [29]. The heme uptake mechanisms are considered to contribute to *V. anguillarum* virulence in fish [29]. However, it is surprising that Nodavirus infection also resulted in the up-regulation of transferrin and ferritin expression, especially within 24 h of infection. The abundance of transferrin transcripts in Nodavirus-infected tissues may not be related to the alteration of the iron metabolism by the pathogen but rather to the ability of enzymatically cleaved forms of this protein to activate fish macrophages [30]. The specific alteration of iron metabolism by *V. anguillarum* infection is also supported by the higher abundance of transcripts coding for hepcidin, a major homeostatic regulator of iron metabolism [31], and for α and β chains of hemoglobin in *V. anguillarum*- than Nodavirus-infected livers (255 clones vs. 1 clone, respectively). The expression of Hepicidin after bacterial infection has been shown in seabass [32] as well as in several other fish species like the striped bass [33], the red sea bream [34], the catfish [35, 36], the Atlantic halibut [37], the zebrafish [38], the rainbow trout [39] and the perch [40]. In this study the qPCR experiments confirmed the up-regulation of hepicidin in *D. labrax* after infection with *V. anguillarum* and showed in addition to this, that the expression of hepcidin might be considered as an excellent marker of bacterial infections, since it was up-regulated in all examined tissues of *V. anguillarum*-infected fish but unaffected in Nodavirus-infected tissues. Another interesting observation of the *in silico* gene expression analysis is the differential abundance of transcripts encoding the isoforms A and B glycolytic/gluconeogenic enzyme fructose-1,6-biphopshate aldolase in bacterial and viral infected tissues. Although the role played by this enzyme in the outcome of these infections is difficult to anticipate due to its dual role in glucose metabolism, these results suggest that the expression ratio between the two enzyme isoforms may be used as a good indicator of the type of infection in the European seabass. Thus, the up-regulation of the B isoform in the spleen exclusively by *V. anguillarum* might be considered another potential marker for this bacterial infection. Similarly, apolipoprotein A1 and 14 kDa apolipoprotein, two major components of high density lipoproteins (HDL) and synthesized in the fish liver [41], also show a differential expression in the liver of fish infected with *V. anguillarum* and Nodavirus following the time course and, therefore, they also may be good candidate indicators of the fish health status and/or the type of infection. The real-time PCR confirmed observations of *in silico* expression analysis and also revealed that the expression of the 14 kDa apolipoprotein and aldolase B in the spleen is an appropriate marker of Nodavirus and *V. anguillarum* infections, respectively. Previous studies in carp and medaka have also shown the involvement of apolipoproteins in the immune response [42, 43]. Finally, the differential expression of one of the clear immune-related genes, the chemokine receptor 4, was also found to be a good putative marker for *V. anguillarum* infection. For assessment of variability of putative markers further studies looking at individuals, exposed to other environmental or pathogenic conditions are needed to exclude possible biological variability caused by infections.

CONCLUSION

In this study we generated a collection of EST sequences from tissues of the European seabass infected with *V. anguillarum* and Nodavirus. We compared gene expression of different tissues after viral and pathogenic bacteria infection. A collection of 3075 unigenes was generated and candidate microsatellite sequences detected. Furthermore, comparisons of *D. labrax* transcripts with zebrafish, human, tetraodon, medaka and stickleback were performed. The majority of putative proteins were located in the centre with a bias towards the right sections, with *D. labrax* as expected being more closely related to the other fish species than to human. Comparison of putative *D. labrax* proteins was also performed among fish species. In this case a slight bias towards stickleback and medaka was observed when comparing medaka, stickleback and tetraodon and a slight bias towards stickleback and medaka was observed when comparing medaka, stickleback and zebrafish. Furthermore, *in silico* analysis of differential gene expression between the two infections based on EST sequences suggests a list of genes with a presumed function in the immune response of *D. labrax*revealing also the importance of looking at "non-classical" immune host proteins and emphasizing the significance of EST sequences generated from cDNA libraries of infected fish tissues. In addition, we show the power of sequencing cDNA sequences for expression analysis by performing real-time PCR experiments for transcripts with high, medium and low R-value. In view of new and high throughput sequence techniques detection of differential expression by measuring *in silico* the abundance of each transcript will enhance significantly the era of functional genomics. Furthermore *in silico* analysis in this study, followed by the confirmation with real-time PCR of potentially interested genes, has revealed some of them as potential biomarkers for bacterial and viral infections in fish.

METHODS

Experimental condition and tissues collection

Two infections, one with Nodavirus strain 475-9/99 isolated from diseased sea bass [from the Instituto Zooprofilattico Sperimentale delle Venezie (Italy) [16]] and one with *V. anguillarum* strain R-82 (serogoup 01) [from the University of Santiago (Spain) [14]] were performed with seabass as previously described [14, 16]. Tissues were taken 4 and 24 h post-infection. Three tissue types (spleen, liver and head kidney) of each experimental condition as well as peritoneal exudate, gill, intestine from *V. anguillarum* infection and brain from Nodavirus infection were selected and immediately frozen with liquid nitrogen. The experiments described comply with the guidelines of the European Union Council (86/609/EU) for the use of laboratory animals and have been approved by the Bioethical Committee of the University of Murcia (Spain) and the CSIC National Committee on Bioethics.

In brief; For Nodavirus infection fish were injected intramuscularly with 100 μl of nodavirus suspension in Minimum Essential Medium (MEM) (5.9×10^6 $TCID_{50}$ ml^{-1}) and placed at 25°C. Mock-infected control fish were injected with the medium alone, and maintained under the same experimental conditions. Three fish from each experimental and control groups were sampled 4 and 24 hours post-infection. Animals were sacrificed by anesthetic (MS-222) overdose and dissected. For the present study brain, spleen, head kidney and liver were sampled. For *V. anguillarum* infection fish were injected intraperitoneally (i.p.) with 1 ml of phosphate-buffered saline (PBS) alone or containing either 2×10^6 live or 10^8 formalin-killed *V. anguillarum* R82 cells (serogroup 01). Under these experimental conditions, about half of the fish were moribund at 24 h post-infection and all of them died within 48 h post-infection. Head kidney (bone marrow equivalent of fish) and peritoneal exudate cells were obtained 4 h and 24 h after bacterial challenge.

RNA extraction

Total RNA was extracted using the NucleoSplin RNA II extraction kit (Machinery Nagel, Dueren, Germany). RNA quality was checked on EtBr stained agarose gels and RNA concentrations and purity were measured using a NanoDrop spectrophotometer. For library construction equal amounts of total RNA extracted out of infected tissues (4 h and 24 h) were pooled. For qPCR experiments total RNA was freshly extracted out of infected tissues originating from three different individuals pooled prior to RNA extraction (liver, spleen and head kidney) with 4 h and 24 h post-infection.

cDNA library construction

All libraries were constructed from total RNA using the Creator SMART cDNA library construction kit (BD Bioscience-Clontech, Mountain View, Canada) using the LD PCR based method. Between 20 and 22 PCR cycles were performed before size separation of inserts. cDNA fragments > 600 bp were selected and directionally ligated at the restriction site Sfi1 of the pDNR-lib vector (BD Clontech) or the pal 32 vector. Plasmids were transformed into *E. coli* strain DH10B (Invitrogen) by electroporation. The libraries were tested for the presence and the size of insert by PCR using two primer pairs. For the libraries constructed with pal 32 vector, the primer pair pal 32 FOR: 5'-CTCGGGAAGCGCGCCATT-3' and pal 32, REV: 5'-TAATACGACTCACTATAGGGC-3' were used. For the libraries constructed with pDNR-lib vector pDNR FOR: 5'-TAAAACGACGGCCAGTA-3' pDNR REV: 5'-GAAACAGCTATGACCATGTTC-3' were used. The products were run on an EtBr stained agarose gel.

DNA sequencing

After plasmid preparation, dideoxy-temination DNA cycle sequencing was performed using the BigDye 3.1 sequencing methodandthepDNRFOR(5'-TAAAACGACGGCCAGTA-3') primer. The sequences were run on an ABI 3730 XL sequencer at MPI Molecular Genetics, Berlin.

Sequence analysis

The raw sequence reads were quality-trimmed and vector- and poly-A-clipped using PREGAP4 [18]. Clustering (grouping of clones related to one another by sequence homology) was performed using the software SeqManII (DNAstar Inc.). After clustering the term 'contig' is used to describe the sequence obtained from one cluster (the sequences of a cluster can be collapsed into a single, non-redundant sequence) and the term 'singleton' describes sequences appearing only once in the entire dataset. The set of sequences obtained by merging contigs and singletons are named as unique sequences.

Simple Sequence Repeats (SSR) in EST sequences

In silico mining for repeat motifs within the obtained unique sequences was perfomed with the programme Msatfinder http://www.genomics.ceh.ac.uk/msatfinder/[19].

Homology search and GO annotation

Gene Ontology (GO) category (Biological process) was assigned after BLASTX search of 3075 unique EST sequences using BLAST2GO. Threshold cutoff was at E-value $1e^{-3}$ and the alignment length of 33 amino acids (aa).

Similarity relationships

The unique sequences from all seabass libraries were submitted to BLASTX similarity searches [20] against the zebrafish, tetraodon, stickleback, medaka and human predicted proteomes (downloadable fromhttp://www.ensembl.org/index.html). For each database the highest BLAST scores (bit score values) in excess of 50 were retained. Relative similarities between triads were visualized as a triangular plot generated by the SimiTri software [21].

Expression analysis

In silico

All sequences of each cDNA library were submitted to BLASTX and BLASTN searches [20]. Transcripts appearing more than once in the cDNA libraries were selected for *in silico* expression analysis after Stekel *et al.* [22]. In brief, this method allows the comparison of gene expression in any number of libraries in order to identify differential expressed genes. The method uses a single statistical test to describe the extent to which a gene is differentially expressed between libraries by a log likelihood ratio statistic and tends asymptotically to a χ^2 distribution [22]. For real-time PCR experiments transcripts with high, medium and low R-value were selected.

Real-time PCR

Gene expression was assessed by real-time PCR (qPCR) in spleen, head kidney and liver at 4 h and 24 h post-infection. RNAs out of three animals pooled prior to RNA extraction were isolated as described above and were used to obtain cDNA by the Superscript II Reverse Transcriptase and oligo (dT)12–18 primer (Invitrogen) following the manufacturer's instructions. Quantitative PCR assays were performed using the 7300 real-time PCR System (Applied Biosystems) with specific primers (Table 3). Each primer (0.5 µl; 10 µM) and the cDNA template (1 µl) were mixed with 12.5 µl of SYBR green PCR master mix (Applied Biosystems) in a final volume of 25 µl. The standard cycling conditions were 95°C for 10 min. followed by 40 cycles of 95°C 15 s and 60°C 1 min. For all reactions two technical duplicates were performed. The comparative CT method (2-ΔΔ CT method) was used to determine the expression level of analysed genes [23]. After evaluation of β-actin as a suitable reference gene for this study in seabass (data not shown) the expression of the candidate genes was normalized. The use of β-actin as a suitable reference gene was also shown in other fish studies [e.g. [24, 25]].

Table 3: Real-time primer sequences

Name	F/R	Sequence 5'-3'
β-actin	Forward	GTGCGTGACATCAAGGAGAA
β-actin	Reverse	GCTGGAAGGTGGACAGAGAG
Apoliprot	Forward	ATACGTCCTGGCACTGATCC
Apoliprot	Reverse	AGCCTGACCTTGCTCACTGT
Chemokin-R4	Forward	TCAAAACGATGACGGACAAG
Chemokin-R4	Reverse	ACACGCTGCTGTACAGGTTG
Transferrin	Forward	CTGGGAAGTGTGGTCTGGTT
Transferrin	Reverse	CAAGACCTCTTGCCCTTCAG
Ferritin_HC	Forward	ATGCACAAGCTCTGCTCTGA
Ferritin_HC	Reverse	TTTGCCCAGGGTGTGTTTAT
Hepcidin-Prec	Forward	CCAGTCACTGAGGTGCAAGA
Hepcidin-Prec	Reverse	TCAGAACCTGCAGCAGACAC
Aldolase-B	Forward	TGACATTGCTCAGAGGATCG
Aldolase-B	Reverse	AGTTGGACATGGAGGGACTG

DECLARATIONS

Acknowledgements

The authors would like to thank Margaret Eleftheriou for carefully proofreading the manuscript. This work was supported by the European Commission's 5th Framework Programme WEALTH (Contract No. 501984, Welfare and health in sustainable aquaculture [WEALTH]).

REFERENCES

1. Chistiakov DA, Hellemans B, Haley CS, Law AS, Tsigenopoulos CS, Kotoulas G, Bertotto D, Libertini A, Volckaert FA: A microsatellite linkage map of the European sea bass Dicentrarchus labrax L. Genetics. 2005, 170: 1821-1826. 10.1534/genetics.104.039719.
2. Chistiakov DA, Hellemans B, Tsigenopoulos CS, Law AS, Bartley N, Bertotto D, Libertini A, Kotoulas G, Haley CS, Volckaert FA: Development and linkage relationships for new microsatellite markers of the sea bass (*Dicentrarchus labrax* L.). Anim Genet. 2004, 35: 53-57. 10.1046/j.1365-2052.2003.01076.x.
3. Chini V, Rimoldi S, Terova G, Saroglia M, Rossi F, Bernardini G, Gornati R: EST-based identification of genes expressed in the liver of adult seabass (*Dicentrarchus labrax*, L.). Gene. 2006, 376: 102-106. 10.1016/j.gene.2006.02.010.
4. Sarropoulou E, Power DM, Magoulas A, Geisler R, Kotoulas G: Comparative analysis and characterization of expressed sequence tags in gilthead sea bream (*Sparus aurata*) liver and embryos. Aquaculture. 2005, 243: 69-81. 10.1016/j.aquaculture.2004.10.012.
5. Sarropoulou E, Franch R, Louro B, Power DM, Bargelloni L, Magoulas A, Senger F, Tsalavouta M, Patarnello T, Galibert F, Kotoulas G, Geisler R: A gene-based radiation hybrid map of the gilthead sea bream *Sparus aurata* refines and exploits conserved synteny with Tetraodon nigroviridis. BMC Genomics. 2007, 8: 44-10.1186/1471-2164-8-44.
6. Martin SA, Caplice NC, Davey GC, Powell R: EST-based identification of genes expressed in the liver of adult Atlantic salmon (*Salmo salar*). Biochem Biophys Res Commun. 2002, 293: 578-585. 10.1016/S0006-291X(02)00263-2.
7. Haugland O, Torgersen J, Syed M, Evensen O: Expression profiles of inflammatory and immune-related genes in Atlantic salmon (*Salmo salar* L.) at early time post vaccination. Vaccine. 2005, 23: 5488-5499. 10.1016/j.vaccine.2005.07.034.
8. Adzhubei AA, Vlasova AV, Hagen-Larsen H, Ruden TA, Laerdahl JK, Hoyheim B: Annotated expressed sequence tags (ESTs) from pre-smolt Atlantic salmon (*Salmo salar*) in a searchable data resource. BMC Genomics. 2007, 8: 209-10.1186/1471-2164-8-209.
9. Hildahl J, Galay-Burgos M, Sweeney G, Einarsdottir IE, Bjornsson BT: Identification of two isoforms of Atlantic halibut insulin-like growth factor-I receptor genes and quantitative gene expression during metamorphosis. Comp Biochem Physiol B Biochem Mol Biol. 2007, 147: 395-401. 10.1016/j.cbpb.2007.02.006.
10. Hildahl J, Sweeney G, Galay-Burgos M, Einarsdottir IE, Bjornsson BT: Cloning of Atlantic halibut growth hormone receptor genes and quantitative gene expression during metamorphosis. Gen Comp Endocrinol. 2007, 151: 143-152. 10.1016/j.ygcen.2006.10.003.
11. Douglas SE, Knickle LC, Kimball J, Reith ME: Comprehensive EST analysis of Atlantic halibut (*Hippoglossus hippoglossus*), a commercially relevant aquaculture species. BMC Genomics. 2007, 8: 144-10.1186/1471-2164-8-144.
12. Douglas SE, Gallant JW, Bullerwell CE, Wolff C, Munholland J, Reith ME: Winter Flounder Expressed Sequence Tags: Establishment of an EST Database and Identification of Novel Fish Genes. Mar Biotechnol (NY). 1999, 1: 458-0464. 10.1007/PL00011802.
13. Bai J, Solberg C, Fernandes JM, Johnston IA: Profiling of maternal and developmental-stage specific mRNA transcripts in Atlantic halibut *Hippoglossus hippoglossus*. Gene. 2007, 386: 202-210. 10.1016/j.gene.2006.09.012.
14. Sepulcre MP, Sarropoulou E, Kotoulas G, Meseguer J, Mulero V: Vibrio anguillarum evades the immune response of the bony fish sea bass (*Dicentrarchus labrax* L.) through the inhibition of leukocyte respiratory burst and down-regulation of apoptotic caspases. Mol Immunol. 2007, 44: 3751-3757. 10.1016/j.molimm.2007.03.021.
15. Ucko M, Colorni A, Diamant A: Nodavirus infections in Israeli mariculture. J Fish Dis. 2004, 27: 459-469. 10.1111/j.1365-2761.2004.00565.x.
16. Poisa-Beiro L, Dios S, Montes A, Aranguren R, Figueras A, Novoa B: Nodavirus increases the expression of Mx and inflammatory cytokines in fish brain. Mol Immunol. 2008, 45: 218-225. 10.1016/j.molimm.2007.04.016.
17. Frerichs GN, Rodger HD, Peric Z: Cell culture isolation of piscine neuropathy nodavirus from juvenile sea bass,*Dicentrarchus labrax*. J Gen Virol. 1996, 77 (Pt 9): 2067-2071. 10.1099/0022-1317-77-9-2067.
18. Staden R, Beal KF, Bonfield JK: The Staden package, 1998. Methods Mol Biol. 2000, 132: 115-130.
19. Thurston MIFD: Msatfinder: detection and characterization of microsatellites. 2005, [http://www.genomics.ceh.ac.uk/msatfinder/]
20. Altschul SF, Madden TL, Schaffer AA, Zhang J, Zhang Z, Miller W, Lipman DJ: Gapped BLAST and PSI-BLAST: a new generation of protein database search programs. Nucleic Acids Res. 1997, 25: 3389-3402. 10.1093/nar/25.17.3389.
21. Parkinson J, Blaxter M: SimiTri – visualizing similarity relationships for groups of sequences. Bioinformatics. 2003, 19: 390-395. 10.1093/bioinformatics/btf870.
22. Stekel DJ, Git Y, Falciani F: The comparison of gene expression from multiple cDNA libraries. Genome Res. 2000, 10: 2055-2061. 10.1101/gr.GR-1325RR.
23. Livak KJ, Schmittgen TD: Analysis of relative gene expression data using real-time quantitative PCR and the 2(-Delta Delta C(T)) Method. Methods. 2001, 25: 402-408. 10.1006/meth.2001.1262.

24. Tang R, Dodd A, Lai D, McNabb WC, Love DR: Validation of zebrafish (*Danio rerio*) reference genes for quantitative real-time RT-PCR normalization. Acta Biochim Biophys Sin (Shanghai). 2007, 39: 384-390. 10.1111/j.1745-7270.2007.00283.x.

25. Ingerslev HC, Pettersen EF, Jakobsen RA, Petersen CB, Wergeland HI: Expression profiling and validation of reference gene candidates in immune relevant tissues and cells from Atlantic salmon (*Salmo salar* L.). Mol Immunol. 2006, 43: 1194-1201. 10.1016/j.molimm.2005.07.009.

26. Harbers M: The current status of cDNA cloning. Genomics. 2008, 91: 232-242. 10.1016/j.ygeno.2007.11.004.

27. Peatman E, Baoprasertkul P, Terhune J, Xu P, Nandi S, Kucuktas H, Li P, Wang S, Somridhivej B, Dunham R, Liu Z: Expression analysis of the acute phase response in channel catfish (*Ictalurus punctatus*) after infection with a Gram-negative bacterium. Dev Comp Immunol. 2007, 31: 1183-1196. 10.1016/j.dci.2007.03.003.

28. Peatman E, Terhune J, Baoprasertkul P, Xu P, Nandi S, Wang S, Somridhivej B, Kucuktas H, Li P, Dunham R, Liu Z: Microarray analysis of gene expression in the blue catfish liver reveals early activation of the MHC class I pathway after infection with *Edwardsiella ictaluri*. Mol Immunol. 2008, 45: 553-566. 10.1016/j.molimm.2007.05.012.

29. Stork M, Di Lorenzo M, Welch TJ, Crosa LM, Crosa JH: Plasmid-mediated iron uptake and virulence in *Vibrio anguillarum*. Plasmid. 2002, 48: 222-228. 10.1016/S0147-619X(02)00111-7.

30. Stafford JL, Belosevic M: Transferrin and the innate immune response of fish: identification of a novel mechanism of macrophage activation. Dev Comp Immunol. 2003, 27: 539-554. 10.1016/S0145-305X(02)00138-6.

31. Ganz T, Nemeth E: Regulation of iron acquisition and iron distribution in mammals. Biochim Biophys Acta. 2006, 1763: 690-699. 10.1016/j.bbamcr.2006.03.014.

32. Rodrigues PN, Vazquez-Dorado S, Neves JV, Wilson JM: Dual function of fish hepcidin: response to experimental iron overload and bacterial infection in sea bass (*Dicentrarchus labrax*). Dev Comp Immunol. 2006, 30: 1156-1167. 10.1016/j.dci.2006.02.005.

33. Lauth X, Babon JJ, Stannard JA, Singh S, Nizet V, Carlberg JM, Ostland VE, Pennington MW, Norton RS, Westerman ME: Bass hepcidin synthesis, solution structure, antimicrobial activities and synergism, and in vivo hepatic response to bacterial infections. J Biol Chem. 2005, 280: 9272-9282. 10.1074/jbc.M411154200.

34. Chen SL, Xu MY, Ji XS, Yu GC, Liu Y: Cloning, characterization, and expression analysis of hepcidin gene from red sea bream (*Chrysophrys major*). Antimicrob Agents Chemother. 2005, 49: 1608-1612. 10.1128/AAC.49.4.1608-1612.2005.

35. Bao B, Peatman E, Li P, He C, Liu Z: Catfish hepcidin gene is expressed in a wide range of tissues and exhibits tissue-specific upregulation after bacterial infection. Dev Comp Immunol. 2005, 29: 939-950. 10.1016/j.dci.2005.03.006.

36. Hu X, Camus AC, Aono S, Morrison EE, Dennis J, Nusbaum KE, Judd RL, Shi J: Channel catfish hepcidin expression in infection and anemia. Comp Immunol Microbiol Infect Dis. 2007, 30: 55-69. 10.1016/j.cimid.2006.10.004.

37. Park KC, Osborne JA, Tsoi SC, Brown LL, Johnson SC: Expressed sequence tags analysis of Atlantic halibut (*Hippoglossus hippoglossus*) liver, kidney and spleen tissues following vaccination against *Vibrio anguillarum* and Aeromonas salmonicida. Fish Shellfish Immunol. 2005, 18: 393-415. 10.1016/j.fsi.2004.10.003.

38. Lin B, Chen S, Cao Z, Lin Y, Mo D, Zhang H, Gu J, Dong M, Liu Z, Xu A: Acute phase response in zebrafish upon Aeromonas salmonicida and Staphylococcus aureus infection: striking similarities and obvious differences with mammals. Mol Immunol. 2007, 44: 295-301. 10.1016/j.molimm.2006.03.001.

39. Gerwick L, Corley-Smith G, Bayne CJ: Gene transcript changes in individual rainbow trout livers following an inflammatory stimulus. Fish Shellfish Immunol. 2007, 22: 157-171. 10.1016/j.fsi.2006.04.003.

40. Rossi F, Chini V, Cattaneo AG, Bernardini G, Terova G, Saroglia M, Gornati R: EST-based identification of genes expressed in perch (*Perca fluviatilis*, L.). Gene Expr. 2007, 14: 117-127. 10.3727/105221607783417600.

41. Kondo H, Kawazoe I, Nakaya M, Kikuchi K, Aida K, Watabe S: The novel sequences of major plasma apolipoproteins in the eel *Anguilla japonica*. Biochim Biophys Acta. 2001, 1531: 132-142.

42. Concha MI, Smith VJ, Castro K, Bastias A, Romero A, Amthauer RJ: Apolipoproteins A-I and A-II are potentially important effectors of innate immunity in the teleost fish *Cyprinus carpio*. Eur J Biochem. 2004, 271: 2984-2990. 10.1111/j.1432-1033.2004.04228.x.

43. Villarroel F, Bastias A, Casado A, Amthauer R, Concha MI: Apolipoprotein A-I, an antimicrobial protein in *Oncorhynchus mykiss*: evaluation of its expression in primary defence barriers and plasma levels in sick and healthy fish. Fish Shellfish Immunol. 2007, 23: 197-209. 10.1016/j.fsi.2006.10.008

CHAPTER 12 FUNCTIONAL GENOMICS REVEALS INCREASES IN CHOLESTEROL BIOSYNTHETIC GENES AND HIGHLY UNSATURATED FATTY ACID BIOSYNTHESIS AFTER DIETARY SUBSTITUTION OF FISH OIL WITH VEGETABLE OILS IN ATLANTIC SALMON (SALMO SALAR)

Michael J Leaver, Laure AN Villeneuve, Alex Obach, Linda Jensen, James E Bron, Douglas R Tocher and John B Taggart

BACKGROUND

FO has been, and still remains, the principal lipid source for fish diets in intensive aquaculture [1]. Because the wild fisheries, from which FO is obtained, are currently fished to the maximal sustainable levels there is pressure to utilise alternative lipid sources in aquaculture diets. Currently the most practical alternatives are VOs and recent studies have demonstrated that VO can be used to replace up to 75% of FO without significant effects on growth in Atlantic salmon [2]. However the effects of VO-based diets on normal metabolism and physiology and ultimately fish health and welfare are not yet understood. A variety of potentially deleterious effects have been reported in VO-fed fish. These include cardiac lesions [3], liver histopathology [4], compromised immune function [5] and disruption of intestinal function [6, 7, 8, 9]. There is good reason to suspect that feeding of VO may have a major impact on fish physiology since VO differ considerably in composition compared to FO. For example VOs are rich in shorter chain, C_{18} polyunsaturated fatty acids (PUFA) and devoid of n-3HUFA. Indeed, fatty acid compositions in fish, including salmon, fed VO are characterised by increased levels of C_{18} PUFA and decreased levels of n-3HUFA, which could compromise their nutritional value to the human consumer [2, 10].

Massively parallel gene expression profiling technologies such as cDNA or oligonucleotide microarrays are powerful tools for discovering genes which change their tissue or cellular expression levels in response to changed conditions and thence enable the physiological mechanisms underlying such changes to be elucidated. Until recently there have been few such resources for commercially important fish species. The recent development of high density cDNA microarrays for Atlantic salmon (GRASP; TRAITS-SGP; [11, 12], the most commercially valuable farmed fish species in Europe and the Americas, has opened the way for fundamental studies on diet-gene interactions and promises to greatly advance understanding of fish nutrition.

The aim of this study was to discover mechanisms for physiological adaptation to VO-based diets in fish. This was

Citation: Michael J Leaver, Laure AN Villeneuve, Alex Obach, Linda Jensen, James E Bron, Douglas R Tocher and John B Taggart, Functional genomics reveals increases in cholesterol biosynthetic genes and highly unsaturated fatty acid biosynthesis after dietary substitution of fish oil with vegetable oils in Atlantic salmon (Salmo salar), doi: 10.1186/1471-2164-9-299 achieved by measuring the effects on hepatic gene expression using a high density cDNA microarray, and by complementary biochemical and compositional assays of Atlantic salmon smolts fed diets in which 100% of the FO was replaced with three VOs, rapeseed oil (RO), linseed oil (LO) or soybean oil (SO).

RESULTS

Growth and Biometry

Slightly, but significantly, lower final weights were obtained with fish fed the LO diets compared to fish fed FO (Table 1). However, no effect of dietary oil was observed in the specific growth rate (SGR) obtained for individually pit-tagged fish. Dietary VOs had no effect on feed efficiency as measured by feed conversion ratio (FCR), hepato-somatic index (HSI) or viscero-somatic index (VSI) or condition factor (Table 1).

Table 1: Growth, biometric parameters and proximate analyses for Atlantic salmon (*Salmo salar*) fed the experimental diets for 16 weeks

	FO	LO	RO	SO
Initial weight (g)	132.0 ± 12.3	132.4 ± 12.6	132.1 ± 12.8	131.3 ± 12.2
Final weight (g)	434.5 ± 64.1[ab]	412.9 ± 61.8[c]	447.4 ± 75.6[a]	419.7 ± 67.8[bc]
SGR	1.03	0.99	1.06	1.01
SGR (Pit-tags)	1.09 ± 0.14	1.07 ± 0.11	1.16 ± 0.10	1.1 ± 0.13
FCR	0.72	0.75	0.71	0.71
VSI	8.60 ± 0.70	8.92 ± 0.86	8.65 ± 0.72	9.06 ± 0.94
HSI	1.19 ± 0.14	1.10 ± 0.16	1.13 ± 0.15	1.14 ± 0.12
CF initial	1.05 ± 0.07	1.06 ± 0.05	1.04 ± 0.06	1.04 ± 0.07
CF final	1.24 ± 0.11	1.25 ± 0.10	1.27 ± 0.12	1.24 ± 0.12
Moisture	65.50 ± 0.80	66.80 ± 1.80	66.50 ± 0.60	68.10 ± 0.80
Protein	51.00 ± 0.50[b]	55.10 ± 1.70[a]	53.90 ± 1.30[a]	55.60 ± 0.30[a]
Lipid	40.50 ± 0.20[a]	35.80 ± 2.50[b]	38.30 ± 1.50[ab]	37.10 ± 0.20[ab]
Ash	5.70 ± 0.10[b]	6.30 ± 0.30[a]	5.90 ± 0.20[ab]	6.30 ± 0.20[a]
P/L Ratio	1.28 ± 0.02[b]	1.40 ± 0.14[ab]	1.32 ± 0.08[ab]	1.49 ± 0.01[a]

Initial and final weights (n = 250), SGR (pit-tags) (n = 20), VSI (n = 12), HSI (n = 12) and CF (n = 12) Moisture, Protein, Lipid, Ash (n = 3) are all means ± SD. Significance of differences between means were determined by one -way ANOVA followed, where appropriate, by Tukey›s multiple comparison test as described in the Materials and Methods. Values within a row with a different superscript letter are significantly different ($P \leq 0.05$). CF, condition factor = 100 × [(body weight (g)/body length (cm)]3; FCR, feed conversion ratio = feed consumed (kg)/weight gain (kg); HSI, hepato-somatic index = 100 × (liver weight × body weight-1); SGR, specific growth rate (%/day) = 100 × [(ln weight final – ln weight initial) × days-1]; VSI, viscero-somatic index = 100 × (carcass weight × body weight-1). P/L, Protein:Lipid ratio.

Proximate, Lipid and Fatty Acid Compositions

All the VO diets significantly increased the relative percentage of protein in the whole fish compared to fish fed the FO diet (Table 1). There was also a tendency for the VO diets to decrease whole body lipid, and increase the protein:lipid ratio although these parameters were only significant in fish fed LO (Table 1). Liver total lipid content was not significantly affected by dietary VOs, but there was a trend towards increased free fatty acids in fish fed LO and increased total neutral lipid and triacylglycerol in fish fed SO and LO (Table 2). Dietary VO had only a few significant effects on the lipid class composition of liver with the most obvious being fish fed LO and SO showing reduced proportions of phosphatidylethanolamine compared to fish fed FO (Table 2). Differences in phosphatidylcholine levels were also observed between RO and LO, although none of the VO diets showed significant differences compared to FO. Increases in total unknown neutral lipid were also apparent in RO and SO fed fish compared to FO. In general the fatty acid compositions of liver and flesh total lipid (Table 3) reflected the fatty acid compositions of the diets. Thus fish fed the FO diet were characterised by high levels of 16:0, 20:5n-3 (eicosapentaenoic acid; EPA) and 22:6n-3 (docosahexaenoic acid; DHA), with 20:4n-6 (arachidonic acid; ARA) as the major n-6 fatty acid. Feeding the VOs resulted in increased proportions of 18:1n-9 and total monoenes, 18:2n-6 and total n-6PUFA, and 18:3n-3, and decreased the proportions of 16:0 and total saturated fatty acids, ARA, EPA and DHA. Specifically, the major fatty acids in fish fed RO, SO and LO were 18:1n-9, 18:2n-6 and 18:3n-3, respectively. Fish fed the vegetable oils also showed increased proportions of the metabolites of these fatty acids, namely 20:3n-3 and 20:4n-3 in fish fed LO, 20:2n-6 and 20:3n-6 in fish fed SO, and 20:1n-9 in fish fed RO.

Table 2: Lipid class composition of liver of Atlantic salmon (*Salmo salar*) fed various dietary oils

Lipid class	FO	LO	RO	SO
Total lipid	4.2 ± 0.2	4.8 ± 0.5	4.4 ± 0.5	4.8 ± 0.8
Phosphatidylcholine	17.7 ± 4.0[ab]	13.3 ± 0.5[b]	19.0 ± 1.6[a]	14.5 ± 0.8[ab]
Phosphatidylethanolamine	10.1 ± 1.5[a]	7.7 ± 1.2[b]	9.4 ± 0.7[ab]	7.8 ± 0.6[b]
Phosphatidylserine	2.6 ± 1.3	3.1 ± 0.5	3.7 ± 0.4	3.1 ± 0.5
Phosphatidylinositol	3.2 ± 1.4	2.8 ± 0.8	4.0 ± 0.3	3.4 ± 0.5
PG/CL	3.2 ± 0.9	2.2 ± 1.1	2.6 ± 0.3	1.8 ± 0.4
Sphingomyelin	1.8 ± 0.3	1.4 ± 0.3	1.8 ± 0.2	1.4 ± 0.3
Lyso-PC	2.8 ± 0.2	2.7 ± 0.4	2.2 ± 0.5	1.9 ± 0.4
Unknown polar lipid	2.0 ± 0.3	2.5 ± 0.2	3.2 ± 0.2	2.8 ± 0.3
Total polar	43.4 ± 6.9[ab]	35.7 ± 0.6[b]	45.8 ± 3.9[a]	36.8 ± 3.2[ab]
Total neutral	56.6 ± 6.9[ab]	64.3 ± 0.6[a]	54.2 ± 3.9[b]	63.2 ± 3.2[ab]
Cholesterol	15.9 ± 0.9	15.7 ± 1.0	16.0 ± 0.8	15.3 ± 0.5
Triacylglycerol	24.8 ± 7.8	29.4 ± 2.2	23.1 ± 3.5	28.3 ± 2.4
Free fatty acid	7.7 ± 2.3	12.1 ± 1.8	8.0 ± 4.0	10.4 ± 0.8
Steryl ester	6.9 ± 2.1	5.6 ± 2.2	3.8 ± 1.3	5.6 ± 1.9
Unknown neutral lipid	1.3 ± 0.9[b]	1.4 ± 0.8[b]	3.3 ± 0.5[a]	3.6 ± 0.6[a]

Values are means ± SD of 4 samples each of tissue pooled from 3 fish. Significance of differences between means were determined by one-way ANOVA followed, where appropriate, by Tukeys multiple comparison post hoc test. Values within a row with a different superscript letter are significantly different ($P \leq 0.05$). CL, cardiolipin; Lyso-PC, lysophosphatidylcholine; PG, phosphatidylglycerol. All values are expressed as percentage total lipid, except total lipid which is expressed as percentage of wet weight.

Table 3: Fatty acid composition (percentage of weight) of total lipid from livers of Atlantic salmon (*Salmo salar*) fed different dietary oils

	FO	LO	RO	SO
14:0	2.7 ± 0.4[a]	0.6 ± 0.0[b]	1.0 ± 0.1[b]	0.6 ± 0.0[b]
16:0	16.7 ± 0.8[a]	10.0 ± 0.8[b]	12.1 ± 1.0[b]	11.4 ± 1.3[b]
18:0	5.9 ± 0.4[a]	6.0 ± 0.4[a]	4.4 ± 0.2[b]	6.1 ± 0.4[a]
Total saturated[1]	25.8 ± 0.5[a]	16.9 ± 1.0[b]	17.9 ± 0.9[b]	18.4 ± 1.2[b]
16:1[2]	3.7 ± 0.6[a]	1.0 ± 0.1[c]	1.8 ± 0.1[b]	0.9 ± 0.1[c]
18:1n-9	9.3 ± 1.1[c]	18.4 ± 1.7[b]	29.1 ± 2.0[a]	18.4 ± 2.2[b]
18:1n-7	2.9 ± 0.1[a]	1.1 ± 0.1[d]	2.4 ± 0.1[b]	1.6 ± 0.1[c]
20:1[3]	2.5 ± 0.4[b]	2.0 ± 0.1[b]	3.8 ± 0.4[a]	2.2 ± 0.2[b]
22:1[4]	1.0 ± 0.4	0.8 ± 0.1	1.0 ± 0.2	0.6 ± 0.1
24:1n-9	1.0 ± 0.1[a]	0.6 ± 0.1[b]	0.8 ± 0.1[ab]	0.6 ± 0.1[b]
Total monoenes	20.3 ± 2.6[b]	24.0 ± 1.9[b]	39.0 ± 2.5[a]	24.3 ± 2.4[b]
18:2n-6	1.9 ± 0.4[c]	9.6 ± 0.3[b]	8.6 ± 0.3[b]	26.4 ± 1.8[a]
20:2n-6	0.5 ± 0.0[c]	1.4 ± 0.1[b]	1.6 ± 0.1[b]	4.3 ± 0.2[a]
20:3n-6	0.3 ± 0.1[d]	0.6 ± 0.1[c]	1.1 ± 0.1[b]	2.5 ± 0.1[a]
20:4n-6	2.8 ± 0.4[a]	0.7 ± 0.1[c]	1.4 ± 0.2[b]	1.6 ± 0.2[b]

Total n-6 PUFA[5]	6.4 ± 0.1^{c}	12.5 ± 0.3^{b}	13.1 ± 0.1^{b}	35.3 ± 1.7^{a}
18:3n-3	0.6 ± 0.2^{c}	20.3 ± 0.9^{a}	2.4 ± 0.2^{b}	2.3 ± 0.1^{b}
18:4n-3	0.6 ± 0.3^{a}	0.8 ± 0.1^{a}	0.3 ± 0.1^{b}	0.2 ± 0.0^{b}
20:3n-3	0.2 ± 0.1^{b}	3.2 ± 0.3^{a}	0.4 ± 0.1^{b}	0.3 ± 0.0^{b}
20:4n-3	1.2 ± 0.3^{b}	3.0 ± 0.0^{a}	1.0 ± 0.1^{b}	0.6 ± 0.1^{c}
20:5n-3	10.7 ± 0.3^{a}	4.0 ± 0.4^{b}	4.8 ± 0.4^{b}	2.6 ± 0.6^{c}
22:5n-3	3.6 ± 0.0^{a}	1.0 ± 0.1^{c}	1.4 ± 0.1^{b}	0.8 ± 0.2^{c}
22:6n-3	30.0 ± 2.7^{a}	14.3 ± 1.6^{c}	19.6 ± 1.8^{b}	15.2 ± 2.3^{c}
Total n-3 PUFA	46.9 ± 2.2^{a}	46.6 ± 1.7^{a}	30.0 ± 1.8^{b}	22.0 ± 3.0^{c}
Total PUFA	54.0 ± 2.2^{b}	59.1 ± 1.6^{a}	43.1 ± 1.7^{c}	57.3 ± 1.5^{ab}

Values are means ± SD of 4 samples each of tissue pooled from 3 fish. Significance of differences between means were determined by one-way ANOVA followed, where appropriate, by Tukey›s multiple comparison test as described in the Materials and Methods. Values within a row with a different superscript letter are significantly different ($P \leq 0.05$). 1, contains 15:0, 20:0 and 22:0, present in some samples at up to 0.4%; 2, predominantly n-7 isomer; 3, predominantly n-9 isomer; 4, predominantly n-11 isomer; 5, totals include 18:3n-6, 22:4n-6 and 22:5n-6 present in some samples at up to 0.5%.

Liver Enzyme Activities

No effect of diet was detected on the activity of liver carnitine palmitoyl transferase 1 (CPT1) activity, although fatty acid β-oxidation was increased in LO and RO fed fish. HUFA synthesis was increased in liver microsomes from fish fed all the VOs compared to fish fed FO. The activity of malic enzyme (ME) was increased in fish fed the RO diet in comparison to fish fed FO (Fig. 1).

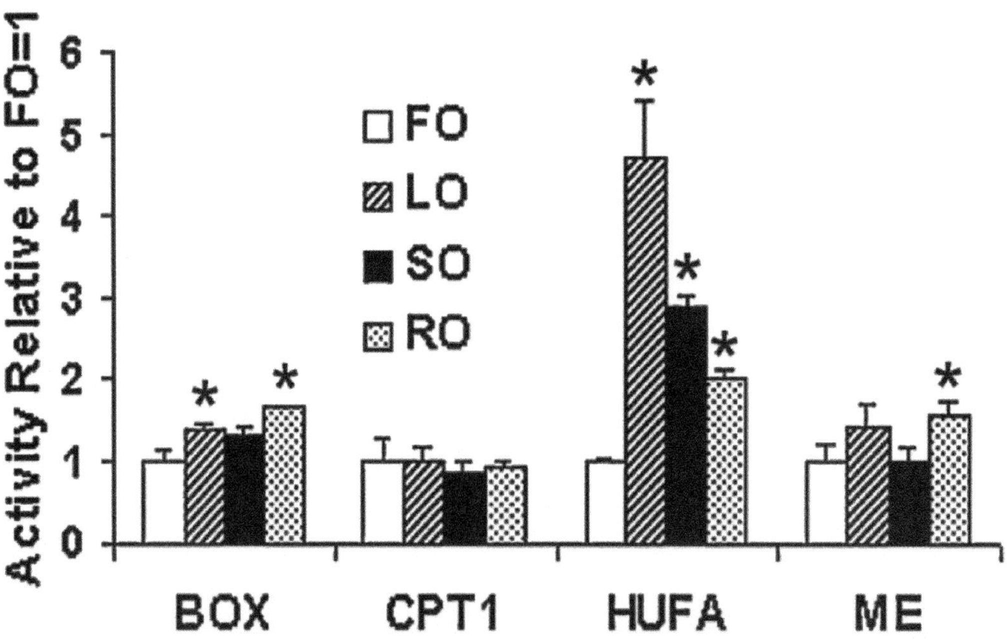

Figure 1: Expression of various enzyme activities in vegetable oil fed Atlantic salmon. Results are expressed relative to fish oil. Those which are significantly different from fish oil ($p \leq 0.05$ t-test) are indicated with an asterisk. BOX, total palmitoyl CoA beta-oxidation; CPT1, carnitine palmitoyl CoA transferase; HUFA, highly unsaturated fatty acid synthesis; ME, malic enzyme. Specific enzyme activities for control fish are: BOX, 33.6 ± 4.7 pmol palmitoyl-CoA oxidised/min/mg protein; CPT1, 56.7 ± 14.8 pmol palmitoyl-carnitine produced/min/mg protein; HUFA, 2.88 ± 0.1 pmol linolenic acid desaturation products/h/mg protein; ME, 69.8 ± 15.2 nmol malate oxidised/min/mg protein.

Microarray gene expression profiling

Comparison of each experimental diet to FO identified 428, 350 and 434 features from RO, LO and SO respectively which were significantly different ($p \leq 0.05$). These lists from each VO were then compared and features which were significantly changed relative to FO in two or more VO diets (T-test, $P \leq 0.05$) were retained. This resulted in a list of 132 features. Features were then eliminated which were changed in different directions in different VO diets, resulting in a list of 128 features, 88 of which could be tentatively identified according to BLAST results. These genes whose expression was changed, and whose identity could be indicated from BLAST similarity, in two or more VO diets compared to FO, are listed in Table 4.

Table 4: Genes in liver changed in two or more vegetable oil diets compared to salmon fed fish oil

PROCESS	ACC. NO.	LO	RO	SO	EST-BLASTX OR EST-TC-BLASTX
HUFA BIOSYNTHESIS	DW588445	**4.131**	**6.1**	**3.529**	(Q9DEX7) Delta-5/delta-6 fatty acid desaturase
	EG354827	**3.814**	**3.864**	0.685	(AAL82631) Salmon delta-5 fatty acyl desaturase, TC51151
	EG647320	**4.286**	**3.37**	1.909	(Q9DEX7) Delta-5/delta-6 fatty acid desaturase
	AAL82631	**5.624**	**11.29**	**8.027**	(Q9DEX7) Delta-5/delta-6 fatty acid desaturase
	AAL82631	**1.84**	**3.657**	**2.932**	(Q9DEX7) Delta-5/delta-6 fatty acid desaturase
	AAR21624	**2.564**	**4.248**	2.688	(Q9DEX7) Delta-5/delta-6 fatty acid desaturase
CHOLESTEROL AND ISOPRENOID BOSYNTHESIS	CK875291	**13.61**	12.23	**13.86**	(O35760) Isopentenyl-diphosphate delta-isomerase 1
	CO471940	**6.282**	**8.078**	**6.958**	(Q4R4W5) Isopentenyl-diphosphate delta-isomerase 1
	EG647637	**14.11**	2.921	**20.59**	(P52019) squalene epoxidase
	CK874989	**4.679**	**4.488**	3.376	(O88822) Lathosterol oxidase
	CK881623	1.132	**0.762**	**0.764**	(O42563) Cytochrome P450 3A27
LIPID METABOLISM AND TRANSPORT	NAN	**1.494**	1.881	**1.613**	(CAA57449) apolipoprotein B [Salmo salar], TC40920
	AM042514	1.085	**1.84**	**1.618**	(Q96BZ4) PLD4 phospholipase D family, member 4, TC47857
	DW589827	**1.403**	0.968	**1.27**	(P59837) Retinol dehydrogenase 12
	BF228485	**0.841**	**0.752**	1.228	(Q9R182) Angiopoietin-related protein 3
	NAN.	**0.489**	**0.709**	**0.652**	(Q9BYV7) BCDO2 Beta-carotene dioxygenase 2
	AJ416951	**0.541**	**0.696**	0.799	(O88275) Peroxisome proliferator-activated receptor gamma
	AM402843	**0.362**	0.67	**0.441**	(Q5TZ29) Novel protein similar to vertebrate apolipoprotein B
CARBOHYDRATE METABOLISM	CK880626	1.093	**1.816**	**2.013**	(P05370) G6PD Glucose-6-phosphate 1-dehydrogenase
	AM083390	**2.685**	1.291	**1.873**	(Q5MIB6) Glycogen phosphorylase
	NAN	**2.316**	**1.609**	1.256	(Q04760) GLO1 glyoxalase 1, TC39739
	EG355381	**0.47**	**0.659**	0.7	(P15121) AKR1B1 aldo-keto reductase 1B1, TC40780
MITOCHONDRIAL FUNCTION	DW589236	**8.96**	**2.524**	1.84	(Q96EL2) MRPS24 mitochondrial ribosomal protein S24, TC34551
	CK891719	**3.013**	**2.495**	**2.212**	(Q99595) Mitochondrial import inner membrane translocase Tim17 A
	NAN	2.067	**1.455**	**1.582**	(Q9EQ20) aldh6a1 aldehyde dehydrogenase 6A1, TC37457
	CK881059	0.97	**0.515**	**0.587**	(Q9H2U2) Inorganic pyrophosphatase 2, mitochondrial
	EG649042	2.754	**0.644**	**0.763**	(P16036) Phosphate carrier protein, mitochondrial
PROTEIN PROCESSING AND TURNOVER	CK898137	2.407	**1.258**	**1.377**	(Q5ZIJ9) Ubiquitin ligase protein MIB2
	CK878932	**2.384**	1.532	**1.599**	(P03974) Transitional endoplasmic reticulum ATPase
	BM414061	**5.083**	**5.299**	**5.287**	(P97571) Calpain-1 catalytic subunit
	CK893924	**1.906**	**1.819**	2.368	(Q99614) Tetratricopeptide repeat protein 1
	DW591059	**1.6**	1.198	**1.72**	(ABQ57500) small heat shock protein HSPB9, TC57595
	AM397488	**0.357**	0.941	**0.517**	(P45878) FK506-binding protein 2, TC38192
RNA PROCESSING	CK888026	1.133	**1.293**	**1.791**	(P09651) heterogeneous nuclear ribonucleoprotein A1, TC61954
	CK892099	**2.289**	1.389	**1.697**	(P28048) La ribonucleoprotein A

	DW588498	1.629	**1.995**	**2.155**	(Q9D198) syf2 SYF2 homolog, RNA splicing factor, TC34249
	CK891976	**1.222**	1.093	**1.373**	(Q6PDM2) sfrs1 splicing factor, arginine/serine-rich 1, DY729372
	DW589408	**0.538**	**0.625**	0.56	(Q9D0E1) heterogeneous nuclear ribonucleoprotein M, TC35947
	CK884048	1.354	**0.792**	**0.874**	(P62317) Small nuclear ribonucleoprotein Sm D2
	AJ425602	**0.526**	**0.481**	0.578	(Q5RF83) Cold-inducible RNA-binding protein
SIGNAL TRANSDUCTION	CK898218	2.04	**1.571**	**1.642**	(Q6PH57) Guanine nucleotide-binding protein G(I)/G(S)/G(T) beta1
	CK881679	**2.84**	1.823	**2.093**	(Q8JGT0) Casein kinase I alpha, TC40170
	NAN	**0.531**	**0.637**	0.812	(Q8BXH3) gucy1b2 guanylate cyclase 1, soluble, beta 2, TC44944
	CK879754	**0.333**	0.627	**0.428**	(Q14511) Enhancer of filamentation 1
	CK877929	0.877	**0.528**	**0.416**	(Q2M5E4) RGS21 regulator of G-protein signaling 21, TC38138
	CK884662	**0.258**	**0.521**	**0.539**	(P62142) Serine/threonine protein phosphatase PP1b catalytic
	AM049833	**0.603**	0.862	**0.689**	(P62155) Calmodulin
	BF228513	0.402	**0.506**	**0.682**	(Q8CFJ8) par-6 partitioning defective 6 homolog gamma, BF228513
TRANSCRIPTION	CN181085	1.588	**1.189**	**1.597**	(Q8BQ46) TAF15 RNA polymerase II, TC53593
	EG648659	**1.637**	**1.396**	1.09	(Q6GPQ6) Endothelial differentiation-related factor 1 homolog
	AM402809	**1.363**	**1.469**	1.242	(Q8BG15) ctdspl2 CTD small phosphatase like 2
	CK898087	**0.698**	**0.627**	0.699	(Q9ERU2) Zinc finger protein 22
TRANSLATION	DW588363	1.032	**1.227**	**1.414**	(P17508) Elongation factor 1-alpha, oocyte form
	EG648663	**1.305**	1.076	**1.433**	(Q9WV70) Nucleolar complex protein 2 homolog
	NAN	**1.974**	1.225	**1.812**	(P18077) RPL35a ribosomal protein L35a, TC60182
	EG647766	**0.695**	**0.74**	0.933	(P47826) 60S acidic ribosomal protein P0
	CK886497	**0.559**	**0.688**	0.854	(P57776) Elongation factor 1-delta
EXTRACELLULAR MATRIX	CK877333	**1.554**	**2.258**	**2.234**	(Q98967) Cystatin
	CK876354	0.869	**0.671**	**0.631**	(Q5D0F2) serine protease inhibitor, Kunitz type, 2, TC52802
	AM042526	**0.534**	0.888	**0.45**	(P48307) TFPI2 tissue factor pathway inhibitor 2, TC51552
	CK885060	**0.386**	**0.419**	**0.476**	(BAB55662) collagen a3(I), TC58897
	AJ424854	**0.432**	**0.483**	0.776	(O93486) Alpha 3 type I collagen, TC43437
	AJ425530	0.191	**0.466**	**0.554**	(Q9JM06) EGF- fibulin-like extracellular matrix protein 2, TC48959
IMMUNE FUNCTION	AM042172	1.456	**0.444**	**0.472**	(AAW66975) Ig mu heavy chain secretory form, TC58483
	AM397498	0.075	**0.667**	**0.446**	(P08603) Complement factor H
ION TRANSPORT	AM402718	**0.464**	**0.561**	1.19	(Q9NYB5) solute carrier organic anion transporter 1C1, TC47786
	CK885259	0.484	**0.625**	**0.746**	(Q9YH26) Sodium/potassium-transporting ATPase alpha-1 chain
	AM402711	0.225	**0.743**	**0.369**	(Q28677) Solute carrier family 12 member 4
APOPTOSIS	CK879455	**0.432**	**0.616**	**0.587**	(O18998) Deoxyribonuclease-1
GROWTH AND PROLIFERATION	EG649002	1.157	**1.746**	**1.278**	(Q5R5D0) Cyclin-G1
	CK876621	0.74	**1.594**	**1.492**	(Q80UU6) ogfr opioid growth factor receptor, TC55122

CYTOSKELETON	EG647322	**4.417**	**4.16**	**2.259**	(P49006) MARCKS-like 1, TC40754
	DW588815	**0.518**	**0.609**	0.831	(P68372) Tubulin beta-? Chain
LYSOSOMAL	AM042088	**0.598**	0.579	**0.583**	(Q90744) Alpha-N-acetylgalactosaminidase
AA BIOSYNTH	CK890942	**0.545**	0.671	**0.727**	(A6H5Y3) mtr methionine synthase, TC52579
MUSCLE	DW591140	1.938	**1.327**	**1.268**	(P06741) MYL1 myosin, light chain 1, TC60893
RESPIRATION	EG648831	**0.467**	**0.426**	0.386	(P02142) Hemoglobin beta-1 subunit
UNKNOWN FUNCTION	EG647376	0.26	**1.808**	**1.739**	(XP_691559) hypothetical protein [Danio rerio], TC35542
	NAN	0.93	**1.98**	**1.353**	(XP_001340266) hypothetical protein [Danio rerio], TC45154
	AM042055	1.284	**1.242**	**1.75**	(Q6I9Y2) THOC7 THO complex 7 homolog, TC54560
	AM402609	**2.644**	**2.732**	**1.84**	(Q56TU0) Type-4 ice-structuring protein precursor, TC59473
	CK898312	1.492	**1.959**	**2.192**	(XP_692475) similar to CG6282-PA, isoform A, TC34514
	CK897369	**0.705**	**0.732**	**0.677**	(Q5JVS0) HABP4 hyaluronan binding protein 4, TC34600
	CK881591	**0.288**	**0.295**	**0.408**	(XP_001338932) hypothetical protein [Danio rerio], TC55348
	BM414217	**0.655**	0.969	**0.709**	(XP_708731) hypothetical protein isoform 2, TC55763
TRANSPOSON	EG647536	0.993	**1.841**	**1.305**	(XP_001341934) similar to ReO_6, retrotransposon, TC63376
	EG647893	**0.318**	0.428	**0.423**	(AAW27349) SJCHGC04625 protein, retrotransposon, TC46087
	BM414246	0.907	**0.525**	**0.54**	(ABV31711) transposase, TC43560
	CK879027	**0.76**	0.985	**0.683**	(Q04202) Transposable element Tcb2 transposase

Numbers indicate expression ratio and a significant change (VO:FO; T-test, $p \leq 0.05$) is indicated by bold type. LO, linseed oil; SO, soyabean oil; RO, rapeseed oil. Accession numbers of array features are provided. TC numbers are also provided, as described in text.

Gene ontology (GO) terms for biological processes were available for 39 of the 88 features with significant BLAST similarities that were changed in VO diets. The biological process GO terms associated with these 39 genes were compared with the total GO list for the TRAITS/SGP, and tested for significant ($p \leq 0.05$) overrepresentation as implemented in GeneSpring GX 7.3.1 (Agilent). Only GO categories represented by more than one gene from the final gene list were considered significant and are reported in Table 5. Overrepresented GO categories included lipid metabolism, sterol biosynthesis, isoprenoid biosynthesis, fatty acid desaturation, mRNA processing, and translational elongation. It should be noted that the current gene ontology annotations (biological process) for the TRAITS/SGP array are only available for 3921 of the 16950 features. Notably amongst those genes in the final list which were overrepresented after GO analysis, were genes showing the greatest increase in expression. These included Δ5/6 fatty acyl desaturases which function in HUFA biosynthesis (enriched GO:6636, fatty acid desaturation) and isopentenyl-diphosphate Δisomerase (IPP), lathosterol oxidase, and squalene epoxidase, all genes of the cholesterol biosynthesis pathway (enriched GO:8299, isoprenoid biosynthesis; GO6695, cholesterol biosynthesis). Levels of other genes associated with lipid metabolism (enriched GO: 6629) or transport were also changed. These included increases in apoliporotein B (apoB) and decreases in an apoB-like gene in LO and SO. Effects were seen in levels of expression of genes involved in transcription, RNA processing and translation (enriched GO:6414).

Table 5: Significantly enriched biological process GO categories in VO fed fish compared to FO fed fish

GO Biological Process Category	Genes with GO term in input list	Genes with GO term in output list	Enrichment p-Value
GO:6629: lipid metabolism	91	5	0.00186
GO:44255: cellular lipid metabolism	72	5	0.000643
GO:6631: fatty acid metabolism	36	3	0.00518
GO:6636: fatty acid desaturation	5	3	8.98E-06
GO:8610: lipid biosynthesis	25	2	0.025

GO:8299: isoprenoid biosynthesis	4	2	0.000571
GO:6695: cholesterol biosynthesis	4	2	0.000571
GO:6720: isoprenoid metabolism	5	2	0.000946
GO:6820: anion transport	54	3	0.0159
GO:6414: translational elongation	30	3	0.00307

Many of the remaining genes from the final list could also be grouped into distinct biological process categories, although they were not identified as such by GO analysis. Changes in carbohydrate metabolism genes included an increase in the expression of glycogen phosphorylase in all VO, and glucose 6-phosphate dehydrogenase in SO and RO. Several changes in expression levels of genes of protein degradation associated with the ubiquitin-proteosome pathway were observed, and also in the protease calpain1 which was upregulated more than five fold in all VO. Signal transduction genes were also changed and MARCKS-like 1 was increased in all VO treatments and protein phosphatase 1β decreased in all VO treatments. Genes for proteins with extracellular locations which changed in VO included the protease inhibitors cystatin, spint2 and tissue factor pathway inhibitor 2, as well as decreased expression of collagen1A3. In addition there were 44 features changed by VO for which no identity could be confirmed by BLAST (not shown).

Quantitative PCR (QPCR)

Results for QPCR are presented in Table 6. Three genes of the cholesterol biosynthesis pathway were upregulated in VO fed fish. One of these, IPP, represented by two microarray features, was selected for QPCR, another 7-dehydrocholesterol reductase was also on the array but was not significantly changed ($p = 0.07$). A further two genes, mevalonate kinase (MK) and SREBP2 were selected for QPCR because they were not present on the microarray, but were identified in BLAST searches of Atlantic salmon EST collections. MK was upregulated in all three VO, whilst SREBP2 mRNA was significantly increased in LO and RO, although not in SO ($p = 0.057$). HMG-CoA reductase (HMGR), another gene in the cholesterol biosynthesis pathway was not changed, either on the basis of microarray data or by QPCR. Δ6 fatty-acyl desaturase mRNA was also confirmed by QPCR to increase in all VO diets. A number of other lipid metabolic genes, selected on the basis of the microarray data were also assayed by QPCR. QPCR confirmed the microarray result that carnitine palmitoyl transferase A (CPT1A) mRNA was not changed, whilst QPCR of apolipoprotein B mRNA (APOB) demonstrated upregulation in LO and RO, whereas the array results indicated upregulation in SO and LO. Three other genes (catalase, PPARβ and short chain alcohol dehydrogenase) whose expression was not different between FO and VO by array analysis were also confirmed not to be changed after QPCR. However PPARγ which showed altered expression levels on micorarray analysis was not changed as measured by QPCR.

Table 6: Gene expression measured by QPCR in the liver of Atlantic salmon fed vegetable oils in comparison to fish oil.

	LO		**SO**		**RO**	
	Expression	**P value**	**Expression**	**P value**	**Expression**	**P value**
Genes changed on array						
isopentenyl-diphosphate Δisomerase	**3.263**	**0.007**	**3.755**	**0.008**	**3.313**	**0.023**
fatty acyl Δ6 desaturase	**2.809**	**0.009**	**3.215**	**0.025**	**3.994**	**0.004**
apolipoprotein B-100	**1.535**	**0.029**	1.137	0.624	**1.416**	**0.036**
PPARγ	1.053	0.805	1.118	0.649	1.122	0.555
Genes not on array (BLAST to ESTs)						
Mevalonate kinase	**2.158**	**0.003**	**2.334**	**0.027**	**2.949**	**0.008**
SREBP2	**2.563**	**0.018**	2.095	0.057	**2.578**	**0.011**
adipophilin	**2.836**	**0.032**	**3.469**	**0.032**	1.338	0.459
acyl-CoA oxidase	**1.432**	**0.034**	1.227	0.343	**1.501**	**0.043**
Reference or no change on array						
β-actin	0.773	0.318	0.908	0.694	0.794	0.291
EF1α	1.199	0.146	1.115	0.564	1.091	0.456
AJ425111	1.079	0.714	0.988	0.939	1.153	0.338
7-dehydrocholesterol reductase	**2.624**	**0.017**	**3.319**	**0.02**	**3.826**	**0.009**
carnitine palmitoyl transferase 1A	1.125	0.411	1.173	0.403	1.099	0.477
catalase	0.919	0.617	0.972	0.892	1.1	0.495
PPARβ1	0.861	0.605	0.932	0.778	0.93	0.78
PPARβ2	0.737	0.228	0.715	0.166	0.756	0.333
HMG-CoA reductase	1.204	0.136	1.132	0.566	1.097	0.516
short chain alcohol dehydrogenase	1.014	0.936	0.806	0.343	1.035	0.865

The identity of the genes tested is indicated. Those genes which were significantly changed (n = 5 fish per treatment, $p \leq 0.05$) in fish fed linseed oil (LO), soyabean oil (SO) or rapeseed oil (RO) compared to fish oil are indicated in bold. Expression numbers indicate fold change relative to fish oil.

DISCUSSION

Although there were some relatively small differences, no great effects of diet on growth were observed in the present trial. Nevertheless, the effects of diet on tissue fatty acid composition are very clear with VO having highly significant effects. These effects are as expected and as reported in several previous trials on salmon [10, 13, 14].

As a means of more fully investigating the possible effects of VO substitution, liver was chosen to undertake a gene expression profiling experiment using the recently developed TRAITS/SGP microarray for Atlantic salmon. Liver integrates a large part of the nutritional uptake from the diet and has particular roles in distributing dietary lipid, and synthesising lipid *de novo* as well as participating in detoxification and excretion pathways. The use of high density cDNA microarrays is now a well established method, and we sought to use the TRAITS/SGP array to discover genes, and indicate biochemical and physiological pathways whose expression patterns might change during adaptation to VO-based diets. In this DNA microarray hybridisation experiment, application of ANOVA indicated that 784 features were changed at $p \leq 0.05$ across all diet groups. However, application of either multiple testing correction (Benjamini and Hochberg) or a post hoc test (Tukeys multiple comparison) reduced this list of features 50 fold. Nevertheless, it was apparent that amongst the features from the list generated by ANOVA which passed filters for multiple testing correction or post hoc testing, were genes for Δ5/Δ6 fatty acyl desaturases. Given that these desaturase genes were known from prior experiments to be up-regulated in VO fed salmon [15] we filtered genes according to the criteria of having expression changed in the same direction in two or more of the VO dietary conditions after individual t-tests. In support of this approach it is notable that of the 132 genes which were scored as being changed in two or more VO diets, only 4 showed expression changes in opposite directions, far less than would be expected by chance alone. We tested the validity of the list of commonly changed candidates by a combination of compositional, biochemical and independent gene expression measurement by QPCR. The overall consistency of gene expression and physiological data demonstrates the validity of our approach to interpreting the microarray results presented here.

The most consistently and highly up-regulated genes that could be identified by homology or had previously been characterised and were clearly overrepresented after GO analysis were those of the Δ5 and Δ6 fatty acyl desaturases. Desaturases are encoded by several genes in salmon and each of them were represented on the array. Every one of these showed significant up-regulation in two or more of the VO diets. QPCR confirmed that desaturase mRNAs were increased and in addition the biosynthesis of HUFA was increased in liver microsomes from VO-fed fish. However, despite this clear increase in desaturase mRNA and HUFA biosynthesis, the amount of HUFA in salmon tissue was still considerably reduced when FO was replaced by VO in the diet. These results confirm previous reports describing the expression of salmon Δ5/Δ6 desaturase, where feeding of VO has been shown to increase desaturase expression and HUFA biosynthesis [15, 16].

In addition to the increase in fatty acyl desaturase expression, increases in mRNA for three enzymes (IPP, squalene epoxidase and lathosterol oxidase) of the cholesterol biosynthetic pathway were also evident from microarray data and from QPCR for IPP. To test the hypothesis that the cholesterol biosynthetic pathway was up-regulated following VO feeding the Atlantic salmon EST collections were searched for genes which are known to participate in cholesterol biosynthesis, but were not present on the TRAITS/SGP array. Two genes were identified and assayed by QPCR, MK and SREBP2. MK is known to be highly regulated by cholesterol level in mammalian liver and SREBP2 is a master transcriptional regulator of both cholesterol and HUFA biosynthesis [17]. Messenger RNA for both of these genes was increased by a similar amount to that seen for IPP. The activity of SREBP2 is controlled by cellular cholesterol status, such that reductions in membrane cholesterol trigger proteolytic cleavage of membrane bound SREBP2 and lead to its translocation to the nucleus where it drives the transcription of the genes for cholesterol and HUFA biosynthesis [17]. Of the 12 genes involved in lipid metabolism which were changed on the microarray after feeding VO, 5 are known to be regulated by SREBPs (Δ5 desaturase, Δ6 deasturase, IPP, lathosterol oxidase and squalene epoxidase) [17]. Another gene in the cholesterol biosynthesis pathway, 7-dehydrocholesterol reductase, which by QPCR was upregulated in all VO, was on present the TRAITS-SGP array, but was not present in the final output gene list. One other gene critical for cholesterol biosynthesis, HMG-CoA reductase was present on the array but remained unchanged and QPCR analysis confirmed that VO had no effect on levels of HMG CoA reductase mRNA. This is surprising since in mammals HMG-CoA reductase is regulated directly by cholesterol levels, by hormonal stimuli, and transcriptionally by SREBPs, and therefore our results suggest some differences in the regulation of this gene between fish and mammals. Nevertheless, taken together the effects on genes of the cholesterol biosynthesis pathway strongly suggests that salmon fed VO respond to reduced cholesterol levels in the diet. Although we did not measure cholesterol contents of the diets, a recent report, which quantifies cholesterol and total sterol levels in the raw ingredients used to formulate the diets, confirms the deficiency of cholesterol in VO diets [18]. Moreover, it is known that in vegetable oils the bulk of sterol is present as sitosterol, stigmasterol, campesterol or brassicasterol depending on source [19]. These plant sterols are not taken up by mammals, and indeed inhibit cholesterol uptake [20] and, although not proven, are unlikely to be taken up by fish. Therefore one of the major responses of salmon liver during adaptation to VO is up-regulation of cholesterol biosynthesis to compensate for deficiencies in the diet. Importantly, and in contrast to the situation with HUFA, lower levels of dietary cholesterol were fully compensated for, since no differences were observed in cholesterol content, in liver or in flesh between fish fed FO or VOs. From the data presented here the compensation for reduced dietary cholesterol is due to up-regulation of hepatic cholesterol biosynthetic pathways, although other pathways such as decreased bile acid synthesis might also contribute. It should be noted, however, that this experiment consisted of a 12 week trial but in a longer term trial, after feeding VO for up to 22 months, both liver cholesterol and plasma LDL cholesterol were lower compared to FO [21].

Further to the up-regulation of cholesterol biosynthesis genes, apoB-100 which is a component of low density lipoprotein (LDL) required for the transport of cholesterol from liver to other tissues, also showed increases in expression level, both from microarray and QPCR measurement. Furthermore, from the microarray results angiopoietin-related protein 3 (ANGPTL3) was down-regulated in VO fed fish. ANGPTL3 is a liver derived factor which suppresses lipid uptake from LDL by tissues and promotes fatty acid release from adipocytes [22]. Thus the reduction in ANGPTL3 observed in salmon fed VO is consistent with the increase in apoB-100 and taken together indicates increases in lipid transport to tissues from liver via LDL. Interestingly an apoB-like protein, which is related but clearly distinct from apoB-100 was down-regulated in VO fed fish, and a protein which is suggested to act as an «anti-freeze» in fish blood [23], but which resembles apoA, another lipid transport molecule, is up-regulated in all VO. Several other genes which showed changes in expression levels in VO fed fish have also been linked to lipid transport. For example cystatin, which was increased in all VO fed fish, is involved in tissue lipid uptake, suggested to be related to its role in controlling elastin and collagen breakdown in the extracellular matrix [24]. Notably in this regard there were also decreases, in two microarray features identified as a collagen subtype, COL1A3, in all VO fish.

Overall these changes indicate a change in the function of the liver with respect to cholesterol synthesis and distribution, with an increase in cholesterol biosynthesis and an increase the transport processes which distribute cholesterol from the liver to other tissues. In animals fed normal cholesterol in the diet, the majority of cholesterol in the liver, from endogenous or exogenous pathways, is transported in the bile, as cholesterol or as bile acids, and thence to the gut where, together with phosphatidylcholine, it assists in the uptake of dietary lipid. In this regard, levels of hepatic phosphatidylcholine and its precursor phosphatidylserine were one of the few lipid compositional measures to vary between dietary treatments. These compositional changes might be related to changes in SO and RO diets of phospholipase D mRNA, which encodes an enzyme which releases choline from PC. Changes in bile transport, if occurring, may have consequences for lipid digestion, uptake and transport in the intestine which might explain the accumulation of lipid droplets in intestinal cells previously been observed in salmonids fed VO [6, 7, 8]. Interestingly, similar intestinal lipid accumulation is observed in rodents with deficiencies in bile transport [25].

Also notable from the microarray data was an increase in mRNA for glycogen phosphorylase (the first step of glycogen utilisation) and also in expression of glucose-6-phosphate dehydrogenase. These systems are involved in NADPH production [26] and suggests that the requirement for reducing power for cholesterol and HUFA biosynthesis is provided by breakdown of glycogen and up-regulation of NADPH generating systems. Another enzyme of importance for biosynthetic NADPH production, ME, was also shown to have increased activity in RO fed fish.

Some reports have indicated an increase in β-oxidation in VO compared to FO fed fish [27, 28]. From our microarray data we did not see increases in the expression of any genes linked to mitochondrial β-oxidation, indeed mRNAs for components of mitochondrial oxidative phosphorylation (pyrophosphates and phosphate carrier proteins) were decreased. Furthermore, QPCR and enzymic activity measurement confirmed that CPT1, an important rate limiting step in mitochondrial β-oxidation, was not significantly altered by VO feeding. However, it is known that peroxisomal β-oxidation predominates in salmon liver [29] and so the mRNA expression level of a critical enzyme of peroxisomal activity, acyl-CoA oxidase [30] was also measured. Acyl-CoA oxidase mRNA was increased in salmon fed the LO and RO diets, and this was consistent with the increased peroxisomal β-oxidation observed in fish fed LO and RO.

Of course this experiment represents a «snapshot» of gene expression 20 hours after final feeding and in liver only. At different prandial states and in different tissues the expression of genes for metabolic enzymes may be different. For example, shortly after feeding, when energy is in abundance, it is possible that the pentose phosphate pathway might be fuelled by gluconeogensis rather than glycogen breakdown, and these differences between FO and VO diets may be masked. With regard to fatty acid β-oxidation, sea bream exhibit differences in mitochondrial and peroxisomal activity depending on whether they are measured at 6 or 24 h after feeding [31].

The microarray data also indicated down-regulation of PPARγ, although, after QPCR no effect of VO feeding was apparent. PPARs are critical transcriptional regulators of lipid metabolism and energy homeostasis, and are encoded by three genes in mammals known as PPARα, PPARβ and PPARγ[32]. However, in salmon at six PPAR genes have so far been discovered and they share considerable homology [33, 34]. Therefore it is possible that the microarray results reflect cross-hybridisation of different, but related transcripts, although a previous microarray study has also reported a decrease in PPARγ mRNA following VO feeding in salmon [35]. In general cDNA microarrays are unlikely to be specific tools for measuring the expression of multiple closely related genes and this problem is exacerbated in Atlantic salmon which have polyploid genomes, often possessing more members within each gene family than the majority of diploid vertebrates. Similar issues been pointed out for the Δ5 and Δ6 fatty acyl desaturase genes which are 95% identical at the nucleotide level in salmon [12].

Increased lipid content, which might be an indication of health problems, in the livers of salmon fed VO has been observed previously [21, 36]. In our study, although there were no significant increases in hepatic lipid, a trend for increased lipid was supported by higher TAG and free fatty acid in VO fed fish and levels of mRNA for a marker for intracellular lipid accumulation, adipophilin [37], were increased in SO and LO, supporting the idea that at least some VO diets cause increases in hepatic lipid accumulation.

Finally from the microarray results, although not directly addressed by QPCR or biochemistry, were changes in the expression of genes involved in RNA and protein processing and turnover. These changes may be a reflection of changes in cholesterol biosynthetic pathways and in lipid transport, reflecting a requirement to export cholesterol to tissues.

CONCLUSION

In conclusion the present study demonstrates the utility of transcriptomic tools for studies on non-model commercially important fish species. Although, there were limitations in the statistical power of the design due largely to the inherent variability amongst recently domesticated farmed Atlantic salmon, by using appropriate filtering of microarray data, independent gene expression tests and biochemical and compositional data we were able to physiologically anchor a significant proportion of the gene expression results. This combined approach leads to the hypothesis that the major lipid metabolic effects of replacing FO with VO in salmon diets are mediated by SREBP2, a transcription factor which responds to reductions in membrane cholesterol level caused by reduced cholesterol in VO based diets. Furthermore, the application of the TRAITS/SGP microarray indicated several genes significantly altered by vegetable oils which suggest mechanisms to explain previously observed health impairment in VO fed salmon.

METHODS

Diets and animals

Four diets (4 mm pellets) with the same basal composition but coated with four different oils were prepared at the Skretting Technology Centre, Stavanger, Norway. The diets were formulated to satisfy the nutritional requirements of salmonid fish (U.S. National Research Council 1993), and to contain 33% fat and 47% protein, consisting of fish meal (55%), corn gluten (10%), wheat (8.3%), oil (26.3%), mineral and vitamin mixes (0.2%) and Carophyl Pink® (0.06%). The oils used were FO (Anchovy oil, Skretting, Stavanger, Norway), RO (Denofa AS, Fredrikstad, Norway), LO (NOBA Vetveredeling B.V., Hamburg, Germany) and SO (Denofa AS, Fredrikstad, Norway). Fatty acid and proximate compositions of the experimental diets are given in Table 7. One thousand Atlantic salmon post-smolts, average weight 132 g, were distributed randomly into four 2 × 2 m tanks (250/tank) at the Nutreco Aquaculture Research Centre, Lerang Research Station, Stavanger, Norway. In order to measure growth in individual salmon, twenty fish per tank were PIT-tagged by implanting a micro-transponder into the peritoneal cavity. The fish were conditioned to the new environment for 3 weeks before feeding the experimental diets. During the conditioning period, the fish received commercial diet containing FO (Atlantic 4 mm; Skretting). Feed intake was monitored during this period, and the trial did not start until appetite was at least 0.8% body weight. After the conditioning period, the fish were fed the experimental diets to satiation according to usual procedures at Lerang Research Station, for a period of 16 weeks. Feed intake was monitored throughout the trial and waste feed was collected from the effluent water from each tank by a wire mesh collector and dried. Feed given, waste feed and the resulting net feed intake were registered daily. The tanks were supplied with sea water at constant temperature (7.9°C ± 0.1°C). The oxygen level varied between 8 and 12 ppm, with an average of 9.9 ppm. A photoperiod of 18 h light and 6 h darkness was applied.

Table 7: Proximate composition. (Percentage of total weight) and fatty acid compositions (percentage of total FA weight) of experimental diets

	FO	LO	RO	SO
Total Protein	46.9	46.0	48.0	47.0
Total Fat	33.2	33.2	34.3	34.1
Ash	6.7	7.6	7.0	7.8
Moisture	6.2	5.8	5.7	4.5
14:0	7.4	0.9	1.2	0.9
16:0	18.7	7.4	7.0	12.6
18:0	3.9	3.5	1.9	3.1
Total saturated[1]	30.9	11.6	10.1	16.6
16:1n-7	7.5	0.9	1.3	0.9
18:1n-9	8.8	17.4	47.2	20.8
18:1n-7	2.7	1.0	2.7	1.7
20:1[2]	3.3	1.5	2.8	1.7
22:1[3]	3.5	2.0	2.5	2.0
24:1n-9	0.5	0.2	0.3	0.2
Total monoenes	26.3	23.0	56.8	27.3
18:2 n-6	3.5	14.2	17.8	44.4
CLA (9c,11t)	0.0	0.0	0.0	0.0
CLA (10t,12c)	0.0	0.0	0.0	0.0
20:4 n-6	0.9	0.1	0.1	0.1
Total n-6PUFA[4]	4.7	14.4	18.0	44.6

18:3 n-3	1.2	44.9	8.1	5.7
18:4 n-3	3.4	0.7	0.8	0.6
20:4 n-3	0.8	0.1	0.1	0.1
20:5 n-3	14.7	2.0	2.5	2.0
22:5 n-3	1.7	0.2	0.2	0.2
22:6 n-3	16.3	2.9	3.4	2.9
Total n-3PUFA	38.1	50.8	15.1	11.5
n-6/n-3	0.12	0.28	1.19	3.89
Total PUFA	42.8	65.2	33.1	56.1

Data are means of duplicate analyses. 1, contains 15:0 and 17:0, present in some samples at up to 0.5%; 2, predominantly n-9 isomer; 3, predominantly n-11 isomer; 4, totals include 18:3n-6, 20:2n-6, 22:4n-6 and 22:5n-6 present in some samples at up to 0.5%; PUFA, polyunsaturated fatty acid.

Sampling Protocols

At the start and end of the trial, all the fish in each tank were anaesthetised with metacain (50 mg/L) and individually weighed and measured. At the end of the trial, the 20 fish per dietary treatment were sampled for compositional, enzymatic and gene expression analyses, with three whole fish frozen immediately on dry ice and subsequently stored at -20°C for whole body compositional (proximate) analyses. The other sampled fish were eviscerated and twelve were used for biometric determinations (hepato- and viscero-somatic indices) and for lipid analyses. Livers were taken from each fish, pooled in four pools of three fish each, and frozen immediately in liquid nitrogen. Samples of liver were dissected from the remaining five fish from each dietary treatment and these samples were individually used for enzymatic, microarray and quantitative PCR (QPCR) analysis. A portion of each liver was taken for enzymatic analyses and frozen in liquid nitrogen. For gene expression analysis, samples of 0.5 g of fresh tissue were rapidly disrupted in 5 ml of TriReagent using an Ultra-Turrax homogeniser (Fisher Scientific, Loughborough, U.K.), and immediately frozen in liquid nitrogen for storage prior to RNA extraction.

Compositional Analyses

Proximate analysis, consisting of moisture content and lipid and protein content, was undertaken as previously described [38]. Lipid analyses were conducted on liver from twelve (four pools of three) fish from each diet and homogenised into pooled "pates". Total lipid was extracted from diets or 1 g portions of tissue pates and separation of lipid classes was performed by high-performance thin-layer chromatography (HPTLC) and fatty acid analysis performed as described previously [38].

Enzymatic and Metabolic Assays

All enzymatic and metabolic assays were carried out on the same five fish randomly selected from each dietary treatment. HUFA biosynthesis was determined in liver microsomes as described by Leaver et al., 2006. Assays of malic enzyme (ME), carnitine acyltransferase I (CPT1) and peroxisomal fatty acid β-oxidation in tissue homogenates were carried out on liver homogenates as previously described [14, 38].

Microarray gene Expression Profiling

Liver was Selected as the tissue on which to focus gene expression studies because of its central role in the integration and control of nutritional physiology and energy homeostasis. Total RNA was extracted from the same five fish from each dietary treatment as used for enzymatic determinations. RNA was prepared from TriReagent homogenates as described in the manufacturers protocol (Sigma). Two separate RNA extracts were prepared from each sample, one for microarray probe preparation and the second for QPCR. Each biological replicate was co-hybridised in a two dye experiment with a single pooled reference sample. The pooled reference sample comprised equal amounts of RNA from each of the 20 biological samples. RNA was reverse transcribed and labelled with either Cy3 (pooled reference) or Cy5 (FO, RO, SO and LO fed fish) fluors using the FAIRPLAY II cDNA indirect labelling kit (Stratagene) according to the manufacturer's instructions. Briefly 20 µg total RNA was reverse transcribed after being primed with oligo dT, which incorporated aminoallyl-dUTP into the newly synthesised cDNA strand. The RNA template was then hydrolysed using 1 M NaOH for 15 min and neutralised with 1 M HCl. The cDNA was precipitated overnight. cDNA pellets were washed in 80% ethanol and air dried before being resuspended in 5 µL 2× coupling buffer (Stratagene FAIRPLAYII reagent). Once the cDNA had fully dissolved (after at least 30 min) 5 µL of either Cy3 or Cy5 dye was added and the samples incubated in the dark for 30 min. The Cy3 and Cy5 dyes (GE Healthcare) were dissolved in DMSO prior to being added to the coupling buffer. To remove unincorporated dye, the labelled cDNA (total volume 10 µL) was passed through a SpinEX column (Qiagen). Dye incorporation was assessed by separating 0.5 µL of the sample on a mini-agarose gel and visualising fluorescent products on a fluorescence scanner (Typhoon Trio, GE Healthcare). No microarray pre-hybridisation step was required. For hybridisation the remainder of each labelled cDNA (7–9 µL; 16–30 pmol each dye) was added to 85 µL hybridisation buffer (UltraHyb, Ambion), 10 µL poly$(A)_{80}$ (10 mg mL^{-1}; Sigma) and 5 µL ultrapure BSA (10 mg mL^{-1}; Ambion). The hybridisation mixture was heated to 95°C for 3 min, then cooled to 60°C before being applied to the microarray (TRAITS/SGP, see section 2.7). Hybridisations were performed on a Gene TAC Hyb Station (Genomic Solutions) for 16 h at 45°C. Slides were then automatically washed with 2×SSC, 0.5% SDS for 10 min at 60°C; 0.2×SSC, 0.5% SDS for 10 min at 42°C; and finally 0.2 × SSC 10 min at 42°C. Following manual rinsing in isopropanol and drying by centrifugation, hybridised slides were scanned at 10 µm resolution using a Perkin Elmer ScanArray Express

HT scanner. BlueFuse software (BlueGnome) was then used to identify and quantify features. Following manual spot editing to remove obvious artefactual features and fusion of duplicate spot data (BlueFuse proprietary algorithm), the resulting intensity values and quality annotations were exported into the GeneSpring GX version 7.3.1 (Agilent Technologies) analysis platform. Data transformation, normalisation and quality filtering were as follows: 1) all raw spot intensity values less than 0.01 were set to 0.01 to facilitate comparisons of transformed data; 2) a ‹per spot per chip› intensity dependent (Lowess) normalisation was undertaken using software defaults (20% smoothing/cutoff 10); 3) data were filtered using a BlueFuse spot confidence value ≥ 0.2 in ≥ 8 slides and BlueFuse spot quality of ≥ 0.5 in ≥ 8 slides to filter out features which were below reasonable quality thresholds Bluefuse uses Bayesian statistical models of the signal and noise distributions on the slide to estimate the confidence that should be placed in each result. Similarly Bluefuse integrates various spot quality parameters (eg circularity, uniformity etc) to provide a spot quality assessment. This gave a final list of 11024 genes which were eligible for statistical analysis and application of ANOVA indicated that 9964 features had sufficient data for testing across all treatments. Experimental annotations complied fully with MIAME guidelines [39] and data has been submitted to the ArrayExpress databank (accession E-TABM-478; [40]).

Atlantic salmon ESTs were tentatively identified using BLASTX to search the Genbank non-redundant protein sequence database (all non-redundant GenBank CDS translations + RefSeq Proteins + PDB + SwissProt + PIR + PRF databases) applying a cut-off value of $\leq e^{-10}$. ESTs which were not assigned using this procedure were then compared to the Atlantic salmon DFCI Atlantic salmon Gene Indexhttp://compbio.dfci.harvard.edu/tgi/tgipage.html comprising ESTs and ESTs assembled into Tentative Consensus (TC) sequences using BLASTN. TCs or ESTs showing 99.9% identity over a continuous overlap of >100 nucleotides were then used to search the Genbank non-redundant protein sequence database applying a cut-off value of $\leq e^{-10}$.

Quantitative PCR (QPCR)

QPCR was performed using an RNA extract from the same five animals that had been taken for enzymic and microarray analysis. Complementary DNA (cDNA) was synthesized from total salmon hepatic RNA using a commercially available kit according the manufacturers instructions (*Reverse*-iT Max 1st strand synthesis kit, ABgene). Briefly each reaction of 20 µL contained 1.5 µg of total RNA, 300 ng of random hexamers and 125 ng of anchored oligo-dT. Following cDNA synthesis at 42°C for 1 hour reactions were stopped by heating at 75°C for 10 min and cDNA diluted to 500 µL total volume with water. Real-time PCR was performed using a Quantica machine (Techne). Quantitative PCR analysis for each gene was performed in triplicate in a total volume of 20 µL containing 5 µL cDNA (equivalent to 15 ng of input RNA), 100 nM of each primer and 10 µL of Absolute QPCR SYBR Green Mix 2× (ABgene). For each target gene, forward and reverse primers were chosen from the available EST sequences by using Primer3 software[41]. Target genes were selected by consideration of microarray data and by BLAST comparison of Atlantic salmon ESTs deposited in Genbank with sequences of mammalian genes of interest. Primer pairs and EST information are provided in Table 8. Thermal cycling was initiated with incubation at 95°C for 15 min in order to activate the Thermo-Start® DNA Polymerase present in the mix. After this initial step, forty-five cycles of PCR were performed. Each PCR cycle consisted of heating for 15 s at 95°C for denaturing, and then for 15 s at 60°C and 30 s at 72°C for annealing and extension. Cycle threshold (CT) values corresponded to the number of cycles at which the fluorescence emission monitored in real time exceeded the threshold limit.

Table 8: Genes, accession numbers and primers used for quantitative PCR

	Acc No	Forward primer	Reverse Primer
Genes from Array			
isopentenyl-diphosphate Δisomerase 1	CK875291	ACAGCCCTATGGTTATGTGTCATCTC	CAAGGTGAGGCGAATGTTTGAAC
7-dehydrocholesterol reductase	DW561983	CTTCTGGAATGAGGCATGGT	ACAGGTCCTTCTGGTGGTTG
HMG-CoA reductase	BE518590	CCTTCAGCCATGAACTGGAT	TCCTGTCCACAGGCAATGTA
Δ6 desaturase	AY458652	CCCCAGACGTTTGTGTCAG	CCTGGATTGTTGCTTTGGAT
apolipoprotein B-100	EG648437	GCTGTGGCTTTAATCCTTGC	CCATGGTGTCAGTGATCAGG
catalase	CK874978	CAACCCCCAGACTCACCTAA	GAAGGTGTGAGAGCCGTAGC
carnitine palmitoyl transferase 1A	AM230810	CCTGTACCGTGGAGACCTGT	CAGCACCTCTTTGAGGAAGG
short chain alcohol dehydrogenase	CK884238	TCTGCACACGAGAGGCATAC	GTCAGGGCAGTCACAGCATA
PPARβ1	AJ416953	GACCACCAACCCCAATGGCTCGGAT	CAGCCCATTCTCAGCCTGGCA-CAAG
PPARβ2	AM229303	CCCCCACCATCTTGGTGGCTCAGAC	TAGACCACTCTCTGCTTGCCA-CAGG
PPARγ	AJ416951	ACCCAGGAACCCAGAGTCAGCGGAC	TTTATAAAGTACTGACCCGCC-GTCA

Genes not on array (BLAST to ESTs)			
Mevalonate kinase	DY708590	CCCTTAATCAGGGTCCCAAT	GGTGCTGGTTGATGTCAATG
SREBP2	DY733476	GACAGGCACAACACAAGGTG	CAGCAGGGGTAAGGGTAGGT
adipophilin	BM413877	GCAGACATGGAAGTCGTTGA	ATGGAAATTTGTGGCTCCAG
acyl-CoA oxidase	DW555420	AAAGCCTTCACCACATGGAC	TAGGACACGATGCCACTCAG
Reference (no change in array ept)			
β-actin	AF012125	ACATCAAGGAGAAGCTGTGC	GACAACGGAACCTCTCGTTA
EF1α	AF321836	CTGCCCCTCCAGGACGTTTACAA	CACCGGGCATAGCCGATTCC
unknown	AJ425111	AGCCTATGACCAACCCACTG	TGTTCACAGCTCGTTTACCG

Melting curve analysis was performed to indicate the production of a single product in these reactions. Standard curves were established for each gene by plotting CT values against the $\log_{10}$ of five different dilutions (in triplicate) of cDNA sample solutions. In addition a subset of randomly selected samples generated by each primer pair was analysed by agarose gel electrophoresis and sequencing to confirm the identity of amplicons. Real-time efficiency was determined for each gene from the slopes given by Quantsoft software, applying the equation $E = 10^{(-1/slope)}$. The calculated relative expression ratio of each gene was based on the PCR efficiency (E) and CT of sample compared with control, and expressed in comparison to the reference genes, glyceraldehyde phosphate dehydrogenase (GAPDH), β-actin, and unidentified EST Acc No. AJ425111 (REST© software, [42]). Statistically significant differences in gene expression between the control (FO) and samples (LO, SO and RO) were evaluated in group means by randomization tests [43] using REST© software. Five thousand random allocations were applied and differences were considered to be significant at $P \leq 0.05$.

Materials

Materials for enzyme assays: [1-^{14}C] Palmitoyl CoA (50–55 mCi/mmol) and [methyl-^{3}H} L-carnitine (60–86 Ci/mmol) were obtained from GE Healthcare. BHT, Carnitine, coenzyme A, DTT, FAF-BSA, glutathione, HEPES, KCN, malate, N-acetylcysteine, NADH, NADP, nicotinamide, palmitoyl-CoA, perchloric acid, silver nitrate and TriReagent were obtained from Sigma.

Materials for lipid analyses: HPTLC (10 cm × 10 cm × 0.15 mm) and TLC (20 cm × 20 cm × 0.25 mm) plates, precoated with silica gel 60 (without fluorescent indicator) were obtained from Merck. All solvents were HPLC grade and were obtained from Fisher Scientific.

Microarrays: the TRAITS/SGP Atlantic salmon microarray comprised 16950 minimally redundant, duplicate cDNA clones and has been described elsewhere [12, 44].

Statistical analysis

Unless otherwise stated, all data are presented as means ± SD (n value as stated). The effects of dietary treatment on biometry, composition and enzyme activity were analysed by one-way analysis of variance (ANOVA) followed, where appropriate, by Tukey›s comparison test. Percentage data and data which were identified as non-homogeneous (Bartlett›s test) were subjected to arcsine transformation before analysis. Differences were regarded as significant when $P \leq 0.05$. Statistical treatment of microarray and QPCR data is described within the specific Methods and Results sections.

The probability that a particular biological process GO term was enriched in the input GO list (All GO annotations in the TRAITS-GS array) compared to the output GO list (all GO annotations in the experimentally altered gene list) was calculated using a hypergeometric distribution model as implemented by GeneSpring GX 7.3.1. The GO annotation of the TRAITS-GS microarray has been described by Taggart et al., 2008 [12]

DECLARATIONS

Acknowledgements

This majority of this work was supported by the European Union (project number: Q5RS-2000-30360) and by a Marie Curie Fellowship to LV (project number: EIF-024589) and also partly by the University of Stirling through provision of a faculty research development fund award and by BBSRC UK (project number: 98/EGA17674).

REFERENCES

1. Sargent JR, Tacon AGJ: Development of farmed fish: a nutritionally necessary alternative to meat. Proceedings of the Nutrition Society. 1999, 58: 377-383.
2. Torstensen BE, Bell JG, Rosenlund G, Henderson RJ, Graff IE, Tocher DR, Lie O, Sargent JR: Tailoring of Atlantic salmon (*Salmo salar* L.) flesh lipid composition and sensory quality by replacing fish oil with a vegetable oil blend. Journal of Agricultural and Food Chemistry. 2005, 53: 10166-10178. 10.1021/jf051308i.
3. Bell JG, McVicar AH, Park MT, Sargent JR: High Dietary Linoleic-Acid Affects the Fatty-Acid Compositions of Individual Phospholipids from Tissues of Atlantic Salmon (*Salmo Salar*) - Association with Stress Susceptibility and Cardiac Lesion. Journal of Nutrition. 1991, 121: 1163-1172.
4. Bell JG, Tocher DR, Farndale BM, McVicar AH, Sargent JR: Effects of essential fatty acid-deficient diets on growth, mortality, tissue histopathology and fatty acid compositions in juvenile turbot (*Scophthalmus maximus*). Fish Physiology and Biochemistry. 1999, 20: 263-277. 10.1023/A:1007743532618.

5. Mourente G, Good JE, Bell JG: Partial substitution of fish oil with rapeseed, linseed and olive oils in diets for European sea bass (*Dicentrarchus labrax* L.): effects on flesh fatty acid composition, plasma prostaglandins E-2 and F-2 alpha, immune function and effectiveness of a fish oil finishing diet. Aquaculture Nutrition. 2005, 11: 25-40. 10.1111/j.1365-2095.2004.00320.x.
6. Olsen RE, Myklebust R, Ringo E, Mayhew TM: The influences of dietary linseed oil and saturated fatty acids on caecal enterocytes in Arctic char (*Salvelinus alpinus* L.): a quantitative ultrastructural study. Fish Physiology and Biochemistry. 2000, 22: 207-216. 10.1023/A:1007879127182.
7. Olsen RE, Dragnes BT, Myklebust R, Ringo E: Effect of soybean oil and soybean lecithin on intestinal lipid composition and lipid droplet accumulation of rainbow trout, *Oncorhynchus mykiss* Walbaum. Fish Physiology and Biochemistry. 2003, 29: 181-192. 10.1023/B:FISH.0000045708.67760.43.
8. Caballero MJ, Obach A, Rosenlund G, Montero D, Gisvold M, Izquierdo MS: Impact of different dietary lipid sources on growth, lipid digestibility, tissue fatty acid composition and histology of rainbow trout, *Oncorhynchus mykiss*. Aquaculture. 2002, 214: 253-271. 10.1016/S0044-8486(01)00852-3.
9. Caballero MJ, Izquierdo MS, Kjorsvik E, Montero D, Socorro J, Fernandez AJ, Rosenlund G: Morphological aspects of intestinal cells from gilthead seabream (*Sparus aurata*) fed diets containing different lipid sources. Aquaculture. 2003, 225: 325-340. 10.1016/S0044-8486(03)00299-0.
10. Bell JG, Tocher DR, Henderson RJ, Dick JR, Crampton VO: Altered fatty acid compositions in Atlantic salmon (*Salmo salar*) fed diets containing linseed and rapeseed oils can be partially restored by a subsequent fish oil finishing diet. Journal of Nutrition. 2003, 133: 2793-2801.
11. von Schalburg KR, Rise ML, Cooper GA, Brown GD, Gibbs AR, Nelson CC, Davidson WS, Koop BF: Fish and chips: Various methodologies demonstrate utility of a 16,006-gene salmonid microarray. BMC Genomics. 2005, 6: 126-10.1186/1471-2164-6-126.
12. Taggart JB, Bron JE, Martin SAM, Seear P, Hoyheim B, Talbot R, Villeneuve L, Sweeney GE, Houlihan DF, Secombes CJ, Tocher DR, Teale AJ: A description of the origin, designs and performance of the TRAITS/SGP Atlantic salmon (*Salmo salar*) cDNA microarray. Journal of Fish Biology. 2008,
13. Rosenlund G, Obach A, Sandberg MG, Standal H, Tveit K: Effect of alternative lipid sources on long-term growth performance and quality of Atlantic salmon (*Salmo salar* L.). Aquaculture Research. 2001, 32: 323-328. 10.1046/j.1355-557x.2001.00025.x.
14. Torstensen BE, Lie O, Froyland L: Lipid metabolism and tissue composition in Atlantic salmon (*Salmo salar* L.) - Effects of capelin oil, palm oil, and oleic acid-enriched sunflower oil as dietary lipid sources. Lipids. 2000, 35: 653-664. 10.1007/s11745-000-0570-6.
15. Zheng XZ, Torstensen BE, Tocher DR, Dick JR, Henderson RJ, Bell JG: Environmental and dietary influences on highly unsaturated fatty acid biosynthesis and expression of fatty acyl desaturase and elongase genes in liver of Atlantic salmon (*Salmo salar*). Biochimica et Biophysica Acta-Molecular and Cell Biology of Lipids. 2005, 1734: 13-24. 10.1016/j.bbalip.2005.01.006.
16. Tocher DR, Bell JG, McGhee F, Dick JR, Fonseca-Madrigal J: Effects of dietary lipid level and vegetable oil on fatty acid metabolism in Atlantic salmon (*Salmo salar* L.) over the whole production cycle. Fish Physiology and Biochemistry. 2003, 29: 193-209. 10.1023/B:FISH.0000045722.44186.ee.
17. Horton JD, Shah NA, Warrington JA, Anderson NN, Park SW, Brown MS, Goldstein JL: Combined analysis of oligonucleotide microarray data from transgenic and knockout mice identifies direct SREBP target genes. Proceedings of the National Academy of Sciences of the United States of America. 2003, 100: 12027-12032. 10.1073/pnas.1534923100.
18. Tocher DR, Bendiksen EA, Campbell PJ, Bell JG: The role of phospholipids in nutrition and metabolism of teleost fish. Aquaculture. 2008,
19. Phillips KM, Ruggio DM, Toivo JI, Swank MA, Simpkins AH: Free and esterified sterol composition of edible oils and fats. Journal of Food Composition and Analysis. 2002, 15: 123-142. 10.1006/jfca.2001.1044.
20. Ostlund RE: Phytosterols and cholesterol metabolism. Current Opinion in Lipidology. 2004, 15: 37-41. 10.1097/00041433-200402000-00008.
21. Jordal AEO, Lie O, Torstensen BE: Complete replacement of dietary fish oil with a vegetable oil blend affect liver lipid and plasma lipoprotein levels in Atlantic salmon (*Salmo salar* L.). Aquaculture Nutrition. 2007, 13: 114-130. 10.1111/j.1365-2095.2007.00455.x.
22. Kersten S: Regulation of lipid metabolism via angiopoietin-like proteins. Biochemical Society Transactions. 2005, 33: 1059-1062. 10.1042/BST20051059.
23. Deng GJ, Laursen RA: Isolation and characterization of an antifreeze protein from the longhorn sculpin, *Myoxocephalus octodecimspinosis*. Biochimica et Biophysica Acta-Protein Structure and Molecular Enzymology. 1998, 1388: 305-314. 10.1016/S0167-4838(98)00180-0.
24. Bengtsson E, To F, Hakansson K, Grubb A, Branen L, Nilsson J, Jovinge S: Lack of the cysteine protease inhibitor cystatin C promotes atherosclerosis in apolipoprotein E-deficient mice. Arteriosclerosis Thrombosis and Vascular Biology. 2005, 25: 2151-2156. 10.1161/01.ATV.0000179600.34086.7d.
25. Voshol PJ, Minich DM, Havinga R, Elferink RPJO, Verkade HJJ, Groen AK, Kuipers F: Postprandial chylomicron formation and fat absorption in multidrug resistance gene 2 P-glycoprotein-deficient mice. Gastroenterology. 2000, 118: 173-182. 10.1016/S0016-5085(00)70426-4.
26. Pollak N, Dolle C, Ziegler M: The power to reduce: pyridine nucleotides - small molecules with a multitude

of functions. Biochemical Journal. 2007, 402: 205-218. 10.1042/BJ20061638.

27. Stubhaug I, Lie O, Torstensen BE: Fatty acid productive value and beta-oxidation capacity in Atlantic salmon (*Salmo salar*L.) fed on different lipid sources along the whole growth period. Aquaculture Nutrition. 2007, 13: 145-155. 10.1111/j.1365-2095.2007.00462.x.

28. Stubhaug I, Tocher DR, Bell JG, Dick JR, Torstensen BE: Fatty acid metabolism in Atlantic salmon (*Salmo salar* L.) hepatocytes and influence of dietary vegetable oil. Biochimica et Biophysica Acta-Molecular and Cell Biology of Lipids. 2005, 1734: 277-288. 10.1016/j.bbalip.2005.04.003.

29. Froyland L, Lie O, Berge RK: Mitochondrial and peroxisomal beta-oxidation capacities in various tissues from Atlantic salmon *Salmo salar*. Aquaculture Nutrition. 2000, 6: 85-89. 10.1046/j.1365-2095.2000.00130.x.

30. Morais S, Knoll-Gellida A, Andre M, Barthe C, Babin PJ: Conserved expression of alternative splicing variants of peroxisomal acyl- CoA oxidase 1 in vertebrates and developmental and nutritional regulation in fish. Physiological Genomics. 2007, 28: 239-252. 10.1152/physiolgenomics.00136.2006.

31. Diez A, Menoyo D, Perez-Benavente S, Calduch-Giner JA, de Celis SVR, Obach A, Favre-Krey L, Boukouvala E, Leaver MJ, Tocher DR, Perez-Sanchez J, Krey G, Bautista JM: Conjugated linoleic acid affects lipid composition, metabolism, and gene expression in Gilthead sea bream (*Sparus aurata* L). Journal of Nutrition. 2007, 137: 1363-1369.

32. Desvergne BA, Michalik L, Wahli W: Transcriptional regulation of metabolism. Physiological Reviews. 2006, 86: 465-514. 10.1152/physrev.00025.2005.

33. Leaver MJ, Boukouvala E, Antonopoulou E, Diez A, Favre-Krey L, Ezaz MT, Bautista JM, Tocher DR, Krey G: Three peroxisome proliferator-activated receptor isotypes from each of two species of marine fish. Endocrinology. 2005, 146: 3150-3162. 10.1210/en.2004-1638.

34. Leaver MJ, Ezaz MT, Fontagne S, Tocher DR, Boukouvala E, Krey G: Multiple peroxisome proliferator-activated receptor beta subtypes from Atlantic salmon (*Salmo salar*). Journal of Molecular Endocrinology. 2007, 38: 391-400. 10.1677/JME-06-0043.

35. Jordal AE, Torstensen BE, Tsoi S, Tocher DR, Lall SP, Douglas SE: Dietary rapeseed oil affects the expression of genes involved in hepatic lipid metabolism in Atlantic salmon (*Salmo salar* L.). J Nutr. 2005, 135: 2355-2361.

36. Ruyter B, Moya-Falcon C, Rosenlund G, Vegusdal A: Fat content and morphology of liver and intestine of Atlantic salmon (*Salmo salar*): Effects of temperature and dietary soybean oil. Aquaculture. 2006, 252: 441-452. 10.1016/j.aquaculture.2005.07.014.

37. Motomura W, Inoue M, Ohtake T, Takahashi N, Nagamine M, Tanno S, Kohgo Y, Okumura T: Up-regulation of ADRP in fatty liver in human and liver steatosis in mice fed with high fat diet. Biochemical and Biophysical Research Communications. 2006, 340: 1111-1118. 10.1016/j.bbrc.2005.12.121.

38. Leaver MJ, Tocher DR, Obach A, Jensen L, Henderson RJ, Porter AR, Krey G: Effect of dietary conjugated linoleic acid (CLA) on lipid composition, metabolism and gene expression in Atlantic salmon (*Salmo salar*) tissues. Comparative Biochemistry and Physiology A-Molecular & Integrative Physiology. 2006, 145: 258-267. 10.1016/j.cbpa.2006.06.034.

39. Brazma A, Hingamp P, Quackenbush J, Sherlock G, Spellman P, Stoeckert C, Aach J, Ansorge W, Ball CA, Causton HC, Gaasterland T, Glenisson P, Holstege FCP, Kim IF, Markowitz V, Matese JC, Parkinson H, Robinson A, Sarkans U, Schulze-Kremer S, Stewart J, Taylor R, Vilo J, Vingron M: Minimum information about a microarray experiment (MIAME) - toward standards for microarray data. Nature Genetics. 2001, 29: 365-371. 10.1038/ng1201-365.

40. ArrayExpress. 2008, [http://www.ebi.ac.uk/arrayexpress]

41. Primer3. 2008, [http://frodo.wi.mit.edu/cgi-bin/primer3/primer3_www.cgi]

42. wwwGene-Quantificationinfo. 2008, [http://www.gene-quantification.info/]

43. Pfaffl MW, Horgan GW, Dempfle L: Relative expression software tool (REST (c)) for group-wise comparison and statistical analysis of relative expression results in real-time PCR. Nucleic Acids Research. 2002, 30:

44. Martin SAM, Taggart JB, Seear P, Bron JE, Talbot R, Teale AJ, Sweeney GE, Hoyheim B, Houlihan DF, Tocher DR, Zou J, Secombes CJ: Interferon type I and type II responses in an Atlantic salmon (*Salmo salar*) SHK-1 cell line by the salmon TRAITS/SGP microarray. Physiological Genomics. 2007, 32: 33-44. 10.1152/physiolgenomics.00064.2007.

CHAPTER 13

A MODERATE INCREASE IN AMBIENT TEMPERATURE MODULATES THE ATLANTIC COD (GADUS MORHUA) SPLEEN TRANSCRIPTOME RESPONSE TO INTRAPERITONEAL VIRAL MIMIC INJECTION

Tiago S Hori, A Kurt Gamperl, Marije Booman, Gordon W Nash and Matthew L Rise

BACKGROUND

The Atlantic cod (*Gadus morhua*) is an important commercial species in several countries including Canada, USA and Norway, whose supply has been threatened by declining wild stocks[1, 2, 3]. In recent years, the aquaculture of cod has emerged as a potential alternative source of fish for these markets[4, 5]. Unfortunately, the development of Atlantic cod aquaculture still faces many challenges, including our incomplete understanding of how changes in the environment (e.g. seawater temperature) impact gadoid culture[2, 6, 7].

Atlantic cod reared in sea-cages are confined within a limited space, and therefore, are more likely than wild cod to be subjected to seasonal fluctuations in temperature. For example, high levels of mortality in farmed cod have been observed when temperatures increase during the summer months[8, 9]. The temperatures to which farmed cod are often exposed in the summer months (e.g. 16°C to 20°C[8]) in themselves, however, are unlikely to be lethal for this species. In fact, previous work from our research group indicates that Atlantic cod juveniles can survive an incremental temperature increase of 1°C every 5 days until temperatures reach ~ 22°C (Gamperl et al., unpublished data).

Elevated temperatures have been shown to modulate the immune response of several commercially important fish species such as the rainbow trout (*Oncorhynchus mykiss*)[10, 11, 12, 13], sea bass (*Dicentrarchus labrax*)[14], orange-spotted grouper (*Epinephelus coioides*)[15], Atlantic salmon (*Salmo salar*)[16] and Atlantic cod[9, 17]. However, the influence of temperature on fish immune function is variable. For example, while constant elevated temperatures can enhance the immune response of salmonids (e.g. improve the protection conferred by vaccines[12]), variable temperatures (i.e. daily fluctuations in temperature) can be immune suppressive in sea bass[14]. Since the thermal regime to which cod are exposed during the spring-summer months in Newfoundland is characterized by a gradual increase in temperature (i.e. variable temperature) and temperatures do not tend to reach the cod's critical thermal maximum (CTM)[9], we hypothesized that a temperature-dependent modulation of the immune system, which could lead to increased susceptibility to pathogens, may be associated with the losses sometimes observed in cod sea-cages.

Microarrays have been widely used to study the immune-relevant gene expression responses of fish and fish cell lines to pathogens and pathogen-associated molecular patterns (PAMPs) (e.g. Rise et al.[18]; Milev-Milovanovic et al.[19]; Workenhe et al.[20]; Booman et al.[21]). In this study we used the 20,000 element (20K) oligonucleotide microarray platform (GEO accession # GPL10532[21]), which includes sequences from both suppression subtractive hybridization (SSH) and normalized libraries enriched for immune and heat-stress responsive transcripts[22, 23, 24, 25, 26] (as well as sequences from several other normalized libraries), and reverse transcription – quantitative polymerase chain reaction

Citation: Tiago S Hori, A Kurt Gamperl, Marije Booman, Gordon W Nash and Matthew L Rise, A moderate increase in ambient temperature modulates the Atlantic cod (Gadus morhua) spleen transcriptome response to intraperitoneal viral mimic injection, doi: 10.1186/1471-2164-13-431

(QPCR), to investigate the effects of an increasing temperature regime (gradually from 10°C to 16°C – Figures1 and2) on the Atlantic cod spleen response to the intra-peritoneal (IP) injection of the viral mimic polyriboinosinic polyribocytidylic acid (pIC). A better understanding of how moderately increased ambient temperature affects the genes and pathways involved in the cod's anti-viral response may shed light on the potential mechanisms of temperature-induced immune suppression[14, 17, 27, 28, 29]. This knowledge will not only be important for the emerging Atlantic cod aquaculture industry, but may also be valuable to our understanding of how accelerated global climate change[30] may impact cold-water marine finfish species.

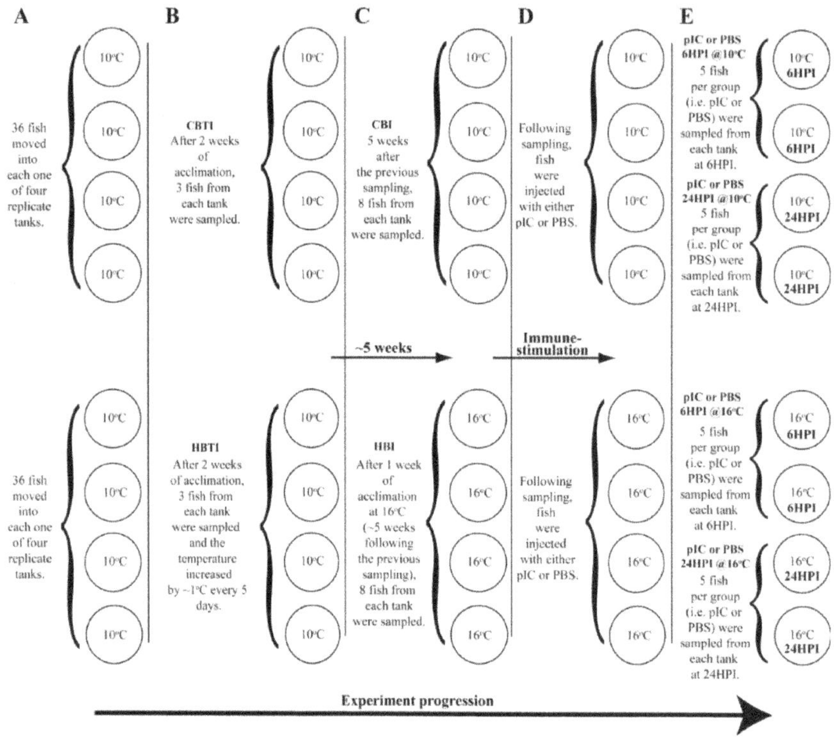

Figure 1: Overview of the experimental design used to expose fish to changes in temperature simulating those observed in Newfoundland sea-cages in the spring-summer, and to investigate the impact of this temperature change on the Atlantic cod's anti-viral response to the viral mimic (pIC). CBTI = Control group with no temperature increase; HBTI = Heat-exposed before temperature increase.

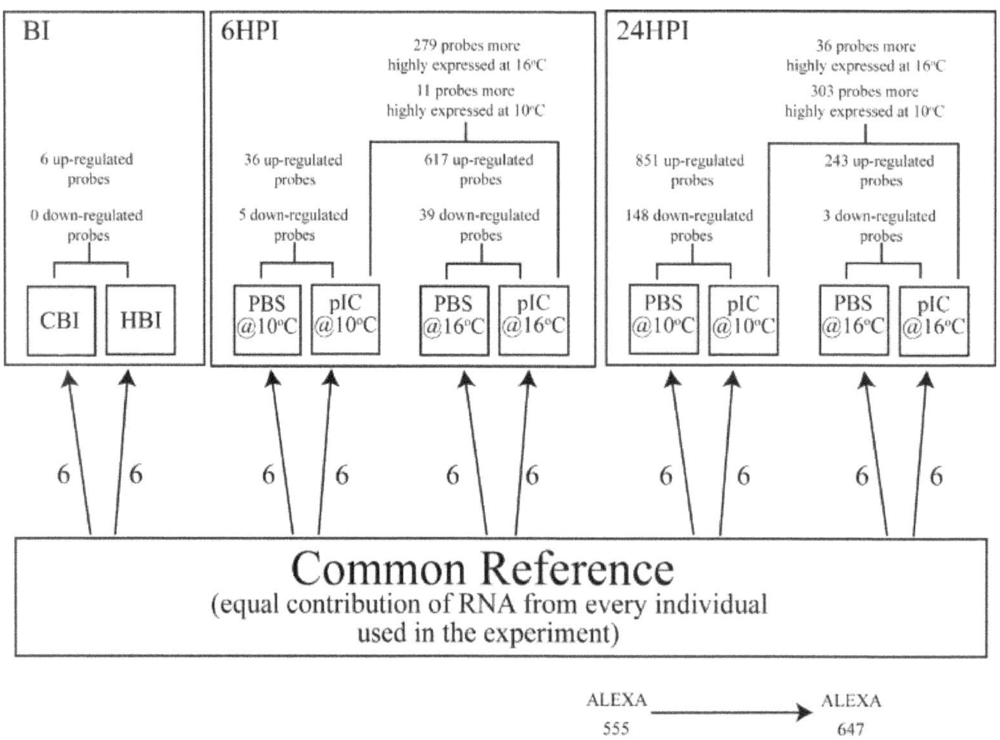

Figure 2: Overview of the microarray experimental design and results. Each bracket connecting 2 boxes represents a direct comparison between these groups using SAM. All genes reported were found to have significant differential expression at FDR = 1%.

RESULTS

Plasma Cortisol

Plasma cortisol averaged (± SE) 15.3 ± 6.4 and 18.2 ± 6.3 ng ml^{-1} in the "control before temperature increase" (CBTI – fish that remained at 10°C throughout the experiment) and "heat-exposed before temperature increase" (HBTI - fish that remained at 10°C up to this sampling point and were subsequently exposed to a gradual increase in temperature) groups, respectively. Cortisol values were approximately two-fold higher ~ 5 weeks later (i.e. one week after temperature had reached 16°C in the heat-exposed tanks) in both sets of fish (Figure1C) with the "control before injection" (CBI) group showing average cortisol levels of 33.1 ± 6.8 ng ml^{-1}and the "heat-exposed before injection" (HBI) group showing average cortisol levels of 37.7 ± 7.5 ng ml^{-1}. However, there were no significant differences in cortisol levels between the groups at either time point, or within the groups between time points.

Impact of non-lethal chronic temperature elevation on the spleen transcriptome

The first analysis carried out using significance analysis of microarrays (SAM) assessed the impact of the temperature increase alone on the cod spleen transcriptome. This analysis was carried out to investigate whether changes in the spleen transcriptome caused by temperature alone could be indicative of immune suppression. We compared 6 individuals from the CBI group to 6 individuals from the HBI group using a modified *t*-test assuming unequal variances in the SAM algorithm[31] as implemented in the Bioconductor package siggenes. The gradual 6°C increase in temperature had a very minor impact on spleen mRNA expression. Only 6 probes (all up-regulated) were identified as differentially expressed (at FDR = 1%) between the CBI and the HBI groups (Figure2 and Additional file1: Table S1).

Impact of pIC injection on the spleen transcriptome

Polyriboinosinic polyribocytidylic acid (pIC) injected fish were also compared to their time- and temperature-matched phosphate buffered saline (PBS) injected controls using SAM. This gave us the following four comparisons: pIC 6HPI vs. PBS 6HPI (at 10 and 16°C) and pIC 24HPI vs. PBS 24HPI (at 10°C and 16°C). We identified 41 (36 up-regulated and 5 down-regulated) differentially expressed probes (at FDR = 1%) between the pIC 6HPI vs. PBS 6HPI at 10°C (Figure2; Table1; Additional file1: Table S2), 656 probes (617 up-regulated and 39 down-regulated) between the pIC 6HPI vs. PBS 6HPI at 16°C (Figure2; Table1; Additional file1: Table S3), 999 probes (851 up-regulated and 148 down-regulated) between pIC 24HPI vs. PBS 24HPI at 10°C (Figure2; Table2; Additional file1: Table S4), and 246 probes (243 up-regulated and 3 down-regulated) between pIC 24HPI vs. PBS 24HPI at 16°C (Figure2; Table2; Additional file1: Table S5). Selected probes having immune-related functional annotations are presented alongside overall fold-change values [i.e. experimental ratio of (ALEXA 647/ALEXA 555) / control ratio of (ALEXA 647/ ALEXA 555)] in Tables1 and2. For complete information on microarray-identified informative probes, including raw p-values, d-values and standard deviations, please refer to Additional file1: Tables S2-S5.

Table 1: Selected probes [1] representing immune-relevant genes that were differentially expressed between fish injected with PBS or pIC and sampled at 6 hours post-injection (6HPI) at 10 or 16°C, or between fish injected with pIC but held at 10 vs. 16°C and sampled at 6HPI

Probes identified as differentially expressed (FDR = 1%) between PBS 6HPI@ 10°C and pIC 6HPI@ 10°C, and between PBS 6HPI@ 16°C and pIC 6HPI@ 16°C[2]				
Probe ID	**Best BLASTx hit[3]**	**Fold-change (pIC @ 16°C / PBS @ 16°C)[4]**	**Fold-change (pIC @ 10°C / PBS @ 10°C)[4]**	**Fold-change (pIC @ 16°C / pIC @ 10°C)[4]**
38613	Interferon stimulated gene 15 [Gadus morhua]	55.97	5.29	8.45
38285	DExD/H box RNA helicase [Paralichthys olivaceus] (Aliases: DHX58, LGP2)	5.13	2.44	2.41
45882	Retinoic acid receptor responder protein 3 [Osmerus mordax] (Alias: HRAS-suppressor like 3)	12.17	5.51	-
38260	Phosphoinositide phospholipase C-eta-2 [Homo sapiens]	5.08	2.19	2.28
44546	Interferon regulatory factor 1 (IRF1) [Gadus mohua]	4.74	1.88	2.48
36426	Cell division cycle 42 [Sus scrofa]	4.00	2.16	1.85
38042	Deoxyribonuclease gamma precursor [Oncorhynchus mykiss]	3.41	1.45	2.18
41397	Prostaglandin E synthase 3 [Osmerus mordax]	2.86	1.87	-

44530	NF-kappa-B inhibitor alpha (IκBα) [Gadus morhua]	2.15	3.12	-
40721	Nostrin protein [Danio rerio]	2.12	2.04	-
39371	Novel immune-type receptor 4 [Oncorhynchus mykiss]	2.03	1.52	-
41181	Ubiquitin carboxyl-terminal hydrolase 5 [Salmo salar]	1.88	2.04	-
36482	CXC chemokine [Psetta maxima]	1.74	1.71	-
Probes identified as differentially expressed (FDR = 1%) between PBS 6HPI@10°C and pIC 6HPI@10°C only [2]				
Probe ID	Best BLASTx hit [3]	Fold-change (pIC @ 16°C / PBS @ 16°C) [4]	Fold-change (pIC @ 10°C / PBS @ 10°C) [4]	Fold-change (pIC @ 16°C / pIC @ 10°C) [4]
46031	Synaptic vesicle glycoprotein 2B [Harpegnathos saltator]	-	3.35	-
44581	NF-kappa-B inhibitor alpha (IκBα) [Gadus morhua]	-	1.46	-
37564	Heat shock protein 90 alpha [Paralichthys olivaceus]	-	-1.50	-
49256	Unknown	-	-10.59	-
Probes identified as differentially expressed (FDR = 1%) between PBS 6HPI@16°C and pIC 6HPI@16°C only [2]				
Probe ID	Best BLASTx hit [3]	Fold-change (pIC @ 16°C / PBS @ 16°C) [4]	Fold-change (pIC @ 10°C / PBS @ 10°C) [4]	Fold-change (pIC @ 16°C / pIC @ 10°C) [4]
44448	Small inducible cytokine SCYA104 [Paralabidochromis chilotes] (Alias: gmSCYA123[a])	24.34	-	7.29
38356	Sacsin [Homo sapiens] (Alias: SACS)	12.54	-	11.14
48390	Endonuclease domain-containing 1 protein precursor [Salmo salar]	9.94	-	-
39328	Zinc finger, NFX1-type containing 1 (ZNFX1) [Homo sapiens]	9.05	-	2.67
43196	VHSV-induced protein [Oncorhynchus mykiss]	8.03	-	3.35
38639	Interleukin-8 variant 5 (IL-8) [Ictalurus punctatus]	7.60	-	5.88
43201	Viperin [Niphon spinosus] (Aliases: Radical S-adenosyl methionine domain containing protein 2, RSAD2)	6.89	-	5.43
50762	Unknown	6.70	-	-
37299	GADD45 alpha [Anoplopoma fimbria]	2.27	-	3.19
38617	Interferon-inducible GTPase a [Salmo salar]	6.53	-	2.80
47384	Unnamed protein product [Tetraodon nigroviridis]	6.50	-	7.71
36121	Anti-apoptotic protein NR-13 [Gadus morhua]	4.44	-	-
37609	Hepcidin precursor [Gadus morhua]	4.30	-	-
36384	CC chemokine type 3 [Gadus morhua]	3.55	-	-
44371	Probable E3 ubiquitin-protein ligase RNF144A-A [Esox lucius]	3.43	-	-
36190	Bloodthirsty [Gadus morhua]	3.38	-	2.05
37904	Type 2 double stranded RNA activated protein kinase [Gadus morhua] (Alias: PKR)	2.98	-	2.52
44579	Mitogen-activated protein kinase kinase 4 [Gadus morhua]	2.76	-	1.90

38599	Interferon regulatory factor 10 (IRF10) [Paralichthys olivaceus]	2.63	-	-
38655	Interferon regulatory factor 7 (IRF7) [Psetta maxima]	2.49	-	2.15
38788	Mannose-specific lectin precursor [Esox lucius]	2.39	-	-
39119	Ras homolog gene family, member T1a [Danio rerio]	2.25	-	1.75
38625	Interleukin 12 receptor beta 2.b [Danio rerio]	2.12	-	-
36407	CD9 antigen [Salmo salar]	2.05	-	1.44
44615	Toll-like receptor 9 (TLR9) [Gadus morhua]	2.03	-	-
37854	Interferon-inducible protein Gig2 [Siniperca chuatsi]	2.01	-	-
35877	Alpha-2-macroglobulin [Epinephelus coioides]	1.94	-	-
41703	Ras-related protein Rab-10 [Salmo salar]	1.90	-	-
36380	CC chemokine type 2 [Gadus morhua]	1.89	-	-
36142	Beta-2-microglobulin [Gadus morhua]	1.73	-	-
38629	Interleukin-1 receptor-associated kinase 4 (IRAK4) [Gadus morhua]	1.67	-	-
38090	GTPase IMAP family member 7 [Salmo salar]	1.64	-	-
36328	FLICE-like inhibitory protein [Oryzias latipes]	1.58	-	1.40
45132	DEAD (Asp-Glu-Ala-Asp) box polypeptide 10-like [Bos taurus]	1.55	-	-
44434	Type 1 death domain-containing protein [Gadus morhua]	1.48	-	-
41613	Dhx33 protein [Xenopus laevis]	1.47	-	-
37498	GTP-binding nuclear protein Ran [Osmerus mordax]	1.43	-	1.36
36122	Novel protein similar to BCL2-related ovarian killer [Danio rerio]	1.43	-	-
44431	Caspase 10 [Gadus morhua]	1.33	-	1.34
37561	Stress-70 protein, mitochondrial precursor [Salmo salar] (Alias: Mortalin)	1.32	-	-
44606	Toll-like receptor 3 (TLR3) [Gadus morhua]	1.32	-	-

[1]Probes presented in this table were selected based on their known roles in vertebrate immune responses. Some genes without a BLASTx hit (e.g. Unknowns) were added because they presented a pIC-induced fold-change higher than 5, and therefore, could represent novel immune-relevant genes. For a complete list of microarray-identified informative probes please refer to Additional file1: Tables S2-S5 and S7-S8.

[2]Fold-change in probe expression is also shown for pIC @ 16°C vs. pIC @ 10°C when significant differential expression was found in the PBS vs. pIC comparisons.

[3]Informative genes were re-annotated using the contig or EST from which the probe printed in the array was designed by comparing the referred sequence to the NCBI's nr database using the BLASTx algorithm as implemented by Blast2GO[67]. The best BLASTx hit (E-value cutoff of 10^{-5}) with an informative name (i.e. not predicted, probable or unnamed protein) is presented in this table. Synonyms were obtained from SWISS-PROT, GENE CARDS and other BLASTx hits as long as they had E-value < 10^{-5}, or from the literature ([a]Borza et al.[58]).

[4]Fold-changes are only reported when probes were differently expressed (FDR = 1%) in a given comparison (e.g. pIC@10°C vs. pIC@16°C) and are shown as outputted by siggenes; otherwise a dash (–) is presented in place of the fold-change. Fold-change values preceded by a minus ("-") sign represent overall fold down-regulation for a particular comparison. For standard deviation, p- and d-values, please refer to Additional file1: Tables S2-S5 and S7-S8.

Table 2: Selected probes [1] representing immune-relevant genes that were differentially expressed between fish injected with PBS or pIC at 24 hour post-injection (24HPI) at 10 or 16°C, or between fish injected with pIC but held at 10 vs. 16°C and sampled at 24HPI

Probes identified as differentially expressed (FDR = 1%) between PBS 24HPI@10°C and pIC 24HPI@10°C, and PBS 24HPI@16°C and pIC 24HPI@16°C[2]				
Probe ID	**Best BLASTx hit[3]**	**Fold-change (pIC @ 16°C / PBS @ 16°C)[4]**	**Fold-change (pIC @ 10°C / PBS @ 10°C)[4]**	**Fold-change (pIC @ 16°C / pIC @ 10°C)[4]**
38604	Interferon stimulated gene 15 (ISG15) [Gadus morhua]	38.35	90.24	-3.05
45882	Retinoic acid receptor responder protein 3 [Osmerus mordax] (Alias: HRAS-suppressor like 3)	31.10	27.01	-
44590	RIG-I C-terminal domain-containing protein 1 [Gadus morhua] (Aliases: DHX58, LGP2)	12.13	30.48	-
48390	Endonuclease domain-containing 1 protein precursor [Salmo salar]	11.53	14.24	-
38356	Sacsin [Homo sapiens] (Alias: SACS)	9.88	17.37	-
44448	Small inducible cytokine SCYA104 [Paralabidochromis chilotes] (Alias: gmSCYA123[a])	7.28	66.64	-2.34
38638	Interleukin-8 variant 5 (IL-8) [Ictalurus punctatus]	6.92	25.33	-5.66
35718	A disintegrin and metalloproteinase domain 8a [Danio rerio]	6.12	3.02	-
47329	Unnamed protein product [Tetraodon nigroviridis]	5.81	5.52	-
41269	Regulator of nonsense transcripts 1-like [Bos taurus]	5.07	13.93	-2.16
44859	Novel protein [Danio rerio]	4.98	11.50	-2.34
37854	Interferon-inducible protein Gig2 [Siniperca chuatsi]	3.41	4.14	-
44324	Aminopeptidase N [Camponotus floridanus]	3.04	3.45	-
44472	CC chemokine type 3 [Gadus morhua]	2.75	1.75	-
55231	Unknown	2.65	5.17	-
47410	Probable E3 ubiquitin-protein ligase RNF144A-A [Salmo salar]	2.64	7.10	-2.83
47613	Unnamed protein product [Tetraodon nigroviridis]	2.48	5.22	-2.72
37542	10 kDa heat shock protein, mitochondrial [Esox lucius]	2.45	1.61	-
44919	Novel protein similar to vertebrate IGSF3 [Danio rerio]	2.43	3.12	-
44371	Probable E3 ubiquitin-protein ligase RNF144A-A [Esox lucius]	2.35	4.56	-1.92
37299	GADD45 alpha [Anoplopoma fimbria]	2.25	4.52	-
44598	STAT1 [Gadus morhua]	2.12	1.98	-
36190	Bloodthirsty [Gadus morhua]	2.11	3.43	-1.78
36612	Complement component C3 [Paralichthys olivaceus]	2.07	2.47	-
43196	VHSV-induced protein [Oncorhynchus mykiss]	2.02	4.04	-
44546	Interferon regulatory factor 1 (IRF1) [Gadus mohua]	1.86	3.35	-1.84
37566	Heat shock protein HSP 90-alpha [Salmo salar]	1.80	1.34	1.55
36407	CD9 antigen [Salmo salar]	1.78	1.93	-
36426	Cell division cycle 42 [Sus scrofa]	1.66	1.93	-
35957	Apolipoprotein A-IV precursor [Salmo salar]	1.60	2.58	-1.53

38545	IgD heavy chain constant region variant b [Gadus morhua]	1.60	1.43	-
44865	Hect domain and RLD 5 [Bos Taurus]	1.52	1.94	-1.40
36923	DnaJ-like subfamily A member 4 [Paralichthys olivaceus]	1.51	1.43	-
Probes identified as differentially expressed (FDR = 1%) between PBS 24HPI@10°C and pIC 24HPI@10°C only [2]				
Probe ID	Best BLASTx hit [3]	Fold-change (pIC @ 16°C / PBS @ 16°C) [4]	Fold-change (pIC @ 10°C / PBS @ 10°C) [4]	Fold-change (pIC @ 16°C / pIC @ 10°C) [4]
38617	Interferon-inducible GTPase a [Salmo salar]	-	8.88	-
49247	Unknown	-	5.74	-2.86
43201	Viperin [Niphon spinosus] (Aliases: Radical S-adenosyl methionine domain containing protein 2, RSAD2)	-	4.68	-2.45
38260	Phospholipase C-eta-2 [Homo sapiens]	-	4.11	-
39368	Novel immune type receptor protein [Danio rerio]	-	3.23	
40397	Deltex-3-like [Salmo salar]	-	2.76	-2.18
39119	Ras homolog gene family, member T1a [Danio rerio]	-	2.66	-
44579	Mitogen-activated protein kinase kinase 4 [Gadus morhua]	-	2.61	-1.66
40174	60 kDa heat shock protein, mitochondrial precursor [Salmo salar]	-	2.46	-
37609	Hepcidin precursor [Gadus morhua]	-	2.44	-
36339	Caspase-1 [Dicentrarchus labrax]	-	2.42	-1.78
38788	Mannose-specific lectin precursor [Esox lucius]	-	2.30	-1.80
36364	Cathepsin L precursor [Anoplopoma fimbria]	-	2.11	-
38599	Interferon regulatory factor 10 (IRF10) [Paralichthys olivaceus]		2.05	-
44503	Type 2 double stranded RNA activated protein kinase [Gadus morhua] (Alias: PKR)	-	1.93	-
45146	DNA (cytosine-5-)-methyltransferase 7 [Danio rerio]		1.89	-1.79
44434	Type 1 death domain-containing protein [Gadus morhua]	-	1.89	-
36797	Cytotoxic and regulatory T cell protein [Oncorhynchus mykiss]	-	1.87	-1.81
40721	Nostrin protein [Danio rerio]	-	1.86	-
44839	Disulfide-isomerase A3 precursor [Salmo salar]	-	1.79	-
44518	Fas [Gadus morhua]	-	1.74	-1.77
36121	Anti-apoptotic protein NR-13 [Gadus morhua]	-	1.67	-
37498	GTP-binding nuclear protein Ran [Osmerus mordax]	-	1.64	-1.54
44550	Interleukin-1 receptor-associated kinase 4 (IRAK4) [Gadus morhua]	-	1.63	-1.37
36328	FLICE-like inhibitory protein [Oryzias latipes]	-	1.59	-
38619	Interferon-inducible protein Gig1 [Psetta maxima]	-	1.55	-
44615	Toll-like receptor 9 (TLR9) [Gadus morhua]	-	1.54	-
42836	Tumor necrosis factor receptor-2 [Paralichthys olivaceus]	-	1.54	-
42266	Strawberry notch homologue 1 [Danio rerio]	-	1.53	-
35877	Alpha-2-macroglobulin [Epinephelus coioides]	-	1.52	-

42501	T-complex protein 1 subunit delta [Takifugu rubripes]	-	1.46	-
41397	Prostaglandin E synthase 3 [Osmerus mordax]	-	1.45	-1.49
36967	E3 SUMO-protein ligase RanBP2 [Homo sapiens]	-	1.42	-
38108	Inhibitor of nuclear factor kappa-B kinase subunit alpha [Danio rerio] (Alias: IKKα)	-	1.41	-
38631	Interleukin-11a [Takifugu rubripes]	-	1.40	-
40138	78 kDa glucose-regulated protein precursor [Salmo salar]	-	1.39	-
37118	Novel protein similar to vertebrate EIF4E [Danio rerio]	-	1.35	1.31
41690	RAS guanyl-releasing protein 1 [Xenopus tropicalis]	-	1.33	-
37568	Glucose-regulated protein 94 [Paralichthys olivaceus]	-	1.33	-
39371	Novel immune-type receptor 4 [Oncorhynchus mykiss]	-	1.31	-1.33
39063	MHC class Ia antigen [Gadus morhua]	-	1.28	-
41711	Rab8b protein [Xenopus tropicalis]	-	1.27	-
35876	Alpha-2-macroglobulin receptor-associated protein [Homo sapiens]	-	1.26	-
41613	Dhx33 protein [Xenopus laevis]	-	1.24	-
45113	Methyltransferase-like protein 2 [Salmo salar]	-	-1.48	1.44
38713	KH domain-containing, RNA-binding, signal transduction-associated protein 1 [Salmo salar]	-	-1.54	-
41688	Novel protein similar to vertebrate IQGAP2 [Danio rerio]	-	-1.65	-
37695	G1/S-specific cyclin-D1 [Salmo salar]	-	-1.68	-
35934	Annexin max3 [Oryzias latipes]	-	-1.83	-
38596	Interferon induced protein 2 [Ictalurus punctatus]	-	-2.38	-
Probes identified as differentially expressed (FDR 1%) between pIC 24HPI@16°C and PBS 24HPI@16°C only [2]				
Probe ID	Best BLASTx hit [3]	Fold-change (pIC @ 16°C / PBS @ 16°C) [4]	Fold-change (pIC @ 10°C / PBS @ 10°C) [4]	Fold-change (pIC @ 16°C / pIC @ 10°C) [4]
44465	CC chemokine type 2 [Gadus morhua]	3.76	-	-
37456	Goose-type lysozyme 2 [Gadus morhua]	3.36	-	-
36381	CC chemokine type 2 [Gadus morhua]	2.98	-	-
42545	Thioredoxin [Oncorhynchus mykiss]	2.35	-	2.17
36388	CEBPA protein [Bos taurus]	1.86	-	-
36468	CCT epsilon subunit [Carassius auratus] (Alias: CCT5)	1.66	-	1.49
38014	Ctssa protein [Danio rerio]	1.66	-	-
35923	Annexin A4 [Ctenopharyngodon idella]	1.62	-	-
42502	Tcp1 protein [Danio rerio] (Alias: CCT1)	1.56	-	-
36813	DEAD (Asp-Glu-Ala-Asp) box polypeptide 23 [Xenopus tropicalis]	1.55	-	-
40508	Glutathione S-transferase Mu 3 [Anoplopoma fimbria]	1.53	-	1.61
45010	Heat shock protein 90 alpha [Paralichthys olivaceus]	1.52	-	1.66
37573	Heat shock protein 90 alpha [Paralichthys olivaceus]	1.42	-	1.65

43222	Warm temperature acclimation protein 65–1 [Dicentrarchus labrax]	1.42	-	-
37574	Heat shock protein 90 beta [Paralichthys olivaceus]	1.37	-	1.52
39904	Peptidylprolyl isomerase A (cyclophilin A) [Danio rerio]	1.33	-	-
Probes identified as differentially expressed (FDR = 1%) between pIC 24HPI @10°C and pIC 24HPI @16°C only [2]				
Probe ID	Best BLASTx hit [3]	Fold-change (pIC @ 16°C / PBS @ 16°C) [4]	Fold-change (pIC @ 10°C / PBS @ 10°C) [4]	Fold-change (pIC @ 16°C / pIC @ 10°C) [4]
37555	Heat shock protein 47 [Oncorhynchus mykiss]	-	-	1.77
40546	Heat shock protein HSP 90-alpha [Harpegnathos saltator]	-	-	1.64
36470	Chaperonin containing TCP1 subunit 6A [Paralichthys olivaceus] (Alias: CCT6)	-	-	1.54
44902	HSP90 multi-domain protein (pfam 00183)	-	-	1.47
37544	Heat shock cognate protein 70 [Pelodiscus sinensis] (Alias: HSC71)	-	-	1.35
42467	TANK-binding kinase 1 [Gadus morhua]	-	-	-1.37
37203	Ferritin heavy subunit [Epinephelus awoara]	-	-	-1.47
41479	Protein kinase C, delta [Danio rerio]	-	-	-1.47

[1]Probes presented in this table were selected based on their known roles in vertebrate immune responses. Some genes without a BLASTx hit (e.g. Unknowns) were added because they presented a pIC-induced fold-change higher than 5, and therefore, could represent novel immune-relevant genes. For a complete list of microarray-identified informative probes please refer to Additional file1: Tables S2-S5 and S7-S8.

[2]Fold-change in probe expression is also shown for pIC @ 16°C vs. pIC @ 10°C when significant differential expression was found in the PBS vs. pIC comparisons.

[3]Informative genes were re-annotated using the contig or EST from which the probe printed in the array was designed by comparing the referred sequence to the NCBI's nr database using the BLASTx algorithm as implemented by Blast2GO[67]. The best BLASTx hit (E-value cutoff of 10^{-5}) with an informative name (i.e. not predicted, probable or unnamed protein) is presented in this table. Synonyms were obtained from SWISS-PROT, GENE CARDS other BLASTx hits as long as they had E-value < 10^{-5}, or from the literature ([a]Borza et al.[58]). Some genes without a BLASTx hit (e.g. Unknowns) were added because they presented a pIC induced fold-change higher than 5, and therefore, could represent novel immune-relevant genes.

[4]Fold-changes are only reported when probes were differently expressed (FDR = 1%) in a given comparison (e.g. pIC@10°C vs. pIC@16°C) and are shown as outputted by siggenes; otherwise a dash (–) is presented in place of the fold-change. Fold-change values preceded by a minus ("-") sign represent overall fold down-regulation for a particular comparison. For standard deviation, p- and d-values, please refer to Additional file1: Tables S2-S5 and S7-S8.

We also compared the PBS-injected fish to the temperature-matched pre-injected fish (i.e. CBI and HBI) to ensure that they were suitable controls. In this analysis, we found 29 probes differentially expressed (with FDR = 1%) between the PBS 6HPI@10°C and CBI groups, 40 differentially expressed probes between the PBS 6HPI@16°C and HBI groups, 18 differentially expressed probes between the PBS 24HPI@10°C and CBI groups, and 97 differentially expressed probes between the PBS 24HPI@16°C and HBI groups (for tables containing information on the above lists see Additional file1: Table S6A-D). Given that only a small overlap (less than 5%) was observed between the probes identified in these comparisons and the pIC vs. PBS comparisons (Additional file1: Tables S6A–D), we concluded that the PBS groups were a suitable control for injection stress.

Impact of Increased Temperature on pIC-induced mRNA Expression

In order to study the effect of the gradual temperature increase on the cod spleen transcript expression response to pIC, we performed two direct comparisons: pIC 6HPI@16°C vs. pIC 6HPI@10°C and pIC 24HPI@16°C vs. pIC 24HPI@10°C. This approach detected 290 differentially expressed probes (with FDR = 1%) (279 more highly expressed at 16°C and 11 more highly expressed at 10°C) between pIC 6HPI@16°C and pIC 6HPI@10°C (Table1; Additional file1: Table S7) and 339 differentially expressed probes (36 more highly expressed at 16°C and 303 more highly expressed at 10°C) between pIC 24HPI@16°C and pIC 24HPI@10°C (Table1; Additional file1: Table S8). Selected microarray-identified probes having immune-related functional annotations are presented with fold-change values in Tables1 and2. For complete information on these probes, including raw p-values, d-values and standard deviations, please refer to Additional file1: Tables S7 and S8.

We further analyzed these genes using gene ontology (GO) and hierarchical clustering analyses. Two distinct analyses were used to look at GO terms mapped to genes differentially expressed between fish injected with pIC at 10 and 16°C. One analysis consisted of plotting level 2 GO terms (belonging to the biological process branch) mapped to the genes differentially expressed between fish injected with pIC at 10 and 16°C (Figure3). The second analysis consisted of using GOSSIP as implemented in Blast2GO to perform Fisher's exact test to check if any GO terms (belonging to any branch) were significantly ($p < 0.01$) enriched in the genes responding to pIC (compared to PBS) at 10 and 16°C for both time points. Figure3 shows four pie charts depicting the proportions of level 2 biological process GO terms of: A) the 279 genes (Figure2) more highly expressed in fish injected with pIC and sampled 6HPI at 16°C in comparison to the time-matched pIC injected fish at 10°C; B) the 11 genes (Figure2) more highly expressed in fish injected with pIC at 10°C and sampled 6HPI in comparison to the time-matched pIC injected fish at 16°C; C) the 36 genes (Figure2) more highly expressed in fish injected with pIC at 16°C and sampled 24HPI in comparison to the time-matched pIC injected fish at 10°C; and D) the 303 genes (Figure2) more highly expressed in fish injected with pIC at 10°C and sampled 24HPI in comparison with the time-matched pIC injected fish at 16°C.

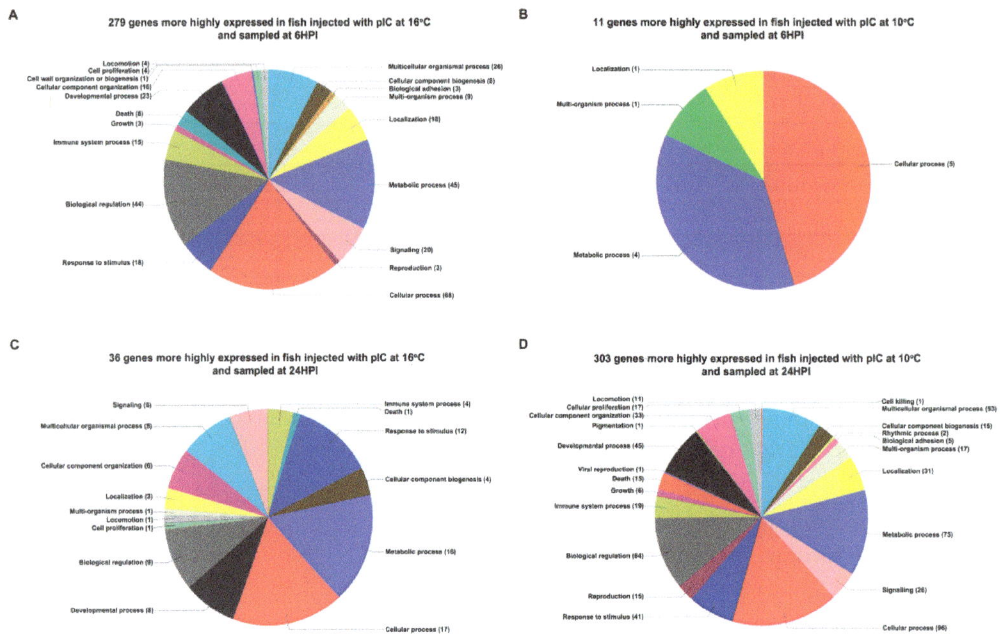

Figure 3: Summary of GO terms belonging to the biological process branch, represented as pie charts, for genes differentially expressed between fish injected with pIC at 10 and 16°C. Genes found to have significant differential expression between pIC@10°C and pIC@16°C at both 6HPI and 24 HPI time points were normalized to GO term hierarchy level 2. GO terms that mapped to the 279 genes more highly expressed in pIC@16°C 6HPI compared to pIC@10°C 6HPI are shown in **A**. The terms that mapped to the 11 genes more highly expressed in pIC@10°C 6HPI compared to pIC@16°C 6HPI are shown in **B**. **C** (36 genes) and **D** (303 genes) represent the same analysis, respectively, for the 24HPI time point.

Several biological processes were highly represented (i.e. by more than 10 sequences) in the 629 genes differentially expressed between fish stimulated with pIC at the different temperatures (i.e. identified in the direct comparisons between fish injected with pIC at 10 and 16°C and sampled at 6HPI and 24HPI). Among these were: cellular component organization, cellular process, developmental process, signaling, biological regulation, response to stimulus, immune system process, and death. Only 4 biological process level 2 GO terms were mapped to the 11 genes more highly expressed in fish stimulated with pIC at 10°C and sampled at 6HPI compared to those stimulated with pIC at 16°C and sampled at the same time point. This is likely a reflection of the low number of probes in this particular gene list. Fifteen of the level 2 biological process GO terms that were mapped to the genes more highly expressed in fish stimulated with pIC at 16°C and sampled at 6HPI were also mapped to the genes that were more highly expressed in fish stimulated with pIC at 10°C and sampled at 24HPI (each compared to its time-matched pIC injected counterpart); we also found that 96 probes were present in both gene lists (Figure4) (Additional file1: Table S9). However, many of the GO terms that were in common between these lists had more sequences mapped to them in the list of probes more highly expressed in fish stimulated with pIC at 10°C and sampled at 24HPI compared to the time-matched fish stimulated with pIC at 16°C. Examples are signaling (20 for pIC 6HPI@16°C and 26 for pIC 24HPI@10°C), immune system process (15 for pIC 6HPI@16°C and 19 for pIC 24HPI@10°C), death (8 for pIC 6HPI@16°C and 15 for pIC 24HPI@10°C), biological regulation (44 for pIC 6HPI@16°C and 64 for pIC 24HPI@10°C), response to stimulus (18 for pIC 6HPI@16°C and 41 for pIC 24HPI@10°C) and cellular process (68 for pIC 6HPI@16°C and 96 for pIC 24HPI@10°C).

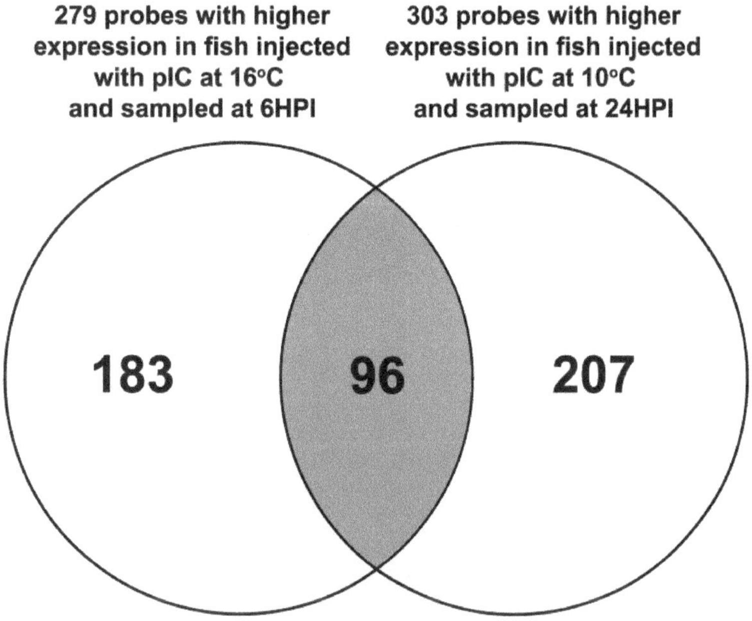

Figure 4: Venn diagram showing the overlap between the genes with significantly higher mRNA expression in pIC@16°C 6HPI compared to pIC@10°C 6HPI and those with significantly higher mRNA expression in pIC @10°C 24HPI compared to pIC@16°C 24HPI.

The GO enrichment analysis also identified several GO terms with significantly different over-/under-representation between the genes up-regulated by pIC (relative to PBS) at 24HPI at the two temperatures. Most GO terms (55 out of 57) identified were over-represented in the genes up-regulated by pIC at 16°C compared with those up-regulated by pIC injection at 10°C (all relative to PBS). Examples of these over-represented GO terms in the 24HPI list of pIC responsive genes at 16°C are: protein folding (GO:0006457), negative regulation of cellular process (GO:0048523), negative regulation of protein metabolic process (GO:0051248), regulation of interferon-gamma-mediated-signaling (GO:0060334), regulation of signaling process (GO:0023051), regulation of signal transduction (GO:0009966), regulation of cytokine-mediated signaling pathway (GO:0001959), unfolded protein binding (GO:0051082) and enzyme regulatory activity (GO:0030234). Two GO terms were under-represented in the 24HPI list of pIC responsive genes at 16°C: macromolecule modification (GO:0043412) and protein modification process (GO:0006464).

We present 2 separate clustering results displayed as heat-maps (Figures5,6 and7). In these heat-maps, all individuals (60) were clustered based on their expression for the genes identified as differentially expressed between the pIC@16°C and pIC@10°C groups at 6HPI (Figure5) or at 24HPI (Figure6 and7). In Figure5, panel A shows the entire heat-map for the 6HPI time point, while panel B shows a sub-tree (marked as a blue bar on the right of panel A) dominated by genes belonging to the interferon (IFN) pathway. Figure6 shows the entire heat-map for the 24HPI time point, while Figure7 shows a sub-tree (marked as a blue bar on the right of Figure6) dominated by genes belonging to the interferon (IFN) pathway. Both clustering results show a clear distinction (with a few outliers) between the groups of fish injected with pIC at the different temperatures and time points. Based on the expression of the genes identified in the above-mentioned comparisons, the pIC, 6HPI @10°C group had an expression profile more similar to that of the non-pIC stimulated fish, while the pIC, 6HPI @16°C group's expression profile was more similar to that of the pIC, 24HPI @10°C group. For the non-pIC stimulated fish (including the PBS-injected and non-injected groups), the clusters were less distinctive.

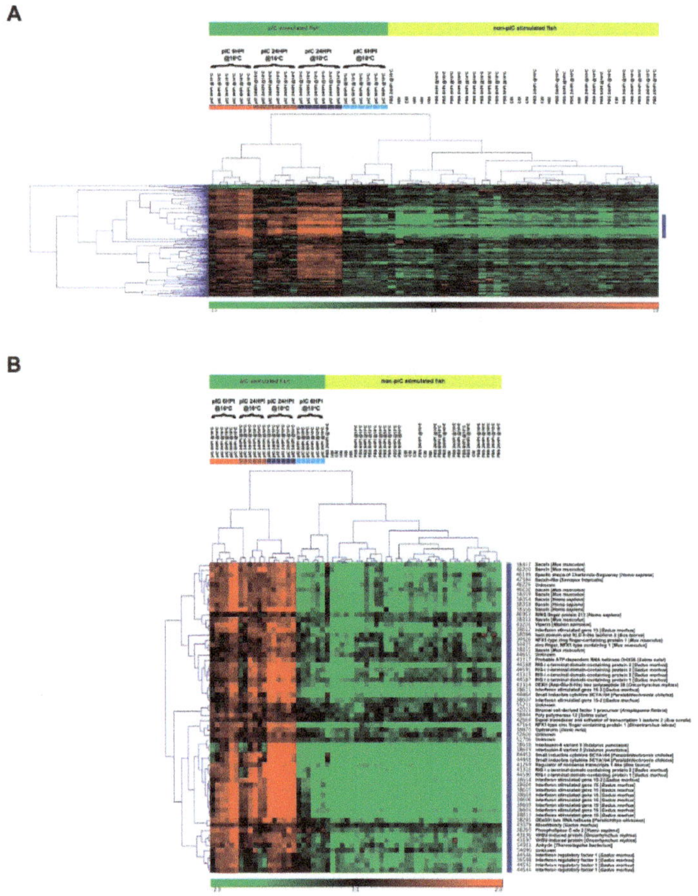

Figure 5: Heat-map depiction of hierarchical clustering of all samples (i.e. microarrays) and the 290 genes significantly differentially expressed between pIC@10°C and pIC@16°C at 6HPI based on their mRNA expression (see Figure2). The colored boxes below the top legend of each panel (**A** and **B**) represent individual fish from the pIC injected groups at the different temperatures and time points. Light red = pIC@16°C 6HPI; light blue = pIC@10°C 6HPI; dark red = pIC@16°C 24HPI; and dark blue = pIC@10°C 24HPI. Panel **A** shows the complete clustering including all genes. Panel **B** shows only the cluster marked in blue on **A**, which is enriched for genes belonging to the interferon pathway.

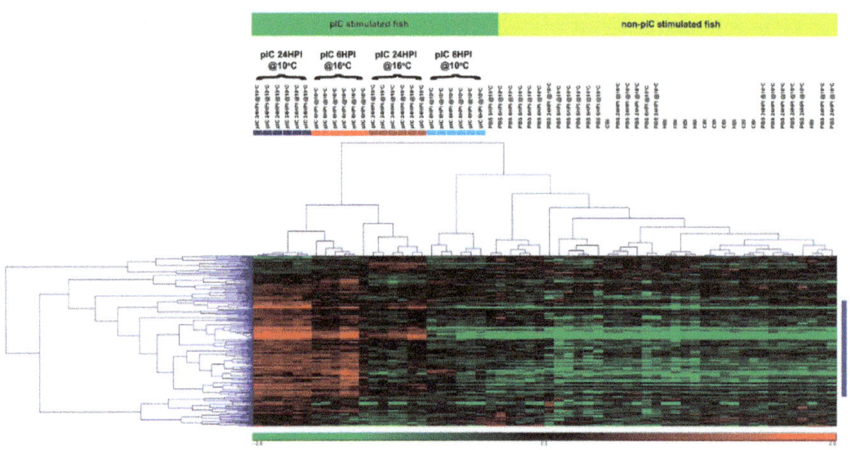

Figure 6: Heat-map depiction of hierarchical clustering of all samples (i.e. microarrays) and the 339 genes differentialy expressed (FDR = 1%) between pIC@10°C and pIC@16°C at 24HPI based on their mRNA expression (see Figure2). The colored boxes below the top legend represent individual fish from the pIC injected groups at the different temperatures and time points. Light red = pIC@16°C 6HPI; light blue = pIC@10°C 6HPI; dark red = pIC@16°C 24HPI; and dark blue = pIC@10°C 24HPI.

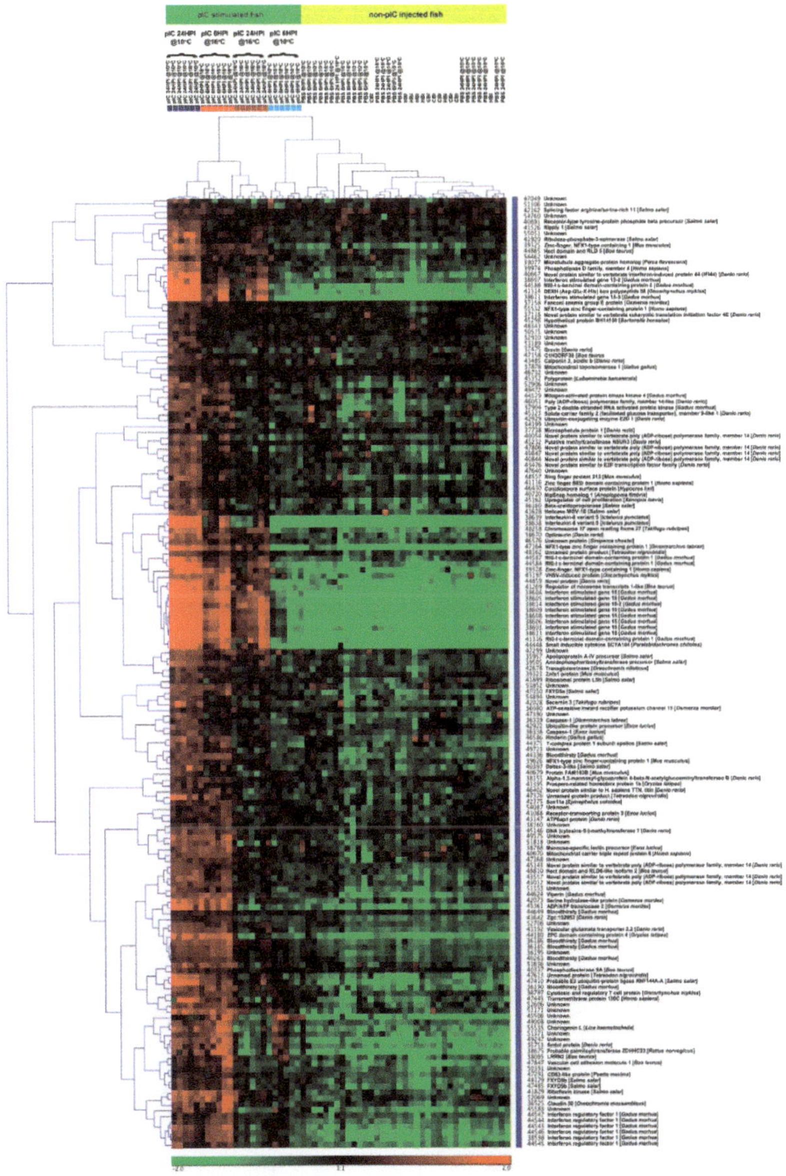

Figure 7: Detail of a sub-tree (marked in blue on Figure 6) from the hierarchical clustering of all samples (i.e. microarrays) and the 339 genes differentialy expressed (FDR = 1%) between pIC@10°C and pIC@16°C at 24HPI based on their mRNA expression (see Figures2and6) enriched for genes belonging to the interferon pathway. The colored boxes below the top legend represent individual fish from the pIC injected groups at the different temperatures and time points. Light red = pIC@16°C 6HPI; light blue = pIC@10°C 6HPI; dark red = pIC@16°C 24HPI; and dark blue = pIC@10°C 24HPI.

QPCR validation of selected immune-relevant genes

Genes of interest (GOI) were selected for QPCR analysis based on their putative roles in pathogen detection, signal transduction/transcription control and as immune effectors; the QPCR data is presented in Figures8,9, and10, respectively, as average $\log_2$ transformed relative quantities (RQs) ± SE. Fifteen of the immune-relevant genes (88%; all except TLR3 and IRAK4) were validated as being significantly ($P < 0.05$) differentially expressed in at least one comparison (e.g. pIC 6HPI@10°C vs. PBS 6HPI@10°C) from which they were identified as informative probes in the microarray experiment. There are several possible explanations for disagreement between microarray and QPCR results (e.g. differences between location of QPCR amplicon and microarray probe or possible misassembly of contigs, for more details see the discussion of Booman et al.[21]). For TLR3 and IRAK4 in the current study, there may have been related sequences (e.g. paralogs) binding to these probe in the array and influencing hybridization results. Also, the fact that 3 of the individuals used in the QPCR experiment were not part of the microarray experiment may have contributed to the disagreement between microarray and QPCR results for these two genes. Of the 4 genes with putative pathogen detection roles that were subjected to QPCR, 3 (TLR9, PKR, and DHX58) were validated (i.e. showed a significant difference in at least one of the group comparisons corresponding to the microarray comparison identifying the genes as differentially expressed). In the QPCR experiment for genes potentially involved in pathogen detection, DHX58 showed the highest fold up-regulation in response to pIC (13.00-fold at 6HPI@16°C and 34.86-fold at 24HPI@10°C) and the highest fold differential expression between fish injected with pIC at different temperatures (at 6HPI DHX58 was 13.64-fold more highly expressed in fish injected with pIC at 16°C, and at 24HPI it was 4.45-fold more highly expressed in fish injected with pIC at 10°C) (Figure8D). PKR showed the same trend but with lower, and yet significant, fold-changes (at 6HPI, 1.56-fold higher in pIC@16°C vs. PBS@16°C and 1.86-fold higher in pIC@16°C vs. pIC@10°C; at 24HPI, 2.73-fold higher in pIC@10°C vs. PBS@10°C and 1.65-fold higher in pIC@16°C vs. PBS@16°C).

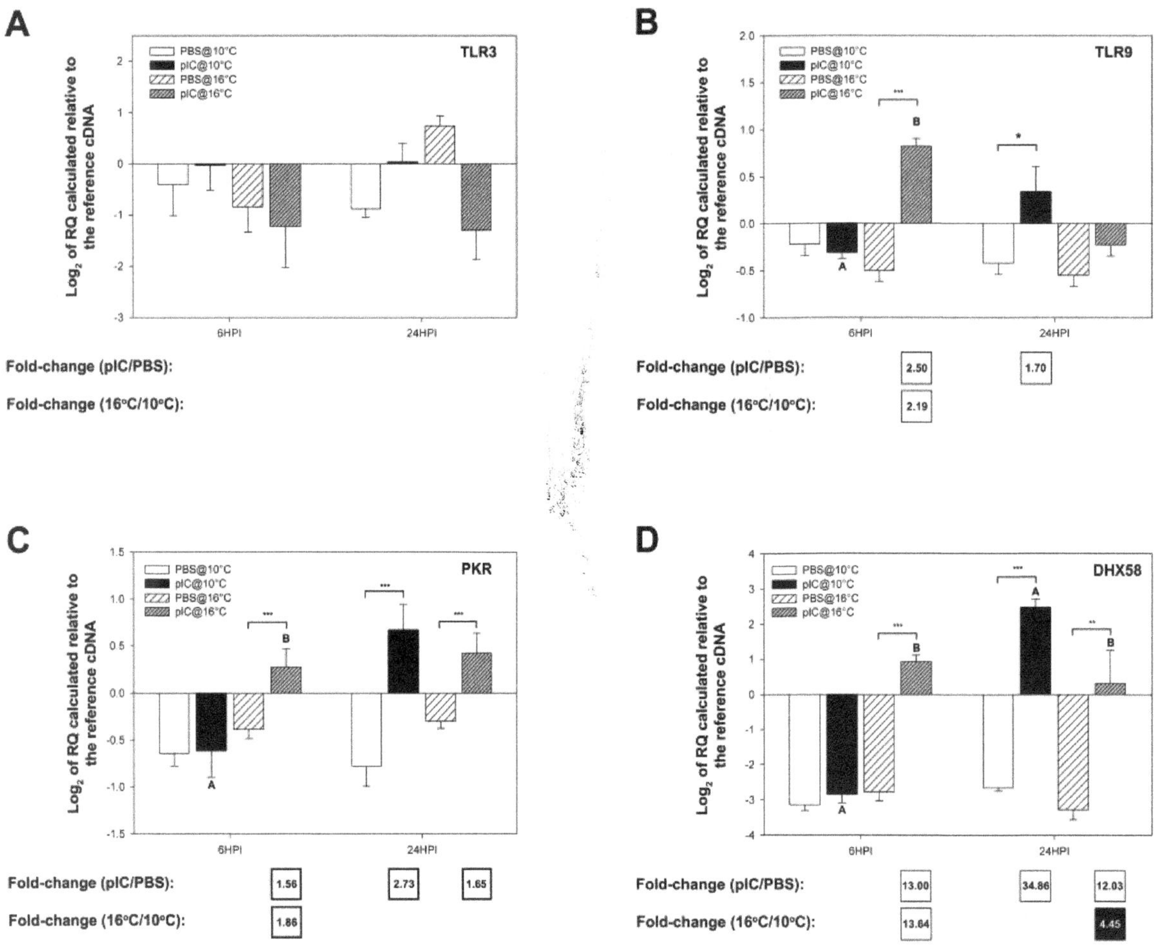

Figure 8: QPCR results for transcripts with putative roles in the detection of a pathogen (see Additional file 2 : Figure S1 schema of the TLR/interferon pathways). Data is presented as mean $\log_2$ transformed data ± SE. Different letters represent significant differences between fish injected with pIC at different temperatures within each time point. An asterisk (*) represents a difference between a given pIC-injected group and the time- and temperature-matched PBS-injected group (* $p < 0.05$; ** $p < 0.01$; *** $p < 0.001$). Fold-changes in white boxes are always pIC/PBS or 16°C/10°C, and are shown in the boxes beneath each panel. Only significant differences are shown. Black boxes show 1/fold-change for comparisons that yielded fold-change values less than one (for more details on fold-change calculations, refer to the methods section). For example, if white boxes (i.e. black numbers on white background) show 16°C/10°C fold change then a black box would show 10°C/16°C fold change. **A**) TLR3; **B**) TLR9; **C**) PKR; **D**) DHX58.

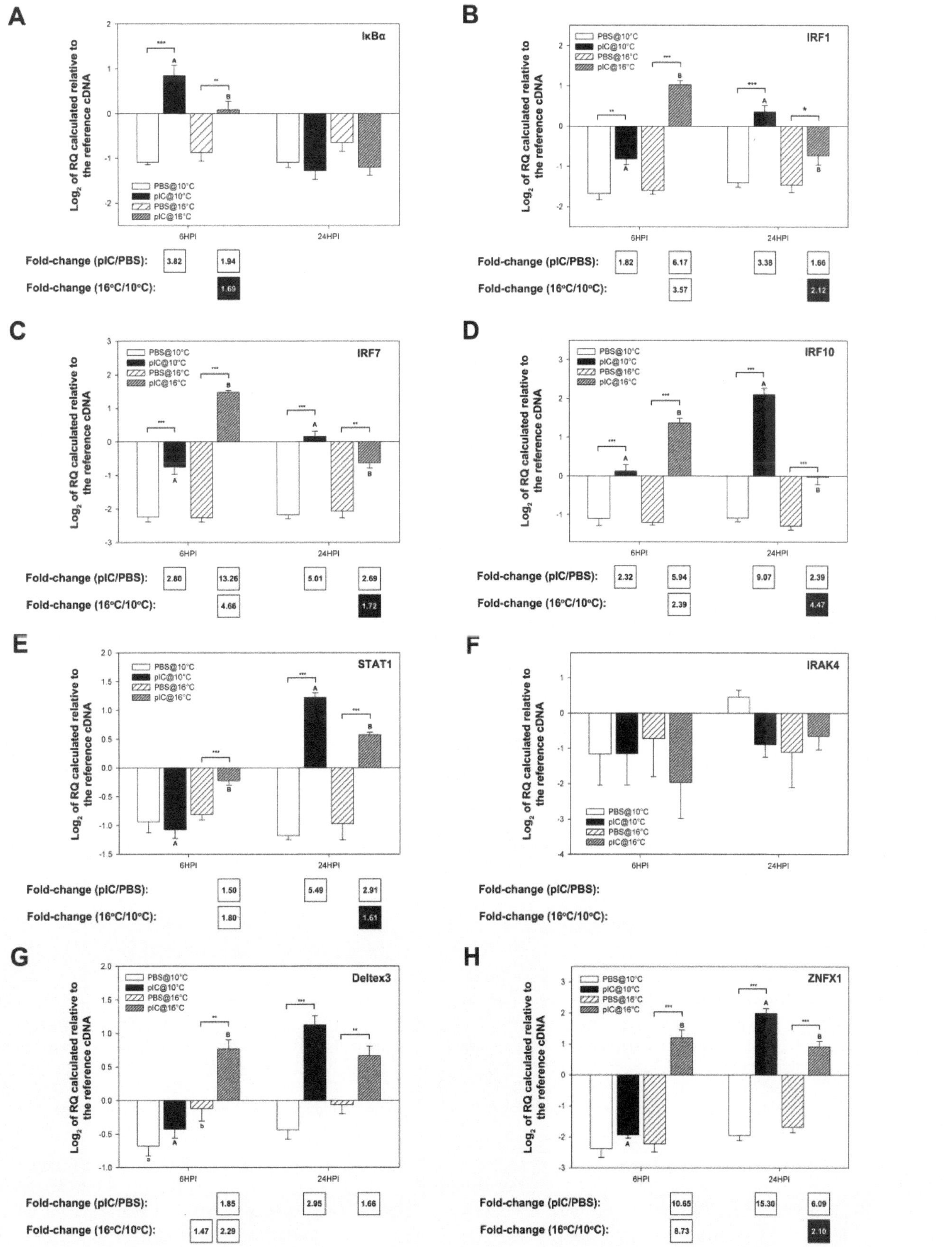

Figure 9: QPCR results for transcripts with putative roles in signal transduction/transcription control (see Additional file 2: Figure S1 schema of the TLR/interferon pathways). Data is presented as mean $\log_2$ transformed data ± SE. Different letters represent significant differences between fish injected with pIC at different temperatures within each time point. An asterisk (*) represents a difference between a given pIC-injected group and the time- and temperature-matched PBS-injected group (* $p<0.05$; ** $p<0.01$; *** $p<0.001$). Fold-changes in white boxes are always pIC/PBS or 16°C/10°C and are shown in the boxes beneath each panel. Only significant differences are shown. Black boxes show 1/fold-change for comparisons that yielded values less than one (for more details on fold-change calculations, refer to the methods section). For example, if white boxes (i.e. black numbers on white background) show pIC/PBS fold change then a black box (i.e. white numbers on black background) would show PBS/pIC fold change. **A)** IκBα; **B)** IRF1; **C)** IRF7; **D)** IRF10; **E)** STAT1; **F)** IRAK4; **G)** Deltex3; **H)** ZNFX1.

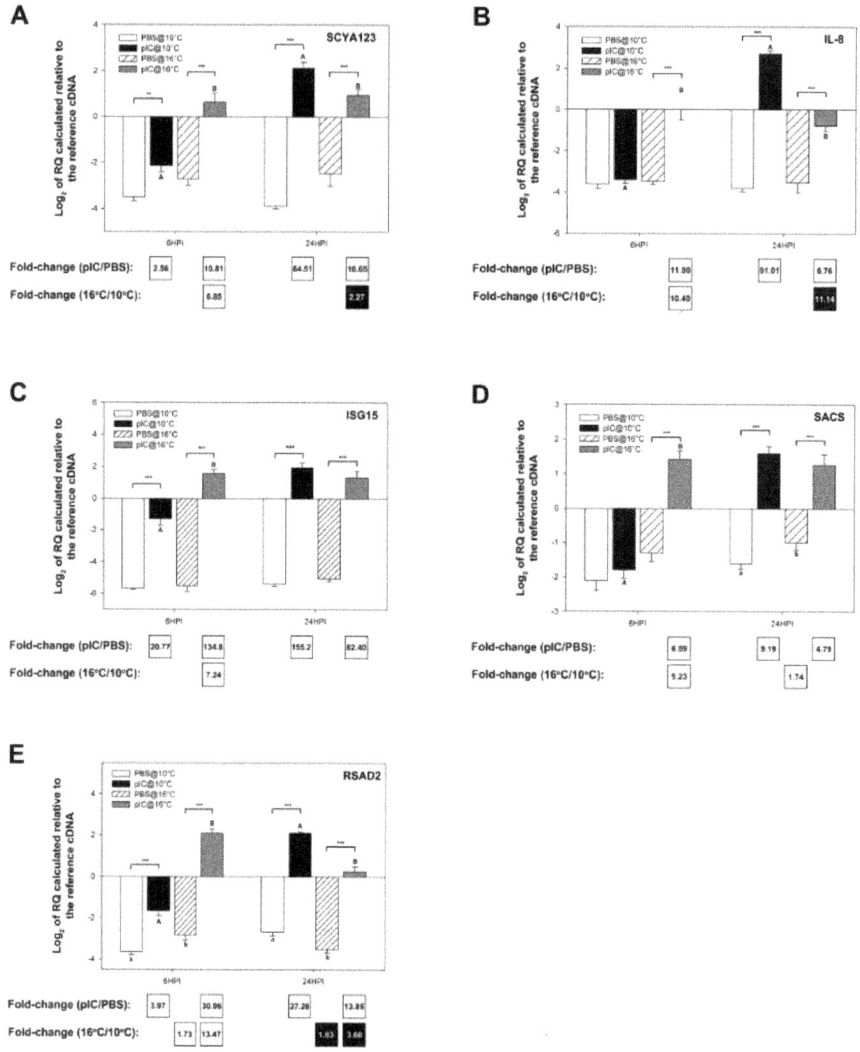

Figure 10: QPCR results for transcripts with putative roles as immune effectors (see Additional file 2 : Figure S1 schema of the TLR/ interferon pathways). Data is presented as mean $\log_2$ transformed data ± SE. Different letters represent significant differences between fish injected with pIC at different temperatures within each time point. An asterisk (*) represents a difference between a given pIC-injected group and the time- and temperature-matched PBS-injected group (* $p<0.05$; ** $p<0.01$; *** $p<0.001$). Fold-changes in white boxes are always pIC/ PBS or 16°C/10°C and are shown in the boxes underneath each panel. Only significant differences are shown. Black boxes show 1/fold-change for comparisons that yielded values less than one (for more details on fold-change calculations, refer to the methods section). **A**) SCYA123; **B**) IL-8; **C**) ISG15; **D**) SACS; **E**) RSAD2.

Of the 8 microarray-identified genes with putative roles in signal transduction/transcription control analyzed with QPCR, 7 (IκBα, IRF1, IRF7, IRF10, STAT1, Deltex3, ZNFX1) were validated. QPCR analysis showed that IκBα mRNA was only differentially expressed between groups at the 6HPI time point (Figure9A: 3.82-fold higher in pIC@10°C vs. PBS@10°C, 1.94-fold higher in pIC@16°C vs. PBS@16°C and 1.69-fold higher in pIC@10°C vs. pIC@16°C). All three studied transcripts encoding interferon regulatory factors (IRFs - 1, 7, and 10, Figure9B, C and D, respectively) showed very similar expression profiles at the mRNA level. These genes responded significantly to pIC at both temperatures and time points, although the magnitude of the fold up-regulation was variable. Of all studied IRFs, IRF7 at the 6HPI time point showed the highest response to pIC (13.26-fold at 16°C) and the highest fold difference between fish injected with pIC at 16 vs. 10°C (4.66-fold). At the 24HPI time point, IRF10 showed the highest response to pIC (9.07-fold at 10°C) and the highest fold difference between fish injected with pIC at 10 vs. 16°C (4.47-fold). Of all the genes with putative roles in signal transduction/transcription control, ZNFX1 showed the highest response to pIC at 10°C (Figure9H: 15.30 fold up-regulation at 24HPI). As previously mentioned, the microarray results for IRAK4 (Tables1 and2) were not validated by QPCR analysis (Figure9F).

All 5 of the putative immune-effector genes studied with QPCR showed significant differential expression at 6HPI and 24HPI between fish injected with pIC vs. PBS at 10 and/or 16°C (Figure10A-E). Further, they showed the highest fold-changes of all the transcripts studied with QPCR. At 6HPI, ISG15 showed the highest response to pIC at both temperatures (20.77-fold at 10°C and 134.8-fold at 16°C), and RSAD2 (alias viperin) (Figure10E) showed the highest differential mRNA expression between fish injected with pIC at the two temperatures (13.47-fold higher in 16°C vs. 10°C). At 24HPI, ISG15 also showed the highest response to pIC at both temperatures (155.2-fold at 10°C and 82.40-fold at 16°C) (Figure10C), while IL-8 presented the highest fold difference between fish stimulated with pIC at 10 vs. 16°C (11.14-fold higher at 10°C) (Figure10B). ISG15 was the only microarray identified, QPCR studied putative immune-effector transcript for which not all differences detected on the microarray were validated by QPCR. While ISG15 was identified as being differently expressed (FDR = 1%) between fish injected with pIC at 10 vs. 16°C at 24HPI in the microarray experiment, only a non-significant change in the same direction was observed in the QPCR at 24HPI.

DISCUSSION

Variations in temperature are an inevitable element of cage-site based aquaculture [8, 32, 33]. We used a regimen of gradual increases in water temperature that simulated the environmental conditions to which the farmed Atlantic cod can be exposed during the spring-summer transition in sea-cages in Newfoundland, and microarrays and QPCR to study the impact of changes in rearing temperature on the spleen anti-viral transcriptome response of cod. Several studies[34, 35, 36] have used functional genomics approaches to study the individual impact of stress or immune stimulation on fish transcriptomes; however, to our knowledge, the current study is the first to use global gene expression analysis to investigate the impact of elevated water temperature on the anti-viral immune transcriptome of a fish species. We were able to demonstrate at the mRNA level that, while a gradual non-lethal increase in water temperature from 10°C to 16°C only caused minor changes in the spleen transcriptome, it had a major influence on the cod spleen gene expression response to IP injection with the viral mimic pIC.

Influence of Elevated Temperature on Plasma Cortisol and the Spleen Transcriptome

In the present study, plasma cortisol levels did not increase significantly when cod were exposed to the increasing temperature regimen. This finding suggests that our juvenile cod were not stressed under these conditions, and that the differences observed in the pIC response of fish exposed to elevated temperatures may not have been the result of transcriptional changes driven by the glucocorticoid receptor (GR). However, this is in contrast to the results of Pérez-Casanova et al.[9] who reported plasma cortisol values of ~ 52 ng ml^{-1}in similar sized cod that had been warmed from 10°C to 16°C using a similar protocol. This discrepancy is likely due to the fact that Pérez-Casanova et al.[9] sampled their fish as temperature reached 16°C, whereas we sampled the cod after they were given a week to acclimate to this temperature. It is probable that the extra week to which our fish were exposed to 16°C allowed their cortisol (stress) level to return towards basal values. This hypothesis is also supported by Pérez-Casanova et al.[9], as these authors reported that cod plasma cortisol levels decreased on day 40 of their experiment (temperature ~ 17°C) and that values remained at pre-stress levels as temperature was increased further.

Increasing water temperature to 16°C alone had a very small impact on the cod spleen transcriptome (only 6 up-regulated genes were detected at FDR = 1%, Figure2). Using microarrays, we were unable to detect any differential expression of β2-microglobulin, MHC-I or IgM heavy chain transcripts due to increases in water temperature alone. This result is again different from the results of Pérez-Casanova et al.[9] who showed, using QPCR with blood cDNA templates, that these genes were differentially expressed between cod at 16°C and at 10°C. Although we cannot rule out tissue-specific differences (spleen vs. blood) in the response of these genes to the temperature increase, it is also possible that the different sampling regimes (see above) could explain these disparate findings. When this data is combined with that for plasma cortisol levels, it suggests that prolonged exposure of Atlantic cod juveniles to sea-cage temperatures of 16°C (in the absence of immune challenge) is unlikely to have negative consequences with regards to their stress state or basal immune transcript expression.

Temperature Effects on the Atlantic Cod Spleen Anti-Viral Transcriptome

Compared to the time- and temperature-matched PBS controls, pIC altered the expression of 41 (37 up-regulated and 4 down-regulated) and 999 (851 up-regulated and 148 down-regulated) transcripts in 10°C fish at 6HPI and 24HPI, respectively, and 656 (617 up-regulated and 39 down-regulated) and 246 (243 up-regulated and 3 down-regulated) transcripts in 16°C fish at 6HPI and 24HPI, respectively (Figure2). Figure3 shows a summary of the GO terms that were associated with the genes differentially expressed between fish injected with pIC at the different temperatures and sampled at 6HPI (Figure3A-B) and 24HPI (Figure3C-D). Not surprisingly, immune system process, signaling and response to stimulus were found in 3 out of the 4 pie charts in relatively high proportions (Figure3A, C and D). These results indicate, as previous reports have shown[7, 11, 12, 14], that elevated temperatures can induce changes in the fish's immune response to an antigen. Furthermore, we suggest that the genes associated with these GO terms (e.g. biological regulation, death, growth, metabolic process; see Figure3) may be good candidates for the development of molecular markers [e.g. exonic, intronic or regulatory region single nucleotide polymorphisms (SNPs)] to be used in marker assisted selection (MAS) programs aimed at developing elite broodstock (e.g. resistant to potentially negative impacts of elevated temperature on immune response) for the Atlantic cod aquaculture industry.

The sample clustering based on genes differentially expressed between the groups injected with pIC and sampled at 6HPI shows that the expression profile of pIC injected fish held at 10°C and sampled at 6HPI was most similar to that of the non-immune stimulated (i.e. non-injected or injected with PBS) fish than to the profile of any other group of fish injected with pIC (i.e. pIC 6HPI@16°C, pIC 24HPI@10°C and pIC 24HPI@16°C) (Figure5A). In addition, clustering based on the 339 genes differentially expressed between fish injected with pIC at the different temperatures and sampled at 24HPI (Figures6 and7), indicates that the mRNA expression profile of pIC-injected fish held at 16°C and sampled 6HPI was most similar to that of fish injected with pIC at 10°C and sampled at 24HPI. Collectively, these data suggest that the moderate increase in water temperature caused a time shift in the transcriptomic response of cod spleen to pIC, with fish held at elevated temperature (16°C) having an earlier maximum response (i.e. at 6HPI) than fish held at optimal temperature (10°C), which had a later maximum response (i.e. at 24HPI). Raida and Buchmann[11] report similar data for mRNA expression of immune-relevant genes in the spleen of rainbow trout injected with a *Yersinia ruckeri* bacterin. These authors found that peak transcript expression of IL-1β, IL-10 and IFN-γ happened earlier in fish stimulated at 15°C or 25°C when compared to fish stimulated at 5°C. Furthermore, it is clear from our heat-maps that, overall, the magnitude of response to pIC for these genes was greatest in the pIC stimulated fish at 10°C and sampled at 24HPI. This result, however, must be interpreted with caution. Our results suggest that holding cod at 16°C may result in a faster, but weaker, immune response to a viral mimic; this hypothesis could be tested in the future by sampling pIC-injected cod exposed to similar temperature regimes at more frequent intervals post-injection. In future research, more frequent sampling (e.g. every 2 hours post-injection) may allow one to determine if the maximum induction of pIC responsive genes that we observed in the fish held at elevated temperature (16°C) and sampled 6 hours after pIC-injection is the peak response, which would indicate that there is indeed a weaker maximal pIC response in these fish compared with fish held at optimal temperature (10°C). However, if peak spleen transcript expression response to pIC occurs before or after 6HPI for fish held at elevated temperature, and before or after 24HPI for fish held at optimal temperature, this could be determined in future studies incorporating more frequent post-injection sampling.

Using the clustering of genes based on their expression profiles for each sample, we were able to identify many different clusters, including one from each time point that is highly enriched for putative members of the interferon pathway. These clusters are marked in blue on Figures5A and6 and shown in detail in Figures5B and7. The interferon pathway is a key part of the fish innate response to viruses[20, 22, 25, 37, 38], and our results indicate that elevated temperature had a considerable impact on the cod innate immune response to a viral mimic.

Impacts of the gradual temperature increase on transcript expression

Genes with putative viral detection roles

One of the key steps in mounting an anti-viral innate immune response (e.g. expression of type I interferons and proinflammatory cytokines) is the detection of an invading pathogen. This recognition often occurs via the detection of a set of pathogen associated molecular patterns (PAMPs). PAMPs bind specifically to germ-line pattern recognition receptors (PRRs), which in turn activate signaling pathways that induce the innate immune response[39, 40, 41]. The immune-stimulant used in this experiment (pIC) is a double-stranded RNA (dsRNA) that mimics the genome and/or RNA intermediates of several viruses and is recognized by PRRs, including the Toll-like receptor 3 (TLR3). Several genes putatively belonging to the TLR pathway (e.g. TLR3) have been identified in fish[36, 41] (see Additional file2: Figure S1 for a schema of a putative type I IFN activation via TLRs pathway in Atlantic cod). While TLR3-like transcript was shown to be slightly up-regulated by pIC in 16°C fish at 6HPI on the microarray (1.32 fold – Table1), the QPCR analysis did not confirm this result (Figure8A). In fact, we did not detect any significant changes in TLR3 transcript due to pIC injection using QPCR. This is not surprising as Rise et al.[22] obtained similar QPCR results for TLR3 (using the same primer pair) in spleens from cod stimulated with pIC. However, it suggests that the differences in response to pIC between fish held at 10 vs. 16°C were not caused by an enhanced sensitivity to double-stranded RNA due to an over-expression of TLR3 in the spleens of fish held at 16°C and injected with pIC. The results of Rodriguez et al.[42] for *in vivo*stimulation of rainbow trout with IP injection of pIC also agree with ours, as these authors detected no significant induction of TLR3 mRNA following injection of this viral mimic. Interestingly, these authors and others[43] have shown up to ~30-fold induction of TLR3 transcripts by pIC in isolated rainbow trout cells/cell culture, and this is similar to what has been observed in mammalian macrophages[44].

In the current study, TLR9 transcripts were found by both the microarray and QPCR analyses to be significantly up-regulated by pIC at 6HPI in 16°C fish, and at 24HPI for fish held at 10°C (Table2; Figure8B); the QPCR analysis showed that mRNA levels of TLR9 were significantly different between 10 and 16°C pIC injected fish sampled at 6HPI (Figure8B). Like in mammals, the main ligand of TLR9 in fish is thought to be viral/bacterial unmethylated CpG DNA[41, 45]. It is unclear what the roles of TLR9 during the host response to viral dsRNA could be, but it is apparent that temperature can also have an impact on the responses of fish to other pathogens such as bacteria by modulating TLR9 mRNA expression[41]. Further research is warranted to elucidate the modulations of TLRs by antigens and/or changes in the environment in teleost fish.

Other important sensors of viruses are cytosolic PRRs, such as the RIG-I like receptors (RLRs) and the dsRNA activated protein kinase (PKR). DHX58 (alias LGP2) is a RLR that was found to be highly induced at the mRNA level by pIC compared to PBS at both 10°C and 16°C and also differentially expressed (at FDR = 1%) between fish injected with pIC at 10 and 16°C at both time points in the QPCR and microarray experiments (Tables1and2, Figure8D). Previous studies have reported up-regulation of DHX58 mRNA by pIC in both cod spleen and brain[22, 25]. In addition, QPCR was used to show that DHX58 mRNA was significantly up-regulated in the brains[25], but not the spleens[22], of asymptomatic high nodavirus carrier Atlantic cod. As in Rise et al.[22, 25] we detected peak up-regulation of DHX58 mRNA at 24HPI for fish injected with pIC at 10°C. In the spleens of fish injected with pIC at 16°C, the fold up-regulation (compared to the PBS control) of DHX58 transcript was similar at 6HPI and 24HPI

(13-fold and 12-fold, respectively) (Figure8D). The roles of DHX58 in the anti-viral response are still unclear[22, 46, 47]. However, there is some experimental evidence that it may induce or repress RIG-I/MDA5 dependent signal transduction in a virus-dependent manner[46]. Thus, it is still not clear what impacts the observed temperature-dependent modulation of DHX58 response to pIC could have in the anti-viral immunity of cod. We observed similar differences in the expression of PKR mRNA (Figure8C). Like TLR9, DHX58, STAT1, Deltex3, ZNFX1, IL-8 and SACS, PKR did not respond to pIC at 6HPI in fish held at 10°C, but it was significantly up-regulated at the same time point in the spleens of fish held at 16°C and injected with pIC. In mammals, the main anti-viral function of PKR was originally thought to be the phosphorylation of eIF2α, leading to reduced protein synthesis[48]. More recently, it has been demonstrated that PKR has roles in modulating apoptosis via NF-κB and the growth-inhibitory activity of IRF1[49]. Therefore the observed influence of pIC, temperature and time on PKR mRNA expression could impact the fish's ability to fight a viral infection.

Genes with Putative Roles in Signal Transduction and Transcription Regulation

In mammals, downstream signaling of TLRs and RLRs involves activation of NF-κB transcription factors[46, 50]. There is some evidence of a similar mechanism in fish, since expression constructs containing a constitutively active form of the zebrafish (*Danio rerio*) TLR3 transfected into ZFL cells induced NF-κB mediated luciferase fluorescence[38]. An important step in TLR signaling in mammals is the activation of the Inhibitor of NF-κB Kinases (IKKs), which in turn phosphorylate the Inhibitor of NF-κB proteins (IκBs). Phosphorylation targets IκBs for degradation, and in their absence, the NF-κB proteins accumulate in the nucleus and regulate mRNA transcription[51, 52]. In higher vertebrates, IκBα is thought to regulate transient activation of NF-κB. As such it is rapidly degraded in response to stressful stimuli, but then quickly re-synthesized due to the presence of an NF-κB response element in its promoter region[51] (see Additional file2: Figure S1 for a putative equivalent Atlantic cod pathway). At 6HPI, both the microarray and the QPCR studies showed an significant up-regulation of IκBα mRNA by pIC at both temperatures (Table1 and Figure9A). This may be an indication of the process of re-synthesis of this protein in the spleen cells following pIC stimulation. However, the QPCR-detected up-regulation of IκBα mRNA by pIC at 6HPI was higher in fish held at 10°C than in fish held at 16°C (3.82-fold vs. 1.94-fold, respectively). Moreover, the difference in mRNA expression of IκBα between 10°C and 16°C fish injected with pIC at 6HPI was also significant (both in the QPCR and the microarray studies). It is possible that the early response of IFN pathway genes observed in the fish held at 16°C (see Figure5B and7) was in part due to a reduced re-synthesis of IκBα, leading to a more pronounced NF-κB-mediated induction of transcription. At 24HPI we detected no significant differences in the mRNA expression of IκBα in fish held at 10°C or 16°C with either the microarray (Table2) or QPCR (Figure9A) studies. The microarray analysis indicated that IKKα was significantly up-regulated by pIC at this point only in the spleens of fish held at 10°C (Table2). Since IKKs are also essential to NF-κB activation, this difference may have contributed to the differences in immune-related gene expression observed at 24HPI (e.g. as shown by the hierarchical clustering; Figures5,6 and7). Collectively, our results suggest that, as in higher vertebrates, timing and intensity of Atlantic cod immune responses are partly regulated by NF-κB transcription factors[51].

The interferon regulatory factor (IRF) genes are also key to the activation of the IFN pathway[53]. In this work all 3 IRFs (i.e. IRF1, IRF7 and IRF10) that had been previously identified as pIC-responsive in cod[22] were shown to be up-regulated in spleen by this viral mimic (Table1 and2, Figure9B, C and D). Moreover, all studied IRFs presented the same overall expression profile as most IFN pathway-related genes identified in this study. This profile is characterized by a higher expression at 6HPI in fish injected with pIC and held at 16°C vs. 10°C, but greater expression in this latter group at 24HPI. In higher vertebrates both IRF1 and IRF7 have been implicated as positive modulators of the type I IFN response to viruses, with IRF7 being essential for interferon-related gene expression[53]. Furthermore, in the current microarray experiment TANK-binding kinase 1 (TBK1), which activates IRF7 by phosphorylation, was found to be more highly expressed in the spleens of fish injected with pIC at 10°C and sampled at 24HPI (1.37 fold – Table2) than in the spleens of fish injected with pIC at 16°C at this time point. It is likely that the differential up-regulation of the IRFs and TBK1 in fish stimulated with pIC at the different temperatures contributed to the global expression profiles (i.e. with several IFN-related genes peaking at different times in fish injected with pIC at 10°C as compared to those injected with pIC at 16°C) depicted in Figures5,6 and7.

Once type I interferons (e.g. IFN-β) are expressed, they interact with specific cell surface receptors (IFN α/β receptors [IFNAR]) triggering the expression of IFN stimulated genes (ISGs)[54]. This signaling cascade occurs partly through the dimerization of the signal transducer and activator of transcription (STAT) 1 and 2 proteins and their interaction with IRF9. The STAT1/STAT2/IRF9 complex then further induces STAT1 expression[54]. In cod, STAT1 mRNA was previously identified as being significantly more highly expressed in the brains of fish with a high nodavirus carrier status compared to fish with no/low nodavirus; STAT1 mRNA was also shown to be significantly up-regulated in cod brain 24 and 72 h after IP injection with pIC[25]. In the current study we found that: 1) at 6HPI STAT1 was only up-regulated in fish injected with pIC at 16°C; 2) as with the IRFs at 24HPI, both pIC-injected groups showed significant STAT1 mRNA up-regulation; and 3) the STAT1 QPCR-detected up-regulation was more pronounced in the 10°C pIC-injected fish and sampled at 24HPI (5.49-fold at 10°C vs. 2.91-fold at 16°C) (Figure9E). This suggests that the modulation of the expression of IRFs at the mRNA level could have affected the production of type I IFN, and together with the observed STAT1 transcript expression profiles may also have contributed to the global gene expression profiles depicted in Figures5,6and7.

Genes with Putative Functions as Immune Effectors

The shift in the timing of the transcriptome response to pIC between fish stimulated at 10°C vs. 16°C (Figures5,6 and7) was also observed for several immune effector genes (Figure10A-E). Of the genes studied with QPCR, IL-8 transcript expression is known to be under the control of NF-κB in mammals as evidenced by marked reductions in its expression in IKKε$^{-/-}$ mouse cells[55] (See Additional file2: Figure S1 for a putative Atlantic cod NF-κB dependent control of immune effector gene transcription). In the present work, IL-8 transcript was 11-fold significantly more highly expressed in the spleens of pIC injected fish held at 16°C and sampled at 6HPI compared with 16°C PBS controls. Further, at 24HPI, IL-8 mRNA was 91-fold up-regulated in the spleens of pIC-injected fish held at 10°C, but only ~ 7-fold up-regulated in the spleens of pIC-injected fish held at 16°C, compared with the time- and temperature-matched PBS controls (Figure10B). Two of the genes identified in the microarray study and subjected to QPCR validation are known ISGs: ISG15 and RSAD2 (alias viperin)[56]. These genes were previously reported to be pIC-responsive in the cod spleen[22] and significantly up-regulated in the brains of cod by asymptomatic high nodavirus carrier state and IP pIC treatment[25]. It is not surprising that, in the current study, these genes presented the same mRNA expression profiles as the IRFs (Figure10C and E). For example, ISG15 was ~ 7 fold significantly more highly expressed in 16°C pIC injected fish at 6HPI and 1.51 fold more highly expressed in 10°C pIC-injected fish sampled at 24HPI when compared to the respective time-matched pIC injected fish (Note: This difference was detected in the QPCR experiment but was only significant in the microarray experiment, possibly due to the fact that our ISG15 QPCR primers only amplified 2 of the 3 known ISG15 paralogs[57]). Further, RSAD2 mRNA was 13.5-fold significantly more highly expressed in 16°C pIC-injected fish at 6HPI compared to 10°C pIC-injected cod at 6HPI and 3.6 fold significantly more highly expressed in 10°C pIC-injected fish at 24HPI compared to 16°C pIC-injected fish at the same time point. However, unlike what was observed for IL-8, the peak fold-change in pIC-stimulated fish for ISG15 and RSAD2 mRNA expression (compared with the time- and temperature-matched PBS controls) was similar for 16°C fish at 6HPI as compared with 10°C cod at 24 HPI (Figure10C and E).

SCYA123[22] and SACS[25] have both been previously identified as up-regulated by pIC in cod, and as expected both SCYA123 and SACS were significantly up-regulated in the spleen of cod injected with pIC at 10 and 16°C (Figure10A and D). However, little is known about the roles of these proteins in the anti-viral response. SCYA123 is a small inducible cytokine, but phylogenetic clustering indicates that it is most related to the CCL19 group[58] of higher vertebrates. In mammals, the CCL19 chemokines are constitutively expressed and thought to have a minor role in immunity (see[58] and references therein). Nonetheless, it is clear that SCYA123 transcripts are up-regulated by pIC and other antigens (e.g. formalin-killed atypical *Aeromonas salmonicida*) in the spleen of cod[21, 22, 23, 58], and therefore, at least in this species, have some role in finfish immunity. Mutations in the SACS gene lead to the human disease spastic ataxia of Charlevoix-Saguenay[59], and it is known that the very large sacsin protein (4579 amino acid residues) encoded by SACS possesses heat shock protein (HSP)-like ATPase domains and HSP-like activity[60]. However, the function of this gene in immunity remains unclear[25]. SACS mRNA was shown by QPCR to be up-regulated by asymptomatic high nodavirus carrier state and IP pIC treatment in brains of cod[25] and by ISA (Infectious Salmon Anemia) virus infection in salmonid TO cells[20]. Moreover, SCYA123 and SACS presented mRNA expression profiles similar to other known immune-related genes reported here (Figures5,6 and7). Interestingly, the transcript expression profiles of SCYA123 and SACS at 6HPI were more similar to those of IL-8 and RSAD2 but at 24HPI was more similar to that of ISG15. Therefore, it is possible that these genes share some of their regulatory mechanisms, and they likely play important roles in Atlantic cod responses to viruses.

Cellular Response to Stress and Innate Immunity

The mild increase in temperature alone did not cause up-regulation of the genes considered to be part of the conserved cellular stress response (e.g. HSP40, HSP70 and peptidyl-prolyl-isomerase[61]). In contrast, several chaperone-like genes were more highly expressed at 24HPI in the fish injected with pIC at 16°C compared to fish injected with pIC at 10°C. In fact, at this time point 2 GO terms related to protein folding (protein folding and unfolded protein binding) were significantly enriched in the list of genes up-regulated by pIC at 16°C when compared to the list of genes up-regulated by pIC at 10°C. In the microarray experiment, chaperone-like genes such as HSP90-alpha, HSP90-beta, HSC71, HSP47, peptidyl-prolyl-isomerase A, CCT1 and CCT6 (Table2) were identified as either being induced by pIC at 24HPI in fish injected at 16°C but not in fish injected at 10°C, or significantly more highly expressed at the mRNA level in the pIC@16°C fish at 24HPI when compared to the pIC@10°C (Table2). Many of these genes (e.g. HSP90-alpha, HSP47 and CCT1) were previously identified as heat shock-responsive in the cod liver[24] and in a multiple-tissue RNA-seq study in rainbow trout[35]. This suggests that even though heat alone did not elicit a robust cellular stress response, the combination of heat and pIC injection did induce molecular biomarkers of cellular stress.

There is evidence from the mammalian literature that HSPs play a key role in modulating the innate immune response[62]. Interestingly at 24HPI, the GO terms "regulation of interferon-gamma-mediated-signaling", "regulation of signaling process", "regulation of signal transduction", and "regulation of cytokine-mediated signaling pathway" were significantly over-represented in the list of genes up-regulated by pIC at 16°C compared to those up-regulated by pIC at 10°C. These GO terms were largely associated with probes representing HSP90-like transcripts or HSP70. It is known that, at least in higher vertebrates, the cellular response to stress prevents innate immune signaling[62], and these effects are to some extent mediated by HSPs. It is possible that the significantly reduced magnitude of induction by pIC observed for some immune-relevant genes (e.g. DHX58, STAT1, SCYA123, IL-8) in Atlantic cod held at 16°C and sampled at 24HPI compared with pIC-injected fish held at 10°C and sampled at 24HPI may have been a result of the activation of the cellular stress response in these animals prior to the 24HPI time point.

CONCLUSIONS

In this study we used microarrays, followed by QPCR, to investigate the interaction between a moderate temperature increase and the innate immune response of cod after pIC stimulation. This is the first time, to our knowledge, that the global gene expression analysis has been used to investigate the interaction between changes in environmental conditions and the antiviral immune response of this commercially important species. We have been able to demonstrate that a moderate increase in seawater temperature, similar to those experienced by farmed Atlantic cod in sea-cages in Newfoundland during the summer, causes a massive shift in the spleen transcriptome response to a viral mimic. Furthermore, we present evidence to suggest that these changes may be, in part, regulated by changes in signal transduction (particularly in the NF-κB and IFN pathways) and by interactions between the cellular stress response and the innate immune response. Collectively, this new information enhances our understanding of the interactions between temperature and the Atlantic cod immune system, and will be valuable in global efforts to improve cod aquaculture practices and broodstock.

METHODS

All experimental procedures were conducted with approval of Memorial University's Institutional Animal Care Committee, and followed the guidelines of the Canadian Council on Animal Care.

Experimental Animals

Passive integrated transponder (PIT)-tagged Atlantic cod belonging to 10 different families (~ 30 fish per family) from the Atlantic Cod Genomics and Broodstock Development Project (CGP) year-class 3 (YC3) (52.6 ± 2.5 g) were obtained from the Huntsman Marine Science Centre in St. Andrew's, New Brunswick. The fish were transported to the Ocean Sciences Centre of Memorial University of Newfoundland in October of 2008, and placed in a 3000 l tank supplied with flow-through seawater (at 10°C and > 90% oxygen saturation). After one month of acclimation, the fish were distributed amongst eight 500 l tanks (~ 36 fish per tank) (Figure1A), each with approximately equal numbers of fish from each family in each tank. Fish were fed to apparent satiation throughout the acclimation period.

Experimental Design

Figure1 provides an overview of the complete experimental design. Four tanks out of the 8 were randomly assigned as temperature control tanks (i.e. kept at 10°C for the entire duration of the experiment), while the remaining 4 tanks were assigned as heat-exposed (Figure1A). After the fish were allowed to acclimate to the 500 l tanks for 2 weeks (Figure1B), 3 fish from each tank (N = 12 per group) had their blood sampled for the determination of plasma cortisol levels. These groups were designated "control before temperature increase" (CBTI) and "heat-exposed before temperature increase" (HBTI). Thereafter, water temperature in the heat-exposed tanks was increased by 1°C every ~ 5 days. These tanks reached 16°C after approximately 1 month, and were held for another ~1 week at that temperature (Figure1C). Fish were fed to apparent satiation during this ~ 5 week period and no significant differences in food consumption were detected between groups (data not shown). At this point, 8 fish per tank were sampled for plasma cortisol, 2 fish per tank had their spleens removed for subsequent microarray and QPCR analyses, and the remaining fish were given an IP injection of either polyriboinosinic polyribocytidylic acid (pIC) (Sigma Co, St. Louis, MO) [2 μg of pIC g^{-1} wet mass, 0.5 μg μl^{-1}in ice-cold 0.2 μm-filtered phosphate buffered saline (PBS)] or an equivalent volume of ice-cold 0.2 μm filtered PBS after recording their PIT tag ID (Figure1D). The fish sampled prior to pIC injection were designated as "control before injection" (CBI) and "heat-exposed before injection" (HBI). Finally, at both temperatures (10 and 16°C), 5 fish from each of two tanks (N = 10 per group / temperature) were sampled at 6 hours post-injection (6HPI) and at 24HPI (Figure1E) to examine the impact of elevated rearing temperature on the spleen's transcriptome response to the viral mimic pIC.

Immune Stimulation

Prior to immune stimulation, fish were starved for 24 hours and feeding was not resumed thereafter. After being lightly anesthetized using a non-lethal dose (100 mg l^{-1}) of tricaine-methane-sulphonate (TMS) (Syndel Laboratories, Qualicum Beach, BC), equal numbers of fish were injected with pIC or PBS. Thereafter, they were allowed to recover (i.e. re-establish normal swimming behavior) in a "recovery-tank" filled with seawater from the same system and temperature as before injection, and then returned to their original tanks. Oxygen levels in the recovery tanks were constantly monitored and maintained above 90% of air saturation.

Tissue and Blood Sampling

At all sampling points, the fish were quickly netted from their tanks and placed in an anaesthetic bath containing a lethal dose of TMS (400 mg l^{-1}). For cortisol analysis, whole blood was quickly obtained from the caudal vein using heparinized (100 U of sodium heparin ml^{-1}) 1 ml syringes (BD Medical Supplies, Franklin Lakes, NJ), centrifuged at 5000 x *g* for 10 minutes (4°C), and the resulting plasma was placed in separate 1.5 ml microcentrifuge tubes and frozen in liquid nitrogen. Spleen samples were collected using standard aseptic molecular biology techniques, placed in certified RNase-free 1.5 ml microcentrifuge tubes and flash-frozen in liquid nitrogen. All spleen samples were stored at −80°C until RNA isolations were performed.

Twelve fish (3 per tank) were sampled for plasma cortisol before the temperature increase (BTI), and compared with values before injection (BI), to allow us to determine the degree of stress induced by the increasing temperature regimen and being held at 16°C for 1 week. Thirty-two fish were sampled prior to the pIC and PBS injections (BI) to allow for correlation analysis of the relationship between temperature-induced plasma cortisol levels and constitutive (pre-injection) spleen immune-related gene expression using QPCR. However, because holding the fish at 10 vs. 16°C only resulted in 6 genes being differentially expressed (see results), this analysis was never performed.

Plasma Cortisol Assay

An enzyme-linked immunosorbent assay (ELISA) kit (Neogen, Lexington, KY) was used to measure plasma cortisol levels, and the methods are described in detail elsewhere[24]. The cortisol data was subjected to a Kolmogorov-Smirnov normality test, and a two-way ANOVA (with temperature and time point as main effects). However, neither of these factors had a significant impact on cortisol data, and thus, no further statistical analyses were carried out.

RNA extractions, DNase-I treatment and column purification

RNA extractions, DNase-I treatment and column purification were performed using the TRIzol reagent (Invitrogen, Carlsbad, CA), RNase-free DNase Set (Qiagen, Valencia, CA) and RNeasy MinElute Clean-up Kit (Qiagen), respectively. All procedures were done according to the manufacturers' instructions with minor changes as described in Hori et al.[24].

Microarray Hybridizations

A schema of the microarray experimental design is presented in Figure2. In summary, 6 individuals from each of the following groups were used for microarray analysis: CBI; HBI; pIC 6HPI@10°C; PBS 6HPI@10°C; pIC 6HPI@16°C; PBS 6HPI@16°C; pIC 24HPI@10°C; PBS 24HPI@10°C; pIC 24HPI@16°C; PBS 24HPI@16°C. In these abbreviations, the "@" followed by a number indicates the temperature at which fish were held, pIC refers to fish immune-stimulated with the viral mimic pIC, PBS refers to sham-injected fish and HPI stands for "hours post-injection".

For all of the above groups, 3 individuals from each of 2 different replicate tanks (N = 6) were selected based on RNA quantity and quality for microarray analysis. Seven µg of spleen DNase I-treated and column-purified total RNA from individuals involved in immune stimulation experiment that included pIC and formalin-killed *Aeromonas salmonicida* injections were pooled to generate a "common reference" sample. The resulting "common reference" pool included all of the individuals used in the present microarray experiment. After pooling, the common reference quantity was measured using spectrophotometry and its quality was checked by 1% agarose gel electrophoresis. For each experimental sample (e.g. pIC 6HPI@10°C, individual 1) or the common reference, 5 µg of DNase I-treated, column-purified RNA were used for target synthesis. Complementary DNA (cDNA) synthesis and labeling were performed using Invitrogen's SuperScript Direct Labeling kit following the manufacturer's instructions. The experimental sample was always labeled with ALEXA 647 fluorophore and the common reference was always labeled with ALEXA 555 fluorophore (Figure2). Labeled cDNA from one experimental sample and the common reference were pooled together and hybridizations were performed at 42°C for ~ 16 hours as described in Booman et al.[21]. Detailed protocols on all hybridizations and washing are available in Booman et al.[21].

Microarray Data Acquisition and Pre-Processing

Microarrays were scanned at 5 µm using a ScanArray Gx Plus scanner and ScanExpress software v4.0 (Perkin Elmer, Woodbridge, ON) with photomultiplier tube (PMT) settings adjusted for each channel of each array and laser power set at 90%. For each channel (i.e. ALEXA 647 or 555) on a given array, the average signal was measured in a circle of diameter ~ 2000 pixels. PMT settings were adjusted in subsequent scans until the average signal was between 600 and 1000 signal units and there was no more than 200 signal units difference between channels. Raw microarray data was saved as TIFF images and probe intensities were extracted using Imagene v7.5 (BioDiscovery, El Segundo, CA). In R, flagged and control spots were removed and the mArray package of Bioconductor was used to $\log_2$ transform and Loess normalize the data by sub-grid (i.e. print-tip Loess). Thresholding and averaging of the probes were performed as described in Booman et al.[21]. The R scripts are described in full in the supplemental materials of Booman et al.[21]. This microarray dataset is described in the Gene Expression Omnibus (GEO) series GSE27299, and individual samples are available under GEO accession numbers GSM675013-GSM675072 (including both pre- and post-normalization data).

Microarray Data Analysis

The Significance Analysis of Microarrays (SAM) algorithm[31] was used to perform two-class comparisons between groups (e.g. pIC 6HPI@10°C vs. pIC 6HPI@16°C) in order to identify differentially expressed genes (at FDR = 1%). Prior to SAM analysis, all probes that were absent in more the 25% of the arrays in the totality of the experiment (i.e. 15 or more out of 60 arrays) were removed from the dataset. This procedure generated a final dataset containing 12,154 probes. Any remaining missing values were imputed using the EM_array method from the LSimpute algorithm[63, 64]. We used the Bioconductor implementation of SAM available in the siggenes package[65] and set our False Discovery Rate (FDR) threshold to 1% to generate gene lists. Detailed R scripts are available in the supplemental materials of Booman et al.[21]. Resulting data was then clustered in Genesis[66] using Pearson uncentered correlation and complete linkage hierarchical clustering.

Resulting gene lists were re-annotated using the expressed sequence tags (ESTs) or contiguous sequences (contigs) from which 50 mer oligonucleotide probes on the array were designed. For details on probe design please refer to Booman et al.[21]. Automated BLASTx alignment of these sequences against the NCBI nr database was performed using the Blast2GO tool[67] with thresholds set as follows: E-value $<10^{-5}$, maximum number of hits = 100 and high scoring pairs (HSP) >33. BLASTx results for each probe were mapped to Gene Ontology (GO) terms using Blast2GO. For detailed information on how BLASTx results are mapped to GO terms, please refer to the Blast2GO manual and Conesa et al.[67]. Biological process GO mapping results for the genes differentially expressed between fish injected with pIC at the different temperatures (i.e. 10 vs. 16°C) were normalized to GO term hierarchy level 2 and plotted as pie charts (Figure3 A-D). GO term enrichment analysis was performed to compare the pIC response at different temperatures using Fisher's exact test as implemented by GOSSIP in the Blast2GO software with a p-value cutoff of <0.01. We compared the distribution of GO terms between lists of genes differentially expressed in the spleens of fish injected with pIC to those injected with PBS at the different temperatures [e.g. (genes up-regulated in pIC injected fish @10°C vs. time-matched PBS injected controls @10°C) compared to (genes up-regulated in pIC injected fish @16°C vs. time-matched PBS-injected controls @16°C)].

QPCR Analysis

From the informative gene lists that were generated using SAM analysis, the following 4 genes of interest (GOI) were selected for QPCR analysis based on their known transcript expression responses in fish exposed to viruses or viral mimics[20, 22, 25]: TLR9 (Toll-like receptor 9), PKR (double-stranded RNA activated protein kinase), IRAK4 (interleukin-1 receptor-associated kinase 4), and IκBα (NF-κB inhibitor alpha). The following 13 microarray-identified genes were selected for QPCR because they were previously identified as up-regulated in the spleen of Atlantic cod injected with pIC and/or in brains of cod with high nodavirus carrier state[22, 25]: Deltex3, DHX58 (DExD/H box RNA helicase; aliases LGP2, RIG-I-c terminal domain containing protein), IL-8 (interleukin-8 variant 5), IRF1 (interferon regulatory factor 1), IRF7 (interferon regulatory factor 7), IRF10 (interferon regulatory factor 10), ISG15 (interferon stimulated gene 15, paralogs 1 and 3), RSAD2 (radical S-adenosyl methionine domain containing protein 2; alias Viperin), SACS (sacsin; alias spastic ataxia of Charlevoix-Saguenay), SCYA123 [small inducible cytokine 123; this follows the nomenclature for cod CC chemokines from Borza et al.[58]], STAT1 (signal transducer and activator of transcription 1), TLR3 (Toll-like receptor 3), and ZNFX1 (zinc finger, NFX1-type containing 1). Nine of the 17 microarray-identified genes subjected to QPCR (DHX58, IL-8, IRAK4, IRF1, ISG15, PKR, RSAD2, SCYA123 and ZNFX1) were amongst the 96 overlapping probes (Additional file1: Table S9) between the list of 279 genes more highly expressed in the pIC 6HPI@16°C vs. pIC 6HPI@10°C comparison and the list of 303 genes more highly expressed in the pIC 24HPI@10°C vs. pIC 24HPI@16°C comparison (Figures2 and4). QPCR primers were either chosen from previous publications from our laboratory[22, 25] (for IRF1, IRF7, IRF10, SACS, Deltex3, STAT1 and ZNFX1) or designed based on sequences representing informative microarray probes using Primer 3 (http://frodo.wi.mit.edu) (for DHX58, IκBα, IL-8, IRAK4, ISG15, PKR, RSAD2, SCYA123, TLR3, TLR9 and ATPS). Every primer set was quality-checked (QC) using reference cDNA templates generated by reverse transcription of the microarray experiment reference RNA. The QPCR primer QC process was carried out as described in Booman et al.[21]. Briefly, a 5-fold dilution (starting with 10 ng of input RNA) standard curve consisting of 5 points was used to calculate amplification efficiencies[68], and melt-curve analysis was carried out to ensure primer sets amplified a single product and there were no detectable primer dimers. The primer sequences, amplification efficiencies and amplicon sizes are shown in Table3.

Table 3: QPCR Primers

Gene		Primer sequence 5' to 3'	Amplification Efficiency (%)	R^2	Amplicon Size (bp)
Deltex3[b]	Forward	TCCACCACAAGACCAGCATCA	98.5	0.99	110
	Reverse	ACTTCACTCGATGCCTTTCGC			
DHX58[c]	Forward	ACAGAAGCCATCGCAGAAAT	92.9	0.99	105
	Reverse	TTTTGCAGCACGAATCAAAC			
IκBα[c]	Forward	GCCAGCAACTGATAAAGCATC	92.1	0.99	132
	Reverse	GGTCACAGAGGGAGACAGAAAA			
IL-8[c]	Forward	GTGTTTCCAGCAGATCACTCG	94.9	0.99	118
	Reverse	TGTTCCACTTGGTGAGGAGTC			
IRAK4[c]	Forward	CGTGGATTACAAGATGGATAAGC	95.8	0.99	102
	Reverse	TCGTCGGGGTCTAAAAAGTC			
IRF1[a]	Forward	AGAAGGACGCCAGTCTGTTCAA	88.5	0.99	100
	Reverse	GCGGAAGTTGGCTTTCCATT			
IRF7[a]	Forward	GGTCGTCGGAGTTCTTGGAGTT	95.1	0.99	102
	Reverse	CCAAACGACAAGGCCAAATG			
IRF10[a]	Forward	CGAGGCGGTAGACCTTGTAG	104.6	0.94	161
	Reverse	GGGCAGGTACAAAGGGAAAT			
ISG15[c]	Forward	AGGACCAACAAAGGCTGATG	94.7	0.99	110
	Reverse	CAGCCGTCCGTTAAGGTAGA			

PKR[c]	Forward	ATTGCATCAGGTTCCCAATC	94.4	0.99	174
	Reverse	GCCGTTACCAGACCAAAATC			
RSAD2[c]	Forward	TGTTTCCACACAGCGAAGAC	95.8	0.99	108
	Reverse	TCCGCCAGAGAAGTTGATCT			
SACS[b]	Forward	CTCCCACTGCCAATGTCATTC	88.5	0.99	102
	Reverse	TCAAGAAAACGTCCCAAGGC			
SCYA123[c]	Forward	GCTCTGGGTCGTGTACCTCT	94.1	1.00	189
	Reverse	TCTCTCTGGACGAACAAGCA			
STAT1[b]	Forward	GCCAATGCCATGTGTTTATG	97.5	0.99	100
	Reverse	ACCTGGAGCAGTTCGTCAGT			
TLR3[c]	Forward	CCTGAAACGCAACTCTATCTCC	90.6	0.99	121
	Reverse	GCCATCAAACATACCCTTCTTT			
TLR9[c]	Forward	TTGCTCGCCAAAACACTATG	99.5	0.99	150
	Reverse	GGAATCCAGTCCCTCTCCTC			
ZNFX1[b]	Forward	ATGCCACTATCGGTGGACAGA	87.3	0.99	108
	Reverse	TCAACAGATTATTGCCCTCGG			
ATPS[c]	Forward	ACATGGATAAATGGCTTTTTGC	99.6	0.98	155
	Reverse	TTGAAGAAGTAGTGTGGCTGGA			

[a]These primers were previously designed and used in Rise et al.[22] but were still quality checked for the present experiment using the reference cDNA as template.

[b]These primers were previously designed and used in Rise et al.[25] but were still quality checked for the present experiment using the reference cDNA as template.

[c]These primers were designed based on ESTs or contigs representing probes that were identified as differentially expressed in the microarray experiment.

Gene expression was normalized to ATPS [ATP synthase H^+ transporting, mitochondrial Fo complex, subunit F2; CGP microarray probe identifier (ID) #36304] mRNA levels. Several steps were taken to ensure that this gene was a suitable normalizer for this QPCR study. Initially, several candidate probes that had overall normalized ratios (i.e. log_2 transformed ALEXA 647/ALEXA 555) close to 0 across all 60 microarrays in this study were selected as potential normalizers. For these, the probe IDs for the ones with the lowest standard deviation were searched in all generated gene lists of differentially expressed genes (with FDR = 1%) (Additional files1: Tables S1-S8), and the ones that were not present in any gene lists were tested with QPCR in a subset of individuals (3 from each condition and time point). ATPS had the lowest threshold cycle (C_T) range (1.2). Lastly, in every multi-plate experiment C_T values for ATPS were always checked; if their range was >1.5 cycles, the samples more distant from the overall C_T mean were removed from the study. For any given GOI, no more than 2 individuals had to be removed from the study.

All cDNAs were reverse transcribed from 1 µg of DNase-I treated, column purified RNA using the RETROScript kit (Ambion, Austin, TX) following the manufacturer's instructions for the protocol that includes template denaturation. Six individuals from each group were analyzed in triplicate reactions. To QPCR validate the microarray experiment-identified GOI, we used 3 individuals that had been used in the microarray and 3 individuals that were not used for microarray hybridizations. Fish included in the QPCR experiment were also selected based on RNA quality and quantity. PCR amplification and cycling were performed as in Booman et al.[21]. In all plates a linker sample consisting of cDNA made from the reference RNA was included for every target (including the normalizer). Prior to analysis, all linker C_T values were checked; inter-plate variation was never >0.5 cycles. All thresholds were set automatically, and relative quantities (RQ) were calculated using actual amplification efficiencies for each primer pair and the linker as a calibrator using the Applied Biosystems (Foster City, CA) 7500 Software Relative Quantification Study Application (v2.0)[69].

Resulting RQs were log_2 transformed in Excel (Microsoft Co., Seattle, WA) and analyzed statistically using Prism (v5.0, GraphPad Software Inc, La Jolla. CA). All groups were initially subjected to a Kolmogorov-Smirnov normality test. For each time point (e.g. 6HPI) a two-way ANOVA with injection (i.e. pIC or PBS) and temperature (10°C or 16°C) as factors was then carried out. When the effects of a given factor were statistically significant ($p < 0.05$), t-tests were carried out to compare pIC-injected groups to their temperature-matched controls (i.e. PBS injected) at a particular time point, PBS-injected fish held at different temperatures, or pIC-injected fish held at different temperatures. Overall fold-change values were calculated as 2^{A-B} as suggested by Cui and Churchill[70], where A is the average log_2 transformed RQ from an experimental group (pIC always considered experimental relative to PBS and 16°C always considered experimental relative to 10°C), and B is the average log_2 transformed RQ from the time- or temperature-matched control.

DECLARATIONS

Acknowledgments

This research was supported in part by Genome Canada, Genome Atlantic, and the Atlantic Canada Opportunities Agency through the Atlantic Cod Genomics and Broodstock Development Project. Funding was also provided by a Canada Research Chair (Tier 2) in marine biotechnology and a Natural Sciences and Engineering Research Council (NSERC) discovery grant to MLR, and the following awards to TSH: the Royal Bank Scholarship in Marine Sciences, the Joe Brown Graduate Research Award; Memorial University of Newfoundland SGS fellowship. The authors would also like to acknowledge the staff of the Dr. Joe Brown Aquatic Research Building (JBARB) for helping with fish husbandry. Lastly, we would like to thank Dr. Stewart Johnson and Dr. Luis Afonso for reading and providing suggestions on earlier versions of this manuscript.

Competing Interests

The authors declare that they have no competing interests.

REFERENCES

1. Myers RA, Hutchings JA, Barrowman NJ: Why do fish stocks collapse? The example of cod in Atlantic Canada. Ecol Appl. 1997, 7 (1): 91-106.
2. Kjesbu OS, Taranger GL, Trippel EA: Gadoid mariculture: development and future challenges. ICES J Mar Sci. 2006, 63: 187-191.
3. Rose GA: Cod: the ecological history of the North Atlantic fisheries. 2007, Breakwater Books, St. John's
4. Brown JA, Minkoff G, Puvanendran V: Larviculture of Atlantic cod (Gadus morhua): progress, protocols and problems. Aquaculture. 2003, 227: 357-372.
5. Roselund G, Halldórsson O: Cod juvenile production: Research and commercial developments. Aquaculture. 2007, 268: 188-194.
6. Roselund G, Skretting M: Worldwide status and perspectives on gadoid culture. ICES J Mar Sci. 2006, 63: 194-197.
7. Pérez-Casanova JC, Afonso LOB, Johnson SC, Currie S, Gamperl AK: The stress and metabolic responses of juvenile Atalntic cod Gadus morhua L. to an acute thermal challenge. J Fish Biol. 2008, 72: 899-916.
8. Gollock MJ, Currie S, Petersen LH, Gamperl AK: Cardiovascular and hematological responses of Atlantic cod (Gadus morhua) to acute temperature increase. J Exp Biol. 2006, 209: 2961-2970.
9. Pérez-Casanova JC, Rise ML, Dixon B, Afonso LO, Hall JR, Johnson SC, Gamperl AK: The immune and stress response of Atlantic cod to long-term increases in water temperature. Fish Shellfish Immunol. 2008, 24: 600-609.
10. Alishahi M, Buchmann K: Temperature-dependent protection against Ichthyophthirius multifiliis following immunisation of rainbow trout using live theronts. Dis Aquat Organ. 2006, 72: 269-273.
11. Raida MK, Buchmann K: Temperature-dependent expression of immune-relevant genes in rainbow trout following Yersinia ruckeri vaccination. Dis Aquat Organ. 2007, 77: 41-52.
12. Raida MK, Buchmann K: Bath vaccination of rainbow trout (Oncorhynchus mykiss) againstYersinia ruckeri: effects of temperature on protection and gene expression. Vaccine. 2008, 26: 1050-1062.
13. Bettge K, Segner H, Burki R, Schmidt-Posthaus H, Wahli T: Proliferative kidney disease (PKD) of rainbow trout: temperature- and time-related changes of Tetracapsuloides bryosalmonae DNA in the kidney. Parasitol. 2009, 136: 615-625.
14. Varsamos S, Flik G, Pepin JF, Wendelaar Bonga SE, Breuil G: Husbandry stress during early life stages affects the stress response and health status of juvenile sea bass, Dicentrarchus labrax. Fish Shellfish Immunol. 2006, 20: 83-96.
15. Cheng A-C, Cheng S-A, Chen Y-Y, Chen J-C: Effects of temperature change on innate cellular and humoral immune responses of the orange-spotted grouper Epinephelus coioides and its susceptibility to Vibrio alginolyticus. Fish Shellfish Immunol. 2009, 26: 768-722.
16. Eggset G, Mikkelsen H, Killie J-EA: Immunocompetence and duration of immunity against Vibrio salmonicida and Aeromonas salmonicida after vaccination of Atlantic salmon (Salmo salar L.) at low and high temperatures. Fish Shellfish Immunol. 1997, 7: 247-260.
17. Magnadottir B, Jonsdottir H, Helgason S, Bjornsson B, Jorgensen TO, Pilstrom L: Humoral immune parameters in Atlantic cod (Gadus morhua L.): I. The effects of environmental temperature. Comp Biochem Physiol B Biochem Mol Biol. 1999, 122: 181-188.
18. Rise ML, Jones SRM, Brown GD, von Schalburg KR, Davidson WS, Koop BF: Microarray analyses identify molecular biomarkers of Atlantic salmon macrophage and hematopoietic kidney responses to Piscirickettsia salmonis infection. Physiol Genomics. 2004, 20: 21-35.
19. Milev-Milovanovic I, Majji S, Thodima V, Deng Y, Hanson L, Arnizaut A, Waldbieser G, Chinchar VG: Identification and expression analyses of poly [I:C]-stimulated genes in channel catfish (Ictalurus punctatus). Fish Shellfish Immunol. 2009, 26: 811-820.
20. Workenhe ST, Hori TS, Rise ML, Kibenge MJ, Kibenge FS: Infectious salmon anaemia virus (ISAV) isolates induce distinct gene expression responses in the Atlantic salmon (Salmo salar) macrophage/dendritic-like cell line TO, assessed using genomic techniques. Mol Immunol. 2009, 46 (15): 2955-2974.
21. Booman M, Borza T, Feng CY, Hori TS, Higgins B, Culf A, Léger D, Chute IC, Belkaid A, Rise M, Gamperl AK, Hubert S, Kimball J, Ouellette RJ, Johnson SC, Bowman S, Rise ML: Development and experimental validation of a 20 K Atlantic cod (Gadus morhua) oligonucleotide microarray based on a collection of over 150,000 ESTs. Mar Biotechnol (NY). 2011, 13 (4): 733-750. Rise ML, Hall J, Rise M, Hori T, Gamperl AK, Kimball J, Hubert S, Bowman S, Johnson SC: Functional genomic analysis

of the response of Atlantic cod (Gadus morhua) spleen to the viral mimic polyriboinosinic polyribocytidylic acid (pIC). Dev Comp Immunol. 2008, 32: 916-971.

22. Feng CY, Johnson SC, Hori TS, Rise M, Hall JR, Gamperl AK, Hubert S, Kimball J, Bowman S, Rise ML: Identification and analysis of differentially expressed genes in immune tissues of Atlantic cod stimulated with formalin-killed, atypical Aeromonas salmonicida. Physiol Genomics. 2009, 37: 149-163.
23. Hori TS, Gamperl AK, Afonso LO, Johnson SC, Hubert S, Kimball J, Bowman S, Rise ML: Heat-shock responsive genes identified and validated in Atlantic cod (Gadus morhua) liver, head kidney and skeletal muscle using genomic techniques. BMC Genomics. 2010, 11: 72
24. Rise ML, Hall JR, Rise M, Hori TS, Browne MJ, Gamperl AK, Hubert S, Kimball J, Bowman S, Johnson SC: Impact of asymptomatic nodavirus carrier state and intraperitoneal viral mimic injection on brain transcript expression in Atlantic cod (Gadus morhua). Physiol Genomics. 2010, 42: 266-280.
25. Bowman S, Hubert S, Higgins B, Stone C, Kimball J, Borza T, Bussey JT, Simpson G, Kozera C, Curtis BA, Hall JR, Hori TS, Feng CY, Rise M, Booman M, Gamperl AK, Trippel E, Symonds J, Johnson SC, Rise ML: An integrated approach to gene discovery and marker development in Atlantic cod (Gadus morhua). Mar Biotechnol (NY). 2011, 13 (2): 242-255.
26. Steinhagen D, Hespe K, Ellmer B, Körting W: Goussia carpelli (Protozoa: Coccidia) infection in stressed and immunosupressed common carp Cyprinus carpio. Dis Aquat Org. 1998, 34: 199-204.
27. Hamel OS: Immunosupression in progeny of Chinook salmon infected with Renibacterium salmoninarum: re-analysis of a brood stock segregation experiment. Dis Aquat Org. 2005, 65: 29-41.
28. Verma AK, Pal AK, Manush SM, Das T, Dalvi RS, Chandrachoodan PP, Ravi PM, Apte SK: Persistent sub-lethal chlorine exposure augments temperature induced immunosupression in Cyprinus carpio advanced fingerlings. Fish Shellfish Immunol. 2007, 22: 547-555.
29. Hansen J, Sato M, Ruedy R, Lo K, Lea DW, Medina-Elizade M: Global temperature change. PNAS. 2006, 103 (39): 14288-14293.
30. Tusher VG, Tibshirani R, Chu G: Significance analysis of microarrays applied to the ionizing radiation response. Proc Natl Acad Sci USA. 2001, 98: 5116-5121.
31. Johansson D, Juell J-E, Oppedal F, Stiansen J-E, Ruohonen K: The influence of the pycnocline and cage resistance on current flow, oxygen flow and swimming behaviour of Atlantic salmon (Salmo salar L.) in production cages. Aquaculture. 2007, 265: 271-287.
32. Oppedal F, Juell J-E, Johansson D: Thermo- and photoregulatory-behaviour influence swimming depth and schooling density of caged Atlantic salmon. Aquaculture. 2007, 265: 70-81.
33. Wang J-T, Li J-T, Zhang X-F, Sun X-W: Transcriptome analysis reveals the time of the fourth round of genome duplication in common carp (Cyprinus carpio). BMC Genomics. 2012, 13: 96
34. Sánchez CC, Weber GM, Gao G, Cleveland BM, Yao J, Rexroad CE: Generation of a reference transcriptome for evaluating rainbow trout responses to various stressors. BMC Genomics. 2011, 12: 626
35. Xiang L-x, He D, Dong W-r, Zheng Y-w, Shao J-z: Deep sequencing-based transcriptome profiling analysis of bacteria-challanged Lateolabrax japonicus reveals insights into the immune-relevant genes in marine fish. BMC Genomics. 2010, 11: 472
36. Robertsen B: The interferon system of teleost fish. Fish Shellfish Immunol. 2006, 20: 172-191.
37. Robertsen B: Expression of interferon and interferon-induced genes in salmonids in response to virus infection, interferon-inducing compounds and vaccination. Fish Shellfish Immunol. 2008, 25: 351-357.
38. Phelan PE, Mellon MT, Kim CH: Functional characterization of full-length TLR3, IRAK-4, and TRAF6 in zebrafish (Danio rerio). Mol Immunol. 2005, 42: 1057-1071.
39. Galligan CL, Murooka TT, Rahbar R, Baig E, Majchrzak-Kita B, Fish EN: Interferon and viruses, signaling for supremacy. Immunol Res. 2006, 35: 27-39.
40. Rebl A, Goldammer T, Seyfert H-M: Toll-like receptor signaling in bony fish. Vet Immunol Immunopathol. 2010, 134: 139-150.
41. Rodriguez MF, Wiens GD, Purcell MK, Palti Y: Characterization of Toll-like receptor 3 gene in rainbow-trout (Oncorhynchus mykiss). Immunogenetics. 2005, 57: 510-519.
42. Chiou PP, Lin C-H, Bols NC, Chen TT: Characterization of virus/double-stranded RNA-dependent induction of the antimicrobial peptide hepcidin in trout macrophages. Dev Comp Immunol. 2007, 31: 1297-1309.
43. Matsumoto M, Seya T: TLR3: Interferon induction by double-stranded RNA including poly(I:C). Adv Drug Deliv Rev. 2008, 60: 805-812.
44. Palti Y: Toll-like receptors in bony fish: from genomics to function. Dev Comp Immunol. 2011, 35 (12): 1263-1272.
45. Kaway T, Akira S: Toll-like receptor and RIG-I-like receptor signaling. Ann NY Acad Sci. 2008, 1143: 1-20.
46. Wilkins C, Gale M: Recognition of viruses by cytoplasmic sensors. Curr Opin Immunol. 2010, 22: 41-47.
47. Gil J, Alcamí J, Esteban M: Activation of NF-κB by the dsRNA-dependent protein kinase, PKR involves the IκB kinase complex. Oncogene. 2000, 19: 1369-1378.
48. García MA, Meurs EF, Esteban M: The dsRNA protein kinase PKR: virus and cell control. Biochimie. 2007, 89: 799-811.
49. Kaway T, Akira S: TLR signaling. Cell Death Differ. 2006, 13: 816-825. Li Q, Verma IM: NF-κB regulation in the immune system. Nature Rev Immunol. 2002, 2 (10): 725-734.
50. Bonizzi G, Karin M: The two NF-κB activation pathways and their roles in innate and adaptive immunity. Trends Immunol. 2004, 25 (6): 280-288.

51. Tamura T, Yanai H, Savitsky D, Taniguchi T: The IRF family transcription factors in immunity and oncogenesis. Annu Rev Immunol. 2008, 26: 535-584.

52. Noppert SJ, Fitzgerald KA, Hertzog PJ: The role of type I interferon's in TLR responses. Immunol Cell Biol. 2007, 85: 446-457

53. Bao X, Indukuri H, Liu T, Liao S-H, Tian B, Brasier AR, Garofalo RP, Casola A: IKKε modulates RSV-induced NF-κB-dependent gene transcription. Virology. 2010, 408: 224-231

54. Liu S-Y, Sanchez DJ, Cheng G: New developments on the induction and anti-viral effectors of type I interferon. Curr Opin Immunol. 2011, 23: 57-64.

55. Furnes C, Kileng Ø, Rinaldo CH, Seppola M, Jensen I, Robertsen B: Atlantic cod (Gadus morhua L) possesses three homologues of ISG15 with different expression kinetics and conjugation properties. Dev Comp Immunol. 2009, 33: 1239-1246

56. Borza T, Stone C, Rise ML, Bowman S, Johnson SC: Atlantic cod (Gadus morhua) CC chemokines: diversity and expression analysis. Dev Comp Immunol. 2010, 34: 904-913.

57. Engert JC, Bérubé P, Mercier J, Doré C, Lepage P, Ge B, Bouchard J-P, Mathieu J, Melançon SB, Schalling M, Lander ES, Morgan K, Hudson TJ, Richter A: ARSACS, a spastic ataxia common in northeastern Québec, is caused by mutations in a new gene encoding a 11.5-kb ORF. Nature Genet. 2000, 24: 120-125.

58. Anderson JF, Siller E, Barral JM: The Sacsin repeating region (SRR): a novel hsp90-related supra-domain associated with neurodegeneration. J Mol Biol. 2010, 400: 665-674.

59. Kultz D: Molecular and evolutionary basis of the cellular stress response. Annu Rev Physiol. 2005, 67: 225-257.

60. Salminen A, Paimel T, Suuronen T, Kaarniranta K: Innate immunity meets with cellular stress at the IKK complex: regulation of the IKK complex by HSP70 and HSP90. Immunol Lett. 2008, 117: 9-15.

61. Bø TH, Dysvik B, Jonassen I: LSimpute: accurate estimation of missing values in microarray data with least squares methods. Nucleic Acids Res. 2004, 32: e34

62. Celton M, Malpertuy A, Lelandais G, de Brevern AG: Comparative analysis of missing value imputation methods to improve clustering and interpretation of microarray experiments. BMC Genomics. 2010, 11: 15

63. Schwender H, Krause A, Ickstadt K: Identifying interesting genes with siggenes. RNews. 2006, 6: 45-50.

64. Sturn A, Quackenbush J, Trajanoski Z: Genesis: cluster analysis of microarray data. Bioinformatics. 2002, 18: 207-208.

65. Conesa A, Götz S, Gracía-Gómez JM, Terol J, Talón M, Robles M: Blast2GO: a universal tool for annotation, visualization and analysis in functional genomics research. Bioinformatics. 2005, 21 (18): 3674-3676.

66. Pfaffl MW: A new mathematical model for relative quantification in real-time RT-PCR. Nucleic Acids Res. 2001, 29: e45-

67. Livak KJ, Schmittgen TD: Analysis of relative gene expression data using real-time quantitative PCR and the 2(−Delta Delta C(T)) method. Methods. 2001, 2: 402-408.

68. Cui X, Churchill GA: Statistical tests for differential expression in cDNA microarray experiments. Genome Biol. 2003, 4: 210

CHAPTER 14

PRODUCTION AND UTILIZATION OF A HIGH-DENSITY OLIGONUCLEOTIDE MICROARRAY IN CHANNEL CATFISH, ICTALURUS PUNCTATUS

Robert W Li and Geoffrey C Waldbieser

BACKGROUND

The channel catfish, *Ictalurus punctatus*, is native to North America and is from the order Siluriformes (superorder Ostariophysi) which is considered more primitive among teleosts [1, 2]. Commercial production of catfish as dietary protein is the leading industry of North American aquaculture, with more than 600 million pounds of catfish produced annually in the United States [3]. One of the most significant factors limiting production is loss to disease, and one of our goals is to improve non-specific immunity through selective breeding. The channel catfish immune system is the best characterized for any fish species, and it is the only fish species where clonal functionally distinct lymphocyte lines can be readily established [4]. However, there is a need to better understand the physiological and immunological pathways controlling host-pathogen interactions *in vivo*.

Improvement of disease resistance in catfish populations depends on an understanding of the genetic control of immune-related pathways, response to pathogens, and correlations with other production traits. Recent developments in genomic technology, particularly high-throughput cDNA sequencing and development of expressed sequence microarrays, have made possible the profiling of global gene expression in experimental fish tissues.

Microarray experiments have been utilized to determine regulation of gene expression during developmental and adaptive processes in several fish species. Using this technology, groups of coordinately expressed genes were identified in zebrafish embryos during development [5, 6]. Microarrays were used to identify changes in gene expression patterns in zebrafish and goby exposed to differing levels of available oxygen [7, 8], in common carp exposed to cold temperature [9], and in rainbow trout exposed to handling stress [10].

Recently, expression microarrays have been used to identify genes involved in host-pathogen relationships. Experimental infection of cells *in vitro* revealed differential gene expression in flounder kidney cells [11], and in rainbow trout monocytes and macrophages [12]. Transcriptional analysis was performed on kidney cells after experimental *in vivo* injection of flounder with a DNA vector expressing the viral hemorrhagic septicemia virus G protein [13], in Atlantic salmon macrophages and anterior kidney tissues after injection of fish with *P. salmonis* [14], on Atlantic salmon liver, spleen, and anterior kidney tissues after cohabitation with *A. salmonicida*-infected fish [15], and from whole zebrafish after infection with *M. marinum* [16]. By measuring changes in gene expression after pathogen challenge, researchers may identify gene expression fingerprints that provide clues to molecular pathways involved in pathogen neutralization and/or removal, identify candidate genes controlling pathogen-specific immunity, and identify heritable differences in gene expression levels that correlate with disease resistance/susceptibility. This data would help in the formulation of a selection index to identify broodstock with superior genetic potential for resistance to disease.

Citation: Robert W Li and Geoffrey C Waldbieser, Production and utilization of a high-density oligonucleotide microarray in channel catfish, Ictalurus punctatus, doi: 10.1186/1471-2164-7-134

While microarray-based transcriptional analysis is a useful tool for functional genomics in fish, to date there has been no microarray platform available for catfish species. Thus, we developed a catfish microarray utilizing existing data from channel catfish expressed sequences obtained in cDNA cloning experiments from several tissues (reviewed in [17] [18]). These sequences have been clustered and annotated in the Catfish Gene Index [19]. The catfish microarray was used to measure differential gene expression in the spleen of lipopolysaccharide-injected and sham-injected catfish.

RESULTS

A high density oligonucleotide microarray was produced that contained 18,989 catfish expressed sequences, each represented by ten perfectly matched and ten mismatched 24-mer oligonucleotides (GEO Database Accession GSE3261). This microarray contained 8057 sequences that matched an annotated sequence in GenBank. Discounting 875 that matched transposons or sequences that were labelled as unknown, unnamed, or hypothetical, there were 7182 annotated sequences (38%) on the microarray. This microarray was used to demonstrate differences in global gene expression in the spleen of LPS-treated catfish. Using Significance Analysis of Microarrays (SAM), we identified 138 sequences that were differentially expressed after LPS treatment

There were 64 genes up-regulated by LPS exposure of which only 26 were annotated based on sequence similarity Among these were cytokines and chemokines, such as IL-1β, CCL4, a small CXC/IL 8-like chemokine, and chondromodulin II. Transcriptional factors such as NF-κB p100 subunit and NF-κB inhibitor alpha-like proteins A and B, interferon regulatory factor I, and AP-1 were also upregulated at least 2-fold. Expression of Toll-like receptor 5 was upregulated at 2–4 h post-exposure. Within the experimental timeframe, almost all induced genes were induced by 4 h post-exposure.

There were 74 genes that were significantly down-regulated in response to LPS, of which only 29 were annotated based on sequence similarity. Among these genes were the immunoglobulin light and heavy chains, MHC class II antigens, invariant chain-like protein 2, an NK lysin-like protein, complement C3, hemoglobin alpha, and a CXC chemokine receptor.

In the present research, we selected nine expressed sequences representing varying levels of expression to verify the microarray results using real-time RT-PCR. Linear regression of log transformed expression data demonstrated a strong positive correlation ($R^2 = 0.87$) between the two technologies (Fig. 1). The microarray results for TNFα failed to pass our quality control criteria, thus no expression ratios were reported (detailed below).

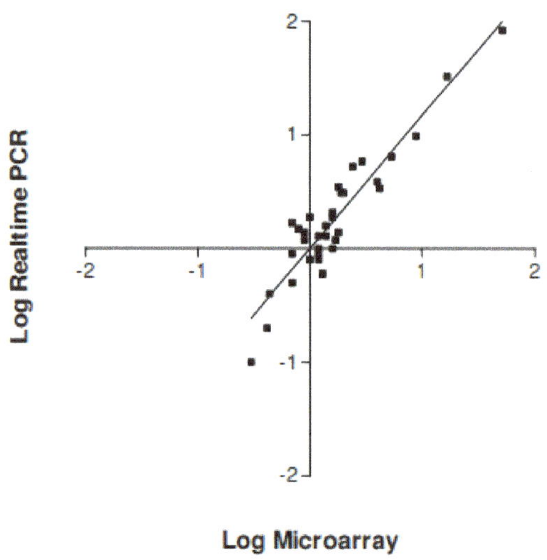

Figure 1: Correlation between levels of gene expression measured by microarray and real-time RT-PCR. The correlation (R^2= 0.87) was calculated using log-transformed values of the fold change obtained for the nine selected genes. Four fish were used per treatment/time combination.

DISCUSSION

Microarray Performance

In our first attempt to monitor the global gene expression profile after LPS stimulation in channel catfish spleen, we have constructed a high-density oligonucleotide microarray containing almost 19,000 catfish unique sequences. In the present research, we used real-time RT-PCR to verify the microarray results. The analysis revealed a strong positive correlation between the microarray and real-time RT-PCR results. The TNFα gene was selected based on its importance in the LPS pathway and also to represent genes that did not make the final "hit" list. The TNFα microarray results failed to pass our quality control criteria, so no ratios were reported. After examination of the oligo sequences for TNFα on the microarray, we found only one of the ten probe sets demonstrated a significant difference in signal intensity between the perfectly matched and mismatched oligos. Five of the TNFα probe sets contained higher signal intensities in the mismatched than in the perfectly matched oligo. The automated oligo design software extracted all ten oligo sets from the most 5' end of the 255 bp TNFα cDNA sequence, so the discrepancy between real-time RT-PCR and microarray results for TNFα was likely due to undesirable oligo design. In fact, the one "good" oligo pair would have identified up-regulation of TNFα by LPS. This observation highlights the importance of high quality oligo design software, and that the ability to identify and eventually remove problematic oligo probes is critical to the collection of robust microarray data and minimization of discordant gene expression data from different technologies.

Differential Expression of Catfish Genes

The evolutionary divergence between mammalian and fish species is considerable, and we do not know if LPS can induce the IKK-NF-κB pathway and MAPK pathways (the ERK, JNK, and p38 pathways) in fish as in mammals. Although LPS-responsive cis-elements have been identified in the 5›-flanking region of many mammalian genes [20], none have been reported in catfish to date. The present research demonstrated increased expression of the p100 subunit of NF-κB at 4 h post-LPS exposure, but a sequence matching NF-κB inhibitor alpha-like proteins A and B were also significantly induced at 2 h post-LPS exposure. NF-κB inhibitor gene expression was also increased in mycobacterium-infected zebrafish [16] and bacterially infected macrophage cells from Atlantic salmon and rainbow trout [12, 14]. There are instances in biological systems that a compound can up-regulate both an enzyme and its inhibitor. For example, the short-chained fatty acid butyrate up-regulated both IGF2 and IGF binding proteins IGFBP2 and IGFBP3 in bovine kidney epithelial cells (R.L., unpublished). An alternative explanation to the present observation is that the inhibitor encoded a protein with a function different from that which was annotated due to sequence similarity.

Toll-like receptor 5 demonstrated significant levels of induction at 4 h post-LPS treatment. In mammals, TLR4 is the extracellular LPS receptor whereas TLR5 recognizes flagellin [21, 22]. The up-regulation of TLR5 by Gram-negative, virulent *Edwardsiella ictaluri* has been independently observed in channel catfish [[23]; Li, unpublished]. This gene was also shown to increase in the liver of Atlantic salmon exposed to the bacterium *A. salmonicida* [15], and in zebrafish exposed to the mycobacterium *M. marinum* [24]. Toll-like receptor 3 was also included on the catfish microarray, but expression was not significantly altered due to LPS exposure. The present data add evidence to the hypothesis that, in fish, microbial products may induce other TLRs than those specific for their own recognition [24].

Interferon regulatory factor 1 (IRF-1) is one of two DNA-binding transcription factors mediating Type 1 interferon gene expression. In rainbow trout gonadal cells, IRF-1 is expressed constitutively and up-regulated by poly I:C but not by LPS [25]. However, microarray experiments using Japanese flounder have demonstrated stimulation of IRF-1 expression in cultured kidney cells after exposure to LPS [11] and in kidney tissue after exposure of fish to the viral hemorrhagic septicemia virus G protein [13].

Apolipoprotein E (ApoE) has been suggested to have a regulatory role in maintaining a critical balance of various proinflammatory cytokines in mice [26] in addition to its key role in lipid metabolism. Expression of ApoE decreases in CaCo-2 cells 24 h after LPS exposure [27], and this was consistent with our observations.

The chemokine receptor CXCR4 gene selected for analysis by quantitative real-time PCR (IpCG06125) was first up-regulated by LPS at 2 h, rapidly down-regulated at 4 h, and remained down-regulated at 8 h and 24 h. A CD14-independent model for LPS recognition has been proposed that includes a LPS activation cluster complex containing CXCR4, heat shock proteins 70 and 90, and growth differentiation factor 5 [28, 29]. In addition, CXCR4 is a functional co-receptor for HIV-1 infection of human macrophages. LPS down-regulates the expression of CXCR4 on monocyte-derived macrophages [30], and the present research is consistent with this observation.

While the majority of the catfish sequences could not be annotated due to sequence identify, several catfish genes could be compared with other fish experiments utilizing microarrays to measure gene expression after experimental infection. The interleukin 1β gene demonstrated the greatest increase in LPS-induced expression in catfish spleen, and this gene was highly induced in flounder kidney cells exposed to LPS [11]. Expression of the immune responsive protein 1 gene increased in catfish spleen, and bacterially infected salmon kidneys and macrophages [14]. Catfish MHC class II gene expression decreased after LPS exposure, and this was also observed in mycobacterium-infected zebrafish [16] and salmon macrophage cells exposed to LPS [12], but LPS-treated flounder kidney cells demonstrated no change in MHC class II gene expression [11]. Immunoglobulin gene expression decreased in catfish spleen after LPS exposure and in bacterially infected Atlantic salmon macrophage cells [14], but it increased in zebrafish after mycobacterial infection [16] and in Atlantic salmon liver and kidney after bacterial infection [14, 15]. Likewise, complement C3 gene expression was decreased in catfish spleen and salmon macrophages [14], but increased in kidneys from infected salmon [14] and in infected zebrafish [16]. While the present experiment demonstrated similarities and differences in species-specific gene expression after bacterial or LPS exposure, the lack of fully annotated sequences on the catfish microarray hindered a more comprehensive comparative analysis. Evaluation of immune responses across species would also enhanced by a reference set of microarray probes on each species-specific microarray.

Overall, there was no correlation between genes that were more abundant in catfish spleen (measured as number of cDNA transcripts in [31]) and genes that demonstrated significant changes in level of expression after LPS stimulation in the present research. While a 2-fold difference in expression is an arbitrary value, there are, undoubtedly, significant physiological and immunological consequences to differences in gene expression less than 2-fold. This differential reflects the inability of current microarray technology to significantly resolve smaller differences in levels of gene expression. The present research exposed difficulties inherent in performing microarray analyses from tissues *in vivo*. Differences observed between individuals within each treatment, due to asynchronized sampling and/or genetic variation, led to averaged effects which could have masked significant responses in some individuals. Nonetheless, microarrays are powerful tools that permit the simultaneous screening of thousands of genes to provide candidates for further study. While the current catfish microarray is limited by the large number of "unknown" candidate sequences, both functional experimentation and comparative genomic analyses should help improve and identification of the large number of catfish ESTs currently without annotation.

From the patterns of gene expression in this experiment, we hypothesize the observed gene expression profiles in LPS-stimulated catfish spleen result mainly from macrophage activation by LPS. However, contributions from other cell types in spleen cannot be discounted. Since macrophages play a key role in a variety of inflammatory diseases and in host defense such as pathogen phagocytosis, antigen presentation to immunocytes, and production of numerous inflammatory mediators, our future work should focus on homogeneous cell populations derived either from cultured cell lines or from macrophages isolated directly from the tissues with techniques such as laser-capture microdissection.

CONCLUSION

The oligonucleotide microarray produced in the present research contained genes expressed in seventeen juvenile tissues and all neonatal tissues and provided a platform for global analysis of catfish gene expression. Levels of expression measured on the microarray were significantly correlated with levels measured by quantitative RT-PCR. Only 38% of the catfish sequences could be presently annotated, but *de novo* synthesis of oligonucleotides on the platform provides flexibility for adding annotated catfish sequences to the microarray as they become available.

The present experiment focused on the identification of differential gene expression in spleen in response to LPS injection. The innate immune response to Gram-negative bacteria is influenced by the ability of the host to respond to the LPS component, the major virulence factor of these bacteria. The present research identified 138 differentially expressed sequences that can provide clues as to the response of the catfish immune system to LPS from gram-negative bacteria, and contribute to discerning the NF-κB pathway in teleosts. Dissecting the LPS signal pathway is important in expanding our knowledge of molecular interactions between invading bacteria and host cells. Understanding the difference in the LPS pathway between fish and mammalian species will also facilitate understanding of the evolution of innate immune systems. Knowledge of the catfish immune system will help researchers identify key pathways in pathogen-induced diseases in catfish.

METHODS

Experimental Fish and LPS Treatment

Channel catfish (USDA103 strain, average weight ~200 g) were maintained in a single 160-liter tank at the Catfish Genetics Research Unit. Fish were fed a commercial diet to apparent satiation daily until 24 h prior to treatment with lipopolysaccharide (LPS). Thirty-two fish were anesthetized with tricaine methanesulfonate (MS222, 100 mg/L) and randomly divided into two groups, control and LPS. There were four individuals in each of time-matched control and LPS treatment time points. Lipopolysaccharide derived from E. coli 0127:B8 (Sigma-Aldrich, St. Louis, MO) was resuspended at 1.0 mg/ml in phosphate buffered saline (PBS, pH 7.0). Fish in the LPS group each received a 0.5 ml intraperitoneal (IP) injection of LPS at a dosage of 2.5 mg/kg body weight. Fish in the control group each received a 0.5 ml IP injection of PBS. Fish were then killed by anesthesia overdose (300 mg/L) at 2, 4, 8, or 24 h post-injection and the spleen was dissected and placed in 1.0 ml Trizol (Invitrogen Corporation, Carlsbad, CA). Samples were snap-frozen in liquid nitrogen and stored at -80°C until RNA isolation. Due to expense, only two fish per treatment/time combination were used for microarray analysis, but all four fish per treatment/time combination were used for real-time RT-PCR analysis. All experimental procedures were conducted in accordance with the principles and procedures approved by the Institutional Animal Care and Use Committee, USDA-ARS Catfish Genetics Research Unit.

Total RNA Purification and Generation of Biotin-Labelled Complementary RNA

Total RNA was extracted using manufacturer's recommendations (Invitrogen), treated with 4–10 units DNase I (Ambion Inc., Austin, TX) per 100 μg total RNA at 37°C for 30 min, then further purified using an RNeasy Mini kit (Qiagen Inc., Valencia, CA). Equal masses of total RNA were pooled from the eight individuals in the control group (two fish in each of four time-matched controls). This pool of RNA was labelled as described below and served as a control in the microarray hybridization. The total RNA from the eight LPS-stimulated fish (two in each of four time points) was not pooled.

Biotin-labelled complementary RNA (cRNA) was generated using a modified procedure for the Superscript Choice System (Invitrogen) for double-stranded (ds) cDNA synthesis followed by *in vitro* transcription. First strand cDNA was synthesized by SuperScript II reverse transcriptase from 2.5 μg of total RNA using 100 pmoles of T7 promoter Oligo dT primer. After ds cDNA synthesis, the cDNA was purified with DNA Clean & Concentrator-5 (Zymo Research, Orange, CA), eluted with 8–16 μl of ddH$_2$O, and concentrated to 3 μl by vacuum centrifugation. Complementary RNA was then synthesized with a MEGAscript *in vitro* Transcription kit (Ambion). The 23.0 μl *in vitro* transcription reaction included the entire 3.0 μl of ds cDNA, 2.3 μl of 10× Ambion reaction buffer, 2.3 μl of 10× Ambion T7 enzyme mix, and 15.4 μl of NTP labelling mix (7.5 mM ATP, 7.5 mM GTP, 5.625 mM UTP, 5.625 mM CTP and 1.875 mM biotin-16-UTP and 1.875 mM biotin-11 CTP). The *in vitro*transcription reaction was incubated at 37°C for 16 hours in a thermal cycler. Transcribed cRNA was then purified with an RNeasy Mini kit. Generally, 40–60 μg of cRNA was obtained from 2.5 μg original RNA. The sample integrity of the cRNA was verified with an Agilent 2100 Bioanalyzer (Agilent Technologies, Inc., Palo Alto, CA). The biotinylated cRNA products were then fragmented to 50–200 bp prior to hybridization by heating cRNA in 1× fragmentation buffer (40 mM Tris-acetate, pH.8.1, 100 mM potassium acetate, 30 mM magnesium acetate) at 95°C for 35 min.

Catfish Oligonucleotide Microarray Production, Hybridization, and Image Acquisition

Published sequences from cDNA libraries were obtained from the TIGR catfish gene index release 5.0 [19]. These ESTs were derived from brain, gill, intestine, liver, kidney (anterior and posterior), macrophage, muscle, olfactory tissue, ovary, peripheral blood leukocytes, pituitary, skin, spleen, stomach, taste tissue, and testis. An additional 9,821 novel sequences were obtained by single-pass directional sequencing of normalized cDNA libraries produced from mRNA obtained from 5 d, 9 d, and 14 d whole fry. The fry and TIGR gene index sequences were clustered using CAP3 software [32] and the resulting contigs and singlets were assigned an IpCG designation (*Ictalurus punctatus* Catfish Genetics). Subsequent to the production of the microarray, the sequence data from the fry cDNA libraries were submitted to GenBank as accessions CV987367 – CV997187 and were incorporated into release 6.0 of the TIGR gene index. The tables were updated to include the release 6.0 Tentative Consensus sequence (TC) designation of these sequences where applicable. The catfish ESTs were compared against the GenBank non-redundant peptide database using BlastX, and only those with E value = 0.005 (minimum bit score = 100) were annotated.

Microarray production, hybridization, and imaging were performed by NimbleGen Systems Inc. (Madison, WI). A total of 18,989 unique catfish ESTs were used to construct the high-density catfish oligonucleotide DNA microarrays. The microarrays were synthesized *in situ* using photo deprotection chemistry with the Maskless Array Synthesizer system [33]. Twenty 24-mer oligonucleotides (features) were designed to represent each EST; ten were perfectly matched oligos positioned throughout the sequence, and these ten were duplicated but with two mismatched bases at positions #13 and #19. The feature size for the microarray was 16 μM × 16 μM, and there were 382,409 features within the 17 mm × 13 mm array area. Of these, 379,652 were catfish-specific, and the remaining 2,757 were used for quality control of oligonucleotide synthesis and hybridization, and for signal normalization.

The microarrays were pre-hybridized with 1X MES hybridization buffer (100 mM MES, 1.0 M Na+, 20 mM EDTA, 0.01% Tween20) plus 40 μg of herring sperm DNA and 200 μg of acetylated BSA at 45°C for 15 min, followed by hybridization in the same buffer with 10 μg of denatured and fragmented cRNA per microarray at 45°C for 16–20 h with constant rotation. After hybridization, the microarrays were immediately washed extensively with non-stringent wash buffer (6X SSPE, 0.01% Tween20) at room temperature, then with stringent wash buffer (100 mM MES salt and free acid solution, 0.1 M Na+, 0.01% Tween20) at 45°C. After final rinsing with non-stringent wash buffer, the microarrays were stained with 1X Stain buffer (100 mM MES, 1 M Na+, 0.05% Tween20, 50 mg/ml of BSA, and 1 mg/ml of Cy3-streptavidin) at room temperature for 25 min. After the stain buffer was removed, the microarrays were rinsed with non-stringent wash buffer and immediately dried under argon gas. The microarrays were scanned with an Axon GenePix 4000B scanner (Molecular Devices Corp., Union City, CA) at 5 μM resolution. The data were extracted from the raw images with NimbleScan software (Nimblegen, Inc.). A total of ten microarrays were used for this project: two replicates for the control pool and one microarray for each of the eight LPS-stimulated individuals.

Data Analysis, Statistics, and Bioinformatics

Relative signal intensity (log2 transformed) for each feature was generated using the Robust Multi-Array Average (RMA) algorithm [34]. The signal intensity was background corrected based on the quantile normalization process [35, 36] for all microarrays from the entire experiment (under uniform conditions) using data from only perfectly matched oligos. Only oligos with a ratio of signal to global average background greater than 2.0, with a perfectly matched signal 150% of the mismatched signal, and with a correlation > 0.9 between biological replicates were included. The normalized data was analyzed using Significance Analysis of Microarrays [37] from the TIGR Multiexperiment Viewer analysis software package [38] with two-class unpaired design for each time point. The RMA value from perfectly matched oligos was used to calculate fold differences as the average signal intensity of treated animals divided by the average signal intensity of control animals. The list of significant genes that changed at least 2-fold was generated with an estimated global false discovery rate <10%. Differentially expressed sequences were compared to the Ensembl peptide databases derived from the zebrafish (v.5) and Tetraodon (v.7) genomes [39] using BlastX (E-value cutoff 0.005, minimum bit score = 100) and annotated accordingly

Real-time RT-PCR

Real-time RT-PCR analysis was carried out with the TaqMan One-Step RT-PCR kit (Applied Biosystems, Foster City, CA) in a 25 μL reaction volume containing 200 nM each amplification primer, 100 nM dual-labelled probe, and 200 ng of total RNA. Products were amplified in an iCycler iQ™ Real Time PCR Detection System (Bio-Rad Laboratories, Hercules, CA) with the following profile: 48°C for 30 min; 95°C for 10 min; 40 cycles of 95°C for 30s, 60°C for 1 min. Expression levels of β-actin were used as endogenous controls within each sample. Relative levels of gene expression were calculated using the $2^{\Delta\Delta CT}$method [40].

DECLARATIONS

Acknowledgements

The authors thank Sylvie Quiniou and Brian Scheffler for critical review of the manuscript, and Beth Flanagan for informatic support. The microarray design and expression data are available as Accession GSE3261 in the Gene Expression Omnibus repository at the National Center for Biotechnology Information. Mention of trade names or commercial products in this publication is solely for the purpose of providing specific information and does not imply recommendation or endorsement by the U. S. Department of Agriculture.

REFERENCES

1. Miya M, Nishida M: Use of mitogenomic information in teleostean molecular phylogenetics: a tree-based exploration under the maximum-parsimony optimality criterion. Molec Phylogenet Evol. 2000, 17: 437-455. 10.1006/mpev.2000.0839.
2. Elmerot C, Arnason U, Gojobori T, Janke A: The mitochondrial genome of the pufferfish, *Fugu rubripes*, and ordinal teleostean relationships. Gene. 2002, 295: 163-172. 10.1016/S0378-1119(02)00688-1.

3. USDA: Catfish Production Report. National Agricultural Statistics Service USDA Washington, D.C, (July 23, 2005), [http://usda.mannlib.cornell.edu/reports/nassr/other/pcf-bbc/2005/cfpd0205.txt]
4. Shen L, Stuge TB, Zhou H, Khayat M, Barker KS, Quiniou SMA, Wilson M, Bengtén E, Chinchar VG, Clem LW, Miller NW: Channel catfish cytotoxic cells: a mini-review. Dev Comp Immunol. 2002, 26: 141-9. 10.1016/S0145-305X(01)00056-8.
5. Ton C, Stamatiou D, Dzau VJ, Liew CC: Construction of a zebrafish cDNA microarray: gene expression profiling of the zebrafish during development. Biochem Biophys Res Commun. 2002, 296: 1134-1142. 10.1016/S0006-291X(02)02010-7.
6. Mathavan S, Lee SG, Mak A, Miller LD, Murthy KR, Govindarajan KR, Tong Y, Wu YL, Lam SH, Yang H, Ruan Y, Korzh V, Gong Z, Liu ET, Lufkin T: Transcriptome analysis of zebrafish embryogenesis using microarrays. PLoS Genet. 2005, 1: 260-76. 10.1371/journal.pgen.0010029.
7. Gracey AY, Troll JV, Somero GN: Hypoxia-induced gene expression profiling in the euryoxic fish *Gillichthys mirabilis*. Proc Natl Acad Sci USA. 2001, 98: 1993-1998. 10.1073/pnas.98.4.1993.
8. Ton C, Stamatiou D, Liew CC: Gene expression profile of zebrafish exposed to hypoxia during development. Physiol Genomics. 2003, 13: 97-106.
9. Gracey AY, Fraser EJ, Li W, Fang Y, Taylor RR, Rogers J, Brass A, Cossins AR: Coping with cold: An integrative, multitissue analysis of the transcriptome of a poikilothermic vertebrate. Proc Natl Acad Sci USA. 2004, 101: 16970-16975. 10.1073/pnas.0403627101.
10. Krasnov A, Koskinen H, Pehkonen P, Rexroad CE, Afanasyev S, Molsa H: Gene expression in the brain and kidney of rainbow trout in response to handling stress. BMC Genomics. 2005, 6: 3-10.1186/1471-2164-6-3.
11. Kurobe T, Yasuike M, Kimura T, Hirono I, Aoki T: Expression profiling of immune-related genes from Japanese flounder*Paralichthys olivaceus* kidney cells using cDNA microarrays. Dev Comp Immunol. 2005, 29: 515-523. 10.1016/j.dci.2004.10.005.
12. Mackenzie S, Iliev D, Liarte C, Koskinen H, Planas JV, Goetz FW, Molsa H, Krasnov A, Tort L: Transcriptional analysis of LPS-stimulated activation of trout (*Oncorhynchus mykiss*) monocyte/macrophage cells in primary culture treated with cortisol. Mol Immunol. 2005, 43: 1340-1348. 10.1016/j.molimm.2005.09.005.
13. Byon JY, Ohira T, Hirono I, Aoki T: Use of a cDNA microarray to study immunity against viral hemorrhagic septicemia (VHS) in Japanese flounder (*Paralichthys olivaceus*) following DNA vaccination. Fish Shellfish Immunol. 2005, 18: 135-147. 10.1016/j.fsi.2004.06.008.
14. Rise ML, Jones SR, Brown GD, von Schalburg KR, Davidson WS, Koop BF: Microarray analyses identify molecular biomarkers of Atlantic salmon macrophage and hematopoietic kidney response to *Piscirickettsia salmonis* infection. Physiol Genomics. 2004, 20: 21-35. 10.1152/physiolgenomics.00036.2004.
15. Ewart KV, Belanger JC, Williams J, Karakach T, Penny S, Tsoi SC, Richards RC, Douglas SE: Identification of genes differentially expressed in Atlantic salmon (*Salmo salar*) in response to infection by Aeromonas salmonicida using cDNA microarray technology. Dev Comp Immunol. 2005, 29: 333-347. 10.1016/j.dci.2004.08.004.
16. Meijer AH, Verbeek FJ, Salas-Vidal E, Corredor-Adamez M, Bussman J, van der Sar AM, Otto GW, Geisler R, Spaink HP: Transcriptome profiling of adult zebrafish at the late stage of chronic tuberculosis due to *Mycobacterium marinum*infection. Mol Immunol. 2005, 42: 1185-1203. 10.1016/j.molimm.2004.11.014.
17. Liu Z: A review of catfish genomics: progress and perspectives. Comp Funct Genom. 2003, 4: 259-265. 10.1002/cfg.265.
18. Nonneman D, Waldbieser GC: Isolation and enrichment of abundant microsatellites from a channel catfish (*Ictalurus punctatus*) brain cDNA library. Anim Biotechnol. 2005, 16: 103-116. 10.1080/10495390500262908.
19. The Institute for Genomic Research. [http://www.tigr.org/tigr-scripts/tgi/T_index.cgi?species=catfish]
20. Sweet MJ, Hume DA: Endotoxin signal transduction in macrophages. J Leukoc Biol. 1996, 60: 8-26.
21. Poltorak A, He X, Smirnova I, Liu MY, Van Huffel C, Du X, Birdwell D, Alejos E, Silva M, Galanos C, Freudenberg M, Ricciardi-Castagnoli P, Layton B, Beutler B: Defective LPS signaling in C3H/HeJ and C57BL/10ScCr mice: mutations in Tlr4 gene. Science. 1998, 282: 2085-2088. 10.1126/science.282.5396.2085.
22. Takeuchi O, Hoshino K, Kawai T, Sanjo H, Takada H, Ogawa T, Takeda K, Akira S: Differential roles of TLR2 and TLR4 in recognition of gram-negative and gram-positive bacterial cell wall components. Immunity. 1999, 11: 443-451. 10.1016/S1074-7613(00)80119-3.
23. Bilodeau AL, Waldbieser GC: Activation of TLR3 and TLR5 in channel catfish exposed to virulent *Edwardsiella ictaluri*. Dev Comp Immunol. 2005, 29: 713-721. 10.1016/j.dci.2004.12.002.
24. Meijer AH, Krens SF, Medina Rodriguez IA, He S, Bitter W, Snaar-Jagalska BE, Spaink HP: Expression analysis of the Toll-like receptor and TIR domain adaptor families of zebrafish. Mol Immunol. 2004, 40: 773-783. 10.1016/j.molimm.2003.10.003.
25. Collet B, Secombes CJ: Type I-interferon signalling in fish. Fish Shellfish Immunol. 2002, 12: 389-397. 10.1006/fsim.2001.0405.
26. Harry GJ, Lefebvre d'Hellencourt C, Bruccoleri A, Schmechel D: Age-dependent cytokine responses: trimethyltin hippocampal injury in wild-type, APOE knockout, and APOE4 mice. Brain Behav Immun. 2000, 14: 288-304. 10.1006/brbi.2000.0606.
27. Ripolles Piquer B, Nazih H, Neunlist M, Huvelin JM, Bard JM: Effect of LPS on basal and induced apo E secretion by 25-OH chol and 9cRA in differentiated CaCo-2. J Cell Biochem. 2004, 91: 786-795. 10.1002/jcb.10786.

28. Triantafilou K, Triantafilou M, Dedrick RL: A CD14-independent LPS receptor cluster. Nat Immunol. 2001, 2: 338-345. 10.1038/86342.

29. Triantafilou M, Triantafilou K: Lipopolysaccharide recognition: CD14, TLRs and the LPS-activation cluster. Trends Immunol. 2002, 23: 301-304. 10.1016/S1471-4906(02)02233-0.

30. Verani A, Sironi F, Siccardi AG, Lusso P, Vercelli D: Inhibition of CXCR4-tropic HIV-1 infection by lipopolysaccharide: evidence of different mechanisms in macrophages and T lymphocytes. J Immunol. 2002, 168: 6388-6395. Kocabas AM, Li P, Cao D, Karsi A, He C, Patterson A, Ju Z, Dunham RA, Liu Z: Expression profile of the channel catfish spleen: Analysis of genes involved in immune functions. Mar Biotechnol. 2002, 4: 526-536. 10.1007/s10126-002-0067-0.

31. Huang X, Madan A: CAP3: A DNA sequence assembly program. Genome Res. 1999, 9: 868-877. 10.1101/gr.9.9.868.

32. Singh-Gasson S, Green RD, Yue Y, Nelson C, Blattner F, Sussman MR, Cerrina F: Maskless fabrication of light-directed oligonucleotide microarrays using a digital micromirror array. Nat Biotechnol. 1999, 17: 974-978. 10.1038/13664.

33. Irizarry RA, Hobbs B, Collin F, Beazer-Barclay YD, Antonellis KJ, Scherf U, Speed TP: Exploration, normalization, and summaries of high density oligonucleotide array probe level data. Biostatistics. 2003, 4: 249-264. 10.1093/biostatistics/4.2.249.

34. Bolstad B, Irizarry R, Astrand M, Speed T: A comparison of normalization methods for high density oligonucleotide array data based on variance and bias. Bioinformatics. 2003, 19: 185-193. 10.1093/bioinformatics/19.2.185.

35. Bioconductor Open Source Software for Bioinformatics. [http://www.bioconductor.org]

36. Tusher VG, Tibshirani R, Chu G: Significance analysis of microarrays applied to the ionizing radiation response. Proc Natl Acad Sci USA. 2001, 98: 5116-5121. 10.1073/pnas.091062498.

37. TIGR Microarray software. [http://www.tigr.org/software/microarray.shtml]

38. Ensembl Genome Browser: Zebrafish (*Danio rerio*) assembly version 5 and Spotted Green Pufferfish (*Tetraodon nigroviridis*) assembly version 7. [http://www.ensembl.org/index.html]

39. Livak KJ, Schmittgen TD: Analysis of relative gene expression data using Real-Time quantitative PCR and the $2^{\Delta\Delta CT}$method. Methods. 2001, 25: 402-408. 10.1006/meth.2001.1262.

CHAPTER 15

EXPRESSED SEQUENCE TAGS (ESTS) FROM IMMUNE TISSUES OF TURBOT (SCOPHTHALMUS MAXIMUS) CHALLENGED WITH PATHOGENS

Belén G Pardo, Carlos Fernández, Adrián Millán, Carmen Bouza, Araceli Vázquez-López, Manuel Vera, José A Alvarez-Dios, Manuel Calaza, Antonio Gómez-Tato, María Vázquez, Santiago Cabaleiro, Beatriz Magariños, Manuel L Lemos, José M Leiroand Paulino Martínez

BACKGROUND

The turbot (*Scophthalmus maximus*; Scophthalmidae; Pleuronectiformes) is a commercially valuable flatfish that has been intensively cultured for the last decade. Its production has steadily increased to 7120 tonnes in 2006 (80% from Spain; FEAP, 2006) and represents one of the most promising aquaculture species in Europe. However, disease outbreaks in turbot have occurred frequently and losses due to infections constitute a serious problem for its culture [1, 2]. The use of antibiotics, vaccines and fish health management practices has partially solved the problem, but the achievement of large-scale production in the highly competitive world market requires enhancing resistance of cultured fish to diseases. Information on the immune response of turbot is still limited, and little is known about host-pathogen interactions in fish species. The screening and identification of immune-relevant genes is essential to analyze the genetic basis for infection, immunity and resistance to pathogens of economic relevance in aquaculture. Expressed sequence tag (EST) analysis is a powerful approach to provide a rapid and efficient method to go from expressed sequences to genes. ESTs are essential for studies of gene function [3, 4], but are also useful to identify polymorphic gene markers, such as microsatellites and single nucleotide polymorphisms (SNPs) [5,6, 7, 8]. These markers are the basis for genetic and physical mapping, and for comparative genome analysis [9, 10, 11]. From a practical perspective, maps can be applied for assisted selection programmes (MAS) and eventually for identification of genes related with quantitative traits (QTL) [12, 13]. In addition, ESTs constitute the basic resources to develop microarrays for functional genomics studies [14].

EST sequence resources are rapidly growing in molecular databases. However, the number of ESTs in fish is generally scarce, excluding some model species and Atlantic salmon among cultured fish [15, 16, 17, 18,19]. The scarcity of EST resources in cultured fish limits the use of modern functional genomic approaches for selective breeding purposes [20]. Among flatfish, aquaculture production has been successfully achieved in turbot, Japanese flounder and Atlantic halibut. Compared to the very large efforts for the development of EST resources in Japanese flounder (8856 ESTs) and Atlantic halibut (17659 ESTs) [21, 22,23, 24, 25, 26, 27, 28], EST resources in turbot are scarce (3171 ESTs). Less than 800 sequences have been deposited to date in the NCBI nucleotide database, most of them from anonymous microsatellite searching [29].

With the aim of increasing the genomic resources in turbot and identifying relevant genes for immunity, three cDNA libraries were constructed from mRNA isolated from immunity-related tissues of turbot (liver, spleen and head kidney) at different times after infection with *Aeromonas salmonicida* and the scuticociliate parasite *Philasterides*

Citation: G Pardo, Carlos Fernández, Adrián Millán, Carmen Bouza, Araceli Vázquez-López, Manuel Vera, José A Alvarez-Dios, Manuel Calaza, Antonio Gómez-Tato, María Vázquez, Santiago Cabaleiro, Beatriz Magariños, Manuel L Lemos, José M Leiro and Paulino Martínez, Expressed sequence tags (ESTs) from immune tissues of turbot (Scophthalmus maximus) challenged with pathogens, doi: 10.1186/1746-6148-4-37 *dicentrarchi*. These pathogens are responsible for important disease outbreaks in turbot, as well as in other culture fish species [30, 31, 32, 33]. Our main goal was to obtain the most accurate information possible to address functional genomic studies on disease resistance. However, the use of non-normalized cDNA libraries made it feasible to get a preliminary picture of the turbot genetic response to pathogens through analyzing transcript distribution among infected *vs.* control libraries. A total of 12584 ESTs were sequenced and compared to GenBank database and a large array of defence or immune-related genes was identified. Also, this large scale EST study increased the number of putative markers for mapping. A total of 191 microsatellites, of which 104 exhibited sufficient flanking sequences for primer design, and 2197 good quality SNPs were identified for the first time in turbot. The cDNA sequences generated will serve as a basis for microarray construction. This first EST study in turbot will provide the support for further research into the genetics, genomics and even proteomics of this important aquaculture species.

RESULTS AND DISCUSSION

cDNA libraries and ESTs

EST analysis is an efficient and fast method for gene discovery [15, 17]. In Pleuronectiformes, this approach has been recently applied in Japanese flounder (*Paralichthys olivaceus*; [21, 22, 24, 25]), winter flounder (*Pseudopleuronectes americanus*; [34]), flounder (*Platichthys flesus*; [35, 36]) and Atlantic halibut (*Hippoglossus hippoglossus*; [26, 27, 28]. However, this order comprises around 600 species, many of them of great economic value both for fisheries and farming. In this study, we have addressed the construction of an EST database for the identification of genes related with immunity and defence in turbot.

The three cDNA libraries constructed held at least 2.5 × 10^6 primary recombinant clones. The directional cloning approach used for construction of cDNA libraries ensured that cDNA inserts appeared mostly in the same orientation within the vector. Clones were sequenced from the 3' end with the vector primer T7 to obtain large gene-specific genomic regions for future oligo-microarray design. The libraries used were non-normalized and, as usually observed [16], substantial redundancy was obtained (around 74%). This approach allowed an analysis of the turbot response to specific pathogens by comparing the amount of transcripts across all genes or groups of genes classified in functional categories. A total of 12584 ESTs were sequenced. After trimming and vector removal 9256 high quality ESTs were obtained (Table 1), showing an average length of 409 bp. Their sequences are available in the dbEST NCBI database under numbers FE943103-FE952358.

Table 1: Summary statistics of ESTs from turbot libraries

	Number	%
Good-quality ESTs	9256	
Redundant sequences	6847	74.0
Unique sequences	3482	36.0
Contigs	1073	30.8
Singletons	2409	69.2
Unique sequences with no BLAST hits	1716	49.3
Unique sequences with BLAST hits	1766	50.7
BLASTN	1091	61.8
BLASTX	675	38.2
Unique sequences with functional annotation	816	23.4
Contigs	489	59.9
Singletons	327	40.1

EST projects generate a large number of redundant sequences due to the random selection of cDNAs from tissue libraries, especially when libraries are non-normalized and a high number of clones are sequenced. Clustering redundant sequences is a critical step to identify genes. The program CAP3http://seq.cs.iastate.edu./. was used to cluster EST sequences using the default parameters. As shown in Table 1, clustering yielded 3482 unique turbot sequences: 2409 singletons (69.2%) and 1073 contigs (30.8%) comprising 6847 ESTs (6.4 sequences/contig) and an average length of 527 bp. Figure 1A shows the histogram distribution of contig sizes. Although most contigs showed two (46.7%) or three-to-five (30.3%) sequences, a small number of highly expressed genes were also detected. As shown in Figure 1B, the beta-globin contig (407 ESTs) and five others contained more than 100 ESTs, which represents a high proportion (1212/9256 = 13.1%) over the total redundancy (74.0%). Most of these ESTs shared homology with genes involved in transport, protein metabolism and response to stress, all of which related to defence. At the other extreme, 77.0% of the 1073 contigs were ≤ 5 times redundant, indicating that most of these unique sequences represent rare mRNAs and that these libraries provide a rich source of sequence information.

A

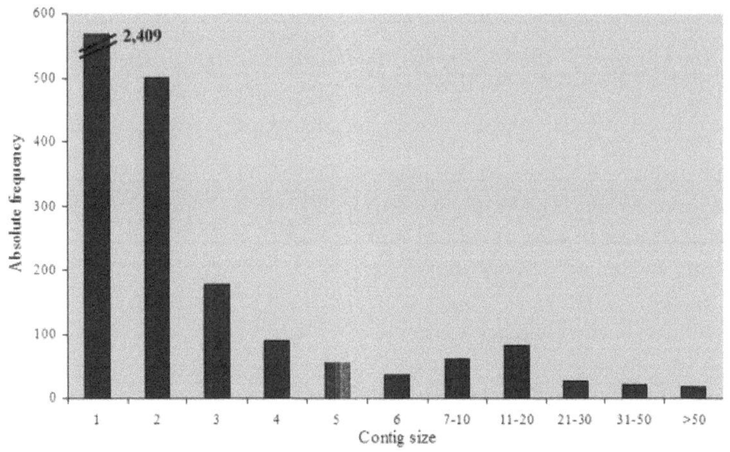

B

Gene	Contig size	e-value	% Hom	Biological process	BLAST
Hepcidin precursor	70	3e-21	83	-	BLASTN
Lysozyme C precursor	76	8e-42	100	Response to stress	BLASTX
Trypsinogen 1 precursor	86	7e-21	67	Proteolysis	BLASTX
Serotransferrin	88	5e-11	85	Ion transport	BLASTN
Trypsinogen 1	114	4e-88	69	Proteolysis	BLASTX
cDNA clone *D. labrax*	148	8e-22	82	-	BLASTN
Ribosomal protein S27a	155	3e-32	68	Prot. metabol. process	BLASTX
Ribosomal protein S2	171	0.0	93	Biosyntethic process	BLASTN
Hemoglobin alpha-1 chain	219	6e-32	86	Transport	BLASTX
Beta type globin	407	3e-127	81	Transport	BLASTN

Figure 1: Sequence prevalence distribution of the identified contigs from turbot libraries. (A) Absolute frequency histogram showing contig size (number of sequences) distribution. (B) Functional and BLAST hits confidence characteristics of the ten largest contigs. Only biological function according to GO terms has been included.

Gene Annotation

ESTs were identified by BLAST searches against nucleotide database at GenBank. Due to the low representation of fish genes, a protein-based homology strategy was also used in the international database searches. Protein sequences have been demonstrated to be more suitable to detect homology over long periods of evolutionary time [37]. Our EST database pprovided a graphical view of all contigs, their PROSITE/PRINTS protein patterns and a search interface by keywords, microsatellite, gene, and UniGene/GO/KEGG information. Tools to search for sequences and markers based on annotations, to perform local BLAST searches, and to select sequences for a prospective microarray were also included. These tools used RepeatMasker http://www.repeatmasker.org/ for masking low-complexity sequences and OligoArray 2.1 http://berry.engin.umich.edu/oligoarray2_1/ to predict secondary structures and potential cross-hybridization.

As shown in Table 1, 1716 (49.3%) unique sequences displayed no significant similarity to known sequences or ESTs in the public databases, whereas the remaining 1766 (50.7%) showed significant matches with e-values ≤ 1e-5. Among these, BLAST database searches allowed assignment of putative function to 816 sequences (23.4%). In spite of their lower frequency among unique sequences, contigs were annotated more frequently (59.9%) than singletons (40.1%). As in other ESTs fish studies, the lower percentage of annotated singletons suggests that these are either novel fish-specific or rapidly evolving genes [38]. Also, it is possible that bioinformatic errors could have a greater impact on singletons, since they are unique sequences whose information cannot be contrasted with other sequences in the database. The lower annotation success regarding similar genomic projects [28, 39] was probably related to the read length (around 500 bp) and specially the direction (from the 3'end) of sequencing. The 3' untranslated region (UTR) is approximately double the length of the 5'UTR according to GenBank fish entries. So, what is gained in specificity for microarray oligo-design is lost for gene annotation.

All unique sequences were annotated based on similarity using BLASTX or BLASTN [40] in the public databases GenBank NR and Unigene. The multiple annotations provided greater assurance about gene description and frequency of annotation than in a single database. The use of consensus sequences allowed sequences without significant similarity regions with a known protein (e.g., 5' or 3' noncoding regions) to be annotated if they were members of an annotated contig. All hits with e-value ≤ 1e-5 and their associated alignments were stored in the database and tracked with any associated functional annotation. We also ran AutoFACT [41] on all sequences in the database. It is interesting to note that while

AutoFACT was able to come up with more function-specific information than our online tool (provided external database hits were found), it was not able to annotate as many sequences as our online custom tool.

Annotated ESTs (816) were classified into functional categories according to GO terms [42] (77.0%), 367 among contigs (75.1%) and 261 among singletons (79.8%). A single sequence very often showed several GO terms in the same ontology, so we try to group them using a single more general category. Overly specific categories were also collapsed into more general terms. Any terms related to defence/immunity were always retained given their interest in this study. Biological process, molecular function and cellular component categories are shown in Figures 2, 3 and 4, respectively. According to these criteria, biological processes associated with proteins (synthesis, metabolism and proteolysis) were the largest annotated categories, though a significant group related to transport and response to stress did appear. In accordance with this, a large proportion of sequences were classified into structural constituent of ribosome and protein-related functions (binding, peptidase activity) regarding molecular function. A remarkably high proportion of annotated sequences were categorized as oxidoreductase activity. Finally, most gene activity was located into the ribosome, followed by the membrane, nucleus and the extracellular and cytoplasm constituents of the cell.

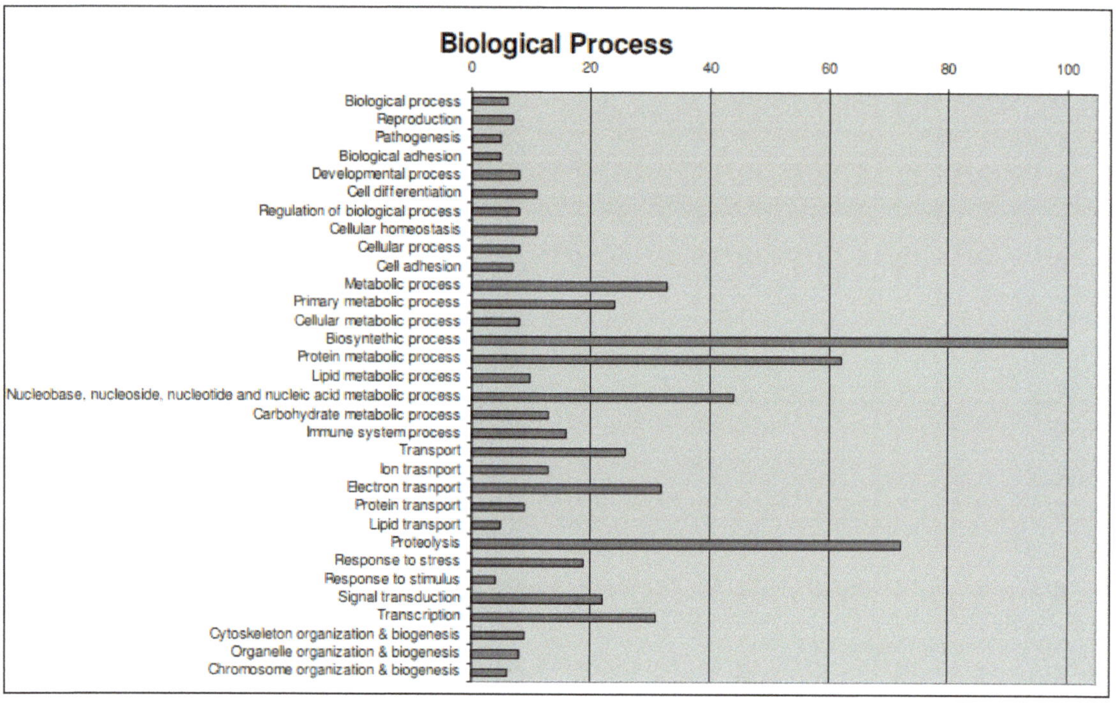

Figure 2: Classification of turbot unique sequences in biological processes categories following Gene Ontology (GO).

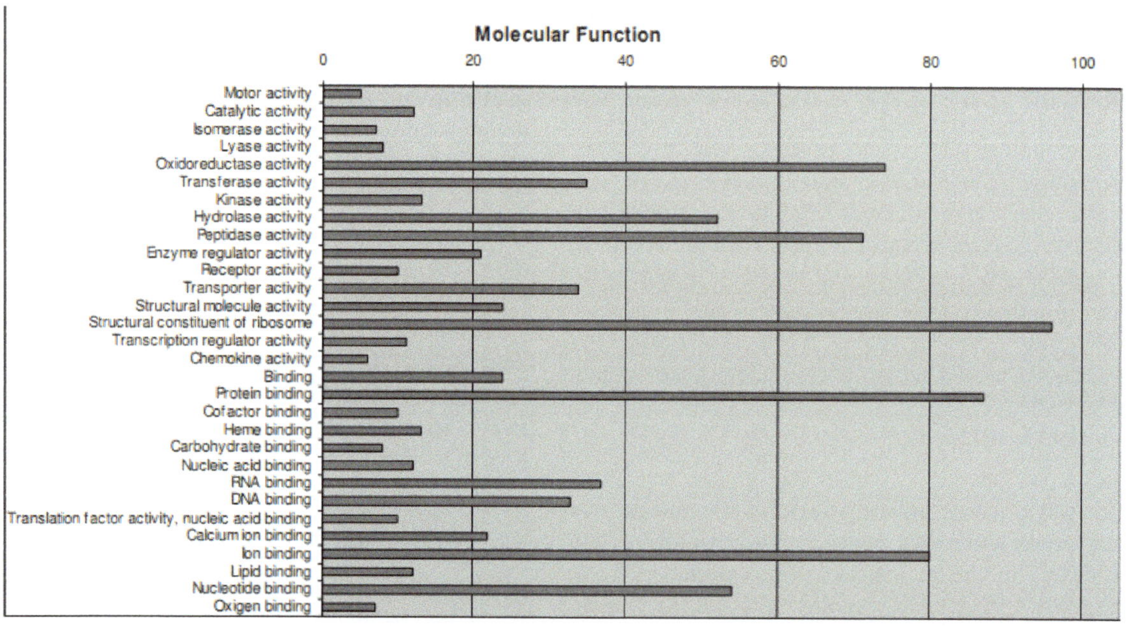

Figure 3: Classification of turbot unique sequences in molecular function categories following Gene Ontology (GO).

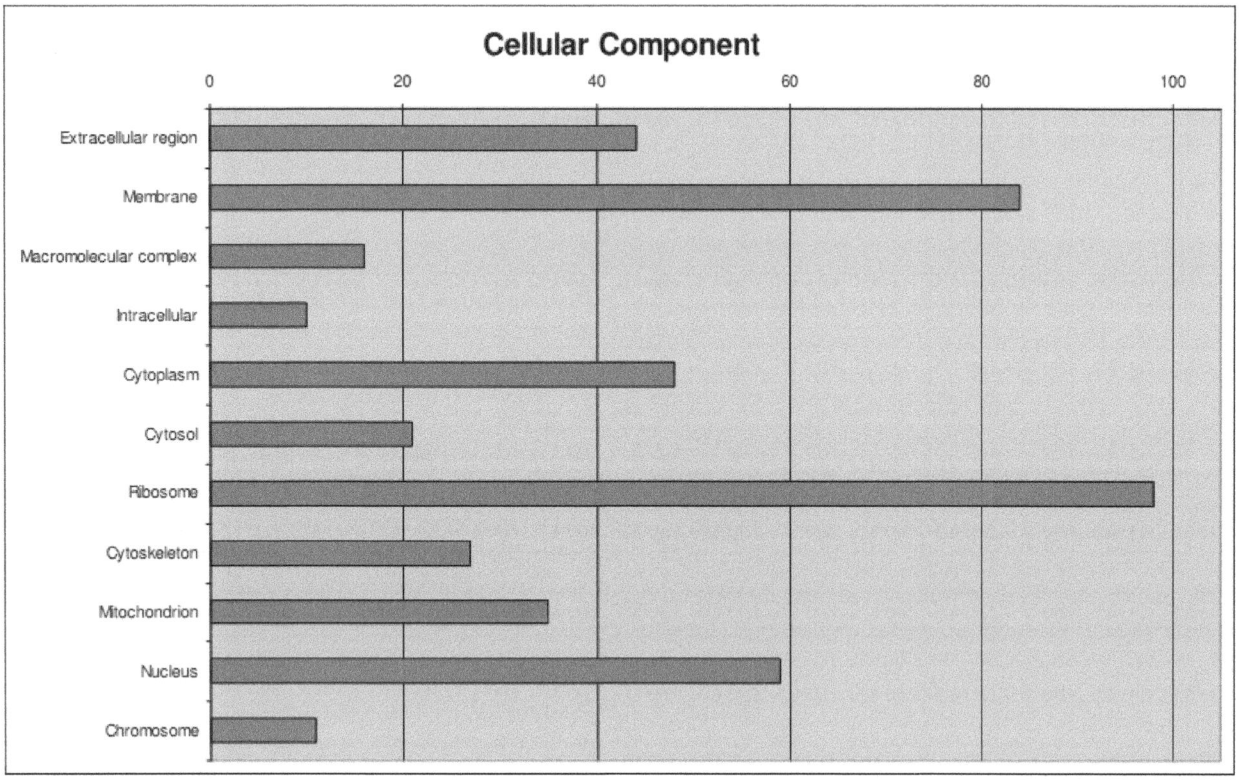

Figure 4: Classification of turbot unique sequences in cellular component categories following Gene Ontology (GO).

Immune Genes

Systematic classification of annotated sequences using available bioinformatics tools (GO, KEGG) provides a useful information to analyze the functional profile of annotated genes. In our study, we were especially interested in identifying immune genes or in a broader sense those genes regulated in response to specific pathogens of turbot. So, we analyzed the relationship of our annotated genes with defence or immunity using scientific literature information available on vertebrates. According to this analysis, 203 genes out of 816 annotated (24.9%) appeared related to defence or immunity in our work (Table 2). This observation is in agreement with the important role of liver, spleen and head kidney in the immune response of fish.

Table 2: Defence and immune-related annotated ESTs from turbot libraries

Genes	No. unique sequences	%
Complement related	16	7.9
Apoptosis related	10	4.9
Immunoglobulin related	8	3.9
Glutathione S-transferase	7	3.5
Elastase	6	2.9
Cytochrome P450	6	2.9
Major histocompatibility complex	5	2.5
Coagulation factor	5	2.5
Interferon related	3	1.5
Perforin	3	1.5
Hepcidin precursor	3	1.5
Nephrosin	3	1.5
Alpha-2-macroglobulin	3	1.5
Other genes	119	58.6
Total	203	24.9

”No. unique sequences” refers to the total amount of the different annotated contigs and singletons for each gene class listed in the Table. “Total percentage” of defence/immune-related genes is referred to the number of unique annotated sequences from turbot libraries (816).

Like in mammals, the immune system of fish is composed of non-specific and specific defence. The innate immune response is an important and highly developed defence mechanism against pathogens in fish [43]. Examples of innate immunity include anatomic barriers, mechanical removal of pathogens, bacterial antagonism, pattern-recognition receptors, antigen-nonspecific defence compounds, complement pathways, phagocytosis, and inflammation [43]. In Table 2, a list of defence/immune-related genes that were found three or more times in our libraries is shown. Complement related genes were predominant (7.9%) followed by apoptosis (4.9%) and immunoglobulin (Ig)-related (3.9%) genes. Glutathione S-transferase, heat shock proteins and cytochrome P450, elastases, major histocompatibility complex (MHC) and coagulation factors involved in innate immunity were also present, as well as others like interferon, perforin, hepcidin, nephrosin or alpha-2-macroglobulin. Interestingly, the majority of our immune-related cDNAs (75.0%) were reported for the first time in turbot, even though some of them have important roles in the immune response like B-cell linker, chemotaxin, complement components, IgD, IgM, interferon stimulated gene 12 (b2), lipopolysaccharide (LPS)-binding protein, natural killer (NK)-lysine type 1, peptidoglycan recognition protein, skin mucus lectin and tumour necrosis factor (TNF) receptor associated factor2, among others. Previously, only a hepcidin [44] and a natural killer cell enhancing factor [45] had been characterized in turbot.

The availability of a large number of sequences from immune-related cDNA libraries, both from non-infected fish (controls) as well as from fish challenged with specific pathogens, suggested a comparison of transcript profiles to identify genes regulated in response to these pathogens. Four or more sequences per gene (contig) are the minimum necessary to get statistical power to check the null hypothesis of even distribution of sequences among the three libraries. Ninety six genes presented four or more sequences in our libraries. However, taking into account the large number of tests performed, we decided to use a more conservative set of genes (with 6 or more sequences) to avoid type I errors. This new set comprised 72 genes, whose distribution among libraries is shown in Additional file 1 together with their probability of departure from the null hypothesis using a chi-square test. To use all information available and to get higher statistical power, we decided to perform an additional analysis by comparing the profiles of all genes grouped according to GO categories (see Additional file 2). Though this approach could be mixing genes with opposite expression patterns, it could provide significant trends associated to specific functions for complementing the “individual-level gene” analysis. Bonferroni correction for multiple tests was considered in this case, because of the higher statistical power achieved with this approach.

As expected, defence/immune-related genes appeared overrepresented in the set of 72 genes (37.5%) with regard to the total number of annotated genes in our libraries (24.9%). Some genes apparently responded in a similar way to both pathogens, being down- (myosin, nephrosin and several peptidases) or up-regulated (actin and lysozyme) regarding control. However, most genes responded to only one pathogen. The different infection profile of the pathogens used in this study, a bacterium (*A. salmonicida*) and a parasite (*P. dicentrarchi*), is expected to stimulate a specific set of genes in the host. So, protein biosynthesis was up-regulated in response to *Aeromonas salmonicida*, as suggested the increased amount or ribosome-related and elongation factor proteins, and their correspondent GO terms (see Additional file 2). Also, a general down-regulation of genes involved in proteolytic activity in response to this bacterium (chymotrypsin B precursor, trypsinogen1, chymotrypsinogen 2, trypsynogen-like serin protease and elastase precursor) was observed, in turn correlated to the lower presence of peptidase and proteolysis GO terms in this library. The possibility that *A. salmonicida* could be blocking these genes cannot be ruled out, since it would facilitate the infection process. Additionally, it was noteworthy that several iron metabolism-related genes turned out to be up-regulated after infection with *A. salmonicida* (haptoglobin fragment 1, globin-related proteins, hepcidin precursor). The high relevance of iron for bacteria infection could explain this observation. The up-regulation of hepcidin after bacteria infection has been recently described in turbot and gilthead seabream [44, 46]. Finally, some other immune relevant genes like MHC II alpha and beta antigens, complement C9, thrombin, chemotaxin, bactericidal permeability-increasing protein, apolipoprotein A-IV3, heat shock protein 90 beta and chemotaxin showed significant up-regulation in response to *A. salmonicida*.

However, the strongest genetic signature was observed in response to *P. dicentrarchi* infection. Specific genes were exclusively detected in this library and sometimes at high frequencies, like those homologous to *Oryzias latipes* and *Paralichthys olivaceous* cDNA clones and to mitochondrial ATP synthase alpha-subunit gene. Functional annotation of these unknown genes now appears relevant to understand the response of turbot to this pathogen. The same was reflected when GO categories were analyzed (see Additional file 2). A large number of differentially regulated gene categories ($P = 0$) appeared associated to *P. dicentrarchi* library according to the different GO criteria. Cytoskeleton organization and biogenesis, as well as carbohydrate metabolic process appeared up-regulated categories among biological processes. Transporter and hydrolase activities, as well as protein, ion and heme binding also were overexpressed regarding molecular function classification. Finally, cytoskeleton and cytosol appeared as the highest active cellular components. Looking at specific categories, a significant increase in the expression of antioxidant genes, like glutathione peroxidase and glutathione-S-transferase, was observed. These genes play a pivotal role to prevent cellular damage due to the increased reactive oxygen species (ROS) during infection [47]. Paramá et al. [48] have recently demonstrated the increase in intracellular ROS by proteases of turbot kidney in response to this pathogen. Tissue trauma or invasion by pathogens induces changes in the quantities of several macromolecules in animal body fluids, which comprise one aspect of the acute phase response (APR). In fish, APR proteins include pentraxins, serum amyloid P, several components of the complement system, transferrin and thrombin. Up-regulation of pentraxin and serotransferrin was observed in response to *P. dicentrarchi* in our study. Transferrin up-regulation could be related to the increase in inflammation and enhanced oxidative stress typical of infections with this parasite [48]. Finally, up-regulation of other important components of the immune response such the profilin and lysozyme was observed. Profilin-like protein has been reported as a toll-like receptor (TLR) 11 ligand in some parasites [49]. TLRs are evolutionary conserved transmembrane proteins that recognize a unique pattern of molecules derived from pathogens or damaged cells, triggering robust but defined innate immune responses [49].

Markers-Containing ESTs

EST studies can also provide resources for the identification of polymorphic DNA markers such as microsatellites and SNPs. In our study, screening of EST sequences for short tandem repeats (2–6 bp) identified 191 microsatellites using a conservative criterion (≥ 6 and ≥ 8 repeats for tri/tetra/penta and dinucleotide motifs, respectively). Of these, 120 had significant hits in BLAST with e-value cut off ≤ 1e-5 and 71 were annotated. Most microsatellites were dinucleotide (128) and trinucleotide (56), while only 6 tetranucleotide and 1 pentanucleotide were found. Among these 191 ESTs sequences, 104 contained sufficient flanking sequences length for primer design. Fifty of these 104 microsatellites-containing ESTs were contigs, therefore the *in silico* comparison of the sequences included in these contigs allowed us to identify 11 putative polymorphic microsatellites.

Because of their abundance within the genome, SNPs are the most common type of genetic markers [50] for studying complex genetic traits and genome evolution [51]. In turbot there have no been reports on SNP identification so far. The use of non-normalized libraries and a large number of individuals in library construction made possible the identification of 2197 good quality SNPs. After the three filters used in the Quality SNP pipeline [52], we finally detected 2197 real and 1158 true SNPs (Table 3), representing a rate of 1.39 and 0.74 SNPs per 100 bp, respectively. Real and true SNPs included 749 and 453 transitions, 974 and 558 transversions and 366 and 125 indels, respectively. With this pipeline, only clusters with at least 4 EST sequences were selected to minimize the detection of SNPs caused by sequencing errors. As shown in Table 4 the majority of SNPs were detected in contigs involving a larger number of sequences, which provides an additional proof of our SNP quality. The identification of microsatellite and SNPs markers within turbot ESTs will contribute to extend the turbot genetic map [10] after linkage analysis in reference families. Since these markers are linked to genes, they will be useful as Type I markers for population genomics scanning in this species and for comparative mapping and fish evolutionary studies.

Table 3: Summary statistics of SNP identification from turbot EST resources

	Real SNPs	**True SNPs andSNPs**
Total sequences analysed	12584	
Number of contigs	257	255
Total SNPs detected	2197	1158
SNP frequency	1.39/100 bp	0.74/100 bp
Total number of transitions	749	453
C/T	556	344
A/G	193	109
Total number of transversions	974	558
A/T	161	87
A/C	352	214
T/G	251	130
C/G	210	127
Total number of indels	366	125
Tri-allelic polymorphisms	99	21
Tetra-allelic polymorphisms	9	1

Real SNPs are those which passed quality filters 1 and 2 using the pipeline QualitySNP and true SNPs are the highest quality SNPs passing the three filters (see Methods).

Table 4: Real and true quality SNP distribution in contigs with 4 or more ESTs

	Number of contigs	Real SNPs	True SNPs
with 4 sequences	22	58	58
with 5–10 sequences	94	295	235
with 11–20 sequences	75	437	266
with 21–30 sequences	28	322	215
with 31–50 sequences	21	289	110
with > 50 sequences	17	796	274
Total	257	2197	1158

Real SNPs are those which passed quality filters 1 and 2 using the pipeline QualitySNP and true SNPs are the highest quality SNPs passing the three filters (see Methods).

CONCLUSION

To our knowledge, this is the first report on a large transcriptional analysis in turbot providing new genomic resources in this important European aquaculture species. This study describes a collection of 9256 ESTs representing 3482 unique sequences obtained from three directionally cloned cDNA libraries from *S. maximus*, all of these being novel ESTs for this species. Therefore, this is a valuable EST collection, which increases genomic resources of turbot and enhances the genomic tools available for non-model fish species. The transcript profile comparison among the three libraries allowed the identification of putative genes generally or specifically related with infections in turbot. In addition, a high number of putative microsatellite and SNP EST-markers are now available for turbot map and highly useful for comparative mapping. These ESTs will be the basis for the development of a turbot microarray, focused on the characterization of the transcriptional response to pathogen exposure.

METHODS

Tissue Source and Challenge

Two batches of 40 individuals (20–30 g each) obtained from a mixture of heterogeneous genetic families were collected at a specialized turbot fish farm. Fish from each batch were challenged intraperitoneally with*A. salmonicida* subsp. *salmonicida* and *P. dicentrarchi*, respectively [53, 54]. The dose was adjusted to obtain around 50% survival (LD50). The challenges were performed at the CETGA facilities in quarantine tanks. Fish were sacrificed prior to organ extraction using a lethal dose of MS222 anesthetic. In order to obtain mRNA representative of both innate and adaptive immune systems across the infection process, liver, spleen and head kidney tissues were collected from 5 sacrificed fish at five sampling points along the infective process for *A. salmonicida* (12 h, 1 day, 3 days, 7 days and 21 days post-inoculation) and at four sampling points for *P. dicentrarchi* (1 day, 3 days, 7 days and 15 days post-inoculation). Analogous batches of control fish were injected with saline serum and sampled at the same days of challenged fish. For each sample time, equal amounts of tissue from liver, kidney and spleen of each fish were pooled and immediately frozen in liquid nitrogen, constituting 15 pools for *A. salmonicida* (3 tissues × 5 sampling points), 12 for *P. dicentrarchi* (3 tissues × 4 sampling points) and the same pools for their respective controls. Each sample of pooled tissues was ground to a fine powder in a mortar and pestled with liquid nitrogen and stored at -80°C until being used for RNA extraction. The use of pools of individuals at each sampling point was planned, both for identifying putative SNPs, as well as for averaging individual effects in future microarray analysis on turbot immune response.

RNA Isolation and cDNA Library Construction

Total RNA was extracted from pooled tissues of control and infected fish using TRIZOL Reagent (Life Technologies) according to manufacturer's recommendations. RNA quality was assessed in a Bioanalyzer (Bonsai Technologies). RNA was quantified using gel and NanoDrop® ND-1000 spectrophotometer (NanoDrop® Technologies Inc) estimations. Poly-A mRNA was isolated using the Dynabeads® mRNA Purification Kit (INVITROGEN). Three cDNA libraries (from *A. salmonicida* and *P. dicentrarchi* infected fish and control) were directionally constructed (5' *Eco*RI, 3' *Xho*I), with equal amounts of RNA from each tissue at each sampling time, using the ZAP-cDNA Library Construction Kit (STRATAGENE) following manufacturer's instructions except for size fractioning. This was performed on cDNAs prior to ligation into vector and carried out with the SizeSep 400 Spun Columns (GE HEALTHCARE). To allow characterization of the inserts in a plasmid system the library was mass excised according to the manufacturer's recommendations, converting it from plaque forming units to a phagemid in *E. coli* SOLR strain. Manually isolated colonies were randomly picked and arrayed into 96-well microtiter plates containing liquid selective medium. Glycerol stocks of overnight cultures were prepared in 96 well plates and stored at -80°C.

Sequencing and Bioinformatics

Plasmid DNA was isolated from around 4000 clones from each library using the DirectPrep® 96 Miniprep kit (QIAGEN) and the Plasmid Miniprep96 kit (MILLIPORE) with a robotic platform (BIOMEK 3000), and checked by electrophoresis in 1% agarose gels. One part of the purified DNA was sequenced following the ABI Prism BigDye™ Terminator v3.1 Cycle Sequencing Kit protocol on an ABI 3100 DNA sequencer

(Applied Biosystems) and the other with the DTCS kit protocol on a CEQ2000 DNA sequencer (BECKMAN COULTER). All clones were sequenced from their 3› ends using a standard T7 primer to obtain the highest specific sequences of genes for oligo-microarray design. Those clones which suffered a systematic drop on sequencing signal after poly-A tails were sequenced from 5› end.

A bioinformatic tool was developed in order to process all data. Basecalling from chromatogram traces was performed by using PHRED [55, 56]. Vector, poly-A tails and low-quality regions were trimmed from EST sequences using a custom Perl script, a local BLAST search engine and the trimmest utility from the EMBOSS suite. High quality ESTs (at least 100 bp and PHRED score ≥ 20 after removal of vector sequence, adapter, and poly-A tail) of both normal and infected libraries were combined and assembled to form clusters using CAP3 http://seq.cs.iastate.edu./. [57] with the overlapping identity percentage and minimum overlapping length parameters set to > 85.0% and 50 bp, respectively, in order to obtain highly reliable contig sequences. These contigs were manually revised to detect possible errors all along the bioinformatics process. ESTs that did not form contigs (singletons) and the contigs resultant of assemblage of multiple sequences were referred to as unique sequences. Singletons and consensus sequences of each contig (unique sequences) were compared against public databases. Unique sequences were searched by both Blastn and Blastx, and the corresponding outputs subsequently parsed using BioPerl for fish-relevant hits and significant UniGene information. GO, KEGG and COG terms were extracted from AutoFACT output, which in turn feeds from Tblastx and RPS-blast output. All results were stored in a my SQL database to be consulted and searched using a custom-designed, friendly AJAX web interface. The e-value cut off was ≤ 1e-5. All annotations thus obtained were complemented with AutoFACT output, whenever available.

Bioinformatic Mining of Microsatellites and SNPs

The set of 3482 unique sequences was searched for microsatellites using the program SPUTNIKhttp://espressosoftware.com/pages/sputnik.jsp. The minimum repeat number used for this search was 8 for dinucleotide and 6 for tri-, tetra-and pentanucleotide microsatellites. Microsatellite-containing ESTs were identified as candidates for marker development if they presented enough flanking sequences on either side of the repeats for primer design. Detection of single nucleotide polymorphisms (SNPs) was done using the pipeline Quality SNP with default settings [52]. This program uses three filters for the identification of reliable SNPs: filter 1 screens for all potential SNPs; filter 2 uses a haplotype-based strategy to detect reliable SNPs; and filter 3 screens SNPs by calculating a confidence score based on sequence redundancy and quality. SNPs that pass filters 1 and 2 are called real SNPs, and those that pass all of the three filters are called true SNPs.

DECLARATIONS

Acknowledgements

This study was supported by a Consellería de Pesca e Asuntos Marítimos and the Dirección Xeral de I+D – Xunta de Galicia project (2004/CP480). Authors wish to thank to Lucía Insua, Sonia Gómez, Ana Teijido and Susana Sánchez for technical assistance.

REFERENCES

1. Toranzo AE, Novoa B, Romalde JL, Núñez S, Devesa S, Mariño E, Silva R, Martínez E, Figueras A, Barja JL: Microflora associated with healthy and diseased turbot (*Scophthalmus maximus*) from three farms in northwest Spain. Aquaculture. 1993, 114: 189-202. 10.1016/0044-8486(93)90295-A.
2. Iglesias R, Paramá A, Álvarez MF, Leiro J, Ubeira FM, Sanmartín ML: *Philasterides dicentrarchi* (Ciliophora: Scuticociliatida) expresses surface immobilization antigens that probably induce protective immune responses in turbot. Parasitology. 2003, 126: 125-134. 10.1017/S0031182002002688.
3. Adams MD, Kelley JM, Gocayne JD, Dubnick M, Polymeropoulos MH, Xiao H, Merril CR, Wu A, Olde B, Moreno RF: Complementary DNA sequencing: expressed sequence tags and human genome project. Science. 1991, 252: 1651-1656. 10.1126/science.2047873.
4. Marra MA, Hillier L, Waterston RH: Expressed sequence tags-ESTablishing bridges between genomes. Trends Genet. 1998, 14: 4-7. 10.1016/S0168-9525(97)01355-3.
5. Liu ZJ, Karsi A, Dunham RA: Development of polymorphic EST markers suitable for genetic linkage mapping of catfish. Mar Biotechnol. 1999, 1: 437-447. 10.1007/PL00011800.
6. Liu ZJ, Li P, Kocabas A, Ju Z, Karsi A, Cao D, Patterson A: Microsatellite-containing genes from the channel catfish brain: evidence of trinucleotide repeat expansion in the coding region of nucleotide excision repair gene RAD23B. Biochem Biophys Res Commun. 2001, 289: 317-324. 10.1006/bbrc.2001.5978.
7. He C, Chen L, Simmons M, Li P, Kim S, Liu ZJ: Putative SNP discovery in interspecific hybrids of catfish by comparative EST analysis. Anim Genet. 2003, 34: 445-448. 10.1046/j.0268-9146.2003.01054.x.
8. Serapion J, Kucuktas H, Feng J, Liu Z: Bioinformatic mining of type I microsatellites from expressed sequence tags of channel catfish (*Ictalurus punctatus*). Mar Biotechnol. 2004, 6: 364-377. 10.1007/s10126-003-0039-z.
9. Murakawa K, Matsubara K, Fukushima A, Yoshii J, Okubo K: Chromosomal assignments of 3Vdirected partial cDNA sequences representing novel genes expressed in granulocytoid cells. Genomics. 1994, 23: 379-389. 10.1006/geno.1994.1514.
10. Bouza C, Hermida M, Pardo BG, Fernández C, Castro J, Fortes G, Sánchez L, Presa P, Pérez M, Sanjuán A, Comesaña S, Álvarez JA, Calaza M, Cal R, Piferrer F, Martínez P: A microsatellite genetic map of the turbot (*Scophthalmus maximus*). Genetics. 2007, 177: 2457-2467. 10.1534/genetics.107.075416.
11. Moen T, Hayes B, Baranski M, Berg PR, Kjøglum S, Koop BF, Davidson WS, Omholt SW, Lien S: A linkage map of the Atlantic salmon (*Salmo salar*) based on EST-

derived SNP markers. BMC Genomics. 2008, 9: 223-10.1186/1471-2164-9-223.

12. Poompuang S, Hallerman EM: Toward detection of quantitative trait loci and marker-assisted selection in fish. Reviews Fish Sci. 1997, 5: 253-277.
13. Liu ZJ, Cordes JF: DNA marker technologies and their applications in aquaculture genetics. Aquaculture. 2004, 238: 1-37. 10.1016/j.aquaculture.2004.05.027.
14. Jørgensen SM, Afanasyev S, Krasnov A: Gene expression analyses in Atlantic salmon challenged with infectious salmon anemia virus reveal differences between individuals with early, intermediate and late mortality. BMC Genomics. 2008, 9: 179-10.1186/1471-2164-9-179.
15. Hirono I, Aoki T: Expressed sequence tags of medaka (*Oryzias latipes*) liver mRNA. Mol Mar Biol Biotechnol. 1997, 6: 345-350.
16. Lo J, Lee S, Xu M, Liu F, Ruan H, Eun A, He Y, Ma W, Wang W, Wen Z, Peng J: 15,000 Unique zebrafish EST clusters and their future use in microarray for profiling gene expression patterns during embryogenesis. Genome Res. 2003, 13: 455-466. 10.1101/gr.885403.
17. Clark MS, Edwards YJ, Peterson D, Clifton SW, Thompson AJ, Sasaki M, Suzuki Y, Kikuchi K, Watabe S, Kawakami K, Sugano S, Elgar G, Johnson SL: Fugu ESTs: new resources for transcription analysis and genome annotation. Genome Res. 2003, 13 (12): 2747-2753. 10.1101/gr.1691503.
18. Rise ML, von Schalburg KR, Brown GD, Mawer MA, Devlin RH, Kuipers N, Busby M, Beetz-Sargent M, Alberto R, Gibbs AR, Hunt P, Shukin R, Zeznik JA, Nelson C, Jones SR, Smailus DE, Jones SJ, Schein JE, Marra MA, Butterfield YS, Stott JM, Ng SH, Davidson WS, Koop BF: Development and application of a salmonid EST database and cDNA microarray: data mining and interspecific hybridization characteristics. Genome Res. 2004, 14: 478-490. 10.1101/gr.1687304.
19. Jaillon O, Aury JM, Brunet F, Petit JL, Stange-Thomann N, Mauceli E, Bouneau L, Fischer C, Ozouf-Costaz C, Bernot A, Nicaud S, Jaffe D, Fisher S, Lutfalla G, Dossat C, Segurens B, Dasilva C, Salanoub M, Levy M, Boudet N, Castellano S, Anthouard V, Jubin C, Castelli V, Katinka M, Vacherie B, Biemont C, Skalli Z, Cattolico L, Poulain J, De Berardinis V, Cruaud C, Duprat S, Brottier P, Coutanceau JP, Gouzy J, Parra G, Lardier G, Chapple C, McKernan KJ, McEwan P, Bosak S, Kellis M, Volff JN, Guigo R, Zody MC, Mesirov J, Lindblad-Toh K, Birren B, Nusbaum C, Kahn D, Robinson-Rechav M, Laudet V, Schachter V, Quetier F, Saurin W, Scarpelli C, Wincker P, Lander ES, Weissenbach J, Roest Crollius H: Genome duplication in the teleost fish *Tetraodon nigroviridis* reveals the early vertebrate proto-karyotype. Nature. 2004, 431: 946-957. 10.1038/nature03025.
20. Chen SL, Xu MY, Hu SN, Li L: Analysis of immune relevant genes expressed in red sea bream (*Chrysophrys major*) spleen. Aquaculture. 2004, 240: 115-130. 10.1016/j.aquaculture.2004.07.008.
21. Inoue S, Nam BH, Hirono I, Aoki T: A survey of expressed genes in Japanese flounder (*Paralichthys olivaceus*) liver and spleen. Mol Mar Biol Biotechnol. 1997, 6: 376-380.
22. Nam BH, Yamamoto E, Hirono I, Aoki T: A survey of expressed genes in the leukocytes of Japanese flounder,*Paralichthys olivaceus*, infected with Hirame rhabdovirus. Dev Comp Immunol. 2000, 24: 13-24. 10.1016/S0145-305X(99)00058-0.
23. Aoki T, Hirono I, Kim M-G, Katagiri T, Tokuda Y, Toyohara H, Yamamota E: Identification of viral induced genes in Ig+ leucocytes of Japanese flounder *Paralichthys olivaceus*, by differential hybridisation with subtracted and un-subtracted cDNA probes. Fish Shellfish Immunol. 2000, 10: 623-630. 10.1006/fsim.2000.0279.
24. Arma NR, Hirono I, Aoki T: Genes expressed in Japanese flounder *Paralichthys olivaceus* spleen: analysis of genes involved in immune function. Fisheries Sci. 2005, 71: 1304-1323. 10.1111/j.1444-2906.2005.01097.x.
25. Aoki T, Hirono I: Immune relevant genes of Japanese flounder, *Paralichthys olivaceus*. Comp Biochem Phys D. 2006, 1: 115-121.
26. Park KC, Osborne JA, Tsoi SCM, Brown LL, Johnson SC: Expressed sequence tags analysis of Atlantic halibut (*Hippoglossus hippoglossus*) liver, kidney and spleen tissues following vaccination against *Vibrio anguillarun* and*Aeromonas salmonicida*. Fish Shellfish Immunol. 2005, 18: 393-415. 10.1016/j.fsi.2004.10.003.
27. Galloway TF, Bardal T, Kvam SN, Dahle SW, Nesse G, Randol M, Kjorsvik E, Andersen O: Somite formation and expression of MyoD, myogenin and myosin in Atlantic halibut (*Hippoglossus hippoglossus* L.) embryos incubated at different temperatures: transient asymmetric expression of MyoD. J Exp Biol. 2006, 209: 2432-2441. 10.1242/jeb.02269.
28. Douglas SE, Knickle LC, Kimball J, Reith ME: Comprehensive EST analysis of Atlantic halibut (*Hippoglossus hippoglossus*), a commercially relevant aquaculture species. BMC Genomics. 2007, 8: 144-10.1186/1471-2164-8-144.
29. Pardo BG, Fernández C, Hermida M, Vázquez A, Pérez M, Presa P, Calaza M, Alvarez-Dios JA, Comesaña AS, Raposo-Guillán J, Bouza C, Martínez P: Development and characterization of 248 novel microsatellite markers in turbot (*Scophthalmus maximus*). Genome. 2007, 50: 329-332. 10.1139/G06-154.
30. Toranzo AE, Barja JL: First report of furunculosis in turbot reared in floating cages in North-West of Spain. Bull Eur Ass Fish Pathol. 1992, 12: 147-149.
31. Björnsdóttir B, Gudmundsdóttir S, Bambir SH, Gudmundsdóttir BK: Experimental infection of turbot, *Scophthalmus maximus* (L.), by *Aeromonas salmonicida* subsp. *achromogenes* and evaluation of cross protection induced by a furunculosis vaccine. J Fish Dis. 2005, 28: 181-188. 10.1111/j.1365-2761.2005.00617.x.
32. Iglesias R, Paramá A, Alvarez MF, Leiro J, Fernandez J, Sanmartin ML: *Philasterides dicentrarchi* (Ciliophora, Scuticociliatida) as the causative agent of scuticociliatosis in farmed turbot *Scophthalmus maximus* in Galicia (NW Spain). Dis Aquat Organ. 2001, 46: 47-55. 10.3354/

dao046047.

33. Paramá A, Arranz JA, Alvarez MF, Sanmartin ML, Leiro J: Ultrastructure and phylogeny of *Philasterides dicentrarchi*(Ciliophora, Scuticociliatia) from farmed turbot in NW Spain. Parasitology. 2006, 132: 555-564. 10.1017/S0031182005009534.

34. Douglas SE, Gallant JW, Bullerwell CE, Wolff C, Munholland J, Reith ME: Winter flounder expressed sequence tags: establishment of an EST database and identification of novel fish genes. Mar Biotechnol. 1999, 1: 458-464. 10.1007/PL00011802.

35. Williams TD, Gensberg K, Minchin SD, Chipman JK: A DNA expression array to detect toxic stress response in European flounder (*Platichthys flesus*). Aquat Toxicol. 2003, 65: 141-157. 10.1016/S0166-445X(03)00119-X.

36. Sheader DL, Gensberg K, Lyons BP, Chipman K: Isolation of differentially expressed genes from contaminant exposed European flounder by suppressive, subtractive hybridisation. Mar Environ Res. 2004, 58: 553-557. 10.1016/j.marenvres.2004.03.044.

37. Pearson WR: Identifying distantly related protein sequences. Comput Appl Biosci. 1997, 13: 325-332.

38. Paschall JE, Oleksiak MF, VanWye JD, Roach JL, Whitehead JA, Wyckoff GJ, Kolell KJ, Crawford DL: FunnyBase: a systems level functional annotation of Fundulus ESTs for the analysis of gene expression. BMC Genomics. 2004, 5: 96-10.1186/1471-2164-5-96.

39. Quilang J, Wang S, Li P, Abernathy J, Peatman E, Wang Y, Wang L, Shi Y, Wallace R, Guo X, Liu Z: Generation and analysis of ESTs from the eastern oyster, *Crassostrea virginica* Gmelin and identification of microsatellite and SNP markers. BMC Genomics. 2007, 8: 157-10.1186/1471-2164-8-157.

40. Altschul SF, Gish W, Miller W, Myers EW, Lipman DJ: Basic local alignment search tool. J Mol Biol. 1990, 215: 403-410.

41. Koski LB, Gray MW, Lang BF, Burger G: AutoFACT: An Automatic Functional Annotation and Classification Tool. BMC Bioinformatics. 6: 151-10.1186/1471-2105-6-151.

42. Gene Ontology: tool for the unification of biology. The Gene Ontology Consortium. Nature Genet. 2000, 25: 25-29. 10.1038/75556.

43. Ellis AE: Innate host defence mechanisms of fish against viruses and bacteria. Dev Comp Immunol. 2001, 25: 827-839. 10.1016/S0145-305X(01)00038-6.

44. Chen SL, Li W, Meng L, Sha ZX, Wang ZJ, Ren GC: Molecular cloning and expression analysis of a hepcidin antimicrobial peptide gene from turbot (*Scophthalmus maximus*). Fish Shellfish Immunol. 2007, 22: 172-181. 10.1016/j.fsi.2006.04.004.

45. Chen Y, Zhang YX, Fan TJ, Meng L, Ren GC, Chen SL: Molecular identification and expression analysis of the natural killer cell enhancing factor (NKEF) gene from turbot (*Scophthalmus maximus*). Aquaculture. 2006, 261: 1186-1193. 10.1016/j.aquaculture.2006.09.034.

46. Cuesta A, Meseguer J, Esteban MA: The antimicrobial peptide hepcidin exerts an important role in the innate immunity against bacteria in the bony fish gilthead seabream. Mol Immunol. 2008, 45: 2333-2342.

47. Valko M, Rhodes CJ, Moncol J, Izakovic M, Mazur M: Free radicals, metals and antioxidants in oxidative stress-induced cancer. Chem Biol Interact. 2006, 160: 1-40. 10.1016/j.cbi.2005.12.009.

48. Paramá A, Castro R, Arranz JA, Sanmartín ML, Lamas J, Leiro J: Scuticociliate cysteine proteinases modulate turbot leucocyte functions. Fish Shellfish Immunol. 2007, 23: 945-956. 10.1016/j.fsi.2007.02.001.

49. Ishii KJ, Uematsu S, Akira S: 'Toll' gates for future immunotherapy. Curr Pharm Des. 2006, 12: 4135-4142. 10.2174/138161206778743484.

50. Picoult-Newberg L, Ideker TE, Pohl MG, Taylor SL, Donaldson MA, Nickerson DA, Boyce-Jacino M: Mining SNPs from EST databases. Genome Res. 1999, 9: 167-174.

51. Syvanen AC: Accessing genetic variation: genotyping single nucleotide polymorphisms. Nat Rev Genet. 2001, 2: 930-942. 10.1038/35103535.

52. Tang J, Vosman B, Voorrips RE, Linden van der CG, Leunissen JAM: QualitySNP: a pipeline for detecting single nucleotide polymorphisms and insertions/deletions in EST data from diploid and polyploid species. BMC Bioinformatics. 2006, 7: 438-10.1186/1471-2105-7-438.

53. Magariños B, Santos Y, Romalde JL, Barja JL, Toranzo AE: Pathogenic activities of live cells and extracellular products of the fish pathogen *Pasteurella piscicida*. J Gen Microbiol. 1992, 138: 2491-2498.

54. Paramá A, Iglesias R, Álvarez MF, Leiro J, Aja C, Sanmartín ML: *Philasterides dicentrarchi* (Ciliophora, Scuticociliatida): experimental infection and possible routes of entry in farmed turbot (*Scophthalmus maximus*). Aquaculture. 2003, 217: 73-80. 10.1016/S0044-8486(02)00523-9.

55. Ewing B, Hillier L, Wendl MC, Green P: Base-calling of automated sequencer traces using phred. I. Accuracy assessment. Genome Res. 1998, 8: 175-185.

56. Ewing B, Green P: Base-calling of automated sequencer traces using phred. II. Error probabilities. Genome Res. 1998, 8: 186-194.

57. Huang X, Madan A: CAP3: a DNA sequence assembly program. Genome Res. 1999, 9: 868-877. 10.1101/gr.9.9.868.

CHAPTER 16

DEEP SEQUENCING-BASED TRANSCRIPTOME PROFILING ANALYSIS OF BACTERIA-CHALLENGED LATEOLABRAX JAPONICUS REVEALS INSIGHT INTO THE IMMUNE-RELEVANT GENES IN MARINE FISH

Li-xin Xiang, Ding He, Wei-ren Dong, Yi-wen Zhang, Jian-zhong Shao

BACKGROUND

Since it is a representative population of lower vertebrates serving as an essential link to early vertebrate evolution, fish is believed to be an important model in various developmental and comparative evolutionary studies [1, 2, 3]. Fish immunogenetics has received considerable attention due to its essential role in understanding the origin and evolution of immune systems [4, 5, 6]. Further, it is also beneficial in the creation of immune-based therapy of severe fish diseases. Great progress in bioinformatics and genome projects in model organisms, including human (*Homo sapiens*), mouse (*Mus musculus*), frog (*Xenopus laevis*), chicken (*Gallus gallus*), and zebrafish (*Danio rerio*), has led to the emergence of studies focusing on the identification and characterization of immune-related genes in teleost fish based on comparative genomics. These have provided preliminary observations on fish immunogenetics and evolutionary history of immune systems from lower vertebrates to mammals [7, 8]. However, large-scale identification of immune-related genes at the genome or transcriptome levels in fish was seen in limited species (e.g. *Danio rerio*) due to the inadequate number of high-throughput deep sequencing technologies available [9, 10]. This is an even more difficult problem in non-model fish species with totally unknown genome sequences.

Recently developed RNA deep sequencing technologies, such as Solexa/Illumina RNA-seq and Digital gene expression (DGE), have dramatically changed the way immune-related genes in fish are identified because these technologies facilitate the investigation of the functional complexity of transcriptomes [11, 12]. RNA-Seq refers to whole transcriptome shotgun sequencing wherein mRNA or cDNA is mechanically fragmented, resulting in overlapping short fragments that cover the entire transcriptome. DGE is a tag-based transcriptome sequencing approach where short raw tags are generated by endonuclease. The expression level of virtually all genes in the sample is measured by counting the number of individual mRNA molecules produced from each gene. Compared with DGE analysis, the RNA-Seq approach is more powerful for unraveling transcriptome complexity, and for identification of genes, structure of transcripts, alternative splicing, non-coding RNAs, and new transcription units. In contrast, the DGE protocol is more suitable and affordable for comparative gene expression studies because it enables direct transcript profiling without compromise and potential bias, thus allowing for a more sensitive and accurate profiling of the transcriptome that more closely resembles the biology of the cell [9, 13, 14, 15, 16, 17]. These two technologies have been used in transcriptome profiling studies for various applications, including cellular development, cancer, and immune defence

Citation: Li-xin Xiang, Ding He, Wei-ren Dong, Yi-wen Zhang, Jian-zhong Shao, Deep sequencing-based transcriptome profiling analysis of bacteria-challenged Lateolabrax japonicus reveals insight into the immunerelevant genes in marine fish, doi: 10.1186/1471-2164-11-472

of various organisms [10, 18, 19, 20, 21, 22, 23, 24, 25, 26, 27, 28, 29]. However, they have not been used in immunogenetic analysis of marine fish species.

Japanese sea bass (*Lateolabrax japonicus*) is an economically important marine species widely cultured in fisheries worldwide. Various diseases caused by bacterial and viral pathogens plague this species [30]. High mortality is associated with infection with *Vibrio harveyi*, a typical gram-negative pathogen of a wide range of marine animals. Infection results in a variety of vibriosis, a common aquatic animal disease associated with high mortality throughout the world [31]. In *L. japonicus*, *V. harveyi* infection leads to bacterial septicaemia with muscle ulcer as well as subcutaneous and gastroenteritic haemorrhage.

The present study is the first to conduct a transcriptome profiling analysis of *V. harveyi*-challenged *L. japonicus*using RNA-seq and DGE to gain deep insight into the immunogenetics of marine fish. Bacteria-challenged *L. japonicus* is expected to be a model system for studying bacterial immunity in marine fish. Further, a global survey of anti-bacterial immune defence gene activities in marine fish can contribute to the in-depth investigation of candidate genes in fish immunity. Results are also expected to improve current understanding of host-pathogen interactions and evolutionary history of immunogenetics from fish to mammals.

RESULTS

Aligning Raw Sequencing Reads To Non-Redundant Consensus

Approximately 34.59 and 33.03 million 75-bp pair end (PE) raw reads (submitted to GEO database, Association No. GSE21721) from the head kidney and spleen tissues of bacteria- and mock-challenged fish, respectively, were generated using Solexa/Illumina RNA-seq deep sequencing analysis. Repetitive, low-complexity, and low-quality reads were filtered out prior to assembly of sequence reads for non-redundant consensus. Using Grape software, reliable reads were assembled into contigs, which were then compared with all PE reads. Overlap of PE reads with two contigs was taken to indicate that the contigs are short segments of a scaffold. Reads were used for gap-filling of these scaffolds to generate final scaffold sequences. Using tgicl and cap3 software programs, scaffold sequences were assembled into clusters that were then analysed for consensus. A total of 150,125 and 140,330 non-redundant consensus sequences, ranging from 100 to 2,000-bp, were generated from each group. Then, consensus sequences were merged for DGE analysis. Removal of partial overlapping sequences yielded 169,950 non-redundant consensus sequences (Table 1). These sequences provide abundant data on healthy and infected conditions, thus allowing for better reference of immune-relevant genes.

Table 1: Distribution of non-redundant consensus sequences

Consensus Length (bp)	Total Number	Percentage
100-500	143865	84.65%
500-1000	14313	8.42%
1000-1500	5475	3.22%
1500-2000	2849	1.68%
≥2000	3448	2.03%
Total	169950	100.00%
Total bp	57817772	
N50	598	
Mean	340	

N50 = median length of all non-redundant consensus sequences

Mean = average length of all consensus sequences

Annotation of all Non-Redundant Consensus Sequences

BLASTX and ESTscan software analysis of 169,950 non-redundant consensus sequences revealed that about 48,987 have reliable coding sequences (CDs) [32, 33]. CD-containing consensus sequences have high potential for translation into functional proteins and most of them translated to proteins with more than 100 aa. Comparison with the Nr and Swissprot databases revealed that 44,842 consensus sequences had good comparability with known gene sequences in existing species. Annotation of the 44,842 sequences using GO and COG databases yielded good results for approximately 16,469 consensus sequences and 9,545 putative proteins [34, 35]. GO-annotated consensus sequences belonged to the biological process, cellular component, and molecular function clusters and distributed among more than 50 categories (Figure 1), including biochemistry, metabolism, growth, development, apoptosis, and immune defence. Similarly, COG-annotated putative proteins were classified functionally into at least 25 molecular families, including cellular structure, biochemistry metabolism, molecular processing, signal transduction, gene expression, and immune defence, that correspond to the categories observed in GO analysis (Figure 2). The KEGG database was used to analyse potential involvement of the consensus sequences in cellular metabolic pathways (Figure 3) [36]. Among the 44,842 consensus sequences, 24,496 can be grouped into seven categories comprised of 219 known metabolic or signalling pathways, including cellular growth, differentiation, apoptosis, migration, endocrine, and various immune-relevant signalling or metabolic pathways (Table 2).

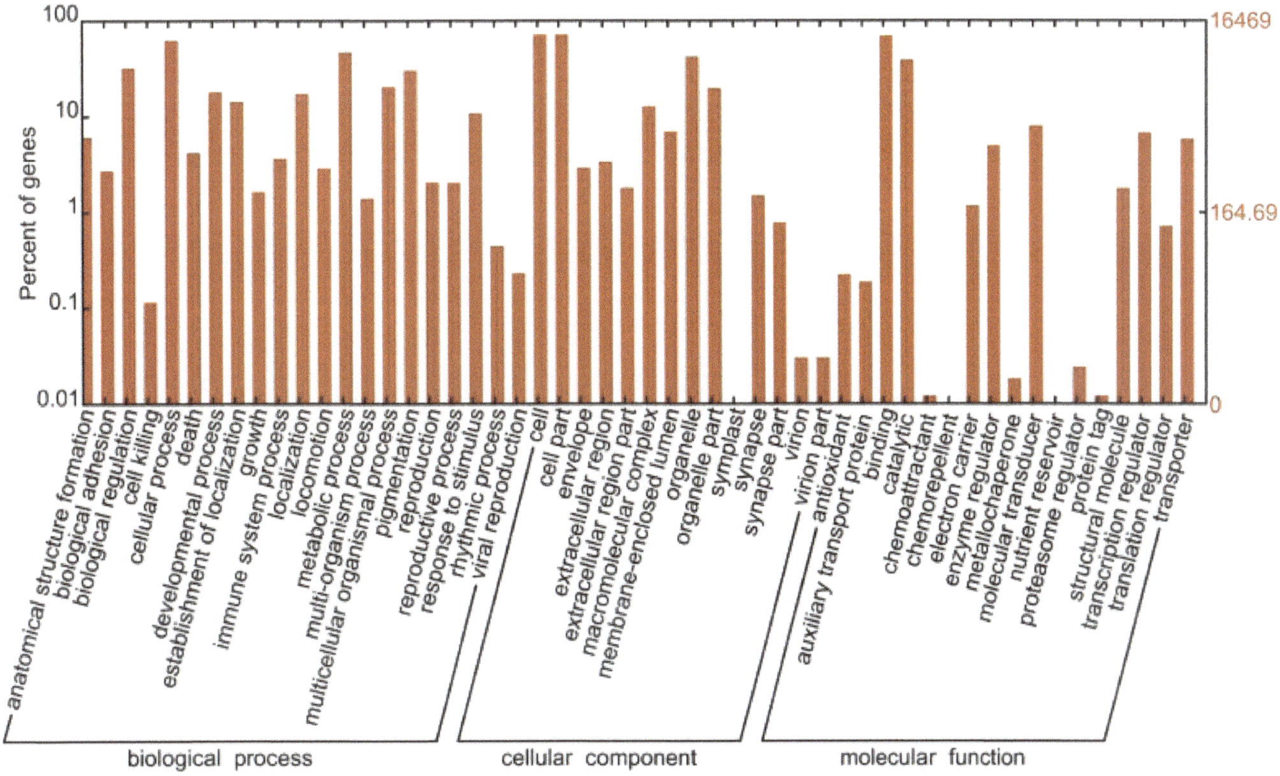

Figure 1: GO annotations of non-redundant consensus sequences. Best hits were aligned to the GO database, and 16,469 transcripts were assigned to at least one GO term. Most consensus sequences were grouped into three major functional categories, namely biological process, cellular component, and molecular function.

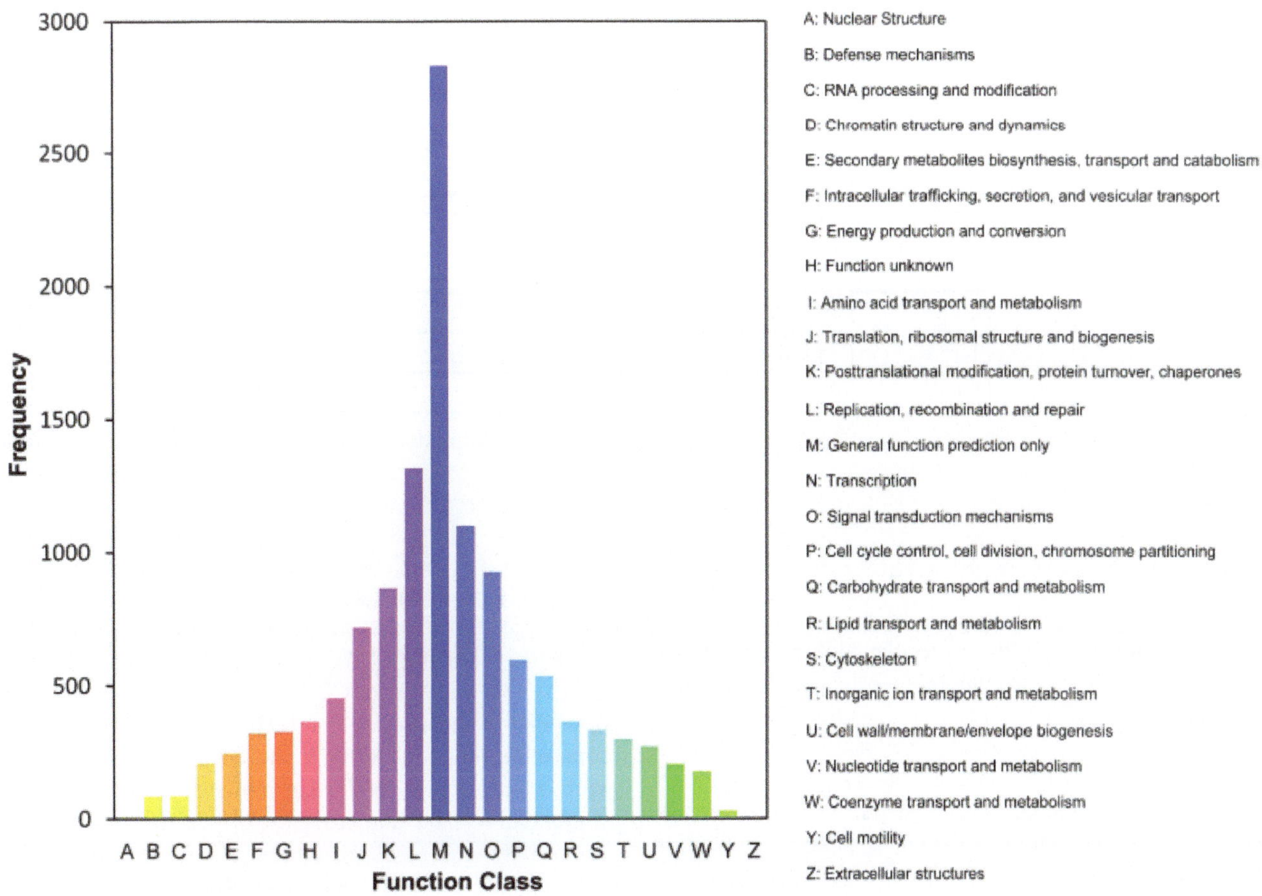

Figure 2: COG annotations of putative proteins. All putative proteins were aligned to the COG database and can be classified functionally into at least 25 molecular families.

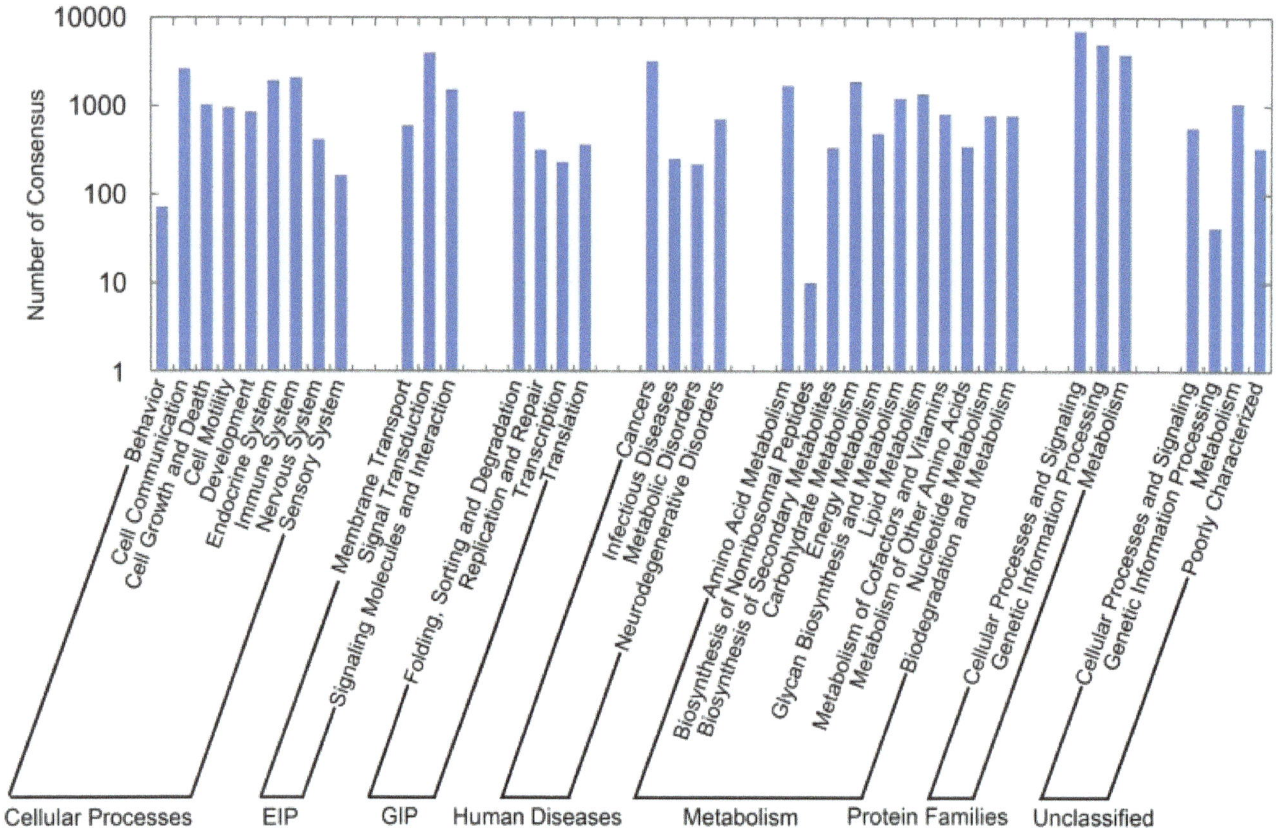

Figure 3: KEGG categories of non-redundant consensus sequences. All non-redundant consensus sequences were annotated using KEGG Automatic Annotation Server for pathway information, and about 24,496 consensus sequences were annotated. The categories GIP and EIP stand for genetic information processing and environmental information processing, respectively.

Table 2: Annotation of non-redundant consensus sequences

Database	Number of annotated consensus sequences	Percentage of annotated consensus sequences
Swissprot	37544	76.64%
Nr	36935	75.40%
GO	16469	33.62%
KEGG	24496	50.01%
COG	9545	19.48%

All 48,987 CD-containing consensus sequences generated by ESTscan were annotated though Swissprot, Nr, GO, KEGG, and COG databases.

Annotation of Immune-Relevant Genes and Pathways

To gain deep insight into the molecular biology of immune systems in *L. japonicus*, the immune-relevant genes, metabolic and signalling pathways were analysed. Approximately 2,673 consensus sequences were found to be homologous to known immune-relevant genes in other species (Table 3, Additional file 1, Table S1), including the most important elements of innate and adaptive immunity, such as pattern recognition receptors (mannose binding proteins, scavenger receptors, Toll-like receptors, and lipopolysaccharide-binding proteins), inflammatory cytokines and receptors (IL-1β, IL-10, IL-12, IL-6, IL-8, TNF-α, IFN-γ, TGF-β, VEGF, IL-1, and TNF receptors), immunoglobins (IgM and IgD), transcriptional factors (NF-kB, Ik-B, AP-1, transcription factor 4, chorion specific transcription factor, and transcriptional enhancer factor 5), complement components (C1q, C1r, C1 s, C2, C4, MASP, Bf, Df, and CR1-5), leukocyte differentiation antigens (CD3, CD4, CD8, and CD80/86), antigen presenting and processing molecules (MHC I, MHC II, TAP1, TAP2, TAP2B, proteosome subunit, PA28α/β, and constant chain protein li), regulatory molecules involved in immune cellular proliferation, differentiation, and apoptosis (growth differentiation factor 11, proliferation associated protein 2G4, apoptotic peptidase activating factor 1, and anti-apoptotic protein NR13), and other molecules involved in immune response (hepcidin, lectin, lysozyme, antimicrobial peptide NK-lysin, chemokines, ferritin, RAG1, and RAG2).

Table 3: Immune-relevant genes/homologues in *L. japonicus*

Putative Gene Catalogs	Consensus Number	Homologs Number in other species	Putative Gene Catalogs	Consensus Number	Homologs Number in other species
Pattern Recognition Genes			***T-cell and B-cell Antigen Activation***		
TLR (1-9,13,14,18,21,23)	44	17	TCR (α,β chain)	6	4
scavenger receptor	19	11	BCL (2,3,6,7,9,10,11) and relevant	44	19
Mannose receptor	49	26	Immunoglobulin and relevant	52	32
MBL	13	9	CD (2-6,8,9,22,40,79,81,97,99,154,166,226,276)	47	29
NOD (1,2)	7	5	BLNK	5	3
C-type lectin (4,6,7,9-12)	36	26	AICDA	3	2
TOLLIP	1	1	CBLB	7	4
LPS-binding/anchor protein	19	3	DOCK2	4	2
HMG (A,B,2L1,20A)	8	6	DRAM	1	1
LRR-containing proteins	168	74	Glomulin	2	2
NALP (1,3,4,6,9,12,13)	21	8	HDAC (1-8,10,11)	40	23
PGRP	1	1	GALNAC4S-6ST	5	3
BTK	2	2	KLF (1-4,6-9,11-13)	29	15
Fibrinogen	4	4	IGBP1	1	1
Complement System			HLA	1	1
C1 (q,r,s,t),C2-C7,C9,C10	40	32	JAG (1,2)	37	6
Complement factor (B,H,I,properdin, DAF)	22	11	KI13B	17	1
Complement receptor (1,2,5)	13	8	LAG	1	1
MASP	4	3	NCK	7	5
RGC	2	2	Homeobox protein (NKX)	4	3
Inflammatory Cytokines and Receptors			PAWR	3	3
IL (1,3,8,10-12,15,16,27,34) and relevant	34	17	Protein kinase C δ	6	3
ILR (1-3,6,7,10,12,15,17,18,20-22,31) and relevant	48	31	Prolactin receptor	5	2
IFN	3	2	RAG (1,2)	3	2
IFN-induced proteins and relevant	59	27	RGS (1-5,7-9,11,12,14,16,18-20)	47	23
IFN (α,β,ω) receptor	2	1	SFTPD/SP-D	1	1
IFN-related regulator	31	16	TYK2	1	1
CRLF (1,3)	3	2	VAV oncogene family (1-3)	13	6
EBI 2	5	2	WAS proteins	23	11
FND (3,C)	12	4	WWP	2	2
TTN/TITIN	5	2	***Antigen Processing and Regulators***		
CXCL(6,12,14),CXC,CXCR(1-5,7)	20	17	MHC I/II	14	5

CCL(2-4,7,14,16,19,C11b),CCR(2-9,11)	33	26	PSMA (1-7)/B/C/D/E(2-4)	93	63
CX3CR	1	1	Legumain	5	3
XCL(1),XCR(1,1a)	6	3	ITCH	5	2
Chemokine receptor-like	4	2	RFX5	4	1
CFLAR	1	1	TAPBP	3	2
EGF	125	50	ASPP (1,2)	24	4
FAM	15	8	ICAM2	1	1
FGF	23	14	RAC1	3	1
FLT3/CD135	2	2	AGT	4	2
GATA (1-3,5,6)	14	8	CDC (2,5-7,16,20,23,27,42,A)	23	15
GDF	3	3	CDK-inhibitor (1-3)	9	8
GFI-1/-1B	13	3	CTLA4	1	1
GPR44	4	2	***Other Genes Related to Immune Cell Response***		
Inhibin β	1	1	CD (33,34,44,45,48,83,151,302)	10	8
Integrin α/β	89	38	Apolipoprotein A	2	2
LIFR	5	3	HELLS	1	1
Lipoteichoic acid	1	1	IMPDH/IMDH	4	3
MAF	2	2	AP-1	20	15
MIF	3	2	c-FOS	2	2
PCGF (1-3, 5, 6, A)	8	6	EGR	11	5
SCYE1/MCA	3	2	ELK4	1	1
SDF2	2	2	IKKA/CHUK, IKKB	8	2
TGF A/B(1-3)	10	9	JAK (1,2)	11	6
TNF and TNR	54	38	JUN oncogene	11	6
Thrombopoietin	1	1	MALT	5	2
VEGF (A,C,D)	5	4	NFAT	32	12
VEGF receptor (1-3)	17	8	NF κ-B inhibitor proteins	11	4
Adapters, Effectors and Signal Transducers			PPM	33	14
ASB (1-3,5-8,12,13)	26	14	RIPK (1-3,5)	7	5
CRADD	2	1	SLC20A-1/2	6	4
DEDD	4	2	Transcription factor Sp and SPT	21	18
EIF 2-α kinase 2	7	2	TANK-binding kinase	9	5
FADD	1	1	TBX	12	7
Heme oxygenase	4	2	TMED (1-5,7-9,A)	12	10
HTR (5A,5B)	22	4	TRIM (1-3,7-9)	20	8
IRAK (1-4)	7	6	Antimicrobial peptide	2	2
CSF	12	5	Co-chaperone	4	3
Myd88	2	1	Ferritin	1	1
NR2C	6	4	Hepcidin	1	1

PEA15	2	1	HSP (7,13,70-72,74,B)	35	20
PELI (1-3)	21	6	Lysozyme	4	3
PHLA (1-3)	4	3	Macroglobulin	9	6
PPARA	2	2	Microtubule-associated	109	45
REL	1	1	Nitric oxide synthase	4	3
RHOA	3	1	Proteolipid protein	1	1
RIP	2	2	Prohibitin	3	2
SARM1	8	2	Selenoprotein	22	11
SIGIRR	2	2	ABCF	8	7
SMAD (1-7,9)	33	16	CAMP	13	4
SOCS (1-7)	18	8	CARD	5	3
TICAM1	1	1	CEBP	7	5
TIRAP	1	1	DMBT	6	4
TRADD	1	1	PTAFR	1	1
TRAF (2-7)	10	7	Protein S100-A	5	4
TANK	2	2	UCRP/CRP	1	1
UBE	47	25	CYBB	1	1
STAT	21	7	Caspase (1-3,6-9)	18	13
NF κ	15	4	Adenosine receptor A (2a,2b)	11	7
MAPK (1-3,5-9,11-15)	47	29	Apoptotic relevant	13	6
MAP2K4(JUKK1)	2	2	IRG1	4	2
MAP3K (2-9,11,14,15)	35	18	Stress response protein	22	6
MAP3K 7-interacting protein (TAB)	4	4	WSC (1,2,4)	8	3
JNK	5	4	OSGI2	1	1

Putative genes/homologues were confirmed based on their similarity to immune-relevant genes in other vertebrate species as predicted by BLAST. Cutoff E-value was set to 1e-10 during BLAST analysis. Characterization of the putative genes is presented in Additional file 1, Table S1.

KEGG analysis revealed that approximately 2,082 consensus sequences were significantly enriched in various known immune-relevant metabolic or signalling pathways (Figure 3). These suggest a considerable conservation of immune-relevant genes and pathways between *L. japonicus* and mammals. Conserved genes and pathway members might include Toll-like receptors (TLR1-9 and TLR13) and corresponding adaptors in mammals (MyD88, TRAM, FADD, CASP8, IRAK4, TRAF6, TAK1, MKK4/7, JNK, and AP-1) and in other fish species (TLR5a/b, TLR14 and TLR21-23) (Figure 4); T cell receptors (TCRα/β) and corresponding signalling transducers (Zap70, Lyn, Fyb, SHP1, CD3, CD28, CD45, LCK, PAK, CDC42, Vav, CARD11, PAG, ITAM, LAT, PIP2, SLP-76, MAP3Ks, IKKs, CBL, NCK, LAT, GRB2, CARMA1, NFAT, MALT1, GRB2, JNK, MKKs, PAK, Ras, Raf1, and MEKK1) (Figure 5); B cell receptors and downstream factors (B-cell antigen receptor complex-associated protein α/β chain, CD81, CD22, and BCL-10); key molecules involved in antigen presenting and processing pathways (MHC class I and II, Hsp70, Hsp90, and Calnexin); members of complement and coagulation pathways (classical pathway members C1q, C1r, C1 s, C2, and C4, MBL pathway members MBL and MASP, alternative pathway members Bf and Df, and complement receptor type 1); and members involved in FcεR I signalling pathway (PI3K and SYK), leukocyte migration (lymphotactin, Rho GTPase-activating protein 5, RhoH, integrin alpha-M, and leukaemia inhibitory factor receptor), and natural killer-mediated cytotoxicity (natural killer cell receptor 2B4, natural killer cell protease 1, natural killer cell enhancement factor, natural killer-tumour recognition sequence, and natural killer cell stimulatory factor 2). Furthermore, a number of consensus genes involved in cellular adhesion, energy production, and amino acid (arginine and praline) metabolisms were also conserved between fish and mammals. These genes are indirectly related to immune responses in mammals. For instance, L-arginine metabolism has been proven to be related to phagocytosis of macrophages, which eventually led to the discovery of NO signalling molecule [37]. Thus, the involvement of these consensus genes in metabolic pathways provides the basis for further identification of the biological functions of candidate genes in fish immune responses.

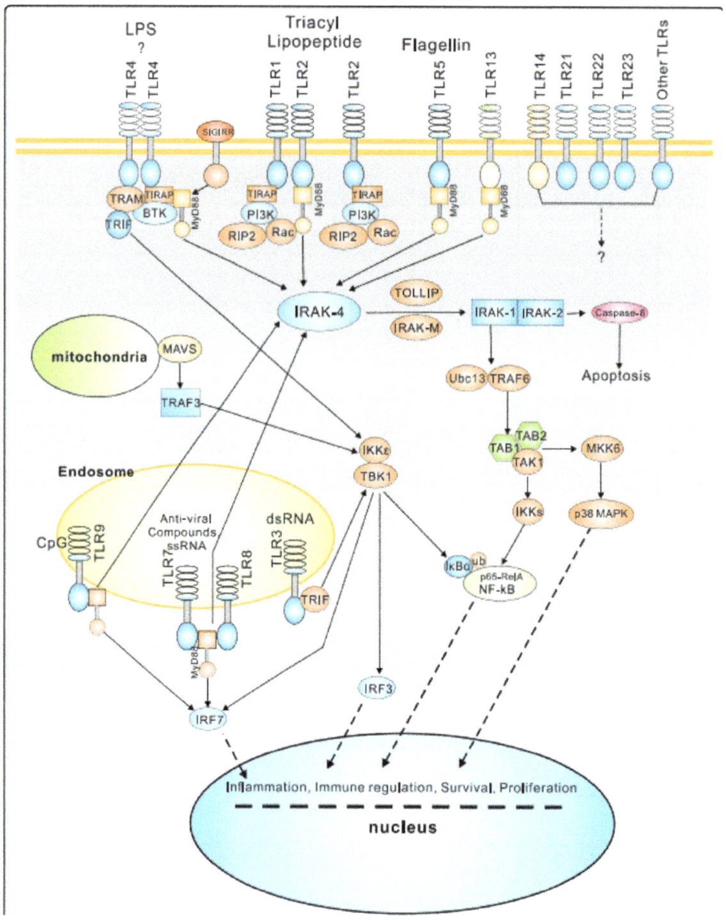

Figure 4: Putative TLR signal pathway. Putative TLR signal pathway of *L. japonicus* was constructed based on knowledge of TLR signalling in mammalian species. However, most interactions have to be confirmed experimentally.

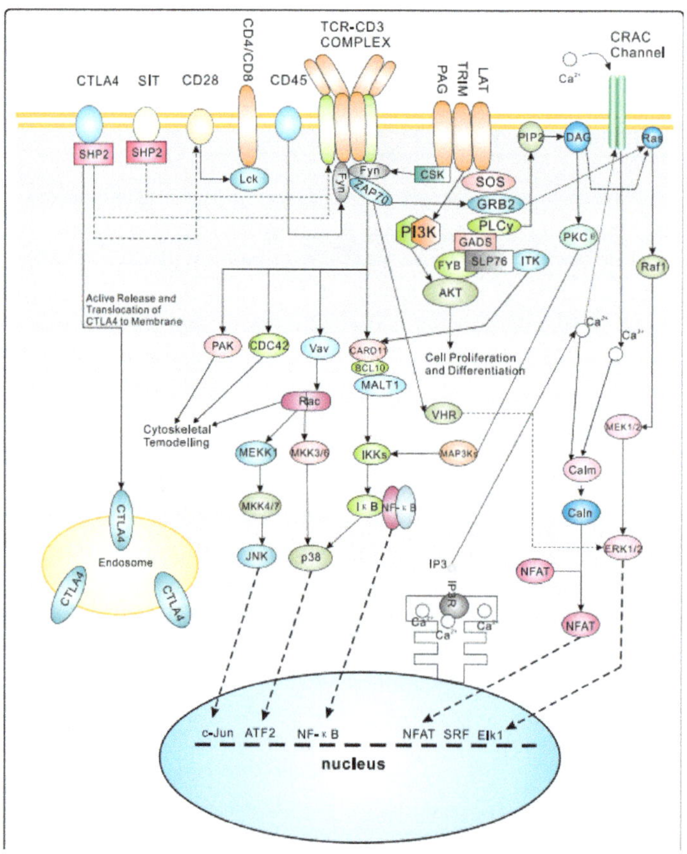

Figure 5: Putative TCR signal pathway. Putative TCR signal pathway of *L. japonicus* was constructed based on knowledge of TCR signalling in mammalian species. However, most interactions require experimental confirmation.

Digital Gene Expression Profile Analysis after Bacterial Challenge

Solexa/Illumina DGE analysis was performed to identify the genes involved in *L. Japonicus* response to bacterial challenge [12]. A total of 3.44 and 3.22 million raw tags (submitted to GEO database, Association No. GSE21712) of the mRNAs extracted from head kidney and spleen of the mock- and bacteria-challenged groups, respectively, were identified by base calling [38, 39]. After transformation of raw sequences into clean tags by data-processing steps using bio-perl scripts, approximately 0.33 and 0.27 million high quality non-redundant tags were obtained in both groups (Figure 6). Gene annotation was performed by tag mapping analysis using the 169,950 non-redundant consensus sequences from RNA-seq-based transcriptome analysis as reference transcript database. Results showed that 71.41% and 74.53% of all distinct tags can be mapped to the entire reference database (sense or anti-sense) in both groups. Out of the 26,394 sense strands and 23,790 anti-sense strands detected in the mock-challenged group, about 36,782 (75.1%) sense or anti-sense strands were mapped by the tags. In contrast, about 34,840 (71.1%) sense or anti-sense genes were mapped out of the 23,359 sense strands and 21,046 anti-sense strands in the infected group (Figure 6). Among the detectable expressed consensus sequences, 9,643 genes had successful annotations. Mapping results are summarized in Additional file 2, Table S2 and Additional file 3, Figure S1.

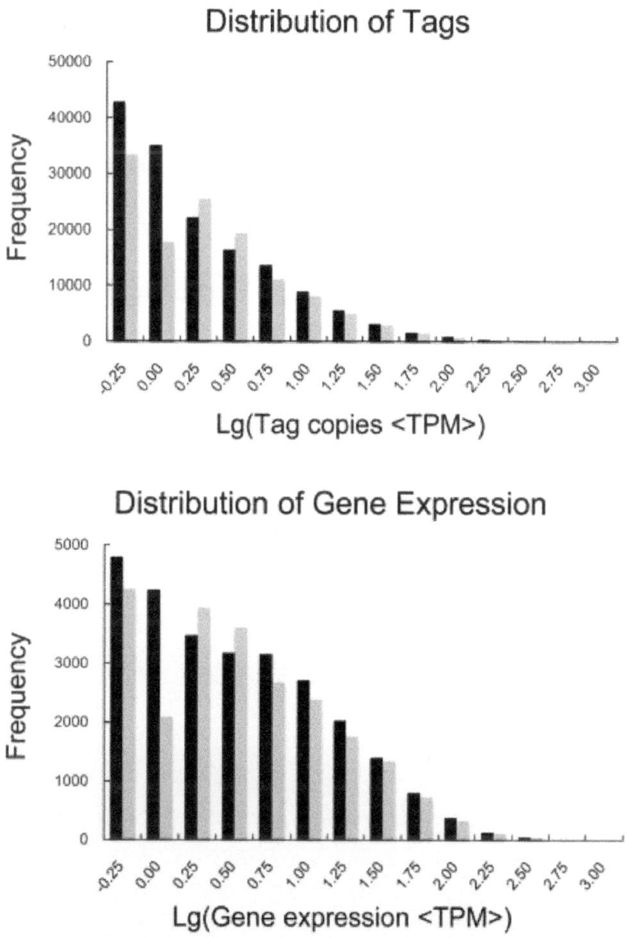

Figure 6: Distribution of tags and gene expression between experimental and control groups. The black and gray columns indicate the distribution of tags and gene expression in bacteria- and mock-challenged groups, respectively. The distribution of tags matches the distribution of genes for both groups. Furthermore, an increase in tags or gene expression is accompanied by a decrease in the frequencies of tags or genes.

Strict Bayesian algorithm was used in differential DGE analysis in order to consider the differences in library size for differential selection between the two differentially expressed gene libraries [40]. Soap2 software was used to map all measured tags to the corresponding assembled consensus sequences [41]. $P \leq 0.01$ and absolute value of $\log_2 \text{Ratio} \geq 1$ were used as the threshold of significant differences in gene expression. Approximately 1,224 CD-containing consensus sequences mapped by 19,548 differentially expressed tags exhibited significant differences after bacterial challenge. Among them, about 376 consensus sequences were significantly up-regulated (including 61 most over-expressed sequences that increased by more than 100-fold), while 848 consensus sequences were significantly down-regulated (including 10 most down-regulated sequences that decreased by 10-100-times) (Figure 7, Additional file 4, Table S3). The distribution of the significant changes detected is illustrated in a volcano plot (Figure 8), where the statistical significance of each consensus was plotted against fold change. Sequences with the highest average differences between the bacteria- and mock-challenged groups (far right and left of the plot) also had the smallest false discovery rate (FDR) values. Analysis of the CD-containing consensus sequences revealed that they all had FDR values less than 0.1, with the highest value being 0.063.

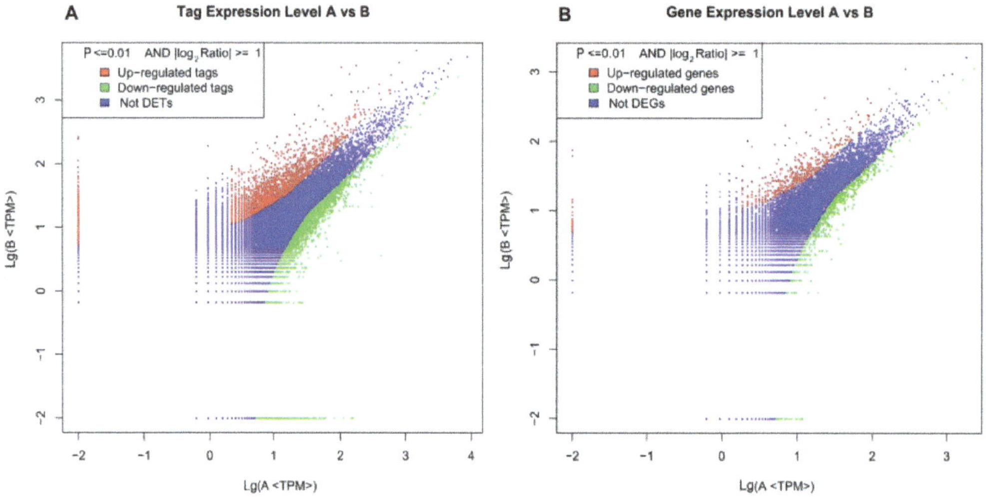

Figure 7: Differential expression analyses of tags and consensus sequences by DGE. The expression level for each tag and consensus was included in the volcano plot (Figures A and B, respectively). 'Not DETs' and 'Not DEGs' indicate 'not detected expression tags' and 'not detected expression genes', respectively. For Figures A and B, the x-axis contains Log_{10} of transcript per million of the bacteria-challenged group and the y-axis indicates Log_{10} of transcript per million of the mock-challenged group. Limitations are based on $P < 0.01$, and the absolute value of Log_2 (B/A) is greater than 1.

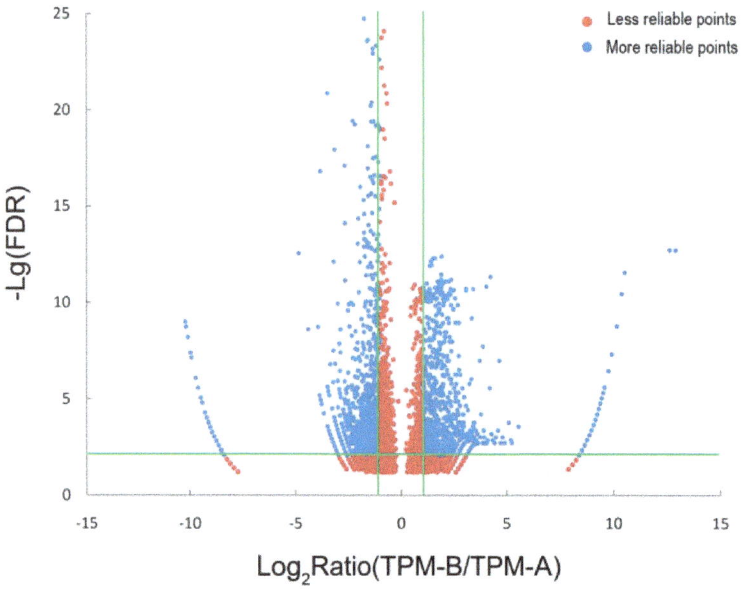

Figure 8: Volcano plot of differentially expressed consensus sequences. For every consensus, the ratio of expression levels in the bacteria-challenged group over that in the mock-challenged group was plotted against the -log error rate. The horizontal line indicates the significance threshold (0.01 FDR), and the vertical lines indicate the twofold change threshold.

Functional annotation of consensus sequences was performed to define the differentially regulated sequences more precisely. Majority of the 1,183 sequences annotated were related to factors involved in various immune-relevant pathways, such as TLR pathways (TLR1/3/13/18/21, BTK, IRAK1, IP2, NR2C2, and UBE2n), TCR signalling pathways (TCR beta chain, Zap70, LCK, SHP1, CARMA1, Vav, NFAT, GRB2, MALT1, NCK, and Raf1), antigen presenting and processing relevant pathways (MHC class II transactivator, CASP1, and CASP8), TGF-β signalling pathway (TGF-β receptor type-2, Serine/threonine-protein phosphatase 4, serine/threonine kinase receptor, BMPs, and SMAD7), and various inflammatory cytokines and receptor relevant pathways (IL18R, IL12R β-2 chain, C-C motif chemokine 7, C-C motif chemokine 19, TGFβ-1-induced transcript 1 protein, TNR superfamily member 1A, and TNFα-induced protein 2). There were annotations in several other biological processes that may indirectly participate in immune response, such as the cell cycle; DNA replication, transcription, and translation; metabolisms of carbohydrates, amino acids, and lipids; and activation of ATPase family members, transcription elongation factor B, membrane transport protein, NADH dehydrogenase, NAD kinase, nucleolar protein 6, tyrosine protein kinase, ribosomal protein L32, nuclear receptor, and replication initiator 1. Among the 61 most over-expressed transcripts and the 10 most down-regulated transcripts, enrichments of factors involved in metabolic or signalling pathways that have not been linked to immune responses previously, such as cytoskeleton regulation, calcium signalling pathway, MAPK signalling pathway, aimnocyl-tRNA biosynthesis, and methionine, glutamate, and aminosugar metabolism, were detected. Highly responsive consensus sequences are shown in Table 4.

Table 4: Representatives of putative immune relevant genes/homologs as predicted by DGE

Catalogs	Consensus ID	Putative Identification	Fold Change
Pattern Recognition Proteins			
TLR and relevant	3365	TLR 1	-6.45
	8296	TLR 21	-374.00
Mannose receptor	158882	Macrophage mannose receptor 1	-2.94
NOD	117946	NOD protein 2	+4.25
C-type lectin	36609	Collectin-12	+3.69
HMG	3367	HMG AT-hook 1	-218.00
LRR-containing proteins	128723	F-box/LRR protein 14	+5.87
	5434	F-box/LRR protein 20	-2.35
	36369	F-box/LRR protein	-249.00
	3391	LRR-containing protein 68	-6.14
	39099	LRR flightless-interacting protein 2	+2.32
NALP	62330	NALP 3	-2.04
	43979	NALP 6	-2.47
	719	NALP 12	-312.00
	14950	NALP 13	-3.77
BTK	37665	BTK	-249.00
Complement System			
C1	39636	C1q-like protein 4	+5.03
Complement factor	58617	Complement factor H-related protein 3	-2.55
Inflammatory Cytokines and Receptors			
IL, ILR, and relevant	7183	IL6R subunit β	-3.06
	4828	IL7R subunit α	-218.00
	29	IL10R α chain	+265.00
	8308	IL12R β-2 chain	-7.09
	70266	Putative IL17R E-like	+2.23
	9160	IL18R	-2.33
	36378	IL18R accessory protein	-2.96
	67818	IL31RA	-2.38
	4392	IL enhancer-binding factor 3	-4.29
IFN and relevant	34	similar to IFN-inducible protein Gig1	-3.76
	39770	IFN-induced helicase C domain-containing protein 1	-2.15
	39343	IFN-induced GTP-binding protein Mx	+2.22
	42748	IFN-induced transmembrane protein 2	-8.97
	41016	IFN regulatory factor 4	-5.98
	43856	IFN regulatory factor 8	-6.14
CXCR	41529	CXC chemokine receptor 5	-2.59
CCL	71546	C-C motif chemokine 19	+2.20
XCR	70282	XC Chemokine receptor 1	+4.03
EGF	8053	SNED	-312.00
FAM	37053	Protein FAM3C	-2.35
FLT	9012	FLT3 ligand cytokine receptor	-405.00
GATA	42892	GATA-binding factor 2	-15.43
GFI	6300	Zinc finger protein GFI-1	+6.58
	16357	Zinc finger protein GFI-1B	-2.01
Integrin	169765	Integrin α-3	+298.00

PCGF	36691	PCGF 2	-2.54
TGF	34228	TGF β-1-induced transcript 1 protein	-4.02
Thrombopoietin	8606	Thrombopoietin receptor	-4.19
TNF and TNR	167272	TNF α-induced protein 8-like protein 2	-6.38
	36673	TNR 1A	-405.00
	2890	TNR 12A	-281.00
	9356	EMILIN 1	-2.40
Adapters, Effectors, and Signal Transducers			
ASB	71979	ASB2	-5.04
DEDD	1437	DEDD	-3.40
Heme oxygenase	71817	Heme oxygenase	+2.48
CSF	50449	Macrophage CSF 1 receptor 1	-2.09
	11081	Macrophage CSF 1 receptor 2	-218.00
	49226	Granulocyte CSF receptor	-3.09
NR2C	42918	NR2C 2-associated protein	+4.22
SMAD	42078	SMAD 1	-437.00
	42615	SMAD 5	-405.00
TRAF	42729	TRAF7	-2.27
	70242	TANK	-2.11
UBE	2069	UBE E2C	+2.07
	9613	UBE E2H	-2.42
NF κ	1073	NF κ-A	+3.18
MAPK	68093	MAPK 3	-9.45
	70484	MAPK 6	-312.00
MAP2K	37130	MAP2K4	-10.39
MAP3K	38157	MAP3K3	-3.26
	7280	MAP3K4	-4.92
T-cell and B-cell Antigen Activation			
CD	60890	CD2	+2.81
	70094	CD22	+4.10
	4032	CD40 (TNR5)	-343.00
	26107	CD79B	-2.48
DOCK2	4848	DOCK2	-2.54
RAG	36208	RGA2	-6.39
RGS	905	RGS12	-2.60
TYK2	9722	Non-receptor TYK2	-2.11
VAV oncogene family	9761	VAV2	-343.00
Antigen Processing and Regulators			
MHC I/II	65343	MHC II transactivator	-218.00
PSM	115644	PSMB3	+3.06
	5004	PSMC2	+2.71
	42169	PSMC6	+3.39
	208	PSMD3	-2.05
	7477	PSMD8	+5.29
TAPBP	10433	TAPBP	-281.00
RAC	36540	RAC1	-2.72
CDC	70253	CDC 16	+2.78
	55784	CDC 42	+331.00
CDK-inhibitor	43078	CDK 4 inhibitor D	+2.58

Other Genes Related to Immune Cell Response			
AP-1	41196	AP-1 complex subunit σ-2	+4.03
	10356	AP1 subunit mu-1	+2.05
JAK	818	JAK2	-312.00
MALT	21139	MALT protein	-6.14
PPM	365	Protein phosphatase 1F	-2.14
RIPK	1908	RIPK1	-7.88
TMED	8436	TMED2	-3.18
Hepcidin	56643	Hepcidin	+6.10
HSP	37584	HSP 71	-2.96
Lysozyme	51	Lysozyme g	+2.18
Macroglobulin	2234	α-2-macroglobulin	+3.31
Microtubule-associated	12594	EMAP 3	-249.00
Proteolipid protein	38379	Proteolipid protein 2	-3.60
Prohibitin	70061	Prohibitin-2	-7.09
Selenoprotein	80124	Selenoprotein K	+3.19
ABCF	3906	ABCF 2	-3.49
CARD	1588	CARD 11	-3.08
Caspase	42397	Caspase 1	-2.05
	31546	Caspase 8	-5.07
Stress response protein	44158	Stress response protein NST1	-2.15

Limitations of all differentially expressed genes are based on $P \leq 0.01$. The absolute value of $Log_2Ratio \geq 1$ and $FDR \leq 0.05$ showed that the genes were significantly altered after bacterial challenge. The cutoff E-value was set to 1e-10 during the gene annotation process. The absolute value of "Fold Change" means magnitude of up- or down-regulation for each gene/homolog after bacteria challenge; "+" indicates up-regulation, "-" indicates down-regulation.

Putative Novel Immune/Stress Response Genes

Among the differentially expressed transcripts, more than 1,183 transcripts were well annotated, whereas approximately 41 transcripts had low sequence homology to known sequences in public databases, suggesting that they might be putative novel immune-relevant genes in *L. japonicus* involved in the response to bacterial challenge. Among these novel sequences, 13 were differentially regulated by more than 100-fold, implying that they were strongly infection-responsive genes. ProDom analysis identified one HSP domain- and one protein kinase domain-containing sequence [42]. SignalP and TMHMM programs distinguished 24 sequences with signal peptides or transmembrane domains, suggesting that they are cytokines and transmembrane proteins, respectively [43, 44]. Observations presented in Additional file 5, Table S4 can provide guidance for further identification. In-depth functional studies of these novel sequences may benefit the exploration of potential marine fish-specific immune-relevant genes for application in the control of fish diseases.

Experimental Validation of Consensus Sequences

To validate the integrity of RNA-seq results, representative consensus sequences with complete encoding regions, such as hepcidin, Myf5, SNARE, and two IL-8-like CXC chemokine family members, were selected for experimental cloning and sequencing analyses by RT-PCR (Additional file 6, Table S5). All experimentally examined genes matched the RNA-seq-generated sequences perfectly. One of the two IL-8-like CXC chemokines was newly discovered by this study. The two IL-8-like CXC chemokine family members were identified through phylogenetic analysis. Both sequences conserved the four cysteine residues (C_{46}-C_{48}-C_{74}-C_{91}) that are the hallmarks of IL-8 CXC chemokines and can be found throughout the vertebrate IL-8 family (Additional file 7, Figure S2 and Additional file 8, Figure S3). This demonstrates the reliability of RNA-seq results and indicates the necessity for further identification of immune-related genes in *L. japonicus*.

DISCUSSION

The transcriptome is the complete repertoire of expressed RNA transcripts in a cell. Its characterization is essential in deciphering the functional complexity of the genome and in obtaining a better understanding of cellular activities in organisms, including growth, development, disease, and immune defence. The definition of the transcriptome has long been a challenging task. Traditionally, global gene expression analysis has relied mostly on several approaches, including RNA hybridisation on high-density arrays, whole-genome tiling arrays, expressed sequence tag (EST), serial analysis of gene expression (SAGE), and SAGE-derived technologies, which include massively parallel signature sequencing (MPSS) and polony multiplex analysis of gene expression. However, these approaches have several inherent limitations. For example, the array-based approaches allow detection of specific sequences only and capture the transcriptome while ignoring splice-junction information or alternative splicing events. The EST approach provides only partial sequences of individual cDNA clones, is sensitive to cloning biases, and is associated with high costs and difficulties in data analysis. SAGE and MPSS are also costly and cannot be used for splicing events [9]. Thus, the newly developed Solexa/Illumina RNA-seq and DGE high-throughput deep sequencing approaches have dramatically changed how functional complexity of the transcriptome can be studied. These approaches overcome many of the inherent limitations of traditional systems, making the detection of alternative splicing events and low-abundance transcripts possible. They have been applied recently to several species, such as yeast, *Arabidopsis*, *Chlamydomonas*, Zebrafish, *Drosophila*, *Caenorhabditis*, and human, for different purposes [9, 10, 18, 19, 20, 21, 22, 23, 24, 25,26, 27, 28, 29].

In this study, the transcriptome profile analysis of bacteria-challenged *L. japonicus* was conducted through these two approaches in an attempt to gain deep insights into the immunogenetics of a marine species. As expected, a large set of transcriptional sequences with complete or differing lengths of encoding regions was generated. KEGG analysis showed that more than 52% of transcripts are enrichment factors involved in approximately 219 known metabolic or signalling pathways, including cellular growth, differentiation, apoptosis, migration, endocrine, and immune system processes. Further, more than 8% of transcripts represent novel fish-specific genes that have never been described previously. Detailed analysis of immune-relevant genes and pathways showed that more than 2,673 transcripts are homologous to known immune-relevant genes, whereas approximately 2,082 transcripts can be enriched in various immune-relevant metabolic or signalling pathways. Challenging the fish with *V. harveyi* resulted in large alterations of the host transcriptome profile, including significant up- or down-regulation of 1,224 transcripts, among which 41 sequences might be novel immune-relevant genes in fish. In addition, several other biological processes that have not been linked to host immunity before, such as the metabolism of carbohydrates, amino acids, and lipids; activation of ATPase, NADH dehydrogenase, NAD kinase, and tyrosine protein kinase; and up-regulation of nuclear receptors, replication initiators, and ribosomal proteins, were found to be dramatically involved in host immune response. These significantly regulated transcripts might represent strong infection-responsive genes in *L. japonicus*, and reflect a number of immune activities during fish defence against bacterial challenge. The transcriptome profiling data sets obtained in this study provide strong basis for future genetic research in marine fish and support further in-depth genome annotation in vertebrates. Future molecular and functional characterisation of infection-responsive genes could lead to global identification of immune-relevant genes and infection markers in marine fish.

At present, transcriptome analysis in fish relies mainly on the EST approach [45, 46]. Although there have been an increasing number of ESTs sequenced in a large number of libraries in various fish species, including rainbow trout (*Oncorhynchus mykiss*), Atlantic salmon (*Salmo salar*), medaka (*Oryzias latipes*), and zebrafish (*Danio rerio*), the immune-relevant transcriptional profiling data sets obtained from fish are still insufficient. Recently, DGE- and microarray-based transcriptome profiling studies performed in zebrafish revealed that zebrafish and its developing embryo are useful *in vivo* models for the identification of host determinants of responses to bacterial infection [9, 10]. However, transcriptional information on immune responses to infection in a non-model marine fish remains elusive. Therefore, the large set of immune-relevant genes and their role in responses to bacterial challenge in *L. japonicus* presented in this study may largely improve knowledge on fish immunogenetics in other analytical systems. The present study also demonstrates the advantages of new deep sequencing approaches for gene discovery, thus providing new leads for functional studies of candidate genes involved in host-bacteria interactions. The RNA-Seq and DGE analyses conducted in this study were found to complement each other well. RNA-Seq was very effective in unravelling transcriptome complexity, and can detect a large set of genes, including numerous low-expressing genes or novel genes. DEG data can be merged with RNA-Seq data sets, indicating an affordable method for comparative gene expression study. Thus, RNA-Seq was initially performed in this study to provide strong reference transcriptome database for subsequent DGE analysis.

Emerging hallmark components and the cells necessary for innate and adaptive immunity in higher vertebrates have been identified in fish [47, 48]. This was the basis for the widely accepted notion that innate and adaptive immunity was established in teleosts about 470 million years ago. However, the exact molecular and cellular basis of immune systems in teleosts remains poorly understood. The precise regulatory mechanisms underlying the innate and adaptive immunity of teleosts remain vague due to the limited immune-relevant genetic information available in fish. The present work on the definition of high-throughput transcriptome data set of the immune system of *L. japonicus* may contribute greatly to better understanding of the molecular and cellular activities involved in fish immunity. Results unexpectedly showed that the fish immune system is more complex than previously thought. On one hand, the substantial amount of immune-relevant genes involved in metabolic and signalling pathways and the induction of genes encoding cell surface receptors, signalling intermediates, transcription factors, and inflammatory mediators show a clear conservation of mechanisms detected in other vertebrate models, including humans. On the other hand, a large set of novel immune response genes and infection markers that have never been linked previously to immune responses in other vertebrate systems was identified in *L. japonicus*, indicating the existence of numerous fish-specific immune activities during early vertebrate evolution.

For instance, the TLR family is the most important class of pattern recognition receptors that play crucial roles in mediating immune responses to pathogenic microorganisms [8, 49, 50]. Triggering of TLRs by ligands leads to the recruitment of adaptor proteins, resulting in the activation of a range of transcription factors, such as NF-κB, activator protein 1 (AP-1), and IFN regulatory factors (IRFs), through distinct signalling pathways. This eventually leads to the downstream activation of proinflammatory cytokines and receptors, such as IFN-α/β, TNF-α, IL-2, IL-6, IL-8, IL10, CD40, CD86, and MIP1α. To date, 13 TLRs (TLR1-13), at least five adaptor proteins (MyD88, Mal/TIRAP, TIR domain-containing adaptor protein, TRIF/TICAM1, TRAM/TICAM2, and SARM), and numerous downstream effectors have been described in mammals and humans. In the present study, a series of TLRs and corresponding adaptor proteins and downstream effectors were identified in *L. japonicus*. The identified TLRs include the majority seen in mammals and humans (TLR1-13), and four TLRs (TLR14, TLR18, TLR21, and TLR23) seen in fish species. Adaptor proteins and downstream effectors identified include the majority known in mammals and humans, including MYD88, BTK, TOLLIP, FADD, HMGB1, HRAS, HSPD1, CASP8, MAPK8IP3, PELI1, RIPK2, SARM1, TICAM2, TIRAP, EIF2AK2, IRAK1, IRAK2, MAP3K7, MAP3K7IP1, NR2C2, PPARA, PRKRA, TRAF6, UBE2N, and UBE2V1. These adaptor proteins and downstream effectors have been found to be well enriched in various known TLR signalling pathways. Downstream transcriptional factors and pro-inflammatory cytokines mediated by these pathways, including NF-κB, JNK/p38, NF/IL6, IRF, IFN-α/β, TNF-α, IL-2, IL-6, IL-8, and IL-10, was also be identified successfully. These suggest that TLR mechanisms are conserved from fish to mammals throughout vertebrate evolution. A putative draft of TLR signalling pathways in *L. japonicus* based on knowledge of TLR signalling in mammalian species was constructed (Figure 4). However, TLR signalling pathways in fish might be more intricate compared with those in mammalian species because of the novel TLRs (TLR14, TLR18, TLR21, and TLR23). An in-depth study of novel TLRs will improve understanding of fish-specific innate immunity in early vertebrates and even the complete evolutionary history of TLR-based innate immunity. DGE analysis revealed that TLR-1, -3, -13, -18, -21 and their signalling intermediates (Rac1, AKT, CASP8, IRAK1, TRAM, IRAK1, IKK alpha/beta, IRF7, and STAT1) were up- or down-regulated dramatically at different levels in the pathway upon bacterial challenge ($P \leq 0.01$). This provides evidence that both conserved (TLR-1, -3, -13) and fish-specific (TLR-18, -21) TLR-based immunity participates in fish defence against bacterial challenge.

The innate immune system is generally believed to represent the evolutionarily ancient aspect of vertebrate immunity. As a representative of lower vertebrates, fish is suggested to possess stronger innate immune responses. However, fish adaptive immunity might be more primitive because of limited immunoglobulins and hallmark components necessary for adaptive immunity identified in this species [7]. In recent years, several hallmarks for T and B cells (TCR, BCR, CD3, CD4, and CD8), antigen presenting and processing molecules (MHCI, MHCII, and DC-SIGN/CD209), co-stimulatory factors (CD80/86, CD83, CD154, and CD40), and immunoglobulins (IgM, IgD, and IgZ/T) have been identified in teleost fish, thus providing preliminary evidence that the adaptive immune system might also be well-established in fish. However, the precise molecular and cellular bases and mechanisms underlying teleost adaptive immunity are still uncharacterised and require further immunogenetic studies. The present study successfully identified a large number of adaptive immune-relevant components homologous to those in higher vertebrates, providing abundant data sets for insights into the characterisation and origin of adaptive immunity in early vertebrates. Data sets imply that adaptive immunity in teleost fish seems to be much more complicated than previously believed. The basic components and signalling pathways necessary for adaptive immunity exist in fish, and a majority showed clear conservation between fish and mammals. For instance, T cell receptor (TCR) signalling pathways regulate T cell activation, one of the most important processes in adaptive immunity [51]. Majority of the four types of TCRs (TCR-α/β, -γδ) and numerous signalling transducers (Zap70, Lyn, LCK, SHP1, CD3, ITAM, LAT, Fyb, SLP-76, CBL, NCK, LAT, GRB2, CARMA1, NFAT, AP1, MALT1, and GRB2) discovered in humans and mammals can be identified in *L. japonicus*. DGE analysis showed that a number of TCR signalling pathway members, including TCR beta chain, Zap70, LCK, SHP1, CARMA1, Vav, NFAT, GRB2, MALT1, NCK, and Raf1, are induced significantly after bacterial challenge ($P \leq 0.01$). These pathway members largely contribute to the proliferation and activation of T cells in mammals, thus suggesting that TCR signalling mechanisms underlying the T cell activation might be conserved between teleost fish and mammals. A putative draft of TCR signalling pathways based on knowledge of pathways known in mammals was constructed (Figure 5). Future studies on these pathways are expected to not only enrich current knowledge on fish immunology but also contribute to better understanding of the evolutionary history of adaptive immunity.

CONCLUSIONS

This study investigated the transcriptome profile of bacteria-challenged *L. japonicus* using Solexa/Illumina RNA-seq and DGE deep sequencing technologies. The substantial amount of transcripts obtained provides a strong basis for future genomic research on marine fish and supports in-depth genome annotation in vertebrates. Globally identified immune-candidate genes, infection markers, and putative signalling pathways in *L. japonicus* revealed that the immune system of fish may be much more complex than previously believed. A considerable amount of immune-relevant genes and pathways in fish showed significant similarity to vertebrate models, suggesting that mechanisms underlying the innate and adaptive immunity in fish may be conserved in higher vertebrates. In addition, a large set of novel immune response genes that have never been linked previously to immune responses in other vertebrate systems indicate the existence of numerous fish-specific immune events during early evolution. This suggests that innate and adaptive immunity might be well established in teleost marine fish. Findings provide deep insight into the immunogenetics of fish species, which can be clinically applied in the therapy of fish diseases. They also contribute to a better understanding of the evolutionary history of innate and adaptive immunity from fish to mammals.

METHODS

Experimental fish

One-year-old Japanese sea bass of both sexes, weighing 48.6 ± 2.5 g, were obtained from the fishery institute of Zhejiang, China. They were kept in running aerated seawater (salinity 30) at 25°C and fed with commercial pellet food at a daily ration of 0.7% body weight. All fish were maintained in the laboratory for at least two weeks prior to experimental use to allow for acclimatisation and evaluation of overall fish health. Only healthy fish, as determined by general appearance and level of activity, were used in the experiment.

Bacterial Strain

Wild-type marine fish-virulent *V harveyi* strain (96-915), a pathogen for bacterial septicaemia in *L. japonicas*, was maintained in the laboratory. It was cultured in Thiosulfate Citrate Bile Salts Sucrose at 27°C overnight. The desired number of cells was adjusted to 5 × 108 CFU/ml. Cells were inactivated with 5% formalin at 27°C overnight before thorough washing with sterile PBS (pH 7.0). They were re-suspended in PBS prior to use.

Bacterial Challenge and RNA Preparation

Fish in the experimental groups were inoculated intraperitoneally with 0.2 ml of *V harveyi* at 1×10^8 CFU per fish. In parallel, fish in the control groups were administrated with 0.2 ml of mock PBS. Both groups were kept under conditions as described above. At seven days post-challenge, fish were sacrificed after anaesthesia, and tissues from the head kidney and spleen were collected. Tissue samples from 15 fishes were mixed for RNA preparation. Total RNA was isolated using a TRIzol reagent (Gibco BRL) following the manufacturer's instructions and treated with RNase free DNase I (Qiagen). RNA concentrations were measured using a spectrophotometer and integrity was ensured through analysis on a 1.5% (w/v) agarose gel.

Sample Preparation for RNA-seq

After RNA extraction, poly-A-containing mRNAs were purified using oligo-dT-attached magnetic beads and fragmented into small pieces using divalent cations under elevated temperature. Cleaved RNA fragments were copied into first strand cDNA using reverse transcriptase and random primers. This was followed by second strand cDNA synthesis using DNA polymerase I and RNase H. These cDNA fragments underwent end repair process, addition of a single 'A' base, and ligation of adapters. Products were subsequently purified and amplified through PCR to create the final cDNA libraries.

Transcriptome Analysis

Transcriptome sequencing was conducted using Solexa/ Illumina RNA-seq. Four fluorescently labelled nucleotides and a specialised polymerase were used to determine the clusters base by base in parallel. The 75-bp raw PE reads (submitted to GEO database, Association No. GSE21721) were generated by the Illumina Genome Analyzer II system. Raw reads were then assembled into non-redundant consensus sequences using Grape, tgicl, and CAP3 softwares [52, 53]. All sequences were examined for possible sequencing errors. Adaptor sequences were trimmed using the Cross_Match software in the Phrap packagehttp://www.phrap.org/[54]. Short sequences (< 100 bp in length) were removed using custom Perl program [55]. The resulting high-quality sequences were assembled into sequence contigs with the TGICL program, which creates an assembly using CAP3. Sequence homology searches were performed using local BLASTall programs against sequences in NCBI non-redundant (nr) protein database and the Swissprot database (E-value < 1e-10). Genes were tentatively identified according to the best hits against known sequences. Assembled consensus sequences were used to determine the GO term, COG term, and were analyzed further using KEGG.

DGE-Tag Profiling

DGE analysis included sample preparation and sequencing. Sequence tag preparation was performed using the Digital Gene Expression Tag Profile Kit (Illumina) according to the manufacturer's instructions. Briefly, 6 μg total RNA was used for mRNA purification using oligo-dT magnetic bead adsorption and oligo-dT was used to guide reverse transcription for double-stranded cDNA synthesis. The generation of 5› ends of tags was done using endonuclease NlaIII, which recognizes and cuts off the CATG sites on cDNA. cDNA fragments with 3› ends were purified through magnetic bead precipitation, and Illumina adapter 1 was added to the 5› ends. The junction of Illumina adapter 1 and CATG site was the recognition site of MmeI, which cuts 17 bp downstream of the CATG site, producing tags with adapter 1. After removal of 3› fragments with magnetic bead precipitation, the 21-bp unique tags with adaptor 1 were purified and ligated to adaptor 2 to form a cDNA tag library. These adapter-ligated cDNA tags were enriched after 15 cycles of linear PCR amplification. The resulting 85-bp fragments were purified by 6% TBE polyacrylamide gel electrophoresis. Fragments were then digested and the single-chain molecules were fixed onto the Solexa Sequencing Chip (flowcell). Sequencing by synthesis was performed using the Illumina Genome Analyzer II system according to the manufacturer›s protocols. Image analysis, base calling, generation of raw 17-bp tags, and tag counting were performed using the Illumina pipeline. Raw data (tag sequences) were deposited in the GEO database under submission number GSE21712.

Aligning DGE Tags to Reference Transcriptome Data Set

Clean tags and count number of DGE libraries from bacteria- and mock-challenged groups were collected and summarised using custom Bio-perl scripts. All tags were mapped to the reference transcriptome generated by RNA-seq. To monitor mapping events on both strands, both sense and complementary antisense sequences were included in the mapping process. Only perfect matches over the entire 21-bp length of the 17-bp tag plus the 4-bp NlaIII recognition site were allowed. This study was limited to tags that mapped to ORFs only and cannot show tags that mapped to mRNA with long 3'UTRs.

Identification of Differentially Expressed Genes

Rigorous algorithms were developed to identify differentially expressed genes between two samples. The correlation of the detected count numbers between parallel libraries was assessed statistically by calculating the Pearson correlation. In addition to the P value, FDR was manipulated to determine differentially expressed genes [56]. Assuming that R differentially expressed genes have been selected, S genes really show differential expression, whereas the other V genes are false positives. If the error ratio $Q = V/R$ must remain below a cutoff (5%), FDR should not exceed 0.05. In this research, $P \leq 0.01$, $FDR \leq 0.1$, and the absolute value of log2Ratio≥1 were used as threshold to assess the significance of gene expression difference. More stringent criteria with smaller FDR and bigger fold-change values can be used to identify differentially expressed genes.

Experimental Validation

Representative consensus sequences with complete ORFs (hepcidin, Myf5, SNARE, syntaxin, and IL-8-like homologues) generated by RNA-seq were selected for experimental cloning and sequencing validation. The cDNAs of these genes were amplified by RT-PCR using the primers shown in Supplemental Table 6. All PCR products were purified using Gel Extraction Kit (Qiagen), cloned into pUCm-T vector (TaKaRa), and sequenced on MegaBACE 1000 Sequencer (GE) using the DYEnamic ET Dye Terminator Cycle Sequencing Kit (Pharmacia). Protein sequence alignments were generated using the Cluster W program (version 1.83). The phylogenies of protein sequences were estimated using MEGA 3.0 with the neighbour-joining method.

ACKNOWLEDGEMENTS

We acknowledge the Beijing Genomics Institute at Shenzhen for its assistance in original data processing and related bioinformatics analysis. We are thankful to professor Guo-liang Wang of Ningbo University for providing us with *Vibrio harveyi* culture. We would also like to thank Guang-ping Liu, Hui-hui Liu, and Jian-qiu Zou for their help in data and figure processing. This work was supported by grants from Hi-Tech Research and Development Program of China (863) (2008AA09Z409), the National Basic Research Program of China (973) (2006CB101805), the National Natural Science Foundation of China (30871936, 30571423), and the Science and Technology Foundation of Zhejiang Province (2006C12038, 2006C23045, 2006C12005, 2007C12011).

Notes

Li-xin Xiang, Ding He and Jian-zhong Shao contributed equally to this work.

REFERENCES

1. Kocher TD: Adaptive evolution and explosive speciation: the cichlid fish model. Nat Rev Genet. 2004, 5 (4): 288-298. 10.1038/nrg1316.
2. Venkatesh B: Evolution and diversity of fish genomes. Curr Opin Genet Dev. 2003, 13 (6): 588-592. 10.1016/j.gde.2003.09.001.
3. Kasahara M, Naruse K, Sasaki S, Nakatani Y, Qu W, Ahsan B, Yamada T, Nagayasu Y, Doi K, Kasai Y: The medaka draft genome and insights into vertebrate genome evolution. Nature. 2007, 447 (7145): 714-719. 10.1038/nature05846.
4. Peatman E, Liu Z: Evolution of CC chemokines in teleost fish: a case study in gene duplication and implications for immune diversity. Immunogenetics. 2007, 59 (8): 613-623. 10.1007/s00251-007-0228-4.
5. Zou J, Tafalla C, Truckle J, Secombes CJ: Identification of a second group of type I IFNs in fish sheds light on IFN evolution in vertebrates. J Immunol. 2007, 179 (6): 3859-3871.
6. Jin HJ, Shao JZ, Xiang LX, Wang H, Sun LL: Global identification and comparative analysis of SOCS genes in fish: insights into the molecular evolution of SOCS family. Mol Immunol. 2008, 45 (5): 1258-1268. 10.1016/j.molimm.2007.09.015.
7. Flajnik MF, Kasahara M: Origin and evolution of the adaptive immune system: genetic events and selective pressures. Nat Rev Genet. 2010, 11 (1): 47-59. 10.1038/nrg2703.
8. Leulier F, Lemaitre B: Toll-like receptors--taking an evolutionary approach. Nat Rev Genet. 2008, 9 (3): 165-178. 10.1038/nrg2303.
9. Hegedus Z, Zakrzewska A, Agoston VC, Ordas A, Racz P, Mink M, Spaink HP, Meijer AH: Deep sequencing of the zebrafish transcriptome response to mycobacterium infection. Mol Immunol. 2009, 46 (15): 2918-2930. 10.1016/j.molimm.2009.07.002.
10. Stockhammer OW, Zakrzewska A, Hegedus Z, Spaink HP, Meijer AH: Transcriptome profiling and functional analyses of the zebrafish embryonic innate immune response to Salmonella infection. J Immunol. 2009, 182 (9): 5641-5653. 10.4049/jimmunol.0900082.
11. Wang Z, Gerstein M, Snyder M: RNA-Seq: a revolutionary tool for transcriptomics. Nat Rev Genet. 2009, 10 (1): 57-63. 10.1038/nrg2484.
12. Anisimov SV: Serial Analysis of Gene Expression (SAGE): 13 years of application in research. Curr Pharm Biotechnol. 2008, 9 (5): 338-350. 10.2174/138920108785915148.
13. Mortazavi A, Williams BA, McCue K, Schaeffer L, Wold B: Mapping and quantifying mammalian transcriptomes by RNA-Seq. Nat Methods. 2008, 5 (7): 621-628. 10.1038/nmeth.1226.
14. Sultan M, Schulz MH, Richard H, Magen A, Klingenhoff A, Scherf M, Seifert M, Borodina T, Soldatov A, Parkhomchuk D: A global view of gene activity and alternative splicing by deep sequencing of the human transcriptome. Science. 2008, 321 (5891): 956-960. 10.1126/science.1160342.
15. Nagalakshmi U, Wang Z, Waern K, Shou C, Raha D, Gerstein M, Snyder M: The transcriptional landscape of the yeast genome defined by RNA sequencing. Science. 2008, 320 (5881): 1344-1349. 10.1126/science.1158441.
16. Morozova O, Marra MA: Applications of next-generation sequencing technologies in functional genomics. Genomics. 2008, 92 (5): 255-264. 10.1016/j.ygeno.2008.07.001.
17. t Hoen PA, Ariyurek Y, Thygesen HH, Vreugdenhil E, Vossen RH, de Menezes RX, Boer JM, van Ommen GJ, den Dunnen JT: Deep sequencing-based expression analysis shows major advances in robustness, resolution and inter-lab portability over five microarray platforms. Nucleic Acids Res. 2008, 36 (21): e141-10.1093/nar/gkn705.
18. Wang B, Guo G, Wang C, Lin Y, Wang X, Zhao M, Guo Y, He M, Zhang Y, Pan L: Survey of the transcriptome of Aspergillus oryzae via massively parallel mRNA sequencing. Nucleic Acids Res. 2010,
19. Levin JZ, Berger MF, Adiconis X, Rogov P, Melnikov A, Fennell T, Nusbaum C, Garraway LA, Gnirke A: Targeted next-generation sequencing of a cancer

transcriptome enhances detection of sequence variants and novel fusion transcripts. Genome Biol. 2009, 10 (10): R115-10.1186/gb-2009-10-10-r115.

20. Tang F, Barbacioru C, Nordman E, Li B, Xu N, Bashkirov VI, Lao K, Surani MA: RNA-Seq analysis to capture the transcriptome landscape of a single cell. Nat Protoc. 2010, 5 (3): 516-535. 10.1038/nprot.2009.236.

21. Asmann YW, Klee EW, Thompson EA, Perez EA, Middha S, Oberg AL, Therneau TM, Smith DI, Poland GA, Wieben ED: 3' tag digital gene expression profiling of human brain and universal reference RNA using Illumina Genome Analyzer. BMC Genomics. 2009, 10: 531-10.1186/1471-2164-10-531.

22. Zhang G, Guo G, Hu X, Zhang Y, Li Q, Li R, Zhuang R, Lu Z, He Z, Fang X: Deep RNA sequencing at single base-pair resolution reveals high complexity of the rice transcriptome. Genome Res. 2010, 20: 646-654. 10.1101/gr.100677.109.

23. David JP, Coissac E, Melodelima C, Poupardin R, Riaz MA, Chandor-Proust A, Reynaud S: Transcriptome response to pollutants and insecticides in the dengue vector Aedes aegypti using next-generation sequencing technology. BMC Genomics. 2010, 11 (1): 216-10.1186/1471-2164-11-216.

24. Liu S, Li D, Li Q, Zhao P, Xiang Z, Xia Q: MicroRNAs of Bombyx mori identified by Solexa sequencing. BMC Genomics. 2010, 11: 148-10.1186/1471-2164-11-148.

25. Yassour M, Kaplan T, Fraser HB, Levin JZ, Pfiffner J, Adiconis X, Schroth G, Luo S, Khrebtukova I, Gnirke A: Ab initio construction of a eukaryotic transcriptome by massively parallel mRNA sequencing. Proc Natl Acad Sci USA. 2009, 106 (9): 3264-3269. 10.1073/pnas.0812841106.

26. Veitch NJ, Johnson PC, Trivedi U, Terry S, Wildridge D, MacLeod A: Digital gene expression analysis of two life cycle stages of the human-infective parasite, Trypanosoma brucei gambiense reveals differentially expressed clusters of co-regulated genes. BMC Genomics. 2010, 11: 124-10.1186/1471-2164-11-124.

27. Morrissy AS, Morin RD, Delaney A, Zeng T, McDonald H, Jones S, Zhao Y, Hirst M, Marra MA: Next-generation tag sequencing for cancer gene expression profiling. Genome Res. 2009, 19 (10): 1825-1835. 10.1101/gr.094482.109.

28. Wang Y, Brahmakshatriya V, Zhu H, Lupiani B, Reddy SM, Yoon BJ, Gunaratne PH, Kim JH, Chen R, Wang J: Identification of differentially expressed miRNAs in chicken lung and trachea with avian influenza virus infection by a deep sequencing approach. BMC Genomics. 2009, 10: 512-10.1186/1471-2164-10-512.

29. Chen X, Li Q, Wang J, Guo X, Jiang X, Ren Z, Weng C, Sun G, Wang X, Liu Y: Identification and characterization of novel amphioxus microRNAs by Solexa sequencing. Genome Biol. 2009, 10 (7): R78-10.1186/gb-2009-10-7-r78.

30. Xie ZY, Hu CQ, Zhang LP, Chen C, Ren CH, Shen Q: Identification and pathogenicity of Vibrio ponticus affecting cultured Japanese sea bass, Lateolabrax japonicus (Cuvier in Cuvier and Valenciennes). Letters in Applied Microbiology. 2007, 45 (1): 62-67. 10.1111/j.1472-765X.2007.02141.x.

31. Austin B, Zhang XH: Vibrio harveyi: a significant pathogen of marine vertebrates and invertebrates. Lett Appl Microbiol. 2006, 43 (2): 119-124. 10.1111/j.1472-765X.2006.01989.x.

32. Iseli C, Jongeneel CV, Bucher P: ESTScan: a program for detecting, evaluating, and reconstructing potential coding regions in EST sequences. Proc Int Conf Intell Syst Mol Biol. 1999, 138-148.

33. Ye J, McGinnis S, Madden TL: BLAST: improvements for better sequence analysis. Nucleic Acids Res. 2006, W6-9. 10.1093/nar/gkl164. 34 Web Server

34. Ye J, Fang L, Zheng H, Zhang Y, Chen J, Zhang Z, Wang J, Li S, Li R, Bolund L: WEGO: a web tool for plotting GO annotations. Nucleic Acids Res. 2006, W293-297. 10.1093/nar/gkl031. 34 Web Server

35. Tatusov RL, Galperin MY, Natale DA, Koonin EV: The COG database: a tool for genome-scale analysis of protein functions and evolution. Nucleic Acids Res. 2000, 28 (1): 33-36. 10.1093/nar/28.1.33.

36. Kanehisa M, Araki M, Goto S, Hattori M, Hirakawa M, Itoh M, Katayama T, Kawashima S, Okuda S, Tokimatsu T: KEGG for linking genomes to life and the environment. Nucleic Acids Res. 2008, D480-484. 36 Database

37. Rassaf T, Kleinbongard P, Kelm M: The L-arginine nitric oxide pathway: avenue for a multiple-level approach to assess vascular function. Biol Chem. 2006, 387 (10-11): 1347-1349. 10.1515/BC.2006.168.

38. Ewing B, Hillier L, Wendl MC, Green P: Base-calling of automated sequencer traces using phred. I. Accuracy assessment. Genome Res. 1998, 8 (3): 175-185.

39. Ewing B, Green P: Base-calling of automated sequencer traces using phred. II. Error probabilities. Genome Res. 1998, 8 (3): 186-194.

40. Audic S, Claverie JM: The significance of digital gene expression profiles. Genome Res. 1997, 7 (10): 986-995.

41. Li R, Yu C, Li Y, Lam TW, Yiu SM, Kristiansen K, Wang J: SOAP2: an improved ultrafast tool for short read alignment. Bioinformatics. 2009, 25 (15): 1966-1967. 10.1093/bioinformatics/btp336.

42. Bru C, Courcelle E, Carrere S, Beausse Y, Dalmar S, Kahn D: The ProDom database of protein domain families: more emphasis on 3D. Nucleic Acids Res. 2005, D212-215. 33 Database

43. Bendtsen JD, Nielsen H, von Heijne G, Brunak S: Improved prediction of signal peptides: SignalP 3.0. J Mol Biol. 2004, 340 (4): 783-795. 10.1016/j.jmb.2004.05.028.

44. Chen Y, Yu P, Luo J, Jiang Y: Secreted protein prediction system combining CJ-SPHMM, TMHMM, and PSORT. Mamm Genome. 2003, 14 (12): 859-865. 10.1007/s00335-003-2296-6.

45. Borchardt T, Looso M, Bruckskotten M, Weis P, Kruse J, Braun T: Analysis of newly established EST databases reveals similarities between heart regeneration in newt

and fish. BMC Genomics. 2010, 11: 4-10.1186/1471-2164-11-4.

46. Lo L, Zhang Z, Hong N, Peng J, Hong Y: 3640 unique EST clusters from the medaka testis and their potential use for identifying conserved testicular gene expression in fish and mammals. PLoS One. 2008, 3 (12): e3915-10.1371/journal.pone.0003915.

47. Lin AF, Xiang LX, Wang QL, Dong WR, Gong YF, Shao JZ: The DC-SIGN of zebrafish: insights into the existence of a CD209 homologue in a lower vertebrate and its involvement in adaptive immunity. J Immunol. 2009, 183 (11): 7398-7410. 10.4049/jimmunol.0803955.

48. Gong YF, Xiang LX, Shao JZ: CD154-CD40 interactions are essential for thymus-dependent antibody production in zebrafish: insights into the origin of costimulatory pathway in helper T cell-regulated adaptive immunity in early vertebrates. J Immunol. 2009, 182 (12): 7749-7762. 10.4049/jimmunol.0804370.

49. Meijer AH, Gabby Krens SF, Medina Rodriguez IA, He S, Bitter W, Ewa Snaar-Jagalska B, Spaink HP: Expression analysis of the Toll-like receptor and TIR domain adaptor families of zebrafish. Mol Immunol. 2004, 40 (11): 773-783. 10.1016/j.molimm.2003.10.003.

50. Oshiumi H, Tsujita T, Shida K, Matsumoto M, Ikeo K, Seya T: Prediction of the prototype of the human Toll-like receptor gene family from the pufferfish, Fugu rubripes, genome. Immunogenetics. 2003, 54 (11): 791-800.

51. Meeker ND, Smith AC, Frazer JK, Bradley DF, Rudner LA, Love C, Trede NS: Characterization of the zebrafish T cell receptor beta locus. Immunogenetics. 62 (1): 23-29. 10.1007/s00251-009-0407-6.

52. Pertea G, Huang X, Liang F, Antonescu V, Sultana R, Karamycheva S, Lee Y, White J, Cheung F, Parvizi B: TIGR Gene Indices clustering tools (TGICL): a software system for fast clustering of large EST datasets. Bioinformatics. 2003, 19 (5): 651-652. 10.1093/bioinformatics/btg034.

53. Huang X, Madan A: CAP3: A DNA sequence assembly program. Genome Res. 1999, 9 (9): 868-877. 10.1101/gr.9.9.868.

54. de la Bastide M, McCombie WR: Assembling genomic DNA sequences with PHRAP. Curr Protoc Bioinformatics. 2007, Chapter 11 (Unit11 14):

55. Stajich JE, Block D, Boulez K, Brenner SE, Chervitz SA, Dagdigian C, Fuellen G, Gilbert JG, Korf I, Lapp H: The Bioperl toolkit: Perl modules for the life sciences. Genome Res. 2002, 12 (10): 1611-1618. 10.1101/gr.361602.

56. Benjamini Y, Drai D, Elmer G, Kafkafi N, Golani I: Controlling the false discovery rate in behavior genetics research. Behav Brain Res. 2001, 125 (1-2): 279-284. 10.1016/S0166-4328(01)00297-

CHAPTER 17

TRANSCRIPTOME ANALYSIS AND SSR/SNP MARKERS INFORMATION OF THE BLUNT SNOUT BREAM (MEGALOBRAMA AMBLYCEPHALA)

Zexia Gao., Wei Luo., Hong Liu, Cong Zeng, Xiaolian Liu, Shaokui Yi and Weimin Wang

INTRODUCTION

Blunt snout bream (*Megalobrama amblycephala* Yih, 1955), commonly known as Wuchang fish, is an endemic species in China. Its natural distribution is restricted to the middle and lower reaches of the Yangtze River, such as Liangzi (LZ), Poyang (PY) and Yuni (YN) Lakes[1], [2]. *M. amblycephala* is a species of ray-finned fish in the genus *Megalobrama*, which includes other three species *M. skolkoii*, *M. hoffmanni* and *M. pellegrini* [3]. Among these four species, *M. amblycephala* has been widely favored for its delicacy and recognized as a main aquaculture species in the freshwater polyculture system since 1960s in China [4]. Due to its successful artificial propagation and high economic value, *M. amblycephala* aquaculture industry has greatly developed in the past decades and its total output had reached 541,115 tons in 2001 [5]. As a consequence of fast domestication and over-fishing, however, germplasm resources of this bream are under threat of recession and mixture due to its artificial breeding [6], [7]. Nowadays, the cultured population of *M. amblycephala* is gradually exhibiting early sexual maturity, low growth rate and disease susceptibility. Therefore, applying genomics tools in the selection of elite broodstock has the potential to enhance the productivity and value of commercial production for this species.

Genetic marker discovery is the first step in the application of genomics to improve broodstock as these markers can be used for the creation of linkage maps and subsequent quantitative trait loci (QTLs) identification. Marker assisted selection (MAS) can be then employed by selecting broodstock based on genotypes at QTL that are relevant to economically important traits such as rapid growth, disease resistance and the control of early maturation. Currently, a limited collection of genetic markers is available for *M. amblycephala*, including mitochondrial DNA (mtDNA) [8], restriction fragment length polymorphisms (RFLP) and microsatellites [9],[10]. Many of the studies using genetic markers to analyze population structure in *M. amblycephala* have employed these genetic markers [2], [8], [11]. So far, only 60 microsatellites have been successfully developed for this species. Single Nucleotide Polymorphisms (SNPs) are highly abundant markers, which are evenly distributed throughout the genome and can be functionally relevant. Because of their potential for high genotyping efficiency, automation, data quality, genome-wide coverage and analytical simplicity [12], SNPs have rapidly become the marker of choice for many applications in genetics and genomics. They are suitable markers for fine mapping of genes and candidate gene association studies aimed at identifying alleles potentially affecting important traits. For instance, simultaneous analyses of thousands of SNPs have enabled genome-wide association studies for complex traits

Citation: Gao Z, Luo W, Liu H, Zeng C, Liu X, Yi S, et al. (2012) Transcriptome Analysis and SSR/SNP Markers Information of the Blunt Snout Bream (Megalobrama amblycephala). PLoS ONE 7(8): e42637. doi:10.1371/journal.pone.0042637

in chicken [13], pig [14], [15], cattle [16]–[18], horse [19] and sheep [20], [21]. However, such studies have not been possible in most aquaculture species because large numbers of SNPs have not been available. Developing a large set of SNP markers for genome analyses in *M. amblycephala* will facilitate fine mapping of QTL, improve the identification and exploitation of genes affecting important traits and enable selective breeding through genomic selection.

Although genome sequencing technology has become progressively more efficient over the past decade, the sequencing of complex genomes remains expensive. Expressed Sequence Tag (EST) sequencing provides an attractive alternative to whole-genome sequencing because this technique produces sequences of the transcribed portions of genes at a fraction of the cost of sequencing complete chromosomes. However, obtaining an EST database by traditional approaches was time-consuming and expensive and consequently restricted to a few model organisms [22], [23]. With the advent of the second-generation sequencing technologies [24], genetic/genomic resources for non-model species have become far more accessible and transcriptome sequencing is becoming one of the most important applications of second-generation sequencing in the field of ecology, evolution and genetics [25], [26]. Among the high throughput sequencing technologies, 454 pyrosequencing [27] allows for rapid transcriptome sequencing at a reasonable cost to any species. It is of particular interest in ecology and evolution [28]–[30] primarily because it yields longer sequencing reads than any other method (up to 600 bp), which allows more accurate de novo sequence assemblies that are often required for non-model organisms. Recently, 454 pyrosequencing was applied for the identification of gene-derived SNPs in a number of species such as *Eucalyptus grandis* [26], butterfly [31], lake sturgeon [32], coral [33], and pine tree [34].

The development of a number of genome resources facilitates both structural and functional analysis of the genome; however, only 20 EST sequences are available for *M. amblycephala*so far. In the present study, we used 454 pyrosequencing technology to perform an EST-based genome scan by sequencing cDNA pools of one-year-old *M. amblycephala* individuals in order to provide the most comprehensive gene sequence resources for *M. amblycephala* hitherto and to gather a large number of genetic markers (microsatellites and SNPs), which would greatly benefit its breeding program and whole genome association studies.

RESULTS AND DISCUSSION

454 pyrosequencing and assembly

We created a normalized cDNA pool using RNA extracted from different tissues of multiple *M. amblycephala* individuals. Pyrosequencing of this cDNA pool on a 454 GS XLR70 Titanium platform produced a total of 1,409,706 reads, with an average sequence length of 411 base pairs (SD=117, range=40–906; Figure 1), approximately 577 Mbp of sequence data. Assuming a similar number of genes and a similar average coding sequence length of 2,000 bp in *M. amblycephala* as in *Danio rerio* (37,855), average transcriptome coverage was estimated at 7.6×. To our knowledge, this is the first comprehensive study on the transcriptome of *M. amblycephala.*

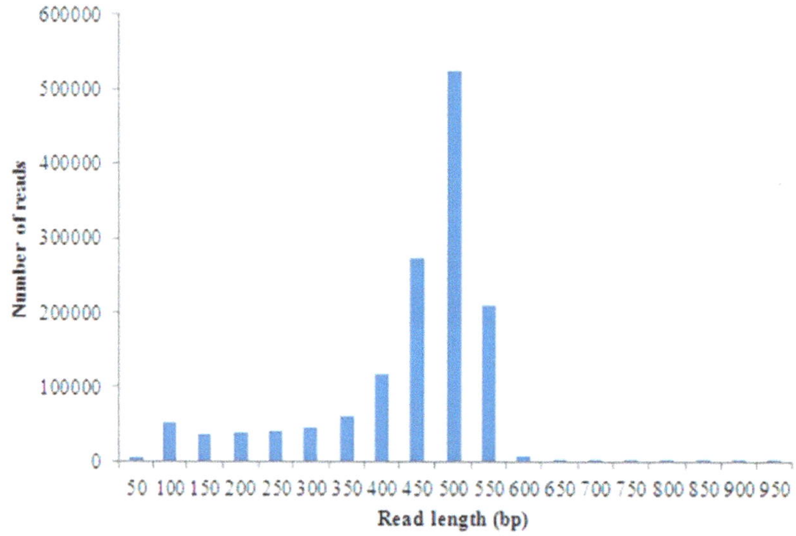

Figure 1: Frequency distribution of 454 sequencing read lengths for *M. amblycephala.* http://dx.doi.org/10.1371/journal.pone.0042637.g001

Clustering of the sequences resulted in 26,802 contigs, with an average length of 730 bp (SD=331 bp). The contigs were consisted of 1,168,029 reads (83%) and the size distribution for these contigs is shown in Figure 2A. The assembly produced a substantial number of large contigs: 19,866 contigs (74.1%) were >500 bp in length, of which 4,685 (17.5%) were >1 kb. The majority of assembled reads were incorporated into the top 14,393 largest contigs, which included 74% of the obtained reads and accounted for 47% of the total assembled length. In addition, the assembly resulted in 73,675 singletons (mean length=374±137 bp, Figure 2B). Singletons represent reads with similarities to other reads, but minor differences resulting in the exclusion from the clusters. The *M. amblycephala* 454 unigenes (contigs and singletons) sequences were uploaded to the website http://wbgd.hzau.edu.cn/additional/. A summary of the 454 sequencing data for *M. amblycephala* was presented in Table 1.

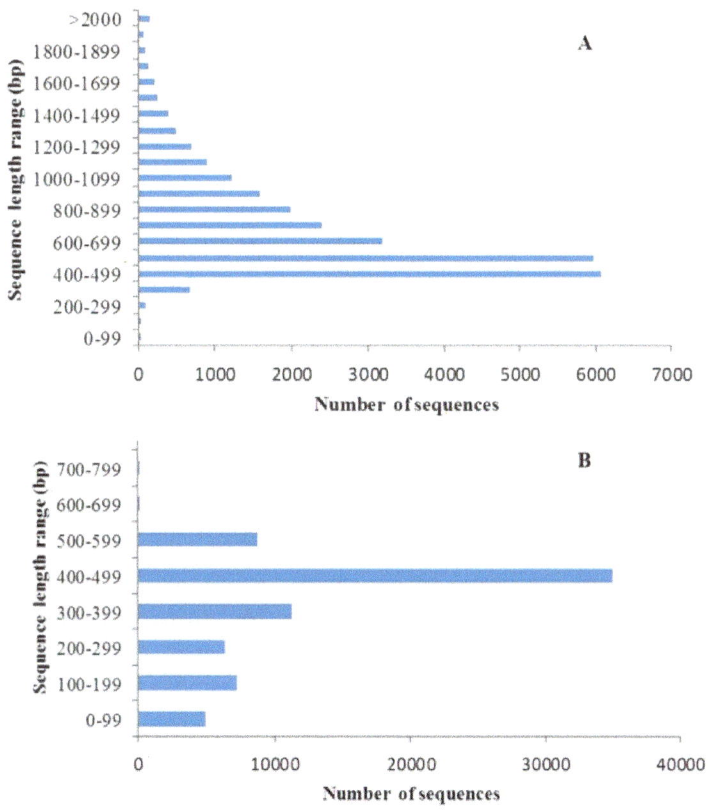

Figure 2: Overview of *M. amblycephala* transcriptome sequence length distributions for all contigs and singletons. (A) Sequence length distribution for all contigs; (B) Sequence length distribution for all singletons. http://dx.doi.org/10.1371/journal.pone.0042637.g002

Table 1: Summary of 454 sequencing, assembly and analysis. http://dx.doi.org/10.1371/journal.pone.0042637.t001

	Sequences (n)
Raw sequencing reads	1,409,706
Reads for assembly	1,280,970
Reads assembled	1,168,029
Contigs	26,802
Singletons	73,675
Total unique sequences	100,477

doi:10.1371/journal.pone.0042637.t001

As compared with other recent studies [31], [33], [34]–[38], our results indicate that short reads from 454 sequencing runs can be effectively assembled and used to readily characterize the gene space of non-model organisms. The large number of reads in good length generated from a full plate pyrosequencing run on a 454 GS XLR70 Titanium instrument resulted in a high expected average transcriptome coverage depth (7.6×) for *M. amblycephala*. As expected with the increased read lengths from the 454 GS XLR70 Titanium instrument, the contigs in our present study were on average larger (mean=730 bp) than those produced in previous studies using earlier 454 technologies with shorter reads (e.g., 197 bp [31]; 440 bp [33]; 500 bp [34]; 422 bp [35]; 300 bp [36]; 521 bp [37]; 547 bp [38]). In addition, a large percentage of our contigs were greater than 500 bp in length and high quality of assembled contigs was indicated by a high proportion of contigs matching to known proteins using BLAST searches.

Gene identification and annotation

Putative gene identification was conducted by either EMBOSS software for identification of open reading frames (ORFs) or by BLASTx similarity search of public protein databases. Of the 26,802 *M. amblycephala* assembled contigs, 5,615 (21%) had an ORF longer than 200 bp, with an average length of 736 bp (SD=244 bp, minimum=305 bp, maximum=3,017 bp; Figure 3).

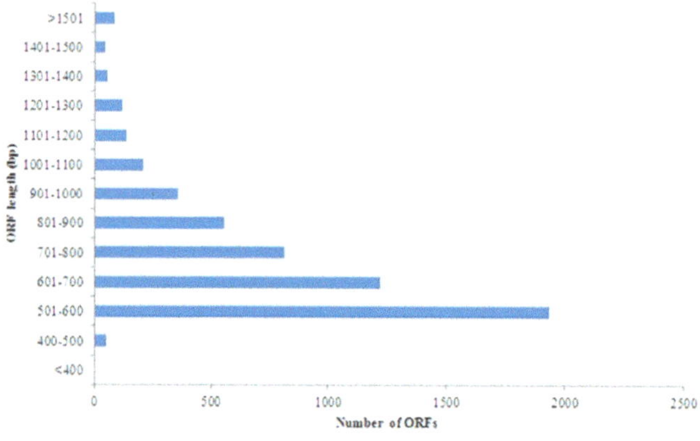

Figure 3: Open reading frame (ORF) length distribution from all contigs of *M. amblycephala*. http://dx.doi.org/10.1371/journal.pone.0042637.g003

BLAST searches for all contigs and singletons obtained from *M. amblycephala* were performed against the nr protein database using the BLASTx algorithm and an arbitrary expectation value of E^{-10}. Among a total 100,477 unigenes (contigs and singletons), 40,687 (40.5%) showed at least one significant alignment to an existing gene model. The assembled transcripts were annotated against UniProt. Of the 26,802 contigs, 17,378 (64.8%) returned an above cut-off Blast hits to the Uniprot database and the average length of these contigs was 803 bp. As for the 19,866 contigs with the length more than 500 bp, 13,915 (70.0%) showed significant hits to the Uniprot database. Of the 73,675 singletons, 24,629 sequences (33.4%) had Blast hits to the Uniprot database and the average read length of these singletons was 424 bp. Altogether, of the 100,477 unigenes, 42,007 (41.8%) had significant BLASTx hits to the Uniprot database and matched 20,674 unique protein accessions (Table 2). Gene ontology terms of biological processes, molecular functions and cellular components were showed in Figure 4. A large number of unigenes were assigned to a wide range of gene ontology categories.

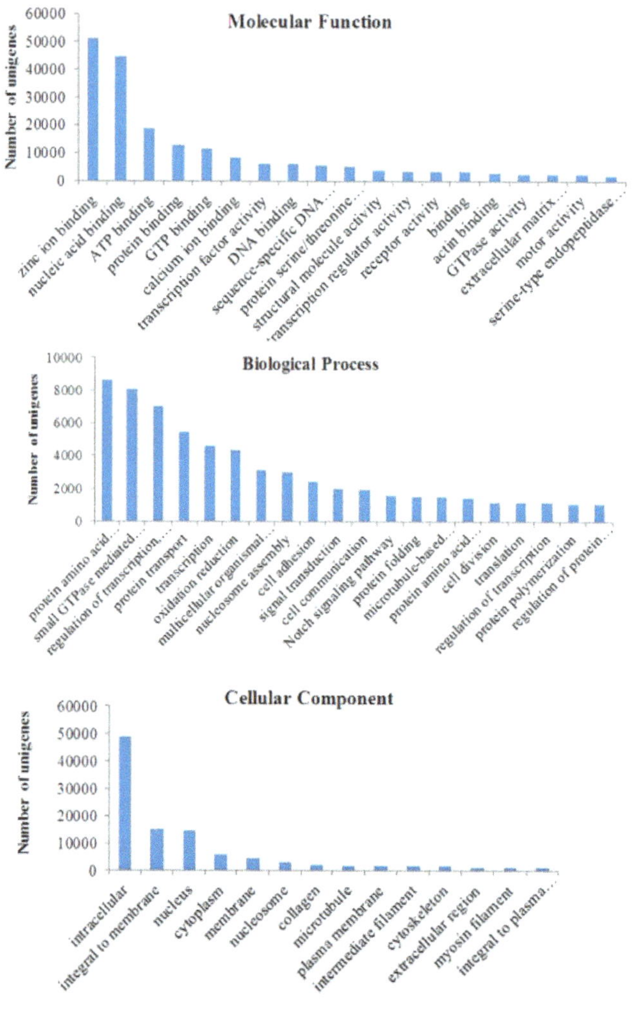

Figure 4: Gene ontology assignments for *M. amblycephala*. The annotated contigs and singletons from *M. amblycephala* 454 sequencing that matched various gene ontology (GO) categories. http://dx.doi.org/10.1371/journal.pone.0042637.g004

Table 2: Summary of BLASTX search analysis of *M. amblycephala* unique sequences. http://dx.doi.org/10.1371/journal.pone.0042637.t002

Dtabase	*M. amblycephala* hits*	Unique protein	% of total unique proteins
NR	40,687	19,230	
Uniprot	42,007	20,674	
Refseq/Ensembl			
Zebrafish	36,179	15,110	55% of 27,251
Medaka	28,740	12,302	50% of 24,661
Tetraodon	28,471	12,053	52% of 23,118
Fugu	29,112	14,890	31% of 47,841
Stickleback	29,736	13,122	48% of 27,576
Human	27,858	11,765	36% of 32,883
Mouse	27,776	11,201	37% of 30,100
Chicken	25,938	10,084	45% of 22,194
Cumulative unique (E^{-10})†	37,289	26,349	

Note:
*Number of significant (E^{-10}) alignments using all *M. amblycephala* unique sequences as queries to search the listed databases.
†Cumulative unique totals were derived from the sum of unique gene/protein identities across all listed species.
doi:10.1371/journal.pone.0042637.t002

Estimating the number of genes and the level of transcript coverage represented in an EST collection is an important issue for transcriptome sequencing projects, but is difficult or impossible without a completely annotated reference genome sequence. We indirectly evaluated transcriptome coverage breadth by determining the number of unique genes detected in our sequence collection using BLAST. The large number of sequences that matched in BLAST searches to unique proteins indicates that our 454 sequencing reads identified a substantial portion of the genes in *M. amblycephala*. A large portion of the *M. amblycephala* unique sequences had best BLAST hits to fish proteins, with a smaller portion (437 genes) hitting to other taxonomic groups. A similarly large number of these were assigned to a wide range of gene ontology categories, indicating that a wide collection of transcripts is represented by our sequence data. Furthermore, many of the contigs and singletons without BLAST hits likely represent additional genes not represented in the annotated protein databases that we searched, or genes that lack BLAST matches due to short length.

Comparative Analysis

In order to assess the level to which the *M. amblycephala* transcriptome has been captured, the unique sequences (100,477) were also searched against the NCBI Refseq and Ensembl databases. A number of significant hits were identified within zebrafish, medaka, tetraodon, fugu, stickleback, human, mouse, and chicken reference protein databases (Table 2). After removal of the redundant protein hits, 10,084–15,110 unique reference proteins were identified within zebrafish, medaka, tetraodon, fugu, stickleback, human, mouse, and chicken databases respectively (Table 2). The *M. amblycephala* 454 unigenes BLAST versus these 8 species were deposited in the website http://wbgd.hzau.edu.cn/additional/. The unique *M. amblycephala* sequences had the most hits (55%) to the unique proteins of zebrafish. A total of 37,289 cumulative unique genes were identified from *M. amblycephala* based on BLASTx searches against the Refseq/Ensembl database and that matched 26,349 unique protein accessions (Table 2). Taken together, these numbers provide strong evidence that this project has captured a majority of the *M. amblycephala* transcriptome.

To assess the evolutionary conservation of the identified unique genes, the number of hits to unique genes in each species of zebrafish, medaka, Tetraodon, Fugu, stickleback, human, mouse, and chicken were compared. A total of 27,488 (65.4% of total number of unique *M. amblycephala* genes) putative known unique genes were found in all eight species; 30,332 (72.2%) were found in all five fish species; and 41,229 (98.1%) were found in at least one of the five fish species (Figure 5), indicating a high level of conservation of gene content among*M. amblycephala* and other teleost fish species.

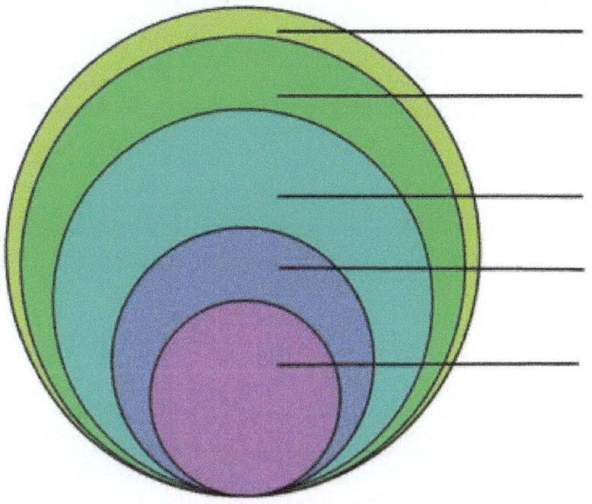

Figure 5: Conservation of *M. amblycephala* gene identities with other species. Number of *M. amblycephala* homologous genes identified from other species using BLASTX searches. http://dx.doi.org/10.1371/journal.pone.0042637.g005

SSR/SNP markers identification

SSR markers.

A total of 4,952 microsatellites were identified from 100,477 assembled unique sequences, including di-, tri-, tetra-, penta- and hexa-nucleotide repeats. Di-nucleotide repeats were the most common type in the EST dataset, with tri- and tetra-nucleotide repeats being present at much smaller frequencies. The characteristics of these microsatellites are summarized inTable 3. The total length of the unique sequences with microsatellites was 2.05 Mb and the average distribution distance of microsatellites was 9.53 Kb. After removing the microsatellites without enough flanking sequence for primer design, 3,372 microsatellites had sufficient flanking sequences (≥50 bp) on both sides to design primers for genotyping, which were distributed within 3,255 contigs. These microsatellites are expected to be useful for genetic linkage mapping and other genetic studies.

Table 3: Summary of microsatellite marker identification from *M. amblycephala*unique sequences. http://dx.doi.org/10.1371/journal.pone.0042637.t003

Type	Number	Proportion in all SSRs	Frequency (%)[1]	Mean distance (kb)[2]	Microsatellites with 50 bp on both sides (Proportion)
Dinucleotide	3,107	62.74%	3.09%	15.18	2,025 (60.05%)
Trinucleotide	1,428	28.84%	1.42%	33.04	1,068 (31.67%)
Tetranucleotide	339	6.85%	0.33%	139.17	224 (6.64%)
Pentanucleotide	61	1.23%	0.06%	773.40	41 (1.22%)
Hexanucleotide	17	0.34%	0.02%	2775.13	14 (0.42%)
Total	4,952	100%	4.92%	9.53	3,372

Note:
[1]Frequency = Total microsatellites number sought out/total number of unique sequences;
[2]Mean distance = Total length of unique sequences/total SSR number.
doi:10.1371/journal.pone.0042637.t003

In order to check the successful amplification proportion of these microsatellites, we randomly designed PCR primers with optimal expected product sizes for 160 microsatellites using Primer3. Of these 160 loci tested for successful PCR amplification in a single trial, 116 (72.5%) loci were successfully amplified in *M. amblycephala*. In addition, 71 (44.4%) of the tested microsatellites were polymorphic across panels of 40 individuals from the Liangzi Lake. The characteristics of these polymorphic microsatellites are listed.

SNP Markers.

Considering the sequencing errors, a sequence variation was declared as a putative SNP or indel (insertion or deletion) when a mismatch was identified in contigs with four or more reads and the minor allele reads existed at least twice within contigs. A total of 56,109 putative SNPs were identified from assembled contigs. These putative SNP types indicated an SNP rate of 1 per 302 bp of the transcribed sequences in *M. amblycephala*. A total of 72,020 insertions and deletions (indels) were discovered, that is, one indel per 158 bp of the transcribed sequences.

Putative SNPs/indels detected may be false positives, potentially arising from sequencing errors or misassembly of paralogous sequence variants or multisite sequence variants [39]. Therefore, in order to select SNPs/indels with high confidence, putative SNPs were screened based on several factors including surrounding sequence quality, absence of additional SNPs/indels in the flanking regions, and minor allele frequency. The setting of a minimum minor allele frequency >15% may help reduce false SNP/indel calling based on sequence errors. Additionally, multiple SNPs/indels located close to one another (<15 bp) often represent sequence errors and prevent the design of primers and probes for SNP/indels genotyping. Therefore, a requirement of no additional SNPs in the 15 bp flanking region around a putative SNP/indels was applied. After filtering, 25,697 SNPs were identified and these SNPs included 17,272 transitions and 8,425 transversions (Figure 6). The filtered SNP frequency in the transcribed sequences was one SNP per 401 bp of the transcribed sequences in *M. amblycephala.* A total of 23,287 filtered indels were identified with one indel per 392 bp of the transcribed sequences. Since the information on minor allele frequency is an important consideration in choosing which SNPs to be used in SNP arrays, the minor allele frequencies of SNPs in the discovery populations were estimated from the sequence data (Figure 7). The average minor allele frequencies were 30.9% in putative filtered SNPs identified for *M. amblycephala*.

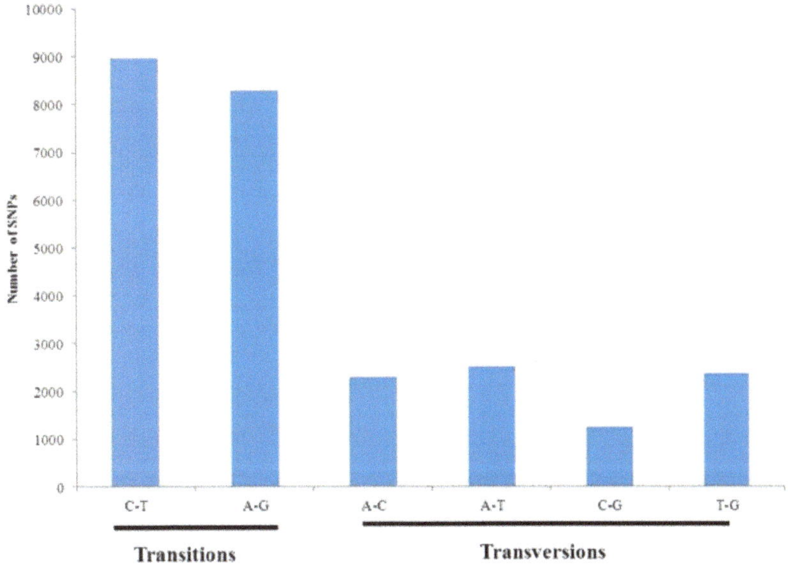

Figure 6: Classification of SNPs identified from 454 sequences of *M. amblycephala*. http://dx.doi.org/10.1371/journal.pone.0042637.g006

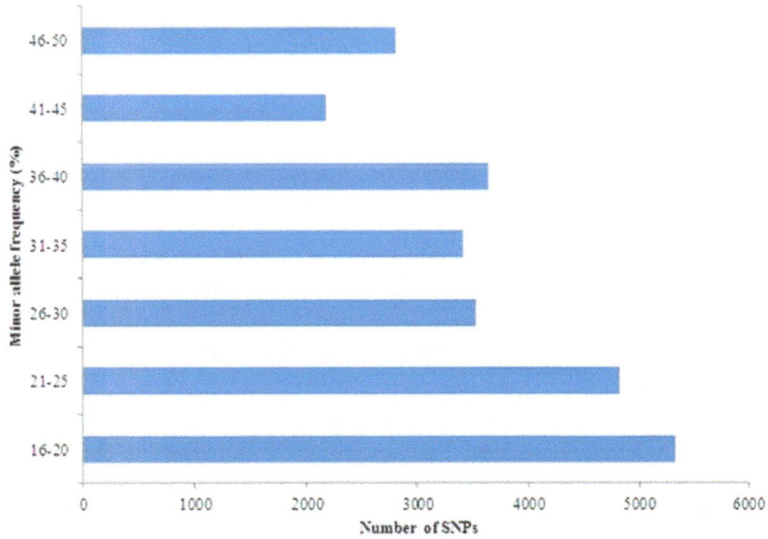

Figure 7: Distribution of minor allele frequencies of filtered SNPs identified for *M. amblycephala* from 454 sequences. The X-axis represents the number of SNPs with given minor allele frequency, while the Y-axis represents the SNP sequence derived minor allele frequency in percentage (>15%). http://dx.doi.org/10.1371/journal.pone.0042637.g007

The number of SNPs and their distribution in genes in a genome is important for the use of SNP in genetic analysis. A total of 12,314 *M. amblycephala* contigs were found to contain filtered SNPs, with the mean number of 2.09 SNPs per contig, while 11,241 contigs were found to contain putative indels with the mean number of 2.07 indels per contig. The number of unique Uniprot accessions hit by contigs containing SNPs was 8,489 in *M. amblycephala*, suggesting that putative filtered SNPs were identified from the vast majority of *M. amblycephala* genes.

Polymorphic genetic markers are important for research involving population genetic structuring, relatedness, the genetic basis of adaptive traits, and genetic selection [40], [41]. The second-generation transcriptome sequencing leads to superior resources for the development of such markers not only because of the enormous amount of sequence data in which markers can be identified, but also because discovered markers are gene-based. Such markers are advantageous on functional variation and the signature of selection in genomic scans or association genetic studies [42], [43]. Currently few genetic marker resources exist for *M. amblycephala*, wherein 454 pyrosequencing in the present study was shown to be an excellent method for large scale prediction of molecular markers for future genetic linkage and QTL analysis in non-model organisms this *M. amblycephala*. Given that these SSR and SNP markers were predicted fromtranscriptomic sequences, they are likely linked to protein-coding genes, and therefore might have substantial physiological implications [44].

Pooling of RNA samples from multiple individuals followed by transcriptome analysis using 454 sequencing provides an excellent opportunity to generate large numbers of SNP markers [23]. Previous studies have used 454 sequencing for finding large numbers of SNPs, which was performed using the whole genome sequence (WGS) as a reference for SNP determination[45]. However, developing such markers using

454 sequencing has also been proved being efficient in species without genomic resources. Beldade et al. [46] used over 20 outbred butterfly (*Bicyclus anynana*) individuals in their wing tissue cDNA library and were able to identify over 14,000 candidate SNPs. Parchman et al. [34] used normalized cDNA collections from multiple tissues and 4 individuals and detected 3,707 high quality SNPs with a minimum coverage depth of 8×. To infer SNP allele frequencies with a minimum reliability, it was recommended that a pool size of at least 10 individuals (20 chromosomes) is used [47]. Lower pool size would result in rather uninformative SNP frequencies estimates [47]. In our study, we used normalized cDNA collections from multiple tissues and 24 individuals with distinct growth performance to sequence, which resulted in a total of 34,771 filtered SNPs. Our results demonstrate that the number of pooling samples is an important factor for generating SNPs through the second-generation sequencing technologies.

CONCLUSIONS

As a native fish species in China, *M. amblycephala* has become a main aquaculture species in the Chinese freshwater polyculture system. The genetic selection of this species is undergoing, which was funded by the earmarked fund for Modern Agroindustry Technology Research System entitled "Staple Freshwater Fishery Industry Technology System". To date, the lack of genomics data available for this species has hampered the progress of its selective breeding. The 100,477 unique sequences in this 454 EST collection represent a major genomic level resource for *M. amblycephala*, and will be useful for comparative genomic studies in the genus *Megalobrama*. The EST contig library developed in this study can be used as a reference transcriptome for analysis of gene expression using cDNA microarray. Our results also demonstrated that the approach to sample individuals with diverse genetic backgrounds for reliable SNP identification allowed the ability to detect many SNPs across the entire genome. The SNPs identified in this study will provide a much-needed resource for genetic studies in *M. amblycephala* and will contribute to the development of high density, cost-effective genotyping platforms. Validation and testing of SNPs using high-density arrays will subsequently lead to the production of a SNP array, which would greatly benefit the breeding program and whole genome association studies for the performance and production traits in *M. amblycephala*.

MATERIALS AND METHODS

Sample and RNA isolation

Twenty-four one-year-old *M. amblycephala* individuals from three natural populations were used for this study, of which 12 individuals have relatively highest growth rates (mean body weight 91.2±10.5 g, 4 individuals from LZ, PY and YN, respectively) and 12 individuals have relatively lowest growth rate (mean body weight 30.6±5.3 g, 4 individuals from LZ, PY and YN, respectively). Samples of 7 tissues including liver, kidney, muscle, testis, ovary, brain and heart were collected. The fish were euthanized with tricaine methanesulfonate (MS 222) at 100 mg/L before tissue collection. All the experimental procedures involving fish were approved by the institution animal care and use committee of Huazhong Agricultural University.

Total RNA was extracted from these samples using TRIzol® reagent (Invitrogen, Carlsbad, CA, USA) according to the manufacturer›s protocol. RNA quality and quantity was measured using the NanoDrop 2000 (Thermo Scientific, Wilmington, DE, USA). All the samples were standardized to 200 ng/μL, and equal volumes of the same tissue samples from different individuals were combined into one pool, making a total of seven RNA pools. The pools were DNase-treated with Turbo DNA-free (Ambion) and purified with RNeasy Mini Kit (QIAGEN, USA) following the manufacturer›s instructions; RNA quality and quantity was determined again at the end of the process.

cDNA synthesis, normalization and 454 sequencing

The seven RNA pools were used for double-strand cDNA synthesis using SMART technology (BD Biosciences Clontech), and then the cDNA was purified using the QIAquick PCR Purification Kit (QIAGEN). Full-length cDNA sample containing equal amounts of cDNA from each library was normalized using the Trimmer kit (Evrogen, Moscow, Russia) according to the manufacturer's instructions. Normalization decreases the prevalence of highly abundant transcripts which should lead to an increase in the number of different transcripts represented in the sequence reads and consequently the number of SNPs available for analysis.

The normalized cDNA pool was sequenced in a single full plate run on the 454 GS-FLX Titanium series pyrosequencer (Roche Applied Science) according to the manufacturer's instructions at the SeqWright Incorporation (http://www.seqwright.com/). The 454 GS-FLX Titanium series reagents has the ability to sequence 400–600 million base pairs per run with 400–500 base pair read lengths.

Sequence Assembly

Prior to assembly, we used the 454 Newbler 2.3 software to perform adaptor, SMART primer and poly-A tail trimming and also quality filtering (threshold quality score=20). The remaining reads were clustered using the MIRA V2.9.26x3 assembler [48] with the "de novo, normal, EST, 454" parameters, specifying a minimum read length of 40 nt, a minimum sequence overlap of 40 nt, and a minimum percentage overlap identity of 80%. Sequences after assembly comprised contigs, singletons and "debris" reads. Contigs based on at least two reads, singletons refer to single reads that clustered with other reads, but were eventually excluded from the final assembly. "Debris" reads may be of high-quality, but display no significant relationships to any other reads during the assembling process, representing very short reads, but also single mRNAs.

ORF Searching and Gene Annotation

All unique sequences obtained after the assembly were analyzed by EMBOSS 6.4.0.1 software [49] to search for ORFs. The putative protein sequences were also generated at the same time by EMBOSS, which could be used to analyze splice variation, determine paralogs, and assess gene families.

Putative gene names, protein domains, and gene ontology terms were assigned to all contigs and singletons (referred to as unigenes) that shared sequence similarity with previously identified genes annotated with those details. For gene name annotation, the unigenes were compared against Uniprot and nr protein sequence databases using BLASTx with a significance threshold of $E<10^{-10}$. Gene names were assigned to each assembled sequence based on the best blast hit annotated with a gene name, as determined by parsing the GenBank record features 'CDS' and 'Protein' for that blast hit. Blast hits annotated with uninformative terms were encountered (e.g., 'unknown', 'uncharacterized', or 'hypothetical') were skipped in this analysis. For computational speed, the search was limited to the first 10 significant hits for each query.

To annotate assembled sequences with Gene Ontology (GO) terms describing biological processes [50], molecular functions, and cellular components, sequences were compared against the UniProt-TrEMBL database of protein sequences [51] using WU-blast. This database was chosen because it has been extensively annotated with GO terms [52]. Each query sequence was assigned GO terms based on those associated with the top blast match for that query, using custom Perl scripts to compare the blast results and the table of annotations. Because the Gene Ontology is hierarchical in structure, the GO terms annotated here can be traced back to one or more parent terms at a given level of detail. The NCBI Refseq protein and Ensemble databases (zebrafish, medaka, Tetraodon, Fugu, stickleback, human, mouse, and chicken) were also used to annotate the unique *M. amblycephala* genes.

Microsatellite and SNP Markers Identification

All the unique sequences were used to search for microsatellite makers using Msatfinder [53]with a repeat threshold of six di-nucleotide repeats or five tri-, tetra-, penta-, or hexa-nucleotide repeats. Unique sequences containing 50-bp sequence on both sides of the microsatellite repeat were considered sufficient for primer design [54].

For putative SNP identification, reads were mapped onto the transcripts we assembled using SSAHA2 [55] with default parameter and then the SNPs were extracted using VarScan under default parameter. A sequence variation was declared as a putative SNP or indel (insertion or deletion) when a mismatch was identified in contigs with four or more sequences and the minor allele sequence existed at least twice within contigs.

In order to identify high quality SNPs, putative SNPs and indels identified as mentioned above were further screened following specific criteria based on the read depth, minor allele frequency, the quality of flanking regions and absence of other SNPs or indels within 15 bp flanking regions. Only those SNPs/indels with minor allele sequences representing more than 15% of the reads aligned at the polymorphic loci were declared as quality SNPs/indels. In addition, no extra SNPs/indels within 15 bp flanking regions were allowed.

Data Deposition

The Roche 454 reads of *M. amblycephala* were submitted to NCBI Sequence Read Archive under the accession number of SRA045792.

REFERENCES

1. Li S, Cai W, Zhou B (1993) Variation in morphology and biochemical genetic markers among populations of blunt snout bream (*Megalobrama amblycephala*). Aquaculture 111: 117–127. doi: 10.1016/b978-0-444-81527-9.50016-6
2. Zhang D (2001) Study on genetic diversity of bluntnose black bream from Yunihu and Liangzi lakes. J China Three Gorges Univ 3: 282–284.
3. Xu W, Xiong BX (2008) Advances in the research on genus *Megalobrama* in China. J Hydroecology 1 (2)7–11.
4. Ke H (1965) The artificial reproduction and culture experiment of *Megalobrama amblycephala*. Acta Hydrobiol Sin 5: 282–283.
5. CAFS (Chinese Academy of Fishery Sciences) (2001) Fishery Statistic Data. Chinese Academy of Fishery Sciences, Beijing.
6. Li S, Yang X (1996) Effects of two-way selection on biochemical genetics of blunt snout bream (*Megalobrama amblycephala*). J Fish China 1996 3: 1–5. doi: 10.1016/b978-0-444-81527-9.50016-6
7. Wang WM (2009) The aquaculture status of blunt snout bream (*Megalobrama amblycephala*). Scientific Fish Farming 2009 4: 44–45. doi: 10.1111/j.1439-0264.2009.00950.x
8. Li HH (2008) MtDNA Sequence Variation and Genetic Structure of the *Megalobrama amblycephala* from Yuni Lake, Liangzi Lake and Poyang Lake. Freshwater Fish 38 (4) 63–65.
9. Li SW, Chang Y, Liang L, Sun X (2006) Rapid isolation of microsatellites from genome of blunt nose black bream (*Megalobrama amblycephala*). J Fish China 13: 187–192.
10. Li WT, Liao XL, Yu XM, Wang D, Tong J (2007) Isolation and characterization of polymorphic microsatellite loci in Wuchang bream (*Megalobrama amblycephala*). Mol Ecol Notes 7: 771–773. doi: 10.1111/j.1471-8286.2007.01689.x
11. Li Y (2010) Genetic structure analysis of three wild *Megalobrama amblycephala*populations and construction of genetic linkage map. PhD dissertation, Huazhong Agricultural University.
12. Morin PA, Luikart G, Wayne RK, Grp SW (2004) SNPs in ecology, evolution and conservation. Trends Ecol Evol 19 (4) 208–216. doi: 10.1016/j.tree.2004.01.009
13. Abasht B, Lamont SJ (2007) Genome-wide association analysis reveals cryptic alleles as an important factor in heterosis for fatness in chicken F-2 population. Anim Genet 38 (5) 491–498. doi: 10.1111/j.1365-2052.2007.01642.x
14. Duijvesteijn N, Knol EF, Merks JWM, Crooijmans RPMA, Groenen MAM, et al. (2010) A genome-wide association study on androstenone levels in pigs reveals a cluster of candidate genes on chromosome 6. BMC Genetics 11: 42. doi: 10.1186/1471-2156-11-42
15. Du ZQ, Zhao X, Vukasinovic N, Rodriguez F, Clutter AC, et al. (2009) Association and haplotype analyses

of positional candidate genes in five genomic regions linked to scrotal hernia in commercial pig lines. PLoS One 4 (3) e4837. doi: 10.1371/journal.pone.0004837

16. Meuwissen THE, Hayes BJ, Goddard ME (2001) Prediction of total genetic value using genome-wide dense marker maps. Genetics 157 (4) 1819–1829.
17. Khatkar MS, Nicholas FW, Collins AR, Zenger KR, Cavanagh JA, et al. (2008) Extent of genome-wide linkage disequilibrium in Australian Holstein-Friesian cattle based on a high-density SNP panel. BMC Genomics 9: 187. doi: 10.1186/1471-2164-9-187
18. Kim ES, Berger PJ, Kirkpatrick BW (2009) Genome-wide scan for bovine twinning rate QTL using linkage disequilibrium. Anim Genet 40 (3) 300–307. doi: 10.1111/j.1365-2052.2008.01832.x
19. Brooks SA, Gabreski N, Miller D, Brisbin A, Brown HE, et al. (2010) Whole-genome SNP association in the horse: identification of a deletion in myosin va responsible for lavender foal syndrome. Plos Genet 6 (4) e1000909. doi: 10.1371/journal.pgen.1000909
20. Kijas JW, Townley D, Dalrymple BP, Heaton MP, Maddox JF, et al. (2009) A genome wide survey of SNP variation reveals the genetic structure of sheep breeds. PLoS One 4 (3) e4668. doi: 10.1371/journal.pone.0004668
21. Becker D, Tetens J, Brunner A, Burstel D, Ganter M, et al. (2010) Microphthalmia in texel sheep is associated with a missense mutation in the paired-like homeodomain 3 (*PITX3*) gene. PLoS One 5 (1)e8689. doi: 10.1371/journal.pone.0008689
22. Weber APM, Weber KL, Carr K, Wilkerson C, Ohlrogge JB (2007) Sampling the arabidopsis transcriptome with massively parallel pyrosequencing. Plant Physiol 144: 32–42. doi: 10.1104/pp.107.096677
23. Wheat CW (2010) Rapidly developing functional genomics in ecological model systems via 454 transcriptome sequencing. Genetica 138: 433–451. doi: 10.1007/s10709-008-9326-y
24. Metzker ML (2010) Sequencing technologies – the next generation. Nat Rev Gen 11: 31–46. doi: 10.1038/nrg2626
25. Rokas A, Abbot P (2009) Harnessing genomics for evolutionary insights. Trends Ecol Evol 24: 192–200. doi: 10.1016/j.tree.2008.11.004
26. Novaes E, Drost DR, Farmerie WG, Pappas GJ, Grattapaglia D, et al. (2008) High-throughput gene and SNP discovery in *Eucalyptus grandis*, an uncharacterized genome. BMC Genomics 9: 312. doi: 10.1186/1471-2164-9-312
27. Margulies M, Egholm M, Altman WE, Attiya S, Bader JS, et al. (2005) Genome sequencing in microfabricated high-density picolitre reactors. Nature 437: 376–380.
28. Schwarz D, Robertson HM, Feder JL, Varala K, Hudson ME, et al. (2009) Sympatric ecological speciation meets pyrosequencing: sampling the transcriptome of the apple maggot *Rhagoletis pomonella*. BMC Genomics 10: 633. doi: 10.1186/1471-2164-10-633
29. Elmer KR, Fan S, Gunter HM, Jones JC, Boekhoff S, et al. (2010) Rapid evolution and selection inferred from the transcriptomes of sympatric crater lake cichlid fishes. Mol Ecol 19: 197–211. doi: 10.1111/j.1365-294x.2009.04488.x
30. Renaut S, Nolte AW, Bernatchez L (2010) Mining transcriptome sequences towards identifying adaptive single nucleotide polymorphisms in lake whitefish species pairs (*Coregonus* spp. Salmonidae). Mol Ecol 19: 115–131. doi: 10.1111/j.1365-294x.2009.04477.x
31. Vera JC, Wheat CW, Fescemyer HW, Frilander MJ, Crawford DL, et al. (2008) Rapid transcriptome characterization for a nonmodel organism using 454 pyrosequencing. Mol Ecol 17 (7) 1636–1647. doi: 10.1111/j.1365-294x.2008.03666.x
32. Hale MC, McCormick CR, Jackson JR, DeWoody JA (2009) Next-generation pyrosequencing of gonad transcriptomes in the polyploid lake sturgeon (*Acipenser fulvescens*): the relative merits of normalization and rarefaction in gene discovery. BMC Genomics 10: 203. doi: 10.1186/1471-2164-10-203
33. Meyer E, Aglyamova GV, Wang S, Buchanan-Carter J, Abrego D, et al. (2009) Sequencing and de novo analysis of a coral larval transcriptome using 454 GSFlx. BMC Genomics 10: 219. doi: 10.1186/1471-2164-10-219
34. Parchman TL, Geist KS, Grahnen JA, Benkman CW, Buerkle CA (2010) Transcriptome sequencing in an ecologically important tree species: assembly, annotation, and marker discovery. BMC Genomics 11: 180. doi: 10.1186/1471-2164-11-180
35. Roeding F, Borner J, Kube M, Klages S, Reinhardt R, et al. (2009) A 454 sequencing approach for large scale phylogenomic analysis of the common emperor scorpion (*Pandinus imperator*). Mol Phylogenet Evol 53: 826–834. doi: 10.1016/j.ympev.2009.08.014
36. Klmer KR, Fan S, Gunter HM, Jones JC, Boekhoff S, et al. (2010) Rapid evolution and selection inferred from the transcriptomes of sympatric crater lake cichlid fishes. Mol Ecol 19 (Suppl. 1) 197–211. doi: 10.1111/j.1365-294x.2009.04488.x
37. Lu FH, Cho MC, Park YJ (2012) Transcriptome profiling and molecular marker discovery in red pepper, *Capsicum annuum* L. TF68. Mol Biol Rep 39 (3) 3327–3335. doi: 10.1007/s11033-011-1102-x
38. Der JP, Barker MS, Wickett NJ, dePamphilis CW, Wolf PG (2011) De novo characterization of the gametophyte transcriptome in bracken fern, *Pteridium aquilinum*. BMC Genomics 12: 99. doi: 10.1186/1471-2164-12-99
39. Liu S, Zhou Z, Lu J, Sun F, Wang S, et al. (2011) Generation of genome-scale gene-associated SNPs in catfish for the construction of a high-density SNP array. BMC Genomics 12: 53. doi: 10.1186/1471-2164-12-53
40. Luikart G, England PR, Tallmon D, Jordan S, Taberlet P (2003) The power and promise of population genomics: from genotyping to genome typing. Nat Rev Genet 4: 981–994. doi: 10.1038/nrg1226
41. Beaumont MA (2005) Adaptation and speciation: what can FST tell us? Trends Ecol Evol 20: 435–440. doi: 10.1016/j.tree.2005.05.017

42. Bouck A, Vision T (2007) The molecular ecologist's guide to expressed sequence tags. Mol Ecol 16 (5) 907–924. doi: 10.1111/j.1365-294x.2006.03195.x

43. Vasemagi A, Nilsson J, Primmer CR (2005) Expressed sequence tag-linked microsatellites as a source of gene-associated polymorphisms for detecting signatures of divergent selection in Atlantic salmon (*Salmo salar* L.). Mol Biol Evol 22 (4) 1067–1076. doi: 10.1093/molbev/msi093

44. Bai X, Rivera-Vega L, Mamidala P, Bonello P, Herms DA, et al. (2011) Transcriptomic Signatures of Ash (*Fraxinus* spp.) Phloem. PLoS One 6 (1) e16368. doi: 10.1371/journal.pone.0016368

45. Barbazuk WB, Emrich SJ, Chen HD, Li L, Schnable PS (2007) SNP discovery via 454 transcriptome sequencing. Plant J 51: 910–918. doi: 10.1111/j.1365-313x.2007.03193.x

46. Beldade P, Rudd S, Gruber JD, Long AD (2006) A wing expressed sequence tag resource for *Bicyclus anynana* butterflies, an evo-devo model. BMC Genomics 7: 130.

47. Pérez-Enciso M, Ferretti L (2010) Massive parallel sequencing in animal genetics: wherefroms and wheretos. Anim Genet 41: 561–569. doi: 10.1111/j.1365-2052.2010.02057.x

48. Chevreux B, Pfisterer T, Drescher B, Driesel AJ, Müller WE (2004) Using the miraEST assembler for reliable and automated mRNA transcript assembly and SNP detection in sequenced ESTs. Genome Res 14: 1147–1159. doi: 10.1101/gr.1917404

49. Rice P, Longden I, Bleasby A (2000) EMBOSS: the European Molecular Biology Open Software Suite. Trends Genet 16 (6) 276–277. doi: 10.1016/s0168-9525(00)02024-2

50. Ashburner M, Ball CA, Blake JA, Botstein D, Butler H, et al. (2000) Gene Ontology: tool for the unification of biology. Nat Genet 25 (1) 25–29. doi: 10.1038/75556

51. UniProt-TrEMBL [http://www.ebi.ac.uk/trembl/FTP/ftp.html].

52. The Gene Ontology [http://www.geneontology.org/GO.downloads.shtml.

53. Thurston MI, Field D (2005) Msatfinder: detection and characterisation of microsatellites. http://www.genomics.ceh.ac.uk/msatfinder/.

54. Rozen S, Skaletsky H (2000) Primer3 on the WWW for general users and for biologist programmers. Methods Mol Biol 132: 365–386. doi: 10.1385/1-59259-192-2:365

55. Ning Z, Cox AJ, Mullikin JC (2001) SSAHA: a fast search method for large DNA databases. Genome Res 11 (10) 1725–1729. doi: 10.1101/gr.194201

INDEX

N

O

P

Q

R

S

T

V

W

Z